Proceedings in Adaptation, Learning and Optimization

Volume 7

Series editors

Yew Soon Ong, Nanyang Technological University, Singapore
e-mail: asysong@ntu.edu.sg

Meng-Hiot Lim, Nanyang Technological University, Singapore
e-mail: emhlim@ntu.edu.sg

Jiuwen Cao · Kezhi Mao
Jonathan Wu · Amaury Lendasse
Editors

Proceedings of ELM-2015 Volume 2

Theory, Algorithms and Applications (II)

 Springer

Editors

Jiuwen Cao
Institute of Information and Control
Hangzhou Dianzi University
Hangzhou, Zhejiang
China

Kezhi Mao
School of Electrical and Electronic
  Engineering
Nanyang Technological University
Singapore
Singapore

Jonathan Wu
Department of Electrical and Computer
  Engineering
University of Windsor
Windsor, ON
Canada

Amaury Lendasse
Department of Mechanical and Industrial
  Engineering
University of Iowa
Iowa City, IA
USA

Proceedings in Adaptation, Learning and Optimization
ISBN 978-3-319-80338-8   ISBN 978-3-319-28373-9   (eBook)
DOI 10.1007/978-3-319-28373-9

Printed on acid-free paper

This Springer imprint is published by SpringerNature
The registered company is Springer International Publishing AG Switzerland

Contents

Large-Scale Scene Recognition Based on Extreme Learning Machines 1
Yuanlong Yu, Lingying Wu, Kai Sun and Jason Gu

Partially Connected ELM for Fast and Effective Scene Classification 19
Dongzhe Wang, Rui Zhao and Kezhi Mao

Two-Layer Extreme Learning Machine for Dimension Reduction 31
Yimin Yang and Q.M. Jonathan Wu

Distributed Extreme Learning Machine with Alternating Direction Method of Multiplier. 43
Minnan Luo, Qinghua Zheng and Jun Liu

An Adaptive Online Sequential Extreme Learning Machine for Real-Time Tidal Level Prediction 55
Jianchuan Yin, Lianbo Li, Yuchi Cao and Jian Zhao

Optimization of Outsourcing ELM Problems in Cloud Computing from Multi-parties. 67
Jiarun Lin, Tianhang Liu, Zhiping Cai, Xinwang Liu and Jianping Yin

H-MRST: A Novel Framework for Support Uncertain Data Range Query Using ELM. 77
Bin Wang, Rui Zhu and Guoren Wang

The SVM-ELM Model Based on Particle Swarm Optimization 93
Miao-miao Wang and Shi-fei Ding

ELM-ML: Study on Multi-label Classification Using Extreme Learning Machine 107
Xia Sun, Jiarong Wang, Changmeng Jiang, Jingting Xu, Jun Feng, Su-Shing Chen and Feijuan He

Sentiment Analysis of Chinese Micro Blog Based on DNN and ELM and Vector Space Model 117
Huilin Liu, Shan Li, Chunfeng Jiang and He Liu

Self Forward and Information Dissemination Prediction Research in SINA Microblog Using ELM 131
Huilin Liu, Yao Li and He Liu

Sparse Coding Extreme Learning Machine for Classification 143
Zhenzhen Sun and Yuanlong Yu

Continuous Top-K Remarkable Comments over Textual Streaming Data Using ELM 155
Rui Zhu, Bin Wang and Guoren Wang

ELM Based Representational Learning for Fault Diagnosis of Wind Turbine Equipment 169
Zhixin Yang, Xianbo Wang, Pak Kin Wong and Jianhua Zhong

Prediction of Pulp Concentration Using Extreme Learning Machine 179
Changwei Jiang, Xiong Luo, Xiaona Yang, Huan Wang and Dezheng Zhang

Rational and Self-adaptive Evolutionary Extreme Learning Machine for Electricity Price Forecast 189
Chixin Xiao, Zhaoyang Dong, Yan Xu, Ke Meng, Xun Zhou and Xin Zhang

Contractive ML-ELM for Invariance Robust Feature Extraction 203
Xibin Jia and Hua Du

Automated Human Facial Expression Recognition Using Extreme Learning Machines 209
Abhilasha Ravichander, Supriya Vijay, Varshini Ramaseshan and S. Natarajan

Multi-modal Deep Extreme Learning Machine for Robotic Grasping Recognition 223
Jie Wei, Huaping Liu, Gaowei Yan and Fuchun Sun

Denoising Deep Extreme Learning Machines for Sparse Representation 235
Xiangyi Cheng, Huaping Liu, Xinying Xu and Fuchun Sun

Extreme Learning Machine Based Point-of-Interest Recommendation in Location-Based Social Networks 249
Mo Chen, Feng Li, Ge Yu and Dan Yang

The Granule-Based Interval Forecast for Wind Speed 263
Songjian Chai, Youwei Jia, Zhao Xu and Zhaoyang Dong

KELMC: An Improved K-Means Clustering Method Using Extreme Learning Machine 273
Lijuan Duan, Bin Yuan, Song Cui, Jun Miao and Wentao Zhu

Wind Power Ramp Events Classification Using Extreme Learning Machines 285
Sujay Choubey, Anubhav Barsaiyan, Nitin Anand Shrivastava, Bijaya Ketan Panigrahi and Meng-Hiot Lim

Facial Expression Recognition Based on Ensemble Extreme Learning Machine with Eye Movements Information 295
Bo Lu, Xiaodong Duan and Ye Yuan

Correlation Between Extreme Learning Machine and Entorhinal Hippocampal System 307
Lijuan Su, Min Yao, Nenggan Zheng and Zhaohui Wu

RNA Secondary Structure Prediction Using Extreme Learning Machine with Clustering Under-Sampling Technique 317
Tianhang Liu, Jiarun Lin, Chengkun Wu and Jianping Yin

Multi-instance Multi-label Learning by Extreme Learning Machine . . . 325
Chenguang Li, Ying Yin, Yuhai Zhao, Guang Chen and Libo Qin

A Randomly Weighted Gabor Network for Visual-Thermal Infrared Face Recognition 335
Beom-Seok Oh, Kangrok Oh, Andrew Beng Jin Teoh, Zhiping Lin and Kar-Ann Toh

**Dynamic Adjustment of Hidden Layer Structure for Convex
Incremental Extreme Learning Machine**...................... 345
Yongjiao Sun, Yuangen Chen, Ye Yuan and Guoren Wang

**ELMVIS+: Improved Nonlinear Visualization Technique
Using Cosine Distance and Extreme Learning Machines** 357
Anton Akusok, Yoan Miche, Kaj-Mikael Björk, Rui Nian,
Paula Lauren and Amaury Lendasse

**On Mutual Information over Non-Euclidean Spaces, Data Mining
and Data Privacy Levels** 371
Yoan Miche, Ian Oliver, Silke Holtmanns, Anton Akusok,
Amaury Lendasse and Kaj-Mikael Björk

Probabilistic Methods for Multiclass Classification Problems 385
Andrey Gritsenko, Emil Eirola, Daniel Schupp, Edward Ratner
and Amaury Lendasse

**A Pruning Ensemble Model of Extreme Learning Machine
with $L_{1/2}$ Regularizer** 399
Bo He, Tingting Sun, Tianhong Yan, Yue Shen and Rui Nian

Evaluating Confidence Intervals for ELM Predictions 413
Anton Akusok, Yoan Miche, Kaj-Mikael Björk, Rui Nian,
Paula Lauren and Amaury Lendasse

Real-Time Driver Fatigue Detection Based on ELM 423
Hengyu Liu, Tiancheng Zhang, Haibin Xie, Hongbiao Chen
and Fangfang Li

**A High Speed Multi-label Classifier Based on Extreme
Learning Machines** 437
Meng Joo Er, Rajasekar Venkatesan and Ning Wang

**Image Super-Resolution by PSOSEN of Local Receptive Fields
Based Extreme Learning Machine** 455
Yan Song, Bo He, Yue Shen, Rui Nian and Tianhong Yan

Sparse Extreme Learning Machine for Regression 471
Zuo Bai, Guang-Bin Huang and Danwei Wang

**WELM: Extreme Learning Machine with Wavelet Dynamic
Co-Movement Analysis in High-Dimensional Time Series** 491
Heng-Guo Zhang, Rui Nian, Yan Song, Yang Liu,
Xuefei Liu and Amaury Lendasse

**Imbalanced Extreme Learning Machine for Classification
with Imbalanced Data Distributions** 503
Wendong Xiao, Jie Zhang, Yanjiao Li and Weidong Yang

Author Index 515

Large-Scale Scene Recognition Based on Extreme Learning Machines

Yuanlong Yu, Lingying Wu, Kai Sun and Jason Gu

Abstract For intelligent robots, scene recognition aims to find a semantic explanation of a scene, i.e., it helps the robots to know where they are. It can be widely applied into various robotic tasks, e.g, topological localization, simultaneous localization and mapping and autonomous navigation. Many of existing methods for scene recognition focused on how to build scene features, such as holistic representations and bags of visual words. However, less attention is put on the classification. Due to the huge number of scene classes in the real world, the variances within each class and the shared features between classes, the classification becomes a challenging issue for scene recognition. This paper proposes an ensemble method for large-scale scene recognition. This proposed method builds a three-level hierarchy for recognizing 397 classes of scenes in the real world. At each level, an ensemble-based classifier is built by using 13 types of features. Extreme learning machine is employed as the basic classifier in each ensemble-based classifier. Experimental results have shown that this proposed method outperforms other state-of-the-art methods in terms of recognition accuracy.

This work is supported by National Natural Science Foundation of China (NSFC) under grant 61473089.

Y. Yu (✉) · L. Wu · K. Sun · J. Gu
College of Mathematics and Computer Science, Fuzhou University,
Fuzhou 350116, Fujian, China
e-mail: yu.yuanlong@fzu.edu.cn

Y. Yu · L. Wu · K. Sun · J. Gu
Department of Electrical and Computer Engineering, Dalhousie University,
Halifax, NS, Canada

J. Gu
e-mail: jgu@dal.ca

1   Introduction

Scene recognition is a fundamental problem in cognitive science, computer vision and robotics. It can be widely applied into computer vision tasks, e.g., object retrieval [1] and 3D reconstruction [2], and robotic tasks, e.g, topological localization, simultaneous localization and mapping (SLAM) [3, 4] and autonomous navigation [5].

During the past decades, the community of robotics proposed several methods, which can recognize where the robot is by identifying landmark objects [6, 7] or segmented regions [5, 8] in the scene. This type of methods is partially inspired by original psychological assumption that scene recognition is a progressive reconstruction of the input from local measurements [9].

But recent psychological experiments suggest that humans have an ability to instantly summarize the characteristics of the whole scene prior to attentional selection [10–14], i.e., local information might be spontaneously ignored during a rapid categorization of scenes. This fact indicates that object variations have less influence on scene recognition even when these variations affect meaningful parts of the scene [15–17]. Inspired by these cognitive findings, a number of holistic representations (e.g., gist [18–21] and bags of features [22–24]) based methods have been proposed for scene recognition in the community of computer vision.

However, less attention is put on the classification issues. The first issue is caused by the large scale of scenes. Large scale indicates that not only the number of scene classes but also the number of training samples are very large. Currently, the 15-categories scene database [25] is widely used for evaluation. But the number of scenes in real environment is very far from 15, e.g., the SUN database [24] used in this paper contains 397 scene classes, as shown in Fig. 1. This fact imposes large burden on classifier design. Some hierarchical methods [24, 25] has been proposed for scene classification, but the ability at identifying multiple classes is still required for each base classifier, especially in large scale cases. Unfortunately, most of existing methods employ one-to-all support vector machine (SVM) [22, 24] or back-propagation based neural network (BP-NN) [21] as base classifiers. Although one-to-all strategy can accommodate multi-class cases, it has to face the problem of serious unbalance between the numbers of positive and negative training samples, especially when the number of negative classes is very large. Although BP-NN is theoretically capable of learning any types of nonlinear classification boundary, it is likely to go into a local optimum and the computational burden for training is also unaffordable.

The third issue is about the inter-class similarity, i.e., different classes share some features. For example, the spatial layouts and contained instruments (e.g., shelves) are similar between bookstore and archive scenes as shown in Fig. 2a, the trees and their layouts are similar between forest path and forest road scenes as shown in Fig. 2c. This issue would lead to serious ambiguity for classification.

This paper attempts to propose a new method for scene classification by solving the above three issues such that recognition accuracy and computational efficiency

Fig. 1 Examples of large-scale scenes in real environment. This images are from the SUN database

for large-scale cases are both improved. This proposed method builds a three-level hierarchy for recognizing 397 types of scenes.

The first contribution of this proposed method is to use extreme learning machine (ELM) [26] as the basic classifier in the hierarchy to address the first issue. ELM has shown great ability to fit nonlinear classification boundaries such that each basic classifier has better predictive performance on multi-class identification compared against SVM based one-to-all strategy. Furthermore, ELM can theoretically obtain a global optimum such that it is unlikely to fall into a local optimum. In terms of computation, the training cost of ELM is much lower compared with BP-NN.

The second contribution is to use the ensemble-based classification strategy [27] to address the second issue. In order to delineate various facets of each scene, a set of training samples obtained in terms of various types of features (e.g., denseSIFT [22] and sparseSIFT based on Hessian-affine interest points [28]) are required. Based on these sets of samples with various features, this proposed method trains ensemble classifiers by using the bagging strategy [29].

Furthermore, the combination of a set of distinct features (totally 13 features used in this proposed method) with ensemble classifiers can also address the inter-class similarity issue (i.e., the third issue) in the sense that the ensemble-based training process favorites those features which can improve classification power.

The remainder of this paper is organized as follows. The details of this proposed method are presented in Sect. 2 and experimental results are given in Sect. 3.

(a)

(b)

(c)

(d)

Fig. 2 Inter-class similarity between scenes. Columns 1 and 2 represent a scene and columns 3 and 4 represent another scene in each row. **a** Bookstore and archive. **b** Golf course and baseball field. **c** Forest path and forest road. **d** Cafeteria and Bistro

2   Exposition of the Proposed Scene Recognition

2.1   *Hierarchy of the Scene Recognition*

The proposed scene recognition method has a three-level hierarchy as shown in Fig. 3. The underlying idea of building this hierarchy includes two facets: First, the relationship between levels and definition of categories are both based on semantic understanding; Secondly, this hierarchy is organized in an ascending order in terms of classification accuracy. SUN database [24] is used as data source to build the hierarchy presented as follows.

Fig. 3 Three-level hierarchy of the proposed scene recognition method

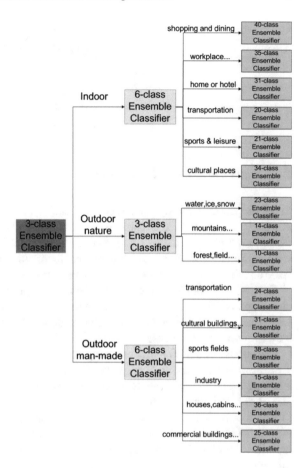

The first level only includes one ensemble-based classifier which identifies the test sample to one of three categories, i.e., indoor, outdoor nature or outdoor manmade category.

The second level is composed of three ensemble-based classifiers, respectively named indoor, outdoor-nature and outdoor-manmade classifiers, each of which corresponds to an output category at the first level. That is, the test sample is sent to the corresponding classifier according to the output at the first level. The indoor classifier assigns the test sample to one of 6 subcategories, the outdoor-nature classifier assigns the test sample to one of 3 subcategories, and the outdoor-manmade classifier assigns the test sample to one of 6 subcategories. Thus the number of output subcategories at the second level is 15.

The third level consists of fifteen ensemble-based classifiers, each of which corresponds to an output subcategory at the second level. Each classifier at this level determines the final label to indicate which scene class the test sample belongs to.

Fig. 4 Bagging strategy for
the ensemble-based scene
classification

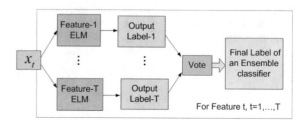

2.2 Ensemble-Based Classifier

In the proposed hierarchy, the structure of each ensemble-based classifier is shown
in Fig. 4. This ensemble consists of a set of basic classifier, called *feature-based
classifiers*, each of which corresponds to a feature type. Each feature-based classifier
is implemented by an ELM. The training and recognition processes of an ELM-based
basic classifier can be seen in Sect. 2.3. The number of feature types is denoted as T.

For each ensemble-based classifier, the training process is to learn all of its basic
classifiers. All types of T features are extracted for each training image. Then each
set of training samples in terms of feature t, denoted as $\{\mathbf{x_{s,t}}\}_{s=1,...S,t=1,...T}$, is used to
train the corresponding basic classifier.

During the recognition process, bagging strategy [29] is used to achieve the final
classification label for each ensemble-based classifier, as shown in Fig. 4. Given
a testing scene image I, all types of features $\{\mathbf{x}_t\}_{t=1,...,T}$ are extracted. Each feature sample $\mathbf{x}_t$ is first sent to the corresponding basic classifier belonging to feature
t to achieve output label. Once all feature samples $\{\mathbf{x}_t\}_{t=1,...,T}$ get their own labels,
these labels ranging from *feature-1* to *feature-T* are voted to get a final label which
indicates the classification output of the ensemble-based classifier for the testing
image I.

2.3 ELM based Basic Classifiers

2.3.1 Structure of an ELM-Based Classifier

ELM is a machine learning algorithm for training single-hidden-layer feed-forward
neural networks whose structure is shown in Fig. 5.

The input layer is connected to the input feature vector $\mathbf{x}$ of a scene image. At
the hidden layer, the number of hidden nodes is denoted as L. The activation func-
tion of each hidden node i is denoted as $g(\mathbf{x}; \mathbf{w}_i, b_i)$, where $\mathbf{w}_i$ is the *input weight*
vector between this hidden node and all input nodes, b_i is the bias of this node
and $i = 1, ..., L$. At the output layer, the number of output nodes is denoted as M.
For classification application, M also indicates the number of classes. The *output
weight* between the i-th hidden node and the j-th output node is denoted as $\beta_{i,j}$, where

Fig. 5 Structure of an
ELM-based classifier

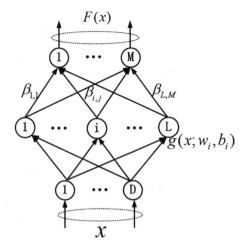

$j = 1, ..., M$. The value of an output node is calculated as shown in (1). Each output node represents a class.

$$f_j(\mathbf{x}) = \sum_{i=1}^{L} \beta_{i,j} \times g(\mathbf{x}; \mathbf{w}_i, b_i) \tag{1}$$

2.3.2 Training Process of an ELM-based Classifier

The supervised training requires N training sample pairs, each of which consists of a feature vector $\mathbf{x}_k$ and its binary class label vector (i.e., ground truth) $\mathbf{d}_k = [d_{k,1}, ..., d_{k,M}]$, where $k = 1, ..., N$. In the label vector, each entry indicates whether or not the sample $\mathbf{x}_k$ belongs to the corresponding class. All labels can form a matrix denoted as $\mathbf{D} = [\mathbf{d}_1, ..., \mathbf{d}_N]^T$.

It can be seen that the training parameters for an ELM include the input weights and biases $\{\mathbf{w}_i, b_i\}_{i=1,...,L}$ as well as the output weight vector β as shown in (4). In the ELM algorithm, the input weights and biases are randomly assigned. Therefore only β is trained.

Let $\mathbf{y}_k$ denote the actual output vector for the input $\mathbf{x}_k$. Taking all training samples $\{x_k\}$ into (1) can form a linear representation:

$$\mathbf{H}\beta = \mathbf{Y} \tag{2}$$

where

$$\mathbf{H} = \begin{bmatrix} \mathbf{h}(\mathbf{x}_1) \\ \vdots \\ \mathbf{h}(\mathbf{x}_N) \end{bmatrix} = \begin{bmatrix} g(\mathbf{x}_1; \mathbf{w}_1, b_1) & \cdots & g(\mathbf{x}_1; \mathbf{w}_L, b_L) \\ \vdots & \vdots & \vdots \\ g(\mathbf{x}_N; \mathbf{w}_1, b_1) & \cdots & g(\mathbf{x}_N; \mathbf{w}_L, b_L) \end{bmatrix} \tag{3}$$

$$\beta = \begin{bmatrix} \beta_1 \\ \vdots \\ \beta_L \end{bmatrix} = \begin{bmatrix} \beta_{1,1} & \cdots & \beta_{1,M} \\ \vdots & \vdots & \vdots \\ \beta_{L,1} & \cdots & \beta_{L,M} \end{bmatrix} \tag{4}$$

and

$$\mathbf{Y} = \begin{bmatrix} \mathbf{y}_1 \\ \vdots \\ \mathbf{y}_N \end{bmatrix} = \begin{bmatrix} y_{1,1} & \cdots & y_{1,M} \\ \vdots & \ddots & \vdots \\ y_{N,1} & \cdots & y_{N,M} \end{bmatrix} \tag{5}$$

The training process aims to minimize the training error $\|\mathbf{D} - \mathbf{H}\beta\|^2$ and the norm of output weight $\|\beta\|$ [26]. So the training process can be represented as a constrained-optimization problem:

$$\begin{array}{ll} \text{Minimize:} & \Psi(\beta, \xi) = \frac{1}{2}\|\beta\|^2 + \frac{C}{2}\|\xi\|^2 \\ \text{Subject to:} & \mathbf{H}\beta = \mathbf{D} - \xi \end{array} \tag{6}$$

where a constant C is used as a regulation factor to control the trade-off between the closeness to the training data and the smoothness of the decision function such that generalization performance is improved.

Lagrange multiplier technique is used to solve the above optimization problem by construct the lagrangian:

$$L_p(\beta, \xi, \alpha) = \frac{1}{2}\|\beta\|^2 + \frac{C}{2}\|\xi\|^2 - \alpha \cdot (\mathbf{H}\beta - \mathbf{D} + \xi) \tag{7}$$

where α is a Lagrange multiplier matrix with dimensions of $N \times M$. Since the constraints in (6) are all equalities, α can be either positive or negative.

Based on KKT optimization conditions, gradients of L_p with respect to β, ξ and α can be obtained as follows:

$$\begin{cases} \dfrac{\partial L_p}{\partial \beta} = 0 & \rightarrow \quad \beta = \mathbf{H}^T \alpha & \text{(8a)} \\[2mm] \dfrac{\partial L_p}{\partial \xi} = 0 & \rightarrow \quad \alpha = C\xi & \text{(8b)} \\[2mm] \dfrac{\partial L_p}{\partial \alpha} = 0 & \rightarrow \quad \xi = \mathbf{D} - \mathbf{H}\beta & \text{(8c)} \end{cases}$$

By substituting (8b) and (8c) into (8a), the above equations can be written as

$$(\frac{\mathbf{I}}{C} + \mathbf{H}^T \mathbf{H})\beta = \mathbf{H}^T \mathbf{D} \tag{9}$$

If matrix $(\frac{\mathbf{I}}{C} + \mathbf{H}^T\mathbf{H})$ is not singular, solution $\hat{\beta}$ can be obtained as

$$\hat{\beta} = (\frac{\mathbf{I}}{C} + \mathbf{H}^T\mathbf{H})^{-1}\mathbf{H}^T\mathbf{D} \qquad (10)$$

By substituting (8a) and (8b) into (8c), the above equations can be also written as

$$(\frac{\mathbf{I}}{C} + \mathbf{H}\mathbf{H}^T)\alpha = \mathbf{D} \qquad (11)$$

If matrix $(\frac{\mathbf{I}}{C} + \mathbf{H}\mathbf{H}^T)$ is not singular, solution $\hat{\beta}$ can be obtained by substituting (11) into (8a)

$$\hat{\beta} = \mathbf{H}^T(\frac{\mathbf{I}}{C} + \mathbf{H}\mathbf{H}^T)^{-1}\mathbf{D} \qquad (12)$$

It can be seen that the dimensions of $(\frac{\mathbf{I}}{C} + \mathbf{H}^T\mathbf{H})$ is $L \times L$ while $(\frac{\mathbf{I}}{C} + \mathbf{H}\mathbf{H}^T)$ is $N \times N$. Therefore, if the number of training samples is huge, the solution in (10) can be used to decrease computational cost; Otherwise, the solution in (12) can be used.

It has been proved that the universal approximation can be satisfied if the activation function g is a nonlinear piecewise continuous function [30–32]. This paper uses sigmoid function as shown in (13).

$$g(\mathbf{x}; \mathbf{w}_i, b_i) = \frac{1}{1 + \exp[-(\mathbf{w}_i \cdot \mathbf{x} + b_i)]} \qquad (13)$$

Algorithm 1 Training routine of an ELM-based classifier

1: Given a set of training samples $\{(\mathbf{x}_k, \mathbf{y}_k)_{k=1,...,N}\}$, activation function g and hidden node number L:
2: Step 1: Randomly assign input weight vector $\mathbf{w}_i$ and bias b_i, where $i = 1, ..., L$, using uniform distribution;
3: Step 2: Normalize values of $\{\mathbf{x}_k\}$ into the range [0, 1] and calculate matrix $\mathbf{H}$;
4: Step 3: Estimate the output weight $\hat{\beta}$ by using (10) or (12).

The training routine for the ELM-based classifier used in this proposed method is shown in Algorithm 1.

Compared with other learning algorithms, e.g. BP algorithm, for neural networks, ELM randomly sets input weights and biases at the hidden layer without training such that the output weights can be quickly estimated based on the least-squares strategy. It can be seen that there are only two tuning parameters: One is the number of hidden nodes (i.e., L) and the other is the regularization factor (i.e., C).

2.3.3 Recognition Process of an ELM-based Classifier

Given the trained output weight $\widehat{\beta}$, the output label vector $\mathbf{F}(\mathbf{x})$ for a test sample $\mathbf{x}$ is obtained by using (14).

$$\mathbf{F}(\mathbf{x}) = \begin{bmatrix} f_1(\mathbf{x}) \\ \vdots \\ f_M(\mathbf{x}) \end{bmatrix} = \widehat{\beta}^T \mathbf{G}(\mathbf{x}) \tag{14}$$

where

$$\mathbf{G}(\mathbf{x}) = \begin{bmatrix} g(\mathbf{x}; \mathbf{w}_1, b_1) \\ \vdots \\ g(\mathbf{x}; \mathbf{w}_L, b_L) \end{bmatrix} \tag{15}$$

The class label of $\mathbf{x}$ is determined as

$$\text{label}(\mathbf{x}) = \arg_{j=1,\ldots,M} \max f_j(\mathbf{x}) \tag{16}$$

3 Experiments

3.1 Experimental Setup

The proposed scene recognition method is tested over the SUN database [24]. This data set has collected 130519 images which cover 397 classes of scenes. The image size ranges from 174×193 to 3648×2736. The images are divided into 10 sets per class and there are 50 images for training and 50 images for test per set. The average recognition rate among these 10 sets is used for performance evaluation.

There are 13 types of features extracted for each image, i.e., the feature number T as shown in Fig. 4 is 13 in our experiments. These features include dense SIFT [22], sparse SIFT [33], histogram of oriented gradients (HOG) [34], gist [18, 21], local binary patterns (LBP) [35], LBP histogram Fourier (LBPHF) [36], self-similarity (SSIM) [23], texton histogram [19], geometry texton histogram [24], geometric classification map [37], geometry color histogram [24], tiny image [38], and line histogram [39].

The performance of this proposed method is compared against SVM based method [24] (as a representative of one-to-all strategy) and BP-NN with two hidden layers (as a representative of nonlinear multi-class identification strategy).

Fig. 6 Average recognition rate curves with respect to parameter L in terms of individual features at each level. **a** Level 1. **b** Level 2. **c** Level 3

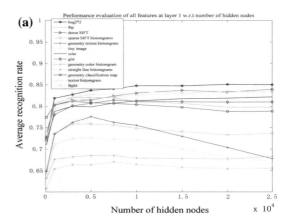

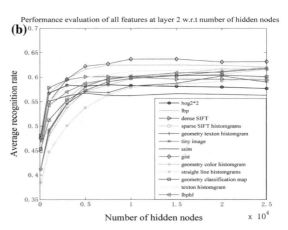

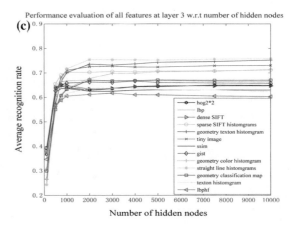

Fig. 7 Average recognition
rate curves with respect to
parameter *C* in terms of
individual features at each
level. **a** Level 1. **b** Level 2.
c Level 3

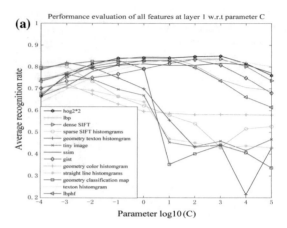

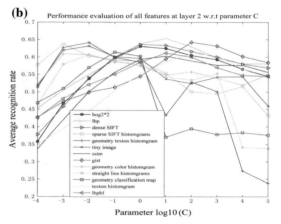

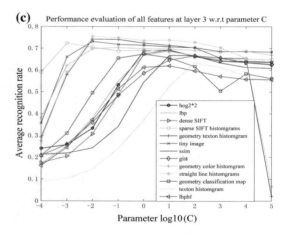

3.2 Tuning Parameters

The ELM based classifier has only two tuning parameters, including number of hidden nodes L and regulation factor C. Our experiments first evaluate performance of each feature-based classifier at individual levels. The average recognition rate curves at three levels with respect to L and C are shown in Figs. 6 and 7 respectively. These two figures illustrate the effects of L and C:

1. With the increase of the number of hidden nodes L, recognition rate goes up for all features. But it becomes stable when L reaches a certain level for most features. The optimal settings of L are dependent on the number of training samples and the number of classes of each basic classifier.
2. With the increase of the regulation factor C, recognition rate goes up for all features. But it goes down slowly or dramatically after C passes over a threshold. The optimal settings of C are very different between features due to the variety in terms of sparsity, data type and value ranges.

Optimal settings of these two tuning parameters for individual features at each level are shown in Tables 1 and 2 respectively.

3.3 Overall Recognition Performance

The overall average recognition rates for 397 classes in terms of individual features and by combination of all features are shown in Table 3. The recognition rates by

Table 1 Optimal setting of tuning parameter L for individual features at each level

L	Level 1	Level 2	Level 3
Gist	7500	15000	4000
LBP	7500	7500	2000
HOG	20000	7500	7500
Dense SIFT	20000	7500	750
Sparse SIFT	5000	25000	10000
Texton histogram	10000	25000	3000
Geometry classification map	5000	20000	5000
LBPHF	25000	25000	3000
Geometry texton histogram	5000	20000	10000
Tiny images	5000	25000	10000
SSIM	25000	5000	1000
Straight line histogram	7500	25000	10000
Geometry color histogram	5000	20000	10000

Table 2 Optimal setting of tuning parameter C for individual features at each level

C	Level 1	Level 2	Level 3
Gist	100	100	100
LBP	10	10	10
HOG	1000	1	1
Dense SIFT	1000	1	10
Sparse SIFT	0.001	0.01	0.001
Texton histogram	100	100	100
Geometry classification map	0.1	0.1	1
LBPHF	10	10	10
Geometry texton histogram	0.001	0.01	0.01
Tiny images	0.001	0.01	0.01
SSIM	1000	10	10
Straight line histogram	0.001	0.01	0.01
Geometry color histogram	0.001	0.001	0.01

Table 3 Overall average recognition rates for 397 classes

	ELM based (%)	SVM based (%)	BP-NN (%)
All features	**42.2**	38.0	8.8
Gist	31.7	16.3	1.8
LBP	31.3	18.0	1.7
HOG	30.2	27.2	0.4
Dense SIFT	30.2	21.5	0.3
Sparse SIFT	29.6	11.5	0.8
Texton histogram	29.2	17.6	0.4
Geometry classification map	28.9	6.1	2.4
LBPHF	28.2	12.8	1.1
Geometry texton histogram	28.0	23.5	1.9
Tiny image	28.0	5.5	7.9
SSIM	27.8	22.5	0.7
Straight line histogram	27.5	5.7	1.2
Geometry color histogram	25.5	9.2	0.4

using SVM based method [24] and BP-NN based method are also shown in this table. It can be seen that the average recognition rate of this proposed ELM-based method by using all features reaches to 42.2 %. It is 4.2 % higher than SVM-based method and much higher than BP-NN based method.

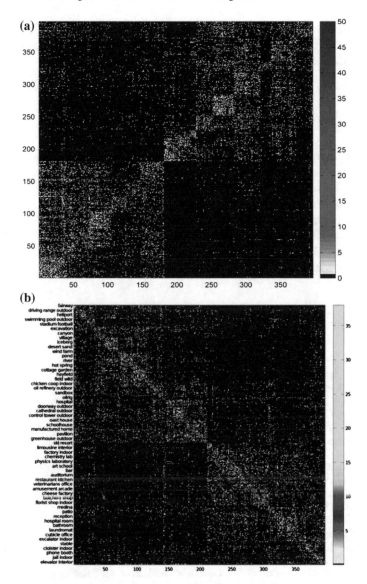

Fig. 8 Patterns of confusion across 397 classes. **a** This proposed ELM based method. **b** SVM based method (obtained from [24])

Figure 8 shows the patterns of confusion across classes obtained by this proposed method and SVM based method. The coordinates in X-axis and Y-axis denote 397 classes. The color at a point with coordinates (x, y) represents the number of test samples whose ground truth label is x while the recognized label is y. It can be seen that SVM based method shows more points in the non-diagonal region while less points along the diagonal line compared with this proposed method. It indicates that there are more true positives while less false positives and false negatives of this proposed method.

4 Conclusion

This paper proposes an ensemble method for large-scale scene recognition. This proposed method builds a three-level hierarchy for recognizing 397 classes of scenes in the real world. At each level, an ensemble-based classifier is built by using 13 types of features. Extreme learning machine is employed as the basic classifier in each ensemble-based classifier. Experimental results have shown that this proposed method outperforms other state-of-the-art methods in terms of recognition accuracy. Future work includes the use of deep neural networks to evaluate the classification performance for the large-scale scenes.

References

1. Arandjelovic, R., Zisserman, A.: Three things everyone should know to improve object retrieval. In: Proceedings of IEEE International Conference on Computer Vision and Pattern Recognition, pp. 1–5 (2012)
2. Agarwal, S., Snavely, N., Simon, I., Seitz, S.M., Szeliski, R.: Building room in a day. In: Proceedings of International Conference on Computer Vision, pp. 1–5 (2010)
3. Johns, E., Yang, G.Z.: Dynamic scene models for incremental, long-term, appearance-based localization. In: Proceedings of International Conference on Robotics and Automation, pp. 1–5 (2011)
4. Cummins, M., Newman, P.: Highly scalable appearance-only slam-fab-map 2.0. In: Proceedings of International Conference on Robotics: Science and Systems, pp. 1–5 (2009)
5. Katsura, H., Miura, J., Hild, M., Shirai, Y.: A view-based outdoor navigation using object recognition robust to changes of weather and seasons. In: Proceedings of IEEE/RSJ International Conference of Inteligent Robots and Systems (IROS), pp. 2974–2979 (2003)
6. Abe, Y., Shikano, M., Fukuda, T., Arai, F., Tanaka, Y.: Vision based navigation system for autonomous mobile robot with global matching. In: Proceedings of International Conference of Robotics and Automation, pp. 1299–1304 (1999)
7. Thrun, S.: Finding landmarks for mobile robot navigation. In: Proceedings of International Conference of Robotics and Automation, pp. 958–963 (1998)
8. Matsumoto, Y., Inaba, M., Inoue, H.: View-based approach to robot navigation. In: Proceedings of IEEE/RSJ International Conference of Inteligent Robots and Systems (IROS), pp. 1702–1708 (2000)
9. Barrow, H.G., Tannenbaum, J.M.: Recovering intrinsic scene characteristics from images. In: Hanson, A., Riseman, E. (eds.) Computer Vision Systems, pp. 3–26. Academic Press, New York (1978)
10. Potter, M.C.: Meaning in visual search. Science **187**(4180), 965–966 (1975)
11. Biederman, I.: Aspects and extension of a theory of human image understanding. In: Pylyshyn, Z. (ed.) Computational Processes in Human Vision: An Interdisciplinary Perspective. Ablex Publishing Corporation, New Jersey (1988)

12. Tversky, B., Hemenway, K.: Categories of the environmental scenes. Cogn. Psychol. **15**, 121–149 (1983)
13. Rensink, R.A., O'Regan, J.K., Clark, J.J.: To see or not to see: the need for attention to perceive changes in scenes. Psychol. Sci. **8**, 368–373 (1997)
14. Sanocki, T., Epstein, W.: Priming spatial layout of scenes. Psychol. Sci. **8**, 374–378 (1997)
15. Oliva, A., Schyns, P.: Colored diagnostic blobs mediate scene recognition. Cogn. Psychol. **41**, 176–210 (2000)
16. O'Regan, J.K., Rensink, R.A., Clark, J.J.: Change-blindness as a result of 'mudsplashes. Nature **398**, 34 (1999)
17. Rensink, R.A.: The dynamic representation of scenes. Vis. Cogn. **7**, 17–42 (2000)
18. Oliva, A., Torralba, A.: Modeling the shape of the scene: a holistic representation of the spatial envelope. Int. J. Comput. Vis. **42**(3), 145–175 (2001)
19. Renninger, L., Malik, J.: When is scene identification just texture recognition? Vis. Res. **44**(19), 2301–2311 (2004)
20. Torralba, A.: Contextual priming for object detection. Int. J. Comput. Vis. **53**(2), 153–167 (2003)
21. Siagian, C., Itti, L.: Rapid biologically-inspired scene classification using features shared with visual attention. IEEE Trans. Pattern Anal. Mach. Intell. **29**(2), 300–312 (2007)
22. Lazebnik, S., Schmid, C., Ponce, J.: Beyond bags of features: spatial pyramid matching for recognizing natural scene categories. In: Proceedings of International Conference on Computer Vision and Pattern Recognition, pp. 2169–2178 (2006)
23. Shechtman, E., Irani, M.: Matching local self-similarities across images and videos. In: Proceedings of IEEE International Conference on Computer Vision and Pattern Recognition, pp. 1–8 (2007)
24. Xiao, J., Hays, J., Ehinger, K.A., Oliva, A., Torralba, A.: Sun database: large-scale scene recognition from abbey to zoo. In: Proceedings of IEEE International Conference on Computer Vision and Pattern Recognition, pp. 1–5 (2010)
25. Li, F.-F., Perona, P.: A bayesian hierarchical model for learning natural scene categories. In: Proceedings of International Conference on Computer Vision and Pattern Recognition, pp. 1–6 (2005)
26. Huang, H.-G., Zhou, H., Ding, X., Zhang, R.: Extreme learning machine for regression and multiclass classification. IEEE Trans. Syst. Man Cybern. Part B: Cybern. **42**(2), 513–529 (2012)
27. Rokach, L.: Ensemble-based classifier. Artif. Intell. Rev. **33**, 1–39 (2010)
28. Matas, J., Chum, O., Urban, M., Pajdla, T.: Robust wide-baseline stereo from maximally stable extremal regions. In: Proceedings of International Conference on Computer Vision and Pattern Recognition, pp. 1–6 (2006)
29. Breiman, L.: Bagging predictor. University of California, Berkeley. Technical Report 421 (1994)
30. Huang, G.-B., Chen, L., Siew, C.-K.: Universal approximation using incremental constructive feedforward networks with random hidden nodes. IEEE Trans. Neural Netw. **17**(4), 879–892 (2006)
31. Huang, G.-B., Chen, L.: Convex incremental extreme learning machine. Neurocomputing **70**(16–18), 3056–3062 (2007)
32. Huang, G.-B., Chen, L.: Enhanced random search based incremental extreme learning machine. Neurocomputing **71**(16–18), 3460–3468 (2007)
33. Mikolajczyk, K., Schmid, C.: Scale and affine invariant interest point detectors. Int. J. Comput. Vis. **60**(1), 63–86 (2004)
34. Dalal, N., Triggs, B.: Histograms of oriented gradients for human detection. In: Proceedings of IEEE International Conference on Computer Vision and Pattern Recognition, pp. 1–8 (2005)
35. Ojala, T., Pietikainen, M., Maenpaa, T.: Multiresolution gray-scale and rotation invariant texture classification with local binary patterns. IEEE Trans. Pattern Anal. Mach. Intell. **24**(7), 971–987 (2002)

36. Ahonen, T., Matas, J., He, C., Pietikainen, M.: Rotation invariant image description with local binary pattern histogram fourier features. Image Anal. Lect. Notes Comput. Sci. **5575**, 61–70 (2009)
37. Hoiem, D., Efros, A.A., Hebert, M.: Recovering surface layout from an image. Int. J. Comput. Vis. **75**(1), 151–172 (2007)
38. Torralba, A., Fergus, R., Freeman, W.T.: 80 million tiny images: A large data set for nonparametric object and scene recognition. IEEE Trans. Pattern Anal. Mach. Intell. **30**(11), 1958–1970 (2008)
39. Kosecka, J., Zhang, W.: Video compass. In: Proceedings of European Conference on Computer Vision, pp. 476–490 (2002)

Partially Connected ELM for Fast and Effective Scene Classification

Dongzhe Wang, Rui Zhao and Kezhi Mao

Abstract Scene classification is often solved as a machine learning problem, where a classifier is first learned from training data, and class labels are then assigned to unlabelled testing data based on the outputs of the classifier. Generally, image descriptors are represented in high-dimensional space, where classifiers such as support vector machine (SVM) show good performance. However, SVM classifiers demand high computational power during model training. Extreme learning machine (ELM), whose synaptic weight matrix from the input layer to the hidden layer are randomly generated, has demonstrated superior computational efficiency. But the weights thus generated may not yield enough discriminative power for hidden layer nodes. Our recent study shows that the random mapping from the input layer to the hidden layer in ELM can be replaced by semi-random projection (SRP) to achieve a good balance between computational complexity and discriminative power of the hidden nodes. The application of SRP to ELM yields the so-called partially connected ELM (PC-ELM) algorithm. In this study, we apply PC-ELM to multi-class scene classification. Experimental results show that PC-ELM outperforms ELM in high-dimensional feature space at the cost of slightly higher computational complexity.

Keywords Scene classification · Extreme learning machine · Support vector machine · Partially connected extreme learning machine

D. Wang (✉) · R. Zhao · K. Mao
School of Electrical and Electronic Engineering, Nanyang Technological University,
50 Nanyang Avenue, Singapore, Singapore
e-mail: DWANG015@e.ntu.edu.sg

R. Zhao
e-mail: RZHAO001@e.ntu.edu.sg

K. Mao
e-mail: EKZMAO@ntu.edu.sg

© Springer International Publishing Switzerland 2016
J. Cao et al. (eds.), *Proceedings of ELM-2015 Volume 2*,
Proceedings in Adaptation, Learning and Optimization 7,
DOI 10.1007/978-3-319-28373-9_2

1   Introduction

In this paper, we consider the problem of scene classification, which is an important issue in many fields such as robotics and Unmanned Aerial Vehicle (UAV). Scene classification is often solved as a machine learning problem. A machine learning-based scene classification system consists of two main components, namely feature extraction and pattern classification. In computer vision studies, a number of feature extraction methods based on bag-of-features (BoF) [1] have been proposed (e.g., SIFT [2], GIST [3], etc.). In scene classification tasks, SVM [5, 6] classifiers are widely used.

One common but intractable problem in scene classification is that there are limited number of labeled image data for training while the dimensionality of image feature space is usually very high. Because of the so called "curse of dimensionality" [4], a low ratio of training sample size to feature dimension may lead to classifier over-fitting. To alleviate the over-fitting problem, support vector machine (SVM) attempts to find a discriminant function that maximizes the margin of separation, i.e., the shortest distance of training samples to the decision boundary. Although SVM has achieved high compatibility and robust performance in high-dimensional domains, it is often complained of its high computational complexity, especially for kernel SVMs.

Recently, a unified single-hidden-layer feedforward neural network namely extreme learning machine (ELM) [7] has been proposed. It has rapidly drawn attentions because of its much faster learning speed than SVMs, expecially in high-dimensional space. This is because the synaptic weights from the input layer to the hidden layer in ELM are randomly generated, without involving any learning procecure. The synaptic weights from the input layer to the hidden layer plays the role of a linear mapping. Thus, the random weights in ELM can be interpreted as the matrix of a random projection (RP).

Inspired by the idea of random projection (RP), Zhao et al. [8] have proposed a Semi-Random Projection (SRP) framework, which takes the advantage of random feature sampling of RP, but employs learning mechanism in the determination of the mapping matrix. The SRP method was applied to ELM architecture to yield the so called "partially connected ELM" (PC-ELM). The supervised learning mechanism within SRP framework is able to find a latent space with large discriminative power and meanwhile to keep the computational complexity low because the semi-random projection is obtained through supervised learning in low dimensional space. Hence, PC-ELM can generates hidden layer nodes with more discriminative power than the original ELM. The main merit of PC-ELM is that it achieves a good balance between computational complexity and generalization performance.

In this work, we present a study of PC-ELM for the scene classification problem. Most image feature extraction methods generate high-dimensional image representations. Regardless of the consistent high learning speed of ELM and robust performance of SVM, the balance of classification performance and computational simplicity is our major concern, especially when the feature space dimensionality is very high. In a sense, PC-ELM manages to achieve this balance by taking the merits of random feature sub-sampling of the SRP algorithm.

2 Semi-random Projection for Extreme Learning Machine

In this section, we first briefly review Extreme Learning Machine (ELM) and the connection between random projection (RP) and ELM. We then discuss the SRP method proposed in [8] and the SRP-based partially connected ELM (PC-ELM).

2.1 Extreme Learning Machine (ELM)

ELM is originally proposed as a generalized single-hidden-layer feedforward neural network whose hidden layer nodes need not be tuned [7]. Given a set of training data $\mathbf{X} = [\mathbf{x_1}, \mathbf{x_2}, \dots, \mathbf{x_N}]$, where $\mathbf{x}_i \in \mathbb{R}^d$ with class label $y_i \in \{1, 2, \dots, N_c\}$, the output of ELM is given as follows:

$$f(\mathbf{x}) = \sum_{i=1}^{M} \beta_i h_i(\mathbf{x}) = \mathbf{h}(\mathbf{x})\boldsymbol{\beta} \tag{1}$$

where $\mathbf{h}(\mathbf{x}) = [h_1(\mathbf{x}), \dots, h_M(\mathbf{x})]$ represents the output of the M hidden layer nodes with respect to the input sample $\mathbf{x}$, and $\boldsymbol{\beta} = [\beta_1, \dots, \beta_M]^T$ denotes the weights connecting the hidden layer and the output layer. By minimizing the training error and the norm of vector $\boldsymbol{\beta}$, the output weights $\boldsymbol{\beta}$ can be found as follows:

$$\boldsymbol{\beta} = \mathbf{H}^T \left(\frac{I}{C} + \mathbf{H}\mathbf{H}^T \right)^{-1} \mathbf{y} \tag{2}$$

where C is the cost parameter, and $\mathbf{H}$ is the hidden-layer output matrix:

$$\mathbf{H} = \begin{bmatrix} \mathbf{h}(\mathbf{x}_1) \\ \vdots \\ \mathbf{h}(\mathbf{x}_N) \end{bmatrix} = \begin{bmatrix} h_1(\mathbf{x}_1) & \cdots & h_M(\mathbf{x}_1) \\ \vdots & \ddots & \vdots \\ h_1(\mathbf{x}_N) & \cdots & h_M(\mathbf{x}_N) \end{bmatrix}. \tag{3}$$

Since the hidden nodes perform nonlinear piecewise continuous activation functions, the output of hidden layer is given as:

$$\mathbf{h}(\mathbf{x}) = [G(\mathbf{w}_1, b_1, \mathbf{x}), \dots, G(\mathbf{w}_M, b_M, \mathbf{x})] \tag{4}$$

where G denotes the activation function. One of the most popular activation functions is the sigmoid function:

$$G(\mathbf{w}_i, b_i, \mathbf{x}) = \frac{1}{1 + \exp(-z)} \tag{5}$$

$$z = \mathbf{w}_i \mathbf{x} + b_i \tag{6}$$

where $\{(\mathbf{w}_i, b_i)\}_{i=1}^M$ are randomly sampled according to any continuous probability distribution.

ELM can be interpreted as the integration of three parts: linear random mapping; nonlinear activation; and linear model learning. The linear random mapping is a kind of random projection (RP). Similar discussions on the relationship between RP and ELM can be found in [12], which combines ELM with RP. Unlike the RP which aims to reduce dimensionality, the linear random mapping in ELM often maps data to even higher dimension. Usually the number of hidden nodes for RP in ELM is larger than the data dimension.

2.2 Random Projection (RP)

Given a set of training data denoted by matrix $\mathbf{X} \in \mathbb{R}^{N \times d}$, where N and d are the data dimension and the sample number of data. We can simply transform the data $\mathbf{X}$ by mapping from the original space to a new space:

$$\mathbf{H} = \mathbf{X}\mathbf{W} \tag{7}$$

where $\mathbf{W} \in \mathbb{R}^{d \times r}$ is the linear transformation matrix, and $\mathbf{H} \in \mathbb{R}^{N \times r}$ denotes the new data representation.

In [9, 10], RP has been proposed as a dimensionality reduction method. RP method is effective to address the computational complexity issue because the transformation matrix $\mathbf{W}$ in RP is generated randomly. However, the randomly generated mapping matrix has some limitations. Without any supervised parameter learning or tuning, RP may not capture the information underlying the data.

The above mentioned limitation of RP is illustrated in Fig. 1. The figure shows samples from two classes in a two-dimensional feature space, which are depicted in red and blue, respectively. As the RP method project the data points onto randomly defined directions, two cases are demonstrated in Fig. 1. We provide an optimal case 2 (separable case), where the projection onto the dash dot line will result in good class separability. However, it is more likely to obtain a result as shown in case 1 (inseparable case) if the data points are projected on the dash line. After all, it is hard to yield ideal projections by solely relying on random projection without using any learning algorithms. The random operation shared by RP and linear random mapping in ELM inspires Zhao et al. [8] to introduce the semi-random projection (SRP) to ELM.

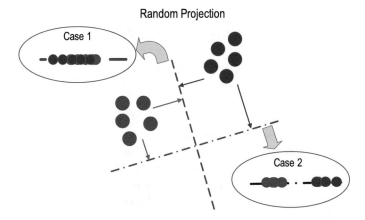

Random Projection

Fig. 1 Illustrations of two possible cases of RP

2.3 Semi-random Projection (SRP)

To address the limitation of RP analysed above, Ref. [8] proposed a novel dimensionality transformation framework called Semi-Random Projection (SRP). In contrast to RP, the main idea of SRP is to learn a latent space where the task-related information can be preserved and meanwhile the learning speed would not drop much. The SRP consists of two parts: random sampling of features (random process) and the transformation matrix learning (non-random process).

Suppose we are given training data $\mathbf{X} \in \mathbb{R}^{N \times d}$, a subset of d_s (d_s is an integer close to $\sqrt{d}$) features is randomly selected. Then the original data matrix is reduced to a sub-matrix $\hat{\mathbf{X}}_i$ with a smaller size of $N \times d_s$ in the ith iteration. The iterative transformation matrix $\hat{\mathbf{w}}_i \in \mathbb{R}^{d_s \times r}$ maps data $\hat{\mathbf{X}}_i$ into the r-dimensional space as follows:

$$\mathbf{h}_i = \hat{\mathbf{X}}_i \hat{\mathbf{w}}_i \tag{8}$$

where $\mathbf{h}_i \in \mathbb{R}^{N \times r}$ is the data representation in ith iteration, in which column is the projection of a sample on the new dimension. $\hat{\mathbf{w}}_i$ can be learned by using the linear discriminant analysis (LDA). The optimization solution for LDA can be derived as follows:

$$\mathbf{W}^* = \underset{\mathbf{W} \in \mathbb{R}^{d \times r}}{\arg\max} \operatorname{Tr} \left(\frac{\mathbf{W}^T \mathbf{L} \mathbf{W}}{\mathbf{W}^T \mathbf{B} \mathbf{W}} \right) \tag{9}$$

where Tr denotes the trace of a matrix. According to the fraction relation in Eq. (9), $\mathbf{L}$ represents the quantity need to be enhanced and $\mathbf{B}$ denotes that need to be suppress. In particular, matrices $\mathbf{L}$ and $\mathbf{B}$ in SRP can be calculated as:

$$\text{SRP}: \begin{cases} \mathbf{L} = \sum_{c=1}^{N_c} n_c (\widehat{\overline{\mathbf{x}}}^c - \widehat{\overline{\mathbf{x}}})(\widehat{\overline{\mathbf{x}}}^c - \widehat{\overline{\mathbf{x}}})^T \\ \mathbf{B} = \sum_{i=1}^{N} (\widehat{\mathbf{x}^i} - \widehat{\overline{\mathbf{x}}^{c_i}})(\widehat{\mathbf{x}^i} - \widehat{\overline{\mathbf{x}}^{c_i}})^T + \eta \mathbf{I}_{d_s} \end{cases} \tag{10}$$

where c_i denotes the class label of the ith sample, $\overline{\mathbf{x}}$ represents the mean vector of the all samples, $\overline{\mathbf{x}}^c$ represents the mean vector of the cth class, N_c and n_c denote the number of classes and the number of samples in the cth class, respectively. $\widehat{*}$ means the corresponding value calculated on the randomly selected feature subset from the original high dimension space. The second term in $\mathbf{B}$ of Eq. (10) represents a regularization term. $\mathbf{I}_{d_s}$ denotes an identity matrix with a size of $d_s \times d_s$ and η represents the regularization weight.

In a N_c-class LDA model, the optimization problem in Eq. (9) can be solved by formulating the following generalized eigenvalue problem:

$$\mathbf{L}\boldsymbol{\varphi} = \lambda \mathbf{B}\boldsymbol{\varphi} \tag{11}$$

where $\boldsymbol{\varphi} = [\boldsymbol{\varphi}_1, \boldsymbol{\varphi}_2, \dots, \boldsymbol{\varphi}_r]$ and $\lambda = [\lambda_1, \lambda_2, \dots, \lambda_r]$ are the eigenvectors and their corresponding eigenvalues. Here, r can be an arbitrary integer from 1 to $N_c - 1$. The optimal transformation matrix can be calculated based on the eigenvectors $\boldsymbol{\varphi}$ with the eigenvalues λ as $\widehat{\mathbf{w}}_i = [\sqrt{\lambda_1}\boldsymbol{\varphi}_1, \sqrt{\lambda_2}\boldsymbol{\varphi}_2, \dots, \sqrt{\lambda_{N_c}}\boldsymbol{\varphi}_r]$.

Next, $\widehat{\mathbf{w}}_i$ can be extended to the original data dimension $\mathbf{w}_i \in \mathbb{R}^{d \times r}$ by interpolating zeros on the unselected positions in every column of $\widehat{\mathbf{w}}_i$ (recall that $\widehat{\mathbf{w}}_i$ is corresponding to the randomly selected feature subset of original d features). This process are repeated o times and a set of transformation sub-matrices will be obtained. Thus, the final transformation matrix $\mathbf{W} \in \mathbb{R}^{d \times (r \cdot o)}$ can be denoted as:

$$\mathbf{W} = [\mathbf{w}_1 | \mathbf{w}_2 | \dots | \mathbf{w}_o] \tag{12}$$

The difference between RP and SRP is that learning is used in SRP for transformation matrix, while random assignment is used in RP.

2.4 Partially Connected ELM (PC-ELM)

Although ELM has achieved success in many applications, its performance in high dimensional space may not as great as in relatively low dimensional data. The limitation of randomised feature mapping from the input layer to the hidden layer has been demonstrated in Sect. 2.2. Inspired by the SRP as depicted in Sect. 2.3, Ref. [8] has implemented parameter-based model learning algorithm associated with linear random mapping in ELM. Specifically, the parameters $\{(\mathbf{w}_i)\}_{i=1}^{M}$ are learned based

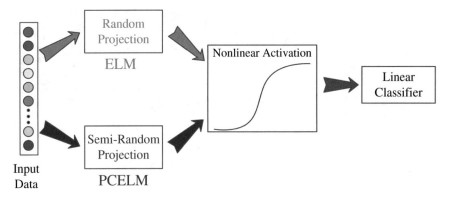

Fig. 2 An overview of ELM and PC-ELM

on SRP instead of random assignment. The application of SRP to ELM yields the so-called partially connected ELM (PC-ELM) as every new dimension of SRP uses a subset of the original feature set. Figure 2 compares the PC-ELM with ELM.

The major difference between ELM and PC-ELM is that ELM assigns random values to the linear feature mapping, while PC-ELM learns $\{(\mathbf{w}_i)\}_{i=1}^{M}$ using SRP. The degree of randomness is thereby suppressed in PC-ELM in contrast to ELM. According to the experimental results in [8], PC-ELM outperforms ELM at the cost of a slight higher computational complexity.

3 Experiments

In this section, we evaluate the application of PC-ELM for scene classification using the benchmark UIUC-Sport events dataset [13]. In order to verify the effectiveness of PC-ELM, we measure the experimental results and compare the classification performance with other state-of-the-art methods including ELM and rbf-SVM.

3.1 Dataset and Experimental Setup

The UIUC-Sport dataset contains 8 sports event categories: rock climbing, badminton, bocce, croquet, polo, rowing, sailing, and snowboarding. The image number in each class ranges from 137 to 250, and there are 1579 images in total. It is noted that the difficulty levels of classification within a category are varying with the distance of the foreground objects. Figure 3 shows some example images in the dataset.

The event recognition task is an 8-class classification problem. Following the experiment setting of [13], 70 local images are randomly selected for training and we test on 60 images. In order to achieve statistically significant experimental results,

Fig. 3 Example images for the UIUC-Sport dataset

we repeat 50 times of the training/ test data random split process and present the averaged results. For evaluation, we perform rbf-SVM, ELM and PC-ELM as the multi-class classifiers on the UIUC-Sports dataset. We measure the experimental results and compare the generalization performance of them in three state-of-the-art image representations including GIST, PHOW [15] (fast dense SIFT). We implement rbf-SVM using LIBSVM [14] with fixed parameter settings, where the hyperparameter C and γ is set to 2 and 0.1, respectively. In order to achieve good performance for ELM/ PC-ELM, we adjust their parameters for individual image representations. Experimental setup is as follows:

- **GIST descriptors.** We resize the images into 256×256; the number of filters is 4; the number of orientations per scale is [8 8 8 8]. For ELM, the cost coefficient C and the number of hidden neurons are set as 1 and 2000, respectively. For PC-ELM, the penalty coefficient C is set to 1, and we define 4 sub-sections LDA (new projection dimension mentioned in Sect. 2.3) with 500 hidden neurons (iteration of feature random selection for each sub-section).
- **PHOW descriptors.** We perform the PHOW descriptors via VLFeat toolbox [16]. Image is resized to 640×480 pixels; regular grids spacing [4 6 8 10]; SIFT descriptors of 6×6 pixels; pyramid level $L=3$; visual vocabulary size $M=200$. For ELM, the cost coefficient C and the number of hidden neurons are set as 0.01 and 4200, respectively. For PC-ELM, the penalty coefficient C is set to 0.01, and we define 7 sub-sections LDA with 600 hidden neurons for each sub-section.

3.2 Results

The first experiment uses the GIST descriptors. We obtain 512-dimension of features for each image samples. We show and compare the average classification performance of rbf-SVM, ELM, and PC-ELM in Fig. 4a, b. As shown in Fig. 4a, PC-ELM outperforms SVM and ELM. Note that the GIST image descriptors present relatively

Fig. 4 Performance results of GIST. The *x*-axis indicates the number of engaged features applied for the experiments. The *y*-axis denotes the average test set classification accuracy results in (**a**). The *y*-axis indicates the training time in (**b**)

(**a**) The average accuracy of 50 trails classification experiments.

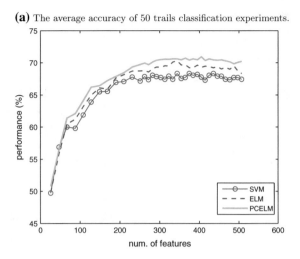

(**b**) The average training time of three classification algorithms.

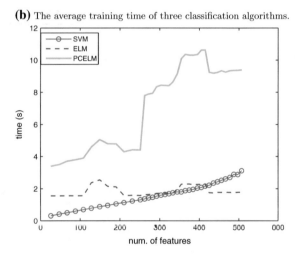

low-dimensional feature space, where the learning speed of PC-ELM is inferior to rbf-SVM and PC-ELM, as depicted in Fig. 4b.

The second experiment uses the dense SIFT image descriptors (PHOW). We obtain 12,600-dimensional feature representation for each image. The second experiment is able to evaluate the performance of the three multi-class classifiers to high-dimensional domain. In Fig. 5a, b, we compare and analyze the average classification performance of rbf-SVM, ELM, and PC-ELM. We observe that PC-ELM produces robust performance and achieves remarkable improvements over ELM and rbf-SVM in Fig. 5a. Meanwhile, the difference of learning speed between the original ELM and PC-ELM is not much. Both ELM and PC-ELM run much faster than the SVM method as the feature dimensionality increases, as illustrated in Fig. 5b.

Fig. 5 Performance results
of PHOW. The *x*-axis
indicates the number of
engaged features applied for
the experiments. The *y*-axis
denotes the average test set
classification accuracy
results in (**a**). In **b**, the *y*-axis
indicates the training time.
Note that this figure is
represented in
semi-logarithmic coordinates

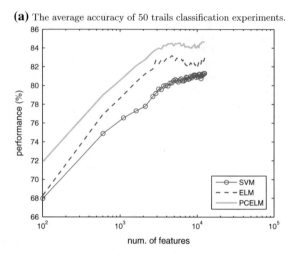

(**a**) The average accuracy of 50 trails classification experiments.

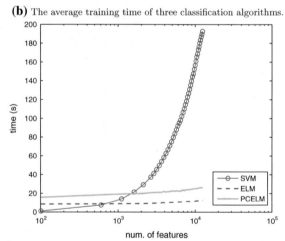

(**b**) The average training time of three classification algorithms.

4   Conclusion

In this paper, we have presented the SRP-based PC-ELM for scene classification. Experimental results on the benchmark dataset show that the PC-ELM network achieves a good balance between learning speed and generalization performance. Experimental results also show that PC-ELM has strong immunity to high-dimensional feature space, which often results in over-fitting to other classifiers. Exploration of the PC-ELM in other applications is undergoing, and results will be reported in our future publications.

References

1. Csurka, Gabriella, et al. "Visual categorization with bags of keypoints." Workshop on statistical learning in computer vision, ECCV. Vol. 1. No. 1–22 (2004)
2. Lowe, David G. "Object recognition from local scale-invariant features." Computer vision, 1999. The proceedings of the seventh IEEE international conference on. Vol. 2, (1999)
3. Oliva, Aude, Torralba, Antonio: Modeling the shape of the scene: A holistic representation of the spatial envelope. International journal of computer vision 42(3), 145–175 (2001)
4. Trunk, G.V.: A problem of dimensionality: A simple example. Pattern Analysis and Machine Intelligence, IEEE Transactions on 3, 306–307 (1979)
5. Boser, Bernhard E., Isabelle M. Guyon, and Vladimir N. Vapnik. "A training algorithm for optimal margin classifiers." Proceedings of the fifth annual workshop on Computational learning theory (1992)
6. Joachims, Thorsten: Text categorization with support vector machines: Learning with many relevant features. Springer, Berlin Heidelberg (1998)
7. Huang, Guang-Bin, Qin-Yu Zhu, and Chee-Kheong Siew. "Extreme learning machine: a new learning scheme of feedforward neural networks." Neural Networks, 2004. Proceedings. 2004 IEEE International Joint Conference on. Vol. 2 (2004)
8. Zhao, Rui, Mao, Kezhi: Semi-Random Projection for Dimensionality Reduction and Extreme Learning Machine in High-Dimensional Space. Computational Intelligence Magazine, IEEE 10(3), 30–41 (2015)
9. Bingham E. and Mannila H., Random projection in dimensionality reduction: Applications to image and text data, in Proc. 7th ACM SIGKDD Int. Conf. Knowledge Discovery Data Mining, pp. 245250 (2001)
10. Li, P, Hastie, T. J., and Church, K. W. Very sparse random projections, in Proc. 12th ACM SIGKDD Int. Conf. Knowledge Discovery Data Mining, pp. 287296 (2006)
11. Huang, Guang-Bin, et al. "Extreme learning machine for regression and multiclass classification." Systems, Man, and Cybernetics, Part B: Cybernetics, IEEE Transactions on 42.2 pp. 513–529 (2012)
12. Cambria, E., Huang, G. B., Kasun, L. L. C., Zhou, H., Vong, C. M., Lin, J., ..., Liu, J. Extreme learning machines [trends & controversies]. Intelligent Systems, IEEE (2013)
13. Li, J-L. and Li, F-F. What, where and who? Classifying event by scene and object recognition. IEEE Intern. Conf. in Computer Vision (2007)
14. Chang, C-C., and Lin, C-J. LIBSVM : a library for support vector machines. ACM Transactions on Intelligent Systems and Technology, 2:27:1–27:27. Software available at http://www.csie.ntu.edu.tw/~cjlin/libsvm (2011)
15. Bosch, A., Zisserman, A., Munoz, X.: Image classifcation using random forests and ferns. In Proc, ICCV (2007)
16. Vedaldi, A., Fulkerson, B. VLFeat: An open and portable library of computer vision algorithms, http://www.vlfeat.org/ (2008)

Two-Layer Extreme Learning Machine
for Dimension Reduction

Yimin Yang and Q.M. Jonathan Wu

Abstract The extreme learning machine (ELM), which was originally proposed for "generalized" single-hidden layer feedforward neural networks (SLFNs), provides efficient unified learning solutions for the applications of regression and classification. It presents competitive accuracy with superb efficiency in many applications. However, due to its single-layer architecture, feature selection using ELM may not be effective for natural signals. To address this issue, this paper proposes a new ELM-based multi-layer learning framework for dimension reduction. The novelties of this work are as follows: (1) Unlike the existing multi-layer ELM methods in which all hidden nodes are generated randomly, in this paper some hidden layers are calculated by replacement technologies. By doing so, more important information can be exploited for feature learning, which lead to a better generalization performance. (2) Unlike the existing multi-layer ELM methods which only work for sparse representation, the proposed method is designed for dimension reduction. Experimental results on several classification datsets show that, compared to other feature selection methods, the proposed method performs competitively or much better than other feature selection methods with fast learning speed.

Keywords Extreme learning machine · Dimension reduction · Feature selection · Multi-layer perceptron · Deep neural network

1 Introduction

During the past years, extreme learning machine (ELM) has been becoming an increasingly significant research topic for machine learning and artificial intelligence, due to its unique characteristics, i.e., extremely fast training, good generalization and universal approximation capability. ELM provide a unified

Y. Yang · Q. M. J. Wu (✉)
Department of Electrical and Computer Engineering, University of Windsor,
Windsor, ON N9B 3P4, Canada
e-mail: jwu@uwindsor.ca

© Springer International Publishing Switzerland 2016
J. Cao et al. (eds.), *Proceedings of ELM-2015 Volume 2*,
Proceedings in Adaptation, Learning and Optimization 7,
DOI 10.1007/978-3-319-28373-9_3

learning framework for "generalized" single-hidden layer feedforward neural networks (SLFNs), including but not limited to sigmoid networks, RBF networks, threshold networks, trigonometric networks, fuzzy inference systems, high-order networks wavelet networks, etc. According to conventional neural network (NN) theories, SLFNs are universal approximators when all parameters of the networks $(\mathbf{a}, b, \beta)$ are allowed to be adjustable [1]. But unlike these learning methods, the ELM [2–6] is a full-random learning method that differs from the usual understanding of learning.

But the original ELM or its variants mainly focuses on regression or/and classifications. Recently multi-layer networks has been becoming an increasingly significant research topic as recent research developments show that multi-layer networks with a suitable learning method can be used for feature extraction. Feature selection (extraction) techniques are designed to reduce dimensionality by finding a meaningful feature subset or feature combinations, because high dimensionality significantly increases the time and space requirements for processing the data [7–9]. Hinton [10] and Vincent [11] show that multi-layer networks with BP learning method (called deep learning (DL), or deep belief netework (DBN)) can be used to reduce the dimensionality of data. The DL considers multi-layer as a whole with unsupervised initialization, and after such initialization the entire network will be trained by BP-based NNs, and all of the layers are "hard coded" together. Meanwhile, another leading trend for multi-layer NN learning is based on the ELM. Charama et al. attempts to develop a multi-layer learning architecture using ELM based autoencoder as its building block. The original inputs are decomposed into multiple hidden layers, and the outputs of the previous layer are used as the inputs of the current one. Tang et al. [12] proposed a new ELM-based hierarchical learning framework for multi-layer perceptron. Different from [13], this new learning system is divided into two main components: unsupervised feature extraction followed by supervised feature classification, and they are bridged by random initialized hidden weights.

In existing ELM based multi-layers method, all the nodes in the feature mapping layer are randomly generated. But if doing so, the "optimize information" may not pass through the purely random hidden layers and the useful features may be destroyed by this purely random layer. In this paper we extend the ELM and propose a new two-layer ELM framework for dimension reduction. Different from current ELM-based multi-layers methods which only work for sparse representation learning, the proposed method is designed for dimension reduction. The learning speed of the proposed method can be several to tens times faster compared to deep networks such as deep belief networks and stacked auto-encoders. Furthermore, our method can provide a much better generalization performance than other feature selection methods such as SNE, SPE, t-SNE, SSNE, DBN, SAE, LGE, etc.

2 Preliminaries and Problem Statement

2.1 Notations

The sets of real numbers is denoted by $\mathbf{R}$. For M arbitrary distinct samples $\{(\mathbf{x}_i, \mathbf{y}_i)\}_{i=1}^{M}$ $(\mathbf{x}_i \in \mathbf{R}^n, \mathbf{y}_i \in \mathbf{R}^m)$, $\mathbf{x}$ denotes the input data and $\mathbf{y}$ denotes the desired output data. g is a sigmoid or sine activation function. Other notations are defined in Table 1.

2.2 Basic-ELM

For M arbitrary distinct samples $(\mathbf{x}_i, \mathbf{t}_i)$, where $\mathbf{x}_i = [x_{i1}, x_{i2}, \dots, x_{in}]^T \in \mathbf{R}^m$ and $t_i \in \mathbf{R}$. ELM is proposed for SLFNs and output function of ELM for SLFNs is

$$f_l(\mathbf{x}) = \sum_{i=1}^{L} \beta_i g(\mathbf{a}_i \cdot \mathbf{x}_j + b_i) = \mathbf{H} \cdot \boldsymbol{\beta} \tag{1}$$

where $\boldsymbol{\beta}$ is the output weight matrix connecting hidden nodes to output nodes, g represents an activation function. ELM theory aims to reach the smallest training error but also the smallest norm of output weights:

$$\text{Minimize} : \|\boldsymbol{\beta}\|^2 + C\|\mathbf{H}\boldsymbol{\beta} - \mathbf{t}\|^2 \tag{2}$$

Table 1 Notations to be used in the proposed method

Notation	Meaning
$(\mathbf{a}, b)$	A hidden node
$\mathbf{a}_f$	Input weights in entrance feature mapping layer $\mathbf{a}_f \in \mathbf{R}^{d \vee m}$
b_f	Bias in entrance feature mapping layer $b_f \in \mathbf{R}$
$(\mathbf{a}_f^j, b_f^j)$	Hidden nodes in the jth entrance feature mapping layer
$(\mathbf{a}_{fi}, b_f)$	The ith hidden node in entrance feature mapping layer
$(\mathbf{a_n}, b_n)$	Hidden nodes in exit feature mapping layer and $\mathbf{a_n} \in \mathbf{R}^{m \times d}$
u_j	Normalized function in the jth general node, $u_j(\cdot) : \mathbf{R} \to (0, 1]$, u_j^{-1} represent its reverse function
$\mathbf{H}_f$	Feature data generated by a entrance feature mapping layer
$\mathbf{H}_f^i$	Feature data generated by the ith feature mapping layer
M	Number of training samples
m	Input data dimension, i.e., $\mathbf{y} \in \mathbf{R}^{m \times N}$
d	Feature data dimension

For L hidden nodes, $\mathbf{H}$ is referred to as ELM feature mapping or Huang's transform:

$$
\mathbf{H} = \begin{bmatrix} g(\mathbf{x}_1) \\ \vdots \\ g(\mathbf{x}_M) \end{bmatrix} = \begin{bmatrix} g_1(\mathbf{x}_1) & \cdots & g_L(\mathbf{x}_1) \\ \vdots & \cdots & \vdots \\ g_1(\mathbf{x}_M) & \cdots & g_L(\mathbf{x}_M) \end{bmatrix} \tag{3}
$$

and $\mathbf{t}$ is the training data target matrix:

$$
\mathbf{t} = \begin{bmatrix} t_1^T \\ \vdots \\ t_M^T \end{bmatrix} = \begin{bmatrix} t_{11} & \cdots & t_{1m} \\ \vdots & \cdots & \vdots \\ t_{M1} & \cdots & t_{Mm} \end{bmatrix} \tag{4}
$$

Huang et al. have proved the following lemma.

Lemma 1. ([2]) *Given any bounded nonconstant piecewise continuous function g : $\mathbf{R} \to \mathbf{R}$, if span $\{G(\mathbf{a}, b, \mathbf{x}) : (\mathbf{a}, b) \in \{\mathbf{R}^d \times R\}$ is dense in L^2, for any target function f and any function sequence randomly generated based on any continuous sampling distribution, $\lim_{n\to\infty} \|f - f_n\| = 0$ holds with probability one if the output weight β_i are determined by ordinary least square to minimize $\|f(x) - \sum_{i=1}^{L} \beta_i g_i(\mathbf{x})\|$.*

The lemma above [14] shows that randomly generated networks with the outputs being solved by least mean square are able to maintain the universal approximation capability, if and only if the activation function g is nonconstant piecewise. Thus the ELM training algorithm can be summarized as follow:

(1) Randomly assign the hidden node parameters, e.g., the input weights $\mathbf{a}_i$ and biases $\mathbf{b}_i$ for additive hidden nodes, $i = 1, \ldots, L$.
(2) Calculate the hidden layer output matrix $\mathbf{H}$.
(3) Obtain the output weight vector.

$$
\boldsymbol{\beta} = \mathbf{H}^\dagger \cdot \mathbf{t} \tag{5}
$$

where $\mathbf{H}^\dagger$ is the Moore-Penrose generalized inverse of matrix $\mathbf{H}$.

3   The Proposed Method

In this section, we propose a new multi-layer ELM autoencoder. The network architecture of the proposed method is to be introduced, and a new learning multi-layer ELM method is also presented.

3.1 The Proposed Framework

The proposed framework is build in a multi-layer mode, as is shown in Fig. 1. Unlike the greedy layer-wise training of the traditional Deep Learning architecture [10, 15], one can see that the proposed method can be divided into two separate phases: unsupervised feature representation and supervised classification. For the former phase, a new ELM-based autoencoder is developed to extract multi-layer related-data features; while for the latter one, the original ELM classifier is implemented for supervised classification.

In the following, we will provide a detailed framework of the proposed method, as well as its advantages against the existing DL and other multi-layer ELM algorithms. The input data first should be transformed into an ELM random feature space, then a replacement learning method is used to update the ELM feature space. Finally after all the parameter in the ELM feature space are fixed, the input raw data will be converted into a low-dimension compact features. There are several steps in this proposed method.

Step 1: Given M arbitrary distinct training samples $\{(\mathbf{x}_k, \mathbf{y}_k)\}_{k=1}^M, \mathbf{x}_k \in \mathbf{R}^n$, which is sampled from a continuous system. Because autoencoder tries to approximate the input data to make the reconstructed outputs being similar to the inputs, here $\mathbf{x} = \mathbf{y}$, then the initial general input weights and biases of the feature mapping layer are generated randomly as:

$$\mathbf{H}_f = g(\mathbf{a}_f \cdot \mathbf{x} + b_f)$$
$$(\mathbf{a}_f)^T \cdot \mathbf{a}_f = \mathbf{I}, (b_f)^T \cdot b_f = 1 \tag{6}$$

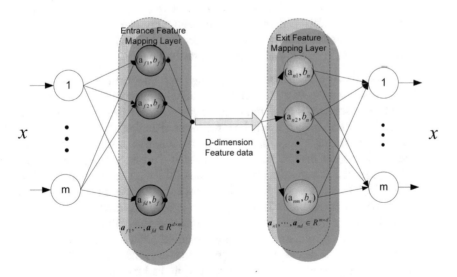

Fig. 1 The structure of our proposed method with two-layer network

where $\mathbf{a}_f \in \mathbf{R}^{d \times n}$, $b_f \in \mathbf{R}$ is the orthogonal random weight and bias of feature mapping layer. $\mathbf{H}_f$ is current feature data.

Step 2: Given a sigmoid or sine activation function g, for any continuous desired outputs $\mathbf{y}$, the optimal parameters of hidden-layer $\{\hat{\mathbf{a}}_h, \hat{b}_h\}$ are obtained as

$$
\begin{aligned}
\mathbf{a}_h &= g^{-1}(u_n(\mathbf{y})) \cdot (\mathbf{H}_f)^{-1}, \quad \mathbf{a}_h \in \mathbf{R}^{d \times m} \\
b_h &= \sqrt{mse(\mathbf{a}_h \cdot \mathbf{H}_f - g^{-1}(u_n(\mathbf{y})))}, \, b_n \in \mathbf{R} \\
g^{-1}(\cdot) &\begin{cases} arcsin(\cdot) & \text{if } g(\cdot) = sin(\cdot) \\ -\log(\dfrac{1}{(\cdot)} - 1) & \text{if } g(\cdot) = 1/(1 + e^{-(\cdot)}) \end{cases}
\end{aligned} \tag{7}
$$

where $\mathbf{H}^{-1} = \mathbf{H}^T(\frac{C}{\mathbf{I}} + \mathbf{H}\mathbf{H}^T)^{-1}$; C is a positive value; u_n is a normalized function $u_n(\mathbf{y}) : \mathbf{R} \to (0, 1]$; g^{-1} and u_n^{-1} represent their reverse function.

Step 3: update $\mathbf{a}_f, b_f$ by

$$
\begin{aligned}
\mathbf{a}_f &= (\mathbf{a}_h)^T \\
b_f &= \sqrt{mse(\mathbf{a}_f \cdot \mathbf{x} - \mathbf{y})}, \, b_f \in \mathbf{R}
\end{aligned} \tag{8}
$$

and update the feature data $\mathbf{H}_f = g(\mathbf{a}_f \cdot \mathbf{x} + b_f)$.

Step 4: Repeat steps 2–3 L times. The parameters $\mathbf{a}_f, b_f, \mathbf{a}_n, b_n$ are obtained and the feature data $\mathbf{H}_f$ equal $g(\mathbf{x}, \mathbf{a}_f, b_f)$.

The above four steps constitute the proposed method, as shown in Algorithm 1 and Fig. 2. The proposed framework is developed based upon random feature mapping and fully exploits the universal approximation capability of ELM, both in feature learning and feature classification. There are three differences between our method and other mutli-layer network feature selection method.

(1) This is completely different from the existing DL frameworks, where all the hidden layers are put together as a whole system, with unsupervised initialization. The whole system need to be retrained iteratively using BP-based NNs. Thus, the training of our method would be much faster than that of the DL.

(2) According to Huang's Transform [16], the randomness in ELM are two folds. First, hidden nodes may be randomly generated. Second, some nodes are randomly generated and some are not, but none of them are tuned. Different from existing ELM methods which hidden nodes in each layer are generated randomly, here we thank that it is impossible to have all the layers randomly generated. If feature mapping layer are randomly generated, the "optimize information" may not pass through the purely random hidden layers or this useful information may be destroyed by the purely random hidden nodes. Thus in our method, hidden nodes are highly correlated with the input data by pulling back the desired output to hidden layers.

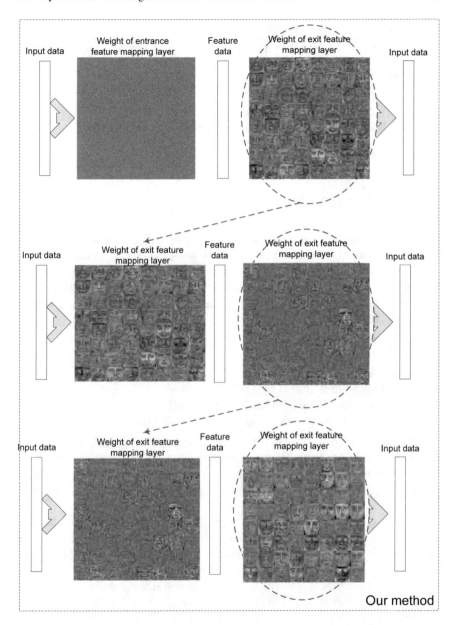

Fig. 2 The learning steps of our proposed method. Input data come from Olive face dataset

---

Algorithm 1 The proposed method

Initialization: Given a training set $\{(\mathbf{x}_i, \mathbf{y}_i)\}_{i=1}^{M} \subset \mathbf{R}^m \times \mathbf{R}^m$, a sine or sigmoid function $g(\cdot)$, maximum loop number L, and $j = 1$.

Learning step:

Step 1) Generate the initial general node $\mathbf{a}_f$ and b_f according to Eq. (6).

while $j < L$ **do**

    Step 2) Calculate $\mathbf{a}_h$ and b_h according to Eq. (7).

    step 3) Set $j = j + 1$, update nodes $\mathbf{a}_f, b_f$ according to Eq. (8), and update $\mathbf{H}_f$.

end while

Obtain feature data $\mathbf{H}_f$.

---

4 Experimental Verification

In this section, aimed at examining the performance of our proposed learning method, we test the proposed method on 9 classification problems. The experiments are conducted in Matlab 2013a with 32 GB of memory and an I7-4770 (3.4G) processor. And we compare the generalization performance of the proposed method with some unsupervised learning methods for dimensionality reduction. The six unsupervised feature selection methods are:

(1) Our proposed algorithm
(2) Deep Belief Network (DBN) [10, 15]
(3) Stacked Auto-Encoders (SAE) [11]
(4) Linear Graph Embedding (LGE) [17]
(5) Orthogonal Linear Graph Embedding (OLGE) [18]
(6) Unsupervised Locality Preserving Projection (LPP) [19]

All databases are preprocessed in the same way (held-out method). Table 2 shows the training and testing data of the corresponding datasets. Permutations of the whole dataset are taken with replacement randomly, and #train (shown in Table 2) of them are used to create the training set. The remaining dataset is used for the test set (#test in Table 2).

The average results are obtained over 10 trials for all problems. For our proposed method, parameter C is selected from $C \in \{2^{-4}, \ldots, 2^8\}$. For DBN, we set epoch at 10 and select the parameter batch size from $[10, \#(\text{Train})/10, \#(\text{Train})/9, \ldots, \#(\text{Train})]$, and the parameter momentum γ from $[0.01, 0.02, \cdots, 0.1]$. The parameter α equals $1 - \gamma$. For SAE, the learning rate equals 1; the parameter batch size is selected from $[10, \#(\text{Train})/20, \#(\text{Train})/18, \ldots, \#(\text{Train})/2]$; and the parameter masked fraction rate δ is selected from $[0.45, 0.46, \ldots, 0.55]$.[1]

All the testing accuracy is obtained following the same steps: first we use these methods to obtain data features, and then an SLFN classifier is used to generate testing accuracy. For DBN and SAE, we first reduce or increase the dimensions of

---

[1] For DBN and SAE, it is impossible for us to set the parameter range too widely as the computational cost of these two methods is very high. If we select these parameters from a wide range, only one trial is needed in more than 10 h.

Table 2 Specification of 15 benchmark datasets

Datasets	#features	#Train	# Test
Acoustic	51	40000	58000
USPS	256	7291	2007
Hill Valley	101	606	606
Protein	357	17766	6621
Gisette	5000	6000	1000
Leu	7129	38	34
Duke	7129	29	15
DNA	180	1046	1186
Satimage	36	4435	2000
S15	43008	1500	2985
Olive Face	4096	320	80

the testing datasets, and then BP and Fuzzy NN networks are used to generate testing accuracy, respectively. For the other methods, we first reduce the dimensions of the dataset, and then a 1000-hidden-nodes ELM network is used to obtain testing accuracy.

To indicate the advantage of the proposed method for unsupervised dimension reduction performance, tests have been conducted of the accuracy of the proposed method compared to other unsupervised dimension reduction methods. For LPP, LGE, OLGE, DBN, SAE, these compared methods can provide mapping function for testing data, i.e., techniques that learn an explicit function between the data space and the low-dimensional latent space. Thus in Table 3 and Fig. 3, training data are firstly used and then low-dimensional testing data are generated by these mapping functions. Table 3 displays the performance comparison of LPP, LGE, OLGE, DBN, SAE, and the proposed method. As we can see, our proposed algorithm consistently outperforms most of the compared feature selection algorithms on these classification problems and image datasets. As seen from these experimental results, the advantage of the proposed method for generalization performance is obvious. Consider Olive face (small number of samples with large input dimensions), and Scene15 (medium number of samples with large input dimensions).

(1) For the Olive face dataset, the testing error of the proposed method is about **4, 3, 14, 16**, and **15** times lower than that of LPP, LGE, OLGE, DBN, and SAE, respectively.
(2) For the Scene15 dataset, the testing accuracy of the proposed method is about **2.5, 2, 2.5, 8**, and **7** times higher than that of LPP, LGE, OLGE, DBN, and SAE, respectively.

Table 3 Generalization Performance Comparison (mean: average testing accuracy)

Datasets	#Dimensionality	LPP		LGE		OLGE		DBN		SAE		Ours	
		Mean (%)	Time	Mean (%)	Time	Mean (%)	Time	Mean (%)	Time	Mean (%)	Time	Mean (%)	Time
Gisstle	4000⟶2	**70.04**	83.88	59.16	78.11	57.00	100.42	50.71	1088.50	50.21	83.76	66.32	**9.46**
Hill	101⟶2	50.52	0.07	51.11	0.05	50.23	0.07	50.23	7.40	88.93	5.40	**91.06**	**0.05**
USPS	256⟶10	90.56	5.36	87.12	5.88	43.75	4.57	**94.31**	80.47	88.75	49.82	91.07	**0.74**
Satimage	36⟶6	**88.41**	1.34	88.20	299.15	86.61	0.62	71.93	23.07	80.95	12.31	85.10	**0.05**
Duke	7129⟶2	45.45	**0.02**	47.27	0.06	54.45	0.03	56.67	9.20	50.00	1.19	**60.55**	0.11
Leu	7129⟶10	52.94	**0.03**	70.59	0.03	60.00	0.04	50.01	15.89	56.25	2.48	**78.86**	0.18
DNA	180⟶2	73.74	0.22	62.90	0.29	59.16	0.18	56.47	16.20	70.70	6.58	**74.15**	**0.14**
Scene15	21504⟶15	28.11	17.56	30.38	2.04	27.39	1.32	9.36	3084.30	10.79	3726.41	**71.05**	**14.47**
Acoustic	50⟶2	54.61	401.53	57.96	412.84	57.45	368.32	62.24	521.67	62.44	391.77	**64.03**	1.51
Protein	357⟶2	47.26	32.17	47.67	32.13	46.36	17.00	46.16	264.67	44.03	83.22	**49.23**	**2.60**
Olive Face	4096⟶40	73.00	**0.12**	80.25	0.13	9.50	0.35	1.24	3.65	2.50	23.08	**95.10**	0.45

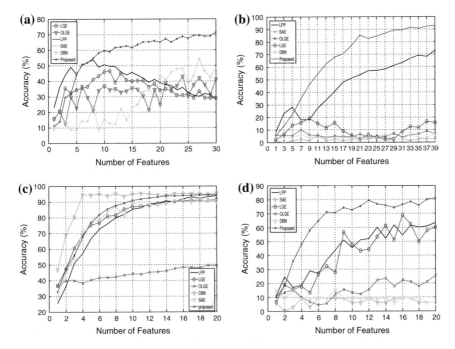

Fig. 3 Average testing accuracy by using LPP, LGE, OLGE, SAE, DBN and the proposed method on S15, Olive face, USPS and Yale face, where the *x*- and *y*-axes show the number of features and average testing accuracy, respectively

5 Conclusion

In this paper, we have proposed a novel multi-layer network training scheme, which is based on the universal approximation capability of the original ELM. The proposed method achieve high low-dimensional representation with layer-wise encoding. Moreover, compared with other feature selection methods, the training of our method is much faster and achieve higher learning accuracy. In these applications, our method functions as a feature extractor and a classier, and it achieve competitively or more better performance than relevant state-of-the-art methods.

References

1. Zhang, R., Lan, Y., Huang, G.-B., Zong-Ben, X.: Universal approximation of extreme learning machine with adaptive growth of hidden nodes. IEEE Trans. Neural Netw. Learn. Syst. **23**(2), 365–371 (2012)
2. Huang, G.B., Chen, L., Siew, C.K.: Universal approximation using incremental constructive feedforward networks with random hidden nodes. IEEE Trans. Neural Netw. **17**(4), 879–892 (2006)

3. Huang, G.B., Zhu, Q.Y., Siew, C.K.: Extreme learning machine: theory and applications. Neurocomputing **70**, 489–501 (2006)
4. Luo, J.H., Vong, C.M., Wong, P.K.: Sparse bayesian extreme learning machine for multiclassification. IEEE Trans. Neural Netw. Learn. Syst. **25**(4), 836–843 (2014)
5. Suresh, S., Dong, K., Kim, H.J.: A sequential learning algorithm for self-adaptive resource allocation network classifier. Neurocomputing **73**(16–18), 3012–3019 (2010)
6. Wang, X.Z., Chen, A.X., Feng, H.M.: Upper integral network with extreme learning mechanism. Neurocomputing **74**(16), 2520–2525 (2011)
7. He, X.F., Ji, M., Zhang, C.Y., Bao, H.J.: A variance minimization criterion to feature selection using laplacian regularization. IEEE Trans. Pattern Anal. Mach. Intell. **33**(10), 2013–2025 (2011)
8. Zhang, B., Li, W., Qing, P., Zhang, D.: Palm-print classification by global features. IEEE Trans. Syst. Man Cybern.: Syst. **43**(2), 370–378 (2013)
9. Bhatnagar, G., Wu, Q.M.J., Senior Member. Biometric inspired multimedia encryption based on dual parameter fractional fourier transform. IEEE Trans. Syst. Man Cybern.: Syst. **44**(9), 1234–1247 (2014)
10. Hinton, G.E., Salakhutdinov, R.R.: Reducing the dimensionality of data with neural networks. Science **313**(5786), 504–507 (2006). [2, 4, 8]
11. Vincent, P., Larochelle, H., Lajoie, I., Bengio, Y., Manzagol, P.A.: Stacked denoising autoencoders: learning useful representations in a deep network with a local denoising criterion. J. Mach. Learn. Res. **11**, 3371–3408 (2010)
12. Tang, J., Deng, C., Huang, G.B.: Extreme learning machine for multilayer perceptron. IEEE Trans. Neural Netw. Learn. Syst. (2015)
13. Huang, G.B., Kasun, L.L.C., Zhou, H., Vong, C.M.: Representational learning with extreme learning machine for big data. IEEE Intell. Syst. **28**(6), 31–34 (2013)
14. Huang, G.B., Zhou, H.M., Ding, X.J., Zhang, R.: Extreme learning machine for regression and multiclass classification. IEEE Trans. Syst. Man Cybern. Part B: Cybern. **42**(2), 513–529 (2012)
15. Salakhutdinov, R., Hinton, G.: An efficient learning procedure for deep Boltzmann machines. Neural Comput. **24**(8), 1967–2006 (2012)
16. Huang, G.-B.: What are extreme learning machines? Filling the gap between Frank Rosenblatt's dream and John von Neumann's Puzzle. Cogn. Comput. **7**(3), 263–278 (2015)
17. Cai, D., He, X.F., Han, J.W.: Spectral regression for efficient regularized subspace learning. In: IEEE 11th International Conference on Computer Vision, vol. 1-6, pp. 214–221 (2007)
18. Cai, D., He, X.F., Han, J.W., Zhang, H.J.: Orthogonal laplacianfaces for face recognition. IEEE Trans. Image Process. **15**(11), 3608–3614 (2006)
19. He, X.F., Niyogi, P.: Locality preserving projections. Adv. Neural Inf. Process. Syst. **16**(16), 153–160 (2004)

Distributed Extreme Learning Machine with Alternating Direction Method of Multiplier

Minnan Luo, Qinghua Zheng and Jun Liu

Abstract Extreme learning machine, as a generalized single-hidden-layer feedforward networks has achieved much attention for its extremely fast learning speed and good generalization performance. However, big data often makes a challenge in large scale learning of ELM due to the limitation of memory of single machine as well as the distributed manner of large scale data storage and collection in many applications. For the purpose of relieving the limitation of memory with big data, in this paper, we exploit a novel distributed extreme learning machine to implement the extreme learning machine algorithm in parallel for large-scale data set. A corresponding distributed algorithm is also developed on the basis of alternating direction method of multipliers which shows effectiveness in distributed convex optimization. Finally, some numerical experiments on well-known benchmark data sets are carried out to illustrate the effectiveness of the proposed DELM method and provide an analysis on the performance of speedup, scaleup and sizeup.

Keywords Extreme learning machine · Neurone work · Alternating direction method of multiplier

1 Introduction

Extreme learning machine (ELM) is a generalized single-hidden-layer feedforward networks, where the parameters of hidden layer feature mapping are generated randomly according to any continuous probability distribution [1] rather than are

M. Luo (✉) · Q. Zheng · J. Liu
SPKLSTN Lab, Department of Computer Science, Xi'an Jiaotong University,
Shaanxi 710049, China
e-mail: minnluo@mail.xjtu.edu.cn

Q. Zheng
e-mail: qhzheng@mail.xjtu.edu.cn

J. Liu
e-mail: liukeen@mail.xjtu.edu.cn

© Springer International Publishing Switzerland 2016
J. Cao et al. (eds.), *Proceedings of ELM-2015 Volume 2*,
Proceedings in Adaptation, Learning and Optimization 7,
DOI 10.1007/978-3-319-28373-9_4

tuned by gradient descent based algorithms. In this case, ELM algorithm achieves extremely fast learning speed and better performance of generalization. The ELM technique has been applied to many applications and performs effectively [2].

It is noteworthy that traditional ELM algorithm is often implemented on a single machine, and therefore it is inevitable to suffer from the limitation of memory with large scale data set. In particular in the era of big data, the data set scale are usually extremely large and the data is often very high-dimensional for detailed information [3]. On the other hand, it is actually necessary to deal with data sets in different machines due to the following two reasons: (1) The data set are stored and collected in a distributed manner because of the large scale of many application; (2) It is impossible to collect all of data together for the reason of confidentiality and the data set can be only accessed on their own machine. Based on the analysis above, how to implement algorithm of ELM with respect to data sets which located in different machines become a key problems.

In previous work, some parallel or distributed ELM algorithms have been implemented to meet the challenge of large-scale data set. For example, Q. He and et. al take advantages of the distributed environment provided by MapReduce [4] and propose an parallel ELM for regression on the basis of MapReduce by designing the proper ⟨*key*, *value*⟩ pairs [5]. X. Wang and et. al did some research on the issue of parallel ELM on the basis of Min-Max Modular network (M^3), namely as M^3-ELM [6]. This approach decomposes the classification problems into small subproblems, then trains an ELM for each subproblem, and in the end ensembles the ELMs with M^3-network. Besides the distributed framework utilized above, the alternating direction method of multipliers (ADMM) that aims to find a solution to a global problem by solving local subproblems coordinately, is also an effective optimization method, and in particular to distributed convex optimization [3]. It is noteworthy that C. Zhang and et. al investigate a large-scale distributed linear classification algorithms on the basis of ADMM framework and achieve a significant speedup over some other classifier [7].

Motivated by the advantages of ADMM in distributed optimization problem, in this paper, we exploit a novel distributed extreme learning machine (DELM) to implement the ELM algorithm on multiple machines in parallel in order to relieve the limitation of memory with large scale data set. It is different from the traditional ELM where all of data are loaded onto one processor and share a common output weight vector of hidden layer. The proposed DELM method allows large-scale data set to been stored in distributed manner; moreover, each processor is associated with one output weight vector β_i, where $i = 1, 2, \ldots, m$ and m denotes the number of processors. Each output weight vector can be determined in parallel, because it depends only on the corresponding sub-dataset. In the framework of ADMM, the shared output weight vector β across all processors is derived by combining all of output weight vectors β_i $(i = 1, 2, \ldots, m)$ with an included regularization vector.

The remainder of this paper is organized as follows. In Sect. 2, notations and preliminaries about ELM are reviewed. Section 3 focuses on the formulated optimization problems of distributed extreme learning machine. A corresponding distributed convex optimization algorithm is developed for the proposed DELM via alternating direction of method of multipliers in Sect. 4. Section 5 presents some experiments on well-known benchmark data sets to illustrate the effectiveness of the proposed DELM method. Conclusions are given in Sect. 6.

2 Principles of Extreme Learning Machine

Extreme learning machine (ELM) refer in particular to a kind of single-hidden-layer feedforward neural networks, where the hidden layer need not be tuned. In this case, the output function of ELM is formulated as

$$f_L(\mathbf{x}) = \sum_{j=1}^{L} \beta_j h_j(\mathbf{x}) = \mathbf{h}(\mathbf{x})\beta, \tag{1}$$

where $\beta = \left(\beta^1, \beta^2, \ldots, \beta^L\right)^{\mathsf{T}} \in \mathbb{R}^L$ is the output weights vector of the hidden layer with L nodes, which needs to be estimated analytically; Feature mapping $\mathbf{h} \colon \mathbb{R}^n \to \mathbb{R}^L$ maps input variable $\mathbf{x} \in \mathbb{R}^n$ to L-dimensional hidden-layer feature space and $\mathbf{h}(\mathbf{x}) = \left(h_1(\mathbf{x}), h_2(\mathbf{x}), \ldots, h_L(\mathbf{x})\right)$ denotes the hidden layer output (row) vector, where the component $h_j(\mathbf{x}) = G(\mathbf{a}_j, b_j, \mathbf{x})$ $(j = 1, 2, \ldots, L)$, as the output function of j-th hidden node, is known to users by generating the parameters $\{(\mathbf{a}_j, b_j) : j = 1, 2, \ldots, L\}$ randomly according to any continuous probability distribution [8]. In general, Sigmoid function

$$G(\mathbf{a}, b, \mathbf{x}) = \frac{1}{1 + \exp\left(-\left(\mathbf{a}^{\mathsf{T}}\mathbf{x} + b\right)\right)},$$

Gaussian function

$$G(\mathbf{a}, b, \mathbf{x}) = \exp\left(-b \|\mathbf{x} - \mathbf{a}\|^2\right)$$

and some other nonlinear activation function are usually used in the framework of ELM.

Given a data set $D = \left\{\left(\mathbf{x}_k^{\mathsf{T}}, t_k\right) : \mathbf{x}_k \in \mathbb{R}^n, t_k \in \mathbb{R}, k = 1, 2, \ldots, N\right\}$, ELM randomly generate the input weights and estimate the output weights vector β by minimizing the training error as well as the norm of output weight for better generalization, i.e.,

$$\beta = \arg\min_{\beta} \frac{1}{2} \|\beta\|_2^2 + \frac{C}{2} \sum_{k=1}^{N} \left(\mathbf{h}(\mathbf{x}_k)\beta - t_k\right)^2$$

$$= \arg\min_{\beta} \frac{1}{2} \|\beta\|_2^2 + \frac{C}{2} \|H\beta - T\|_2^2 \tag{2}$$

where C is the trade-off between the training error and the regularization; $T = (t_1, t_2, \ldots, t_N)^T \in \mathbb{R}^N$ denotes the actual output vector; H represents the hidden layer output matrix, i.e.,

$$H = \begin{pmatrix} \mathbf{h}(\mathbf{x}_1) \\ \mathbf{h}(\mathbf{x}_2) \\ \vdots \\ \mathbf{h}(\mathbf{x}_N) \end{pmatrix} = \begin{pmatrix} h_1(\mathbf{x}_1) & h_2(\mathbf{x}_1) & \ldots & h_L(\mathbf{x}_1) \\ h_1(\mathbf{x}_2) & h_2(\mathbf{x}_2) & \ldots & h_L(\mathbf{x}_2) \\ \vdots & \vdots & \vdots & \vdots \\ h_1(\mathbf{x}_N) & h_2(\mathbf{x}_N) & \ldots & h_L(\mathbf{x}_N) \end{pmatrix} \in \mathbb{R}^{N \times L}.$$

It is evident that the closed solution of optimization (2) can be derived as $\beta = H^T \left(\frac{I}{C} + HH^T \right)^{-1} T$ and the output of ELM satisfy

$$f_L(\mathbf{x}) = \mathbf{h}(\mathbf{x}) H^T \left(\frac{I}{C} + HH^T \right)^{-1} T. \tag{3}$$

Moreover, if the feature mapping $\mathbf{h}$ is not known to users, a kernel matrix $\Omega = (\Omega_{i,j}) \in \mathbb{R}^{N \times N}$ with respect to data set D can be defined as

$$\Omega_{i,j} = h(\mathbf{x}_i) \cdot h(\mathbf{x}_j) = K(\mathbf{x}_i, \mathbf{x}_j),$$

where kernel function $K : \mathbb{R}^n \times \mathbb{R}^n \to \mathbb{R}$ is usually used as Gaussian function

$$K(\mathbf{x}_i, \mathbf{x}_j) = G(\mathbf{x}, \mathbf{x}_j, \sigma) = \exp\left(-\frac{\|\mathbf{x} - \mathbf{x}_j\|^2}{\sigma^2} \right).$$

By Eq. (3), the output of ELM with kernel function K is formulated as

$$f_K(\mathbf{x}) = \mathbf{h}(\mathbf{x}) H^T \left(\frac{I}{C} + HH^T \right)^{-1} T$$

$$= (K(\mathbf{x}, \mathbf{x}_1), K(\mathbf{x}, \mathbf{x}_2), \ldots, K(\mathbf{x}, \mathbf{x}_N)) \left(\frac{I}{C} + \Omega \right)^{-1} T. \tag{4}$$

It is noteworthy that the traditional ELM algorithm is often implemented on a single machine, and therefore it is inevitable to suffer from the limitation of memory with large scale data set. In particular in the era of big data, the data set scale are usually extremely large and the data is often very high-dimensional for detailed information. On the other hand, it is actually necessary to deal with data sets in different machines for the following two reasons: (1) The data set are stored and collected in a distributed manner because of the large scale of many application; (2) It is impossible to collect all of data together for the reason of confidentiality and the data set can be only accessed on their own machine. Based on the analysis above, how to implement algorithm of ELM with respect to data sets which located in different machines become a key problems.

In addition, traditional ELM is computation-intensive with large scale set as well as large number of nodes in hidden layer because of the high complexity for inverse conversion of the large hidden layer output matrix H (see (3)). This situation will be worse in the case of kernel matrix $\Omega \in R^{N \times N}$ since the number of nodes is equal to the number of samples with large scale data set.

3 Distributed Extreme Learning Machine (DELM)

In this section, we extend the traditional ELM algorithm and propose a new distributed extreme learning machine (DELM) on the basis of alternating direction method of multipliers (ADMM) [3], where the data sets to be processed are located in different machines. Subsequently, we first introduce the optimization problem formulated for DELM, and then exploit an distributed algorithm for the proposed optimization problem on the basis of ADMM in the next section.

Given a dataset $D = \left\{ \left(\mathbf{x}_k^{\mathsf{T}}, t_k \right) : \mathbf{x}_k \in \mathbb{R}^n, t_k \in \mathbb{R}, k = 1, 2, \ldots, N \right\}$, we let $\{D_1, D_2, \ldots, D_m\}$ be a partition of all data indices $\{1, 2, \ldots, N\}$ and each data set D_i $(i = 1, 2, \ldots, m)$ is located in different machines. Then, the optimization problem of DELM is formulated as

$$\min_{\beta_1, \beta_2, \ldots, \beta_m, \mathbf{z}} \quad \frac{1}{2} \|\mathbf{z}\|_2^2 + \frac{C}{2} \sum_{i=1}^{m} \sum_{j \in D_i} \left(h(\mathbf{x}_j)\beta_i - t_j \right)^2 \tag{5}$$

$$\text{subject to} \quad \beta_i = \mathbf{z} \quad (i = 1, 2, \ldots, m)$$

where, different from the traditional ELM where all of samples in data set D share a common output weight vector, we associate each data set D_i with a local variable $\beta_i = (\beta_{i1}, \beta_{i2}, \ldots, \beta_{iL})^{\mathsf{T}} \in \mathbb{R}^L$ $(i = 1, 2, \ldots, m)$ which represents the corresponding output weight vector. It is noteworthy that the samples in data set D_i are just processed in the i-th processor, and therefore the output weight vector β_i can be determined in parallel. Moreover, a common global variable $\mathbf{z}$ is included in the optimization of DELM to integrate all of the output weight vectors β_i $(i = 1, 2, \ldots, m)$. In this case, it guarantees to learn a shared output weight vector with respect to data set D (see Fig. 1).

In addition, similar to the traditional ELM, the parameters of hidden node output function shared by all of data set D_i $(i = 1, 2, \ldots, m)$ are generated randomly according to any continuous probability distribution in the framework of DELM. For better representation, optimization (5) can be reformulated as

$$\min_{\beta_1, \beta_2, \ldots, \beta_m, \mathbf{z}} \quad \frac{1}{2} \|\mathbf{z}\|_2^2 + \frac{C}{2} \sum_{i=1}^{m} \|H_i \beta_i - T_i\|_2^2 \tag{6}$$

$$\text{subject to} \quad \beta_i = \mathbf{z} \quad (i = 1, 2, \ldots, m)$$

Fig. 1 Distributed extreme
learning machine

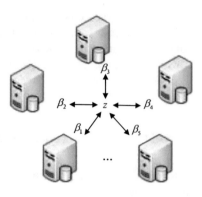

where H_i is the hidden-layer output matrix with respect to the data set D_i, i.e.,

$$H_i = \begin{pmatrix} \mathbf{h}(\mathbf{x}_1) \\ \mathbf{h}(\mathbf{x}_2) \\ \vdots \\ \mathbf{h}(\mathbf{x}_{N_i}) \end{pmatrix} = \begin{pmatrix} h_1(\mathbf{x}_1) & h_2(\mathbf{x}_1) & \cdots & h_L(\mathbf{x}_1) \\ h_1(\mathbf{x}_2) & h_2(\mathbf{x}_2) & \cdots & h_L(\mathbf{x}_2) \\ \vdots & \vdots & \vdots & \vdots \\ h_1(\mathbf{x}_{N_i}) & h_2(\mathbf{x}_{N_i}) & \cdots & h_L(\mathbf{x}_{N_i}) \end{pmatrix} \in \mathbb{R}^{N_i \times L}; \qquad (7)$$

$T_i \in \mathbb{R}^{N_i}$ denotes the actual output of data set D_i and N_i refers to the number of samples in data set D_i, i.e., $\sum_{i=1}^{m} N_i = N$ with $|D_i| = N_i$ for $i = 1, 2, \ldots, m$.

Associated with the sub-datasets D_i $(i = 1, 2, \ldots, m)$, the corresponding m hidden-layer output matrixes $H_i \in \mathbb{R}^{N_i \times L}$ $(i = 1, 2, \ldots, m)$ are utilized instead of H with respect to the whole data set D in traditional ELM. In this case, the problem of inverse conversion for large matrix H is avoided to some extent. At the same time, the proposed DELM method makes it possible to implement ELM algorithm in parallel.

4 ADMM Based Algorithm for DELM

In this section, we first introduce the notations and preliminaries of ADMM, and then exploit an ADMM based distributed algorithm for the formulated optimization problem of DELM.

4.1 ADMM

ADMM is an effective optimization method to solve the following composite optimization problem [3]

$$\min \ f(\mathbf{x}) + g(\mathbf{z}) \qquad (8)$$
$$\text{s.t.} \quad A\mathbf{x} + B\mathbf{z} = \mathbf{c}$$

where functions f and g are all convex. With the augmented Lagrangian for (8),

$$\mathscr{L}_\rho(\mathbf{x}, \mathbf{z}, \lambda) = f(\mathbf{x}) + g(\mathbf{z}) + \lambda^T (A\mathbf{x} + B\mathbf{z} - \mathbf{c}) + \frac{\rho}{2} \|A\mathbf{x} + B\mathbf{z} - \mathbf{c}\|_2^2,$$

the algorithm of ADMM consists of iterations

$$\mathbf{x}^{k+1} = \arg\min_{\mathbf{x}} \mathscr{L}_\rho(\mathbf{x}, \mathbf{z}^k, \lambda^k) \tag{9}$$

$$\mathbf{z}^{k+1} = \arg\min_{\mathbf{z}} \mathscr{L}_\rho(\mathbf{x}^{k+1}, \mathbf{z}, \lambda^k)) \tag{10}$$

$$\lambda^{k+1} = \lambda^k + \rho \left(A\mathbf{x}^{k+1} + B\mathbf{z}^{k+1} - \mathbf{c} \right) \tag{11}$$

where $\rho > 0$ and k denotes the iteration time. Moreover, this algorithm iterates satisfy the following theoretical guarantees of convergence:

1. $\mathbf{r}^k = A\mathbf{x}^k + B\mathbf{z}^k - \mathbf{c} \to 0$ as $k \to +\infty$;
2. $f(\mathbf{x}^k) + g(\mathbf{z}^k)$ converges to the optimal objective function as $k \to +\infty$.

It is evident that ADMM is a simple but powerful algorithm that aims to find a solution to a global problem by solving local subproblems coordinately [3].

Remark 1 Under the framework of ADMM, $\mathbf{x}^{k+1}$ is often added with the previous value of $\mathbf{z}^k$ in optimizations (10) and (11) for fast convergence, i.e.,

$$\hat{\mathbf{x}}^{k+1} = t\mathbf{x}^{k+1} + (1 - t)\mathbf{z}^k$$

where $t \in [1.5, 1.8]$ is usually used [9]. This technique makes a further improvement and affect efficiency of ADMM significantly.

4.2 Algorithms for DELM

For the optimization problem (6), let $\beta := \{\beta_1, \beta_2, \ldots, \beta_m\}$ and $\lambda := \{\lambda_1, \lambda_2, \ldots, \lambda_m\}$ with $\lambda_i \in \mathbb{R}^L$ for $i = 1, 2, \ldots, m$. We consider the corresponding augmented Lagrangian function given by

$$\mathscr{L}_\rho(\beta, \mathbf{z}, \lambda) = \frac{1}{2} \|\mathbf{z}\|_2^2 + \frac{C}{2} \sum_{i=1}^m \|H_i\beta_i - T_i\|_2^2 + \sum_{i=1}^m \left(\lambda_i^\mathsf{T} (\beta_i - \mathbf{z}) + \frac{\rho}{2} \|\beta_i - \mathbf{z}\|_2^2 \right) \tag{12}$$

Starting from some initial vector $\beta^0 = \{\beta_1^0, \beta_2^0, \ldots, \beta_m^0\}$, $\lambda^0 = \{\lambda_1^0, \lambda_2^0, \ldots, \lambda_m^0\}$, $\mathbf{z}^0$, the variables at iteration $k \geq 0$ are updated in the framework of ADMM as

$$\beta^{k+1} = \arg\min_{\beta} \mathscr{L}_{\rho}\left(\beta, \mathbf{z}^k, \lambda^k\right) \tag{13}$$

$$\mathbf{z}^{k+1} = \arg\min_{\mathbf{z}} \mathscr{L}_{\rho}\left(\beta^{k+1}, \mathbf{z}, \lambda^k\right) \tag{14}$$

$$\lambda_i^{k+1} = \lambda_i^k + \rho\left(\beta_i^{k+1} - \mathbf{z}^{k+1}\right) \quad i = 1, 2, \dots, m. \tag{15}$$

Moreover, because the Lagrangian $\mathscr{L}_{\rho}$ is separable in β_i, we can solve the optimization problem (13) in parallel, i.e.,

$$\beta_i^{k+1} = \arg\min_{\beta_i} \frac{C}{2}\left\|H_i\beta_i - T_i\right\|_2^2 + (\lambda_i^k)^{\mathsf{T}}\left(\beta_i - \mathbf{z}^k\right) + \frac{\rho}{2}\left\|\beta_i - \mathbf{z}^k\right\|_2^2 \tag{16}$$

for $i = 1, 2, \dots, m$. Let $\lambda_i = \rho\mathbf{u}_i$ $(i = 1, 2, \dots, m)$ for better representation, we write down an equivalent problem of optimization (16) as

$$\beta_i^{k+1} = \arg\min_{\beta_i} \mathscr{A}_i = \frac{C}{2}\left\|H_i\beta_i - T_i\right\|_2^2 + \frac{\rho}{2}\left\|\beta_i - \mathbf{z}^k + \mathbf{u}_i^k\right\|_2^2, \tag{17}$$

The necessary condition for the minimum of optimization problem (17) is derived by setting the partial derivatives with respect to β_i to zero, i.e.,

$$\frac{\partial\mathscr{A}_i}{\partial\beta_i} = CH_i^{\mathsf{T}}\left(H_i\beta_i - T_i\right) + \rho\left(\beta_i - \mathbf{z}^k + \mathbf{u}_i^k\right) = 0. \tag{18}$$

Therefore, we have

$$\beta_i^{k+1} = \left(\frac{\rho}{C}I + H_i^{\mathsf{T}}H_i\right)^{-1}\left[\frac{\rho}{C}\left(\mathbf{z}^k - \mathbf{u}_i^k\right) + H_i^{\mathsf{T}}T_i\right].$$

On the other hand, with $\lambda_i = \rho\mathbf{u}_i$ $(i = 1, 2, \dots, m)$, the optimization problem (14) is rewritten as

$$\mathbf{z}^{k+1} = \arg\min_{\mathbf{z}} \mathscr{B} = \frac{1}{2}\left\|\mathbf{z}\right\|_2^2 + \sum_{i=1}^{m}\frac{\rho}{2}\left\|\beta_i^{k+1} - \mathbf{z}\right\| + \rho\left(\mathbf{u}_i^k\right)^{\mathsf{T}}\left(\beta_i^{k+1} - \mathbf{z}\right). \tag{19}$$

Thus, the closed form solution for optimization problem (19) is obtained by setting the partial derivatives with respect to $\mathbf{z}$ to zero; and we have

$$\mathbf{z}^{k+1} = \frac{\sum_{i=1}^{m}\left(\beta_i^{k+1} + \mathbf{u}_i^k\right)}{m + 1/\rho}.$$

It is evident that the update of $\mathbf{z}^{k+1}$ is indeed an averaging step over the present local output weight vectors β_i and $\mathbf{u}_i$ $(i = 1, 2, \dots, m)$ to some extent [3].

Based on the analysis above, we summarize the algorithm for DELM on the basis of ADMM in Algorithm 1 which is also visualized by Fig. 2 for better understanding.

Fig. 2 Distributed extreme learning machine with ADMM

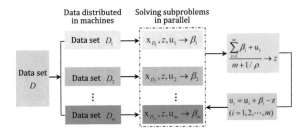

Algorithm 1 Distributed extreme learning machine with ADMM

Input: Data set $D = \left\{ \left(\mathbf{x}_j^\mathsf{T}, t_j \right) : \ \mathbf{x}_j \in \mathbb{R}^n, t_j \in \mathbb{R}, j = 1, 2, \ldots, N \right\}$ with a partition of all data indices $\{D_1, D_2, \ldots, D_m\}$.

Initialization: $\mathbf{z}^0 = 0, \mathbf{u}_i^0 = 0 \ (i = 1, 2, \ldots, m), \rho, t.$

1: **for** $k = 0, 1, 2, \ldots$ **do**

2:      **Update** β_i^{k+1} **in parallel:**

$$\beta_i^{k+1} = \left(\frac{\rho}{C} I + H_i^\mathsf{T} H_i \right)^{-1} \left[\frac{\rho}{C} \left(\mathbf{z}^k - \mathbf{u}_i^k \right) + H_i^\mathsf{T} T_i \right]$$

$$\hat{\beta}_j^{k+1} = t \beta_i^{k+1} + (1 - t) \mathbf{z}^k$$

     $(i = 1, 2, \ldots, m).$

3:      **Update** $\mathbf{z}^{k+1}$: $\mathbf{z}^{k+1} = \frac{\sum_{i=1}^m (\hat{\beta}_i^{k+1} + \mathbf{u}_i^k)}{m + 1/\rho}$;

4:      **Update** $\mathbf{u}_i^{k+1}$: $\mathbf{u}_i^{k+1} = \mathbf{u}_i^k + \hat{\beta}_i^{k+1} - \mathbf{z}^{k+1} \ (i = 1, 2, \ldots, m).$

5: **end for**

Output: Output weights vector $\mathbf{z}$ with respect to data set D.

---

5 Experiments

In this section, some numerical experiments on well-known benchmark data sets are carried out to illustrate the effectiveness of the proposed DELM method. We first demonstrate a significant comparable (even better) performance over the traditional ELM with single processor. We then show that the proposed DELM algorithm is effective to handle optimization problem with a large scale data set and provide an analysis on the performance of speedup.

In order to illustrate regression accuracy of the proposed DELM, the data sets of $sinc - train$ and $sinc - test$ stock from the home page of Extreme Learning Machine (http://www.ntu.edu.sg/home/egbhuang) are utilized in this experiments to compare with the traditional ELM. Each of data set $sinc - train$ and $sinc - test$ composes of 5000 instances with one dependent variable and one independent variable.

For fair comparison, let the number of hidden nodes be 20, the output function of each hidden node be "Sigmoid" and the value of parameter C be 1 in the framework of both traditional ELM and the proposed DELM. Besides, we assign the value of parameter $\rho = 1$ and $t = 1.6$ in DELM. In the conditions above, the number of

Fig. 3 The objective
function of DELM with
respect to different number
of processors

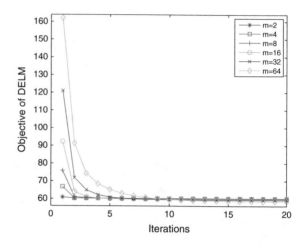

processors $m = 2, 4, 8, 16, 32, 64$ are respectively utilized in DELM and the training
data are split and distributed evenly among the processors. With respect to these different number of processors, we plot the corresponding values of objective function
in Fig. 3 as the increase of iteration times. It indicates that the objective functions for
different number of processors converges within no more than 10 times of iteration
though slightly more number of iteration are required with respect to more number
of processors. Therefore, the proposed DELM achieves fast convergence on the basis
of ADMM. We also show the values of root mean square error (RMSE) associated
with training data and testing data as the increase of iteration times in Fig. 4(a) and
Fig. 4(b), respectively; moreover, the corresponding exact values of RMSE after 20
times iteration are written down in Table 1. It indicates that, on one hand, the result
of DELM is comparable with the results of the traditional ELM (training RMSE

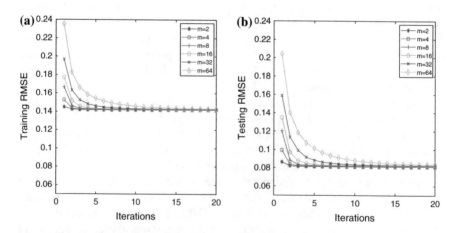

Fig. 4 The values of RMSE with respect to different number of processors. **a** Training. **b** Testing

Table 1 The values of RMSE with 20 times iteration

Number of processors	2	4	8	16	32	64
Training RMSE	0.1409	0.1419	0.1420	0.1421	0.1425	0.1432
Testing RMSE	0.0813	0.0813	0.0814	0.0817	0.0824	0.0836

0.1424 and testing RMSE 0.0824); on the other hand, it is evident that the proposed DELM has better robustness on regression accuracy with respect to different number of processors.

6 Conclusion

In this paper, we develop an effective distributed extreme learning machine to implement the traditional ELM algorithm in parallel for large-scale data set. This method takes advantage of ADMM algorithm and find a solution to a global problem by solving local subproblems coordinately. In this case, the problem of inverse conversion for large matrix in traditional ELM is avoided to some extent by implementing the corresponding sub-problem in parallel.

Acknowledgments This work was supported in part by National Science Foundation of China (Grant No. 91118005, 91218301, 61502377, 61221063, 61428206), Ministry of Education Innovation Research Team (IRT13035), Key Projects in the National Science and Technology Pillar Program of China (2013BAK09B01) and the National Science Foundation (NSF) under grant CCF-1500365.

References

1. Huang, G.-B., Chen, L., Siew, C.K.: Universal approximation using incremental constructive feedforward networks with random hidden nodes. IEEE Trans. Neural Netw. **17**(4), 879–892 (2006)
2. Huang, G.-B., Zhu, Q.Y., Siew, C.K.: Extreme learning machine: theory and applications. Neurocomputing **70**(1), 489–501 (2006)
3. Boyd, S., Parikh, N., Chu, E., Peleato, B., Eckstein, J.: Distributed optimization and statistical learning via the alternating direction method of multipliers. Found. Trends Mach. Learn. **3**(1), 1–122 (2010)
4. Dean, J., Ghemawat, S.: Mapreduce: simplified data processing on large clusters. In: Sixth Symposium on Operating System Design and Implementation (2004)
5. Wang, X., Chen, Y., Zhao, H., Lu, B.: Parallel extreme learning machine for regression based on mapreduce, Neurocomputing **102**
6. He, Q., Shang, T., Zhuang, F., Shi, Z.: Parallelized extreme learning machine ensemble based on min-max modular network, Neurocomputing **128**
7. Zhang, C., Lee, H., Shin, K.G.: Efficient distributed linear classification algorithms via the alternating direction method of multipliers. In: The 15th International Conference on Aritificial Intellienge and Statistics (AISTATS)

8. Huang, G.-B., Zhou, H., Ding, X., Zhang, R.: Extreme learning machine for regression and multiclass classification. IEEE Trans. Syst. Man Cybern. Part B Cybern. **42**(2), 513–529 (2012)
9. Eckstein, J.: Parallel altrenating direction multiplier decomposition of convex programs. J. Optim. Theory Appl. **80**(1), 39–62 (1994)

An Adaptive Online Sequential Extreme Learning Machine for Real-Time Tidal Level Prediction

Jianchuan Yin, Lianbo Li, Yuchi Cao and Jian Zhao

Abstract An adaptive variable-structure online sequential extreme learning machine (OS-ELM) is proposed by incorporating a hidden nodes pruning strategy. As conventional OS-ELM increases network dimensionality by adding newly-received data samples, the resulted dimension would expand dramatically and result in phenomenon of "dimensionality curse" finally. As the measurement samples may come endlessly, there is a practical need to adjust the dimension of OS-ELM not only by adding hidden units but also by pruning superfluous units simultaneously. To evaluate the contribution of existing hidden units and locate the superfluous units, an index is implemented referred to as normalized error reduction ratio. As the OS-ELM adds new samples in hidden units, those existing units contribute less to current dynamics would be deleted from network, thus the resulted parsimonious network can represent current system dynamics more efficiently. This online dimension adjustment approach can handle samples which are presented one-by-one or chuck-by-chuck with variable chuck size. The adaptive variable-structure OS-ELM was implemented for online tidal level prediction purpose. To evaluate the efficiency of the adaptive variable structure OS-ELM, tidal prediction simulations was conducted based on the actual measured tidal data and meteorological data of Old Port Tampa in the United States. Simulation results reveal that the proposed variable-structure OS-ELM demonstrates its effectiveness in short term tidal predictions in respect of accuracy and rapidness.

Keywords Online sequential extreme learning machine · Pruning strategy · Tidal prediction

J. Yin (✉) · L. Li · Y. Cao · J. Zhao
Navigation College, Dalian Maritime University, Dalian 116026, China
e-mail: yinjianchuan@gmail.com

© Springer International Publishing Switzerland 2016
J. Cao et al. (eds.), *Proceedings of ELM-2015 Volume 2*,
Proceedings in Adaptation, Learning and Optimization 7,
DOI 10.1007/978-3-319-28373-9_5

1   Introduction

Tidal prediction is an important issue in areas of oceanographic and coastal engineering. It also concerns much to the tidal energy utilization as well as marine safety
and efficiency. Precise tidal prediction can be a vital issue for navigational safety
when a ship sails through shallow waters; it is also can be used for ships to carry
more cargo with extra water depth brought by tide. Tidal level changes periodically
under the forces caused by the movements of celestial bodies such as moon, earth
and sun. Its dynamics is also influenced by environmental factors such as barometer
pressure, water temperature, wind, current, rainfall and ice, etc. [1]. Therefore, the
dynamic changes of tidal level is a complex time-varying process which is hard to be
predicted precisely in practice. The conventional harmonic analysis method is still
the most widely used one and is still the basis for long-term tidal prediction [2]. The
harmonic method is able to represent the influences on tide caused by celestial bodies
and coastal topography. However, changes of tidal level are not only influenced by
periodical celestial movements, but also by time-varying meteorological factors [3].
The changes of these factors and their influences on tidal change are complex and
hard to be represented by strictly constructed model. Therefore, to generate precise
tidal predictions in real time, there is a practical need to construct an adaptive model
to consider the influences of above-mentioned time-varying environmental changes.

The development of intelligent computation techniques such as neural network,
fuzzy logic and evolutionary computations have been widely applied in vast areas [3,
4]. In recent years, a novel sampling and learning scheme is presented for neural networks referred to as sequential learning scheme. Sequential learning is an adaptive
learning strategy which processes samples sequentially and tunes network accordingly [5, 6]. Sequential learning is initiated by Platt [5]; Lu developed RAN by incorporating pruning strategy in RAN, and the resulted minimal RAN (MRAN) is more
parsimonious than RAN and demonstrate more computational efficiency [6]. However, there are too much parameters in MRAN which increases the manual work burden. Another type of sequential learning mode is online sequential extreme learning
machine (OS-ELM) [7, 8], which is derived from the theory of extreme learning
machine (ELM) [9, 10]. OS-ELM improved ELM from batch learning to be able
to handle data which arrives one-by-one or chunk-by-chunk [7], facilitate its online
applications.

In this paper, a pruning strategy is incorporated in OS-ELM by deleting the obsolete hidden units. Contributions of each unit are evaluated adaptively and those units
contribute less to the output would be deleted from network. The resulted variable-
structure OS-ELM is implemented for online tidal level prediction. The increasing
and deleting of hidden units are conducted during identification process and the
achieved network can be implemented simultaneously. The identification and prediction processes are performed in one step. To evaluated the effectiveness of the
proposed method, tidal prediction simulations are conducted based on the measurement data of Old Port Tampa in the United States.

2 Online Sequential Extreme Learning Machine (OS-ELM)

In this part, we skip the rigorous proof for ELM [9, 11, 12]. The main idea of ELM is that for N arbitrary distinct samples $(\mathbf{x}_k, t_k)$, in order to obtain arbitrarily small non-zero training error, one may randomly generate $\tilde{N}(\leq N)$ hidden nodes (with random parameters $\mathbf{a}_i$ and b_i). Under this assumption, $\mathbf{H}$ is completely defined. Then, Eq. (9) becomes a linear mapping and the output weights $\boldsymbol{\omega}$ are estimated as

$$\hat{\boldsymbol{\omega}} = \mathbf{H}^\dagger \mathbf{T} = \left(\mathbf{H}^{\mathrm{T}} \mathbf{H}\right)^{-1} \mathbf{H}^{\mathrm{T}} \mathbf{T}, \tag{1}$$

where $\mathbf{H}^\dagger$ is the Moore-Penrose generalized inverse of the hidden layer output matrix $\mathbf{H}$. Calculation of the output weights can be done in a single step. This avoids any lengthy training procedure to choose control parameters (learning rate and learning epochs, etc.), thus enables its extreme processing speed. Universal approximation capability of ELM has been analyzed in [9], which indicated that SLFNs with randomly generated additive or radial basis function (RBF) nodes can universally approximate any continuous target function on any compact subspace of $\mathbf{R}_n$.

As training data may be presented one-by-one or chunk-by-chunk in real-time applications, the ELM is modified so as to make it suitable for online sequential computation [7]. Suppose a new chunk of data is given, it results in a problem of minimizing

$$\left\| \begin{bmatrix} \mathbf{H}_0 \\ \mathbf{H}_1 \end{bmatrix} \omega - \begin{bmatrix} \mathbf{T}_0 \\ \mathbf{T}_1 \end{bmatrix} \right\|. \tag{2}$$

When a new sample arrives or a chunk of samples arrive, the connecting weight ω becomes

$$\omega^{(1)} = \mathbf{K}_1^{-1} \begin{bmatrix} \mathbf{H}_0 \\ \mathbf{H}_1 \end{bmatrix}^{\mathrm{T}} \begin{bmatrix} \mathbf{T}_0 \\ \mathbf{T}_1 \end{bmatrix}, \tag{3}$$

where $\mathbf{K}_1 = \begin{bmatrix} \mathbf{H}_0 \\ \mathbf{H}_1 \end{bmatrix}^{\mathrm{T}} \begin{bmatrix} \mathbf{H}_0 \\ \mathbf{H}_1 \end{bmatrix} = \mathbf{K}_0 + \mathbf{H}_1^{\mathrm{T}} \mathbf{H}_1$.

For the efficiency of sequential learning, it is reasonable to express $\omega^{(1)}$ as a function of $\omega^{(0)}$, $\mathbf{K}_1$, $\mathbf{H}_1$ and $\mathbf{T}_1$, which is independent of the original data set.

$$\begin{bmatrix} \mathbf{H}_0 \\ \mathbf{H}_1 \end{bmatrix}^{\mathrm{T}} \begin{bmatrix} \mathbf{T}_0 \\ \mathbf{T}_1 \end{bmatrix} = \mathbf{H}_0^{\mathrm{T}} \mathbf{T}_0 + \mathbf{H}_1^{\mathrm{T}} \mathbf{T}_1 \tag{4}$$

$$= \mathbf{K}_1 \omega^{(0)} - \mathbf{H}_1^{\mathrm{T}} \mathbf{H}_1 \omega^{(0)} + \mathbf{H}_1^{\mathrm{T}} \mathbf{T}_1$$

$\omega^{(1)}$ can be expressed as follows by combining (3) and (4):

$$\omega^{(1)} = \mathbf{K}_1^{-1} \begin{bmatrix} \mathbf{H}_0 \\ \mathbf{H}_1 \end{bmatrix}^{\mathrm{T}} \begin{bmatrix} \mathbf{T}_0 \\ \mathbf{T}_1 \end{bmatrix} = \omega^{(0)} + \mathbf{K}_1^{-1} \mathbf{H}_1^{\mathrm{T}} \left(\mathbf{T}_1 - \mathbf{H}_1 \omega^{(0)}\right). \tag{5}$$

Iteratively, when the $(k+1)$th new chunk of data arrives, the recursive method is implemented for acquiring the updated solution. $\omega^{(k+1)}$ can be updated by

$$\omega^{(k+1)} = \omega^{(k)} + \mathbf{K}_{k+1}^{-1}\mathbf{K}_{k+1}^{\mathrm{T}}\left(\mathbf{T}_{k+1} - \mathbf{H}_{k+1}\omega^{(k)}\right). \tag{6}$$

with

$$\mathbf{K}_{k+1}^{-1} = \mathbf{K}_k^{-1} - \mathbf{K}_k^{-1}\mathbf{H}_{k+1}^{\mathrm{T}}\left(\mathbf{I} + \mathbf{H}_{k+1}\mathbf{K}_k^{-1}\mathbf{H}_{k+1}^{\mathrm{T}}\right)^{-1} \times \mathbf{H}_{k+1}\mathbf{K}_k^{-1}. \tag{7}$$

The detailed statement of the recursive least square method can be found in [7] and will not be repeated here.

3 Dimension Adjustment Strategy for OS-ELM

To achieve an OS-ELM with adaptive variable structure, the pruning strategy is incorporated in this study. The pruning operation is conducted to those hidden units which possess less representing ability to current system dynamics by contributing less to the current input-output mapping. The current dynamics is embodied by a sliding data window, and the contribution of the existing hidden units is measured by an index referred to as normalized error reduction ratio (*nerr*). We employ in the pruning strategy based on a sliding data window [13, 14]. When sample is received one by one, the sliding window is updated accordingly by adding the new one in the hidden layer and discarding the foremost one. When the samples are presented in chuck-by-chuck way, assume that the chuck comprise of l samples, the foremost l samples would be removed from network. Under extreme conditions, when a large amount of samples are received at a time, or the number of received samples exceed the size of sliding data window, l samples are selected from the samples randomly, with $l \leq N$.

A sliding data window is expressed by incorporating input and output pairs:

$$W_{SD} = [(x_{t-N+1}, y_{t-N+1}), \dots, (x_t, y_t)], \tag{8}$$

where N denotes the width of W_{SD}, that is the number of sample pairs included in the window; t is the index of time and in this study t denotes the index of the newest sample, whether it is presented one-by-one or chuck-by-chuck. The determination of N is based on the change rates of dynamics systems. Theoretically a complete cycle of system changes is needed.

Sliding data window is composed by input matrix X and output matrix Y for represent system mapping, where $X \in R^{n \times N}$ is the input of sliding data window, and n is the dimension of input variables, $Y \in R^{N \times m}$ is the output of the sliding data window, and m is the dimension of output variables.

By incorporating the subset selection scheme of orthogonal least squares (OLS) algorithm [15] into a sequential learning mode, we can adjust dimension of OS-ELM by adding new sample as a hidden unit directly, as well as pruning units which consecutively contribute little to output. For implementing pruning strategy, the contribution of each existing hidden units is calculated based on an index referred to normalized error reduction ratio, which is a generalization form of error reduction ratio (err) [15, 16].

The learning procedure begins with random allocation of hidden units in OS-ELM and increase the dimension of OS-ELM by allocating newly received samples in hidden layer. When a new observation arrives, the observation x_i is added as a new hidden unit directly as the new sample conveys more information on current system dynamics. The hidden layer is constructed by the new hidden units and the original units together. The dimension of constructed hidden units is $R^{n \times M}$, where M is the number of hidden units, n is the dimension of the candidate hidden units, which is the same as the the dimension of input variables.

We choose the Gaussian functions as the active functions of the hidden units. The response matrix is expressed by $\Phi \in R^{N \times M}$, with the individual element is expressed as:

$$\phi_{j,k} = \exp\left(-\frac{\|x_j - c_k\|^2}{2\sigma^2}\right) \quad 1 \le j \le N, 1 \le k \le M, \tag{9}$$

where c_k is the center location of kth hidden unit, σ is a width constant, and $\| \cdot \|$ denotes the Euclidean norm. It should be noticed that the dimension of Φ is $N \times M$, which is different from the square response matrix in OLS algorithm.

As the time series samples of a practical system is usually correlated and the response matrix is highly coupled, the orthogonal method is utilized to distinguish the contribution of each hidden units [15]. By utilizing Gram-Schmidt method, the set of basis vectors Φ can be translated into orthogonal basis vectors by $\Phi = WA$.

The space spanned by the set of w_k is the same as that spanned by set of ϕ_k. W is composed by orthogonal columns. According to vector space theory, $\sum_{k=1}^{M} \cos^2 \theta_{k,l} = 1$ hold valid only in single-output conditions. Different from square response matrix generated in OLS algorithm, the response matrix in this condition is generally not square as the number of hidden units and the number of samples in the sliding data window is usually not coincide.

As the result of $\sum_{k=1}^{M}[err]_k \ne 1$, it is impossible to set a criterion for subset selection. To evaluate the contribution of existing hidden units by one same criterion, the index of normalized error reduction ratio (nerr) is designed by:

$$[nerr]_k = \frac{[err]_k}{\sum_{k=1}^{M}[err]_k}. \tag{10}$$

The summation of nerr is then normalized and $\sum_{k=1}^{M}[nerr]_k = 1$, thus the nerr can be directly used as a criterion for evaluating the contribution of each existing hidden unit.

To realize the pruning of hidden units, those units whose summation of error reduction ratio falls below the preset accuracy threshold ρ are selected at each step.

The units with smallest *nerr* is firstly selected whose $[nerr]_{k_1} = \min\{[nerr]_k, 1 \leq k \leq M\}$. After the first unit, the second unit is selected: $[nerr]_{k_2} = \min\{[nerr]_k, 1 \leq k \leq M, k \neq k_1\}$ if $[nerr]_{k_1} \leq \rho$. Similar selection procedure continues until their sum reaches an accuracy threshold $\sum_{k=k_1}^{k=k_S+1} [nerr]_k > \rho$.

Thus $k_1, \ldots, k_S$ and corresponding hidden units are selected. Similar selection procedure is conducted at each step. If certain hidden units are selected for p consecutive observations, the units can be considered as obsolete for current system dynamics and would be pruned from the network. That is, units in the intersection I will be eliminated from the network.

$$I = \left\{ S_k \bigcap S_{k-1} \bigcap \cdots \bigcap S_{k+p-1} \right\} \tag{11}$$

This operation of consecutive selection avoid the misdeletion of useful units and guarantee the stability of the resulted variable OS-ELM consequently. After hidden units being added or pruned at each step, the existing parameters of the network are adjusted. Under condition that no hidden units are incorporated in I and none was pruned, as the new sample or new chuck of samples still be added into the hidden layer, the connecting parameters are updated using the conventional recursive least square method in conventional OS-ELM method:

$$\omega^{(k+1)} = \omega^{(k)} + \mathbf{K}_{k+1}^{-1} \mathbf{K}_{k+1}^{\mathrm{T}} \left(\mathbf{T}_{k+1} - \mathbf{H}_{k+1} \omega^{(k)} \right). \tag{12}$$

Under conditions where both the number of hidden units and the number of existing hidden units are not very large, the parameters are determined by linear least mean squares estimation (LLSE) method. Comparing with other sequential learning algorithms, the VS-OSELM is featured by its adaptive learning scheme and small number of tuning parameters. At each step, VS-OSELM adds new sample directly as a hidden unit, and remove units consecutively align poor with output. This learning scheme highly reduces the computational burden of sequential learning. And there are only 3 parameters (N, ρ and p), which is much less than sequential learning algorithms of RAN and MRAN. These features facilitate its practical identification and prediction applications.

4 Simulation of Tidal Level Prediction Based on Variable-Structure OS-ELM

4.1 Structure of Online Tidal Prediction Scheme

The change of tidal level is a complex process which is affected not only by movement of celestial bodies but also by environmental changes. These factors made it a difficult task to generate precise tidal level predictions by a strictly founded model.

OS-ELM has demonstrated its efficiency in describing nonlinear processes. By incorporating the pruning strategy in OS-ELM, the achieved VS-OSELM possesses compact network structure. In this study, the VS-OSELM is implemented in tidal level prediction simulation.

The OS-ELM-based prediction is a kind of sequential learning scheme whose hidden units are added and pruned sequentially in this study. During the simulation, both processes of identification and prediction are conducted at each step. The prediction is performed once the hidden units and connecting parameters are determined.

The tidal level y is considered as the effects of periodical movement of celestial bodies as well as time-varying environmental influences of u and other unmodeled factors. As the OS-ELM is suitable for representing the nonlinear time-varying dynamics of tidal changes, in this study, OS-ELM is used for online tidal prediction of y based on nonlinear autoregressive with exogenous inputs (NARX) model:

$$y(t) = f(y(t-1), \ldots, y(t-n_y), u(t-1), \ldots, u(t-n_u)), \tag{13}$$

where the y and u are system output and input, with n_y and n_u denote orders of y and u, respectively. In this study, u in (13) contains environmental factors including water temperature T and air pressure P. That is, $u(t-1), \ldots, u(t-n_u)$ contains

$$T(t-1), \ldots, T(t-n_T), P(t-1), \ldots, P(t-n_P), \tag{14}$$

where n_T and n_P are orders of the T and P in prediction model, respectively.

The 1-step-ahead prediction is realized by replacing the t in (13) with $t+1$:

$$y(t+1) = f(y(t), \ldots, y(t-n_y+1), u(t), \ldots, u(t-n_u+1)), \tag{15}$$

After the VS-OSELM is online adjusted by learning data pairs of y_R in current sampling pool of sliding data window, it is applied for tidal level prediction immediately.

For m-steps-ahead prediction, the processes of identification and prediction are expressed as follows:

$$y(t) = f(y_R(t-m), \ldots, y(t-m-n_y+1), u(t-m), \ldots, u(t-m-n_u+1)), \tag{16}$$

and

$$y(t+m) = f(y_R(t), \ldots, y(t-n_y+1), u(t-m), \ldots, u(t-m-n_u+1)). \tag{17}$$

Once the identification process is completed, currently available information of y, P and T are then set as input according to (17) and the $y(t+m)$ is the m-steps-ahead prediction of the tidal level.

4.2   Real-Time Tidal Prediction Simulation

Simulation of online tidal level prediction was conducted to verify the feasibility and efficiency of the proposed VS-OSELM. The simulation was conducted based on the measured hourly tidal data of Old Port Tampa in Florida, USA. The hourly tidal level samples are measured from GMT0000 January 1 to GMT2300 June 30, 2015, 4344 samples in total. To evaluate the efficiency of the proposed online prediction method, 4000 steps of prediction was conducted. All the measurements of tidal level, air pressure, water temperature and the parameters of harmonic constants in this study are achieved from web site of American National Oceanic and Atmospheric Administration: http://co-ops.nos.noaa.gov. In the online tidal prediction simulations, both the identification and prediction processes are performed in each step. Simulations are processed in MATLAB 7.4 environment running at 2.40 GHz (CPU) and 1.92 GB memory (RAM).

The measured tidal data of Port of Old Port Tampa is shown in Fig. 1. The prediction result achieved by using conventional harmonic method is also shown in the figure.

The coefficient of the measured tidal level and the predicted value by Harmonic method is 0.879529. As the harmonic method only takes consider of the period influences of celestial bodies and ignores the influences of the environmental changes, there exist time-varying errors in the prediction results. It can be noticed that the prediction error are large in January to March. It reach its highest to 0.6710 m at 1355-th hour. Ship voyage plan stipulated based on such prediction may cause accidents of grounding or collision when ship sails through shallow water or under bridge. Therefore, there is a practical need to reduce predictive tidal error and give accurate predictions.

In this study, the index of root mean square error (RMSE) is utilized to evaluate the performances of prediction. Based on the measured data, the $RMSE_p$ of tidal prediction using harmonic method is 0.149933 m. These diverges are mostly caused

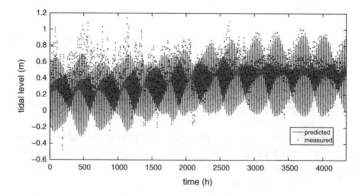

Fig. 1   Measured tidal level and the predicted results by using harmonic method (king point)

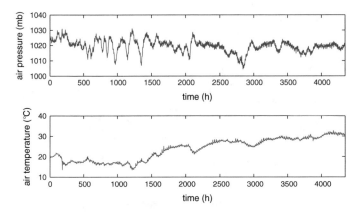

Fig. 2 Changes of air pressure and water temperature

by environmental factors such as air pressure and water temperature, which is shown in Fig. 2.

It can be notice by comparing Fig. 2 that the rising of water temperature coincides with the increase of mean sea level; and the variations of the air pressure corresponds to the fluctuations of tidal level from January to March, 2015. The VS-OSELM was implemented for tidal prediction based on the measurement data of Old Port Tampa. In the study, the running step is set as 4000, and the parameters for NARX model are set as $n_y = 12$, $n_p = 1$ and $n_T = 1$. For VS-OSELM, the first 72 samples are set as sample pool for ELM to select and the number of hidden neurons assigned to the ELM is 18, Gaussian function was selected as activation function. The parameters for the pruning strategy are set as $N = 72$, $\rho = 0.0001$ and $p = 2$. The prediction error of 1-step-ahead tidal level prediction is shown in Fig. 3.

It can be noticed that the predicted tidal levels track the real ones well even when the tidal level fluctuates form January to March, 2015. It is shown that the errors

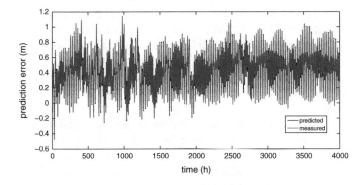

Fig. 3 Prediction result for 1-step-ahead tidal prediction by VS-OSELM

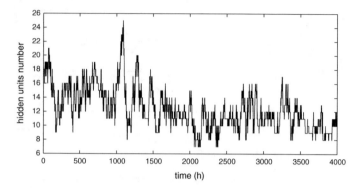

Fig. 4 Scatter diagram of measured tidal level and prediction result by VS-OSELM

are round zero, which is different from the result acquired by the harmonic method. The RMSE of prediction by using VS-OSELM is 0.0364727 m, which is much less than the result achieved by the harmonic method.

The corresponding correlation coefficient is 0.995032, which is higher than that achieved by harmonic method. As the hidden units in VS-OSELM is adjustable by adding received new ones and discarding the obsolete ones, the number of hidden units is variable. The evolution of the number of hidden units is depicted in Fig. 4.

It is noticed by Fig. 4 that the employed hidden units number is quite small in contrast to the conventional OS-ELM which expands its dimension gradually and would reach a considerably large number. As the adjustment of hidden units is a dynamic process, the hidden units number changes almost in every step. So that we can only measure the dimension of the network by an index referred to as average hidden units number (AHUN). The larger AHUN means a relatively big dimension and a smaller one implies a compact network structure. The overall processing time for the VS-OSELM is 17.898055 s, which enable it for online applications such as online prediction and control.

For comparison purpose, the conventional OS-ELM is also conducted and the prediction result is shown in Fig. 5. The samples are presented to OS-ELM sequentially, and the models of identification and prediction are the same as that of the VS-OSELM. For conventional OS-ELM, the first 72 samples are set as sample pool for ELM to select and the number of hidden neurons assigned to the ELM is set as 24 to get optimal prediction accuracy. Altogether 50 times of simulation is conducted and the average prediction result is depicted in Fig. 5 together with the actual measured ones.

It is noted by comparing Figs. 3 and 5 that the prediction results coincide with the actual ones better than that by using conventional OS-ELM, especially in tracking of the fluctuant tidal levels. It is noticed that error level is higher than that by using VS-OSELM and the prediction result depicts more fluctuations even abrupt changes of. The maximum prediction error reaches 0.5, which is too large for practical navigational applications. The predictive RMSE for OS-ELM prediction is 0.0503m, which

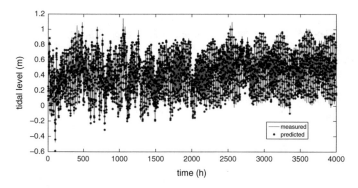

Fig. 5 Prediction result for 1-step-ahead tidal prediction by conventional OS-ELM

is larger than the 0.0365 m by using VS-OSELM. Both the predictive RMSE of OS-ELM and VS-OSELM are smaller than 0.149933 m by using harmonic method. The correlation coefficient between the actual levels and predicted ones is 0.9791.

It can be noticed from above simulation results that for short-term prediction, the VS-OSELM performs superior to conventional OS-ELM both in predictive accuracy and processing speed. By incorporating the pruning strategy in OS-ELM, the resulted VS-OSELM demonstrate better prediction accuracy and faster processing time than conventional OS-ELM. As the resulted network possesses parsimonious structure, the processing burden for matrix calculation is drastically released and the processing speed is highly accelerated accordingly. Notwithstanding the reduced network dimension, the generalization ability of the achieved VS-OSELM is not deteriorated much because the scattered hidden units with limited number can represent the current system dynamics well. Especially for time-varying systems, the redundant network will deteriorate the adaptive ability of network instead. The conclusion is based on the prediction simulation setting that the system is a time-varying system whose dynamics changes quickly with variations of environmental changes.

5 Conclusions

A variable structure OS-ELM is proposed based by incorporating the pruning strategy in the adjustment of hidden units. The computational burden is alleviated by reducing the dimension of hidden layer. Furthermore, the adaptive adjustment strategy for hidden units enable the existing hidden units represent current system dynamics, which facilitate its online application for short-term prediction of time-varying dynamics. The hidden units adjustment strategy for controlling the dimension of hidden layer should be further studied to ensure the stability of the achieved network. The variable structure OS-ELM can also be implemented to other online applications such as online identification, prediction and control of time-varying systems.

Acknowledgments This work is supported by National Natural Science Foundation of China (Grant No. 51279106), the Applied Basic Research Fund of the Chinese Ministry of Transport (Grant No. 2014329225010) and Fundamental Research Funds for the Central Universities (Grant No. 3132014028).

References

1. Fang, G.H., Zheng, W.Z., Chen, Z.Y.: Analysis and Prediction of Tide and Tidal Current. Ocean Press, Beijing (1986)
2. Lee, T.: Back-propagation neural network for long-term tidal predictions. Ocean Eng. **31**, 225–238 (2004)
3. Liang, S., Li, M., Sun, Z.: Prediction models for tidal level including strong meteorologic effects using a neural network. Ocean Eng. **35**, 666–675 (2008)
4. Lee, T., Jeng, D.: Application of artificial neural networks in tide-forecasting. Ocean Eng. **29**, 1003–1022 (2002)
5. Platt, J.: A resource allocating network for function interpolation. Neur. Comput. **3**(2), 213–225 (1991)
6. Lu, Y.W., Sundararajan, N., Saratchandran, P.: A sequential learning scheme for function approximation using minimal radial basis function neural networks. Neur. Comput. **9**, 461–478 (1997)
7. Liang, N.Y., Huang, G.B., Saratchandran, P., Sundararajan, N.: A fast and accurate online sequential learning algorithm for feedforward networks. IEEE Trans. Neur. Netw. **17**(6), 1411–1423 (2006)
8. Sun, Y.J., Yuan, Y., Wang, G.R.: An OS-ELM based distributed ensemble classification framework in P2P networks. Neurocomputing **74**(16), 2438–2443 (2011)
9. Huang, G.B., Zhu, Q.Y., Siew, C.K.: Extreme learning machine: theory and applications. Neurocomputing **70**, 489–501 (2006)
10. Huang, G., Huang, G.B., Song, S.J., You, K.Y.: Trends in extreme learning machines: a review. Neural Networks **61**, 32–48 (2015)
11. Huang, G.B., Ding, X.J., Zhou, H.M.: Optimization method based extreme learning machine for classification. Neurocomputing **74**, 155–163 (2010)
12. Huang, G.B., Zhou, H.M., Ding, X.J., Zhang, R.: Extreme learning machine for regression and multiclass classification. IEEE Trans. Syst. Man Cy. B **42**(2), 513–529 (2012)
13. Li, J.M., Chen, X.F., He, Z.J.: Adaptive stochastic resonance method for impact signal detection based on sliding window. Mech. Syst. Signal Pr. **36**, 240–255 (2013)
14. Chen, C.Y., Li, T.H.S., Yeh, Y.C., Chang, C.C.: Design and implementation of an adaptive sliding-mode dynamic controller for wheeled mobile robots. Mechatronics **19**(2), 156–166 (2009)
15. Chen, S., Cowan, C.F.N., Grant, P.M.: Orthogonal least squares learning algorithm for radial basis function networks. IEEE Trans. Neural Netw. **2**(2), 302–309 (1991)
16. Yin, J.C., Wang, L.D., Wang, N.N.: A variable-structure gradient RBF network with its application to predictive ship motion control. Asian J. Contr. **14**(3), 716–725 (2012)

Optimization of Outsourcing ELM Problems in Cloud Computing from Multi-parties

Jiarun Lin, Tianhang Liu, Zhiping Cai, Xinwang Liu and Jianping Yin

Abstract In this letter, we introduce a secure and practical multi-parties cooperating mechanism of outsourcing extreme learning machines (ELM) in Cloud Computing. This outsourcing mechanism enables original ELM to perform over large-scale dataset in which multi-parties are involved. We propose a optimized partition policy in Cloud Computing to significantly improve the training speed and dramatically reduce the communication overhead. According to the partition policy, cloud servers are mainly responsible for calculating the inverse of an intermediate matrix derived from the hidden layer output matrix, which is the heaviest computation. Although most of the computation is outsourced in Cloud Computing, the confidentiality of the input/output is assured because the randomness of the hidden layer is fully exploited. Theoretical analysis and experiments have shown that the proposed multi-parties cooperating mechanism for outsourcing ELM can effectively release customers from heavy computations.

1 Introduction

In the era of Big Data and Cloud Computing, computation outsourcing has drawn considerable attention in both academic and industry community. Traditional machine learning can no longer efficiently process big data which is giant-volume, fast-velocity and intensely various. Fortunately, Extreme Learning Machines (ELM), provides some tips upon how to effectively address the challenges.

ELM was originally proposed by G-B Huang et al. [1–3], as a fast learning algorithm for Single-hidden Layer Feedforward Neural Networks (SLFNs), in which the input weights and bias attached with the hidden layer are randomly assigned and

The authors are with the College of Computer Science, National University of Defense Technology, 410073, Changsha, P.R. China.

J. Lin (✉) · T. Liu · Z. Cai · X. Liu · J. Yin
College of Computer, National University of Defense Technology, Changsha,
People's Republic of China
e-mail: nudtjrlin@gmail.com

© Springer International Publishing Switzerland 2016
J. Cao et al. (eds.), *Proceedings of ELM-2015 Volume 2*,
Proceedings in Adaptation, Learning and Optimization 7,
DOI 10.1007/978-3-319-28373-9_6

67

the output weights are analytically determined. ELMs have been extended for "generalized" SLFNs which may not be neuron alike. To improve the training speed, especially over large-scale datasets, researchers have proposed many enhanced ELM variants [4], such as Parallel ELM [5] based on MapReduce, GPU-accelerated and parallelized ELM ensembles for large scale regression.

To avoid committing any large capital expenditures, users would like to outsource the ELM tasks into Cloud Computing in a pay-per-use manner at relatively low price. However, Outsourcing is also giving up users' direct control of their data and application, in which sensitive information might be involved [6], such as business financial records, proprietary research data, etc. To ensure the data security and fight against unauthorized information leakage, sensitive data is frequently desired to be encrypted in advance so as to provide end-to-end data confidentiality assurance in the cloud and beyond. Moreover, the requirements of security are indispensable juristically in some practical application systems, such as the protection of Personal Health Information (PHI) according to the Health Insurance Portability and Accountability Act (HIPAA) [7].

In our previous work [4], we proposed a secure and practical mechanism, named Partitioned ELM, to outsource ELM in Cloud Computing to enable ELM to be applied in big data applications. As the parameters of the hidden nodes need not be tuned but can be randomly assigned, they claimed that ELM is well suited to be outsourced for improving training speed. The mechanism explicitly decomposes ELM algorithm into a public part and a private part. The assignments of random nodes and the calculation of the hidden layer's output matrix is kept locally in private part. Only the output matrix and its pseudo inverse are transmitted between users and cloud. The public part is executed in the cloud and is mainly responsible for the calculation of pseudo inverse. The pseudo inverse also serves as the correctness and soundness proof for result verification. Cloud servers cannot mine out sensitive information from the hidden layer's output matrix. Therefore, the customer can perform the ELM tasks with improved training speed while preserving the confidentiality of the input and the output.

However, our previous work focused on outsourcing original ELM in Cloud Computing from a very single party, and the partition policy is based on that the pseudo inverse was calculated through Singular Value Decomposition (SVD), which is native and extreme expensive computationally. In this paper, we will discuss outsourcing ELM in Cloud Computing from multiple parties while preserving the privacy and discuss another efficient partition policy for outsourcing ELM in Cloud Computing. Besides, we discuss the optimization of outsourcing batches of ELM problems in cloud computing through pipelien and parallelization. To the best of our knowledge, we are the first to outsource ELM in Cloud Computing from multiple parties to improve the training speed while assuring the confidentiality of the input/output.

2 Preliminary

Given a set of training data $(\mathbf{X}, \mathbf{T})$, in which $\mathbf{x_i} = [x_{i1}, x_{i2}, ..., x_{in}]^T \in \mathbf{R}^n$ and $\mathbf{t_i} = [t_{i1}, t_{i_2}, ..., t_{i_m}]^T \in \mathbf{R}^m$, $i = 1, 2, ..., N$, ELM can train the SLFN at an extremely fast speed thanks to one of the salient feature of ELM - the hidden layer need not be tuned. ELM trains SLFNs with randomly assigned hidden nodes, i.e., the bias and weights between the input layer and hidden layer $(\mathbf{w}, \mathbf{b})$ are assigned randomly, given the activation functions are infinitely differentiable. $w_i = [w_{i1}, w_{i2}, ..., w_{in}]^T \in R^n$ is the input weight vector connecting the ith hidden node with the input nodes [3]. From the other perspective, we can reform the input weights matrix. We use $\mathbf{a_{j_i}} = [a_{j1}, a_{j2}, ..., a_{jM}]$ to denote the weight vector (row) between jth input node and the M hidden nodes. The input weights $\mathbf{w}$ can be also expressed as $\mathbf{w} = [\mathbf{a_1}, \mathbf{a_2}, ..., \mathbf{a_n}]$, i.e., $w_{ij} = a_{ji}$. After the hidden nodes are assigned, the hidden layer output matrix $\mathbf{H}$ of the neural network is determined.

The architecture of SLFNs trained by ELM is shown in Fig. 1. It can be mathematically modeled as

$$\sum_{i=1}^{M} \beta_i h_i(x) = \mathbf{h}(\mathbf{x})\beta, \tag{1}$$

where $\mathbf{H} = \mathbf{h}(\mathbf{x}) = [h_1(x), h_2(x), ..., h_M(x)]$ is the output (row) vectors of the hidden layer with respect to the input $\mathbf{x}$ and $\beta = [\beta_1, \beta_2, ..., \beta_M]$ is the output weights vector between M hidden nodes and m output nodes. The pseudo inverse of $\mathbf{H}$ is denoted as $\mathbf{H}^\dagger$. β can be determined analytically by

$$\beta = \mathbf{H}^\dagger T. \tag{2}$$

Fig. 1 Training SLFN with ELM

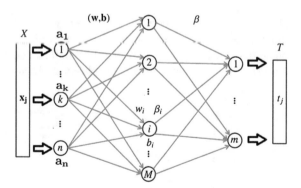

3 Multi-parties Cooperating Outsourcing of ELM

3.1 Threat Model of Cloud Computing

There are two kinds of entities involved in the mechanism of outsourcing ELMs in Cloud Computing: cloud customers and cloud servers. The former entity has several computationally expensive large scale ELM problems to outsource in Cloud Computing, and the latter one has literally unlimited resources and provides utility computing services.

The threats of security and privacy mainly stem from the cloud servers who may behave in "honest-but-curious" model, which is also called semi-honest model that was assumed in many previous researches. The cloud servers would persistently be interested in analyzing the data to mine more information, either because it is compromised or it intends to do so. More worse, the cloud servers may behave beyond semi-honest model, i.e., they may cheat the customer to save power or reduce executing time while expecting not to be caught at the same time. In this paper, we firstly assume that the cloud server performs the computation honestly and discuss the correctness and soundness in the later subsection.

To focus on outsourcing ELM in Cloud Computing, we omit the authentication processes, assuming that communication channels are reliably authenticated and encrypted, which can be achieved in practice with little overhead [8].

3.2 Outsourcing ELM from Multi-parties Cooperating in Different Ways

Different parties may cooperate in different ways. The basic architecture is shown in Fig. 2. Mainly, there are two ways of cooperation among multi-parties. The first one is that different parties will contribute different samples with the same features. For example, the same sensors are deployed at different spots to collect same kinds

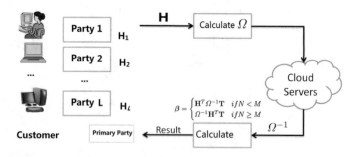

Fig. 2 Architecture of outsourcing multi-parities ELM

of data for different subjects. The second one is that each party contributes different features of a very sample, i.e., different functions are featured in different parties.

We model each sample as a column vector. From the perspective of the aggregated training data across different paries, the data contributed in the first way is horizontally partitioned while that of the second way is vertically partitioned. So we will discuss different outsourcing mechanisms over horizontally partitioned data and vertically partitioned data in the following two subsections, respectively.

3.2.1 Horizontally Partitioned Data

In this case, we firstly delegate a trusted third-party or vote a party as the Primary Party according the power of CPU, memory, network bandwidth, etc. The details of how to delegate or vote a primary party and how to authenticate different parties are out of the focus of this paper. While most of public cloud servers are out of the trust domain, the trusted party could be a trusted third-party or be voted among the involved parties, which are probably in the same trust domain. So it is acceptable to delegate a trusted party for generating parameters. The primary party will take the responsibility of generating and distributing the hidden layer's parameters. As soon as the lth party P_l receives them, it begins to calculate the corresponding part $\mathbf{H}_l$ of the hidden layer's output matrix with its own data records. Thereafter, $\mathbf{H}_l$ is sent to the cloud servers. After the cloud servers calculate and feedback the pseudo inverse $\mathbf{H}^\dagger$, each participated party can calculate the desired output weights β according to Eq. 1, as well as perform the testing of the trained SLFNs with the trained parameters.

Furthermore, there is another way of calculating the output weights. The cloud servers only send it to the primary party. The primary party is in charge of calculating and distributing the output weights, as well as the testing of trained SLFNs.

3.2.2 Vertically Partitioned Data

Assuming that, there are L parties and k_l features contributed by lth party P_l. n is the number of features for the aggregated samples, $\sum_{l=1}^{L} k_l = n$. The input weights between input layer and the hidden layer can be partitioned according to the feature distribution among different parities. To outsource ELM from multiple parties in Cloud Computing without leaking information to the other parties, different parts of the hidden layer parameters are randomly assigned by related parties, i.e., ith party is responsible for $\mathbf{a}_j$ if jth feature is contributed by party P_i. When the data is vertically partitioned, there is no need to delegate a primary party as long as the testing data is previously collected.

Worth noting that, different parties are responsible for corresponding parts of output matrix of the hidden layer, and then send them to cloud server. In this way, cloud servers can know how many features of the data record are contributed by corresponding parties, i.e., the distribution of features among different parties.

To avoid the leakage of information of the distribution, the same way with the horizontally partitioned data can be employed, in which the primary party firstly assembles the different parts of $\mathbf{h}(\mathbf{x})$ from corresponding parties, and interacts with the cloud servers.

3.3 Improved Partitioned Policy for Outsourcing ELM

In this subsection, we will discuss an improved partitioned policy for outsourcing ELM in Cloud Computing based on the architecture shown in Fig. 2.

As shown in [4], the heaviest computation of ELM is the calculation of pseudo inverse of the hidden layer's output matrix $\mathbf{H}$. And the matrices transmitted between customers and cloud servers are $\mathbf{H}$ and $\mathbf{H}^{\dagger}$, whose size are $N \times M$ and $M \times N$, respectively. Assuming that the bandwidth between the cloud customer and the cloud servers is B bytes per second. We use T_d to denote the communication delay, it would be $T_d = 2 \times \frac{8 \times N \times M}{B}$ if each data of the matrix is a double.

In big data applications, N is usually pretty huge and great larger than M. The communication overhead would be a horrible disaster and the time saving from computation outsourcing would be killed. Therefore, we would like to find another partitioned policy so that the outsourcing mechanism cannot only improve training speed but also reduce communication overhead.

Different methods can be used to calculate pseudo inverse of a matrix: orthogonal projection method, orthogonalization method, iterative method, and singular value decomposition (SVD), etc. We further discuss using orthogonal projection method to calculate pseudo inverse in this paper. To improve generalization performance and make the solution more robust, we add a regularization term as shown elsewhere [3],

$$\beta = \left(I/C + \mathbf{H}^{\mathbf{T}}\mathbf{H}\right)^{-1}\mathbf{H}^{\mathbf{T}}\mathbf{T} \tag{3}$$

We define an intermediate matrix $\Omega = I/C + \mathbf{H}^{T}\mathbf{H}$ firstly. Then the output weights of trained neural network are

$$\beta = \Omega^{-1}\mathbf{H}^{T}\mathbf{T} \tag{4}$$

We observe that the size of Ω and Ω is only $M \times M$, which is significantly smaller than that of $\mathbf{H}$ due to the fact that in most of cases $N \geq M$. More specifically, the calculation of Ω's inverse ($O(M^3)$) is the heaviest operation. Therefore, we would like to outsource the inverse calculation in Cloud Computing. The communication delay is changed to $T_d = 2 \times \frac{8 \times M \times M}{B}$. Through outsourcing, the training speed is significantly improved and the communication overhead is dramatically reduced.

As Ω is directly derived from $\mathbf{H}$, it is impossible to calculate it without the whole knowledge of $\mathbf{H}$. The confidentiality of input and output is still guaranteed as cloud servers cannot dig out information from Ω. However, a primary party should be voted or delegated to take charge of calculating both $\mathbf{H}$ and Ω, over either horizontally partitioned data or vertically partitioned data.

To further improve the speedup of the training of SLFNs using ELM, we multiply $\mathbf{H}^T\mathbf{T}$ firstly, and then multiply it with Ω^{-1} when determine the desired output weights β analytically.

3.4   Security Analysis and Result Verifications

The ELM is instinctively suitable to be outsourced in Cloud Computing while assuring the confidentiality of the training samples and the desired parameters of neural networks. In the private part, the parameters $(\mathbf{w},\mathbf{b})$ are assigned randomly which are a part of the desired parameters of the training SLFNs. These parameters must be assigned by the cloud customer but not the cloud server. The feature mapping $\mathbf{h}(\mathbf{x})$ is also unknown to the cloud servers. Without any knowledge of the feature mapping functions and the randomly generated parameters, cloud servers cannot obtain any knowledge about the exact X or $(\mathbf{w}, \mathbf{b})$ from $\mathbf{H}$ or Ω.

Till now, we have been assuming that the server is honestly performing the computation, while being interested in learning information. However, the cloud server might behave unfaithfully. Therefore, the customer must be enabled to verify the correctness and soundness of the results.

In our mechanism, the pseudo inverse itself from the cloud server can also serve as the verification proof. It is plainly easy to verify the result through the definition of matrix inverse. Therefore, the correctness and soundness of the results can be verified while incurring few computation overhead and no extra communication overhead.

4   Performance Evaluation

We test the new mechanism over a large-scale dataset named CIFAR-10 [9] which consists of 50000 training color images (32×32) and 10000 testing images in 10 classes. To reduce the attribute number, we firstly transform the color images into gray images. We conduct 5 trials for each M, and randomly choose two classes as the training samples and testing samples. The customer computations are conducted on a common workstation (Intel Core i5-3210M CPU, 2.50GHz, 4GB RAM). To evaluate the overall speedup, the cloud server computations are separately conducted on two workstations with different resources: a workstation with an Intel Core i5-3470 CPU (3.20GHz, 6GB RAM) and a more powerful workstation with an Intel Core i7-4770 CPU (3.40GHz, 16GB RAM).

Through outsourcing the calculation of the pseudo inverse from a common workstation to another workstation with much more computing power, we can evaluate the training speedup of the proposed mechanism without a real cloud environment. Our proposed mechanism focuses on improving the training speed through outsourcing while the training accuracy and testing accuracy are not affected.

We use t_0 to denote the training time of the original ELM in which the pseudo inverse is calculated using orthogonal projection method. In the new Partitioned ELM, the time cost at the local customer side and at the cloud server side are denoted as t_l and t_c, respectively. Then, we define the *asymmetric speedup* of the proposed mechanism as $\lambda = t_0/t_l$, which physically means the savings of the computing resources for the customer and is independent on how resourceful the cloud server is and directly related with the size of ELM problems. The *overall speedup* is defined as $\eta = t_0/(t_l + t_c)$. Obviously, the more powerful the cloud server is, the higher the overall speedup.

The results are listed in Table 1. t_{c1} and η_1 means the time cost and overall speedup at the first cloud, respectively. Memory is becoming the dominant computing resource when solving the ELM problem with the increasing M. Therefore, different with experiments in [4], a powerful enough laptop with large enough memory serves as a client so that ELM would not terminate due to memory limit.

As illustrated in Table 1, λ is increasing along M, which means the larger the problems' overall size, the larger speedups the proposed mechanism can achieve. In [4], λ of the original Partitioned ELM would increase to greater than 50 when the number of hidden nodes is larger than 3000. However, in the new proposed mechanism, it increases steadily with the increase of M, slower than that of the original Partitioned ELM. The reason is that SVD was used to calculate the pseudo inverse of the $\mathbf{H}$ in [4], while we use the orthogonal projection method in this paper, which is much faster and requires much less memory. λ is independent with the cloud server, while η is directly related, which higher when the destination cloud server is more powerful.

Over two random classes of CIFAR-10 dataset, the training accuracy (from 83 to 95 %) and testing accuracy (from 80 to 84 %) is also increasing steadily from with the number of hidden nodes. It is worth noting that, the experiment is conducted

Table 1 Performance over 2 random classes of CIFAR-10 dataset

M	$t_0(s)$	$t_l(s)$	$t_{c1}(s)$	$t_{c2}(s)$	λ	η_1	η_2
1000	3.84	3.56	0.19	0.12	1.08	1.03	1.04
2000	11.65	8.50	1.31	1.12	1.37	1.19	1.21
3000	21.39	12.48	3.95	3.00	1.71	1.30	1.38
4000	47.64	24.57	7.86	6.74	1.94	1.47	1.52
5000	72.18	29.87	15.13	13.12	2.42	1.60	1.68
6000	103.55	37.38	26.21	20.62	2.77	1.63	1.79
7000	152.15	48.38	40.61	30.97	3.15	1.71	1.92
8000	210.88	59.94	57.75	46.52	3.52	1.79	1.98
9000	293.25	77.63	78.59	65.41	3.78	1.88	2.05
10000	385.17	92.40	105.82	85.44	4.17	1.94	2.17

on gray images without feature extraction in purpose of evaluating the efficiency of the proposed outsourcing mechanism and optimization policy only. The accuracy of Extreme Learning Machine is not affected after outsourcing in cloud computing.

5 Conclusion

In this paper, we improve Partitioned ELM which was proposed in [4] by different partition policy and extend it to multiple parties. Different parties may contribute the training samples in vertical way or horizontal way which is more common in practice. Through outsourcing the calculation of inverse of a intermediate matrix which is derived from the hidden layer's output matrix, the computation time cost at customer side would be significantly reduced and the communication between customers and cloud servers can be dramatically reduced while the confidentiality of input/output is still guaranteed. The high physical saving of customers' computing resources and the literally unlimited resources in Cloud Computing enable ELM to be applied to big data applications, even when sensitive data is involved.

Acknowledgments This work was supported by the National Natural Science Foundation of China (Project No.61379145, 61170287, 61232016, 61070198, 61402508, 61303189).

References

1. Huang, G.B., Zhu, Q.Y., Siew, C.K.: Extreme learning machine: a new learning scheme of feedforward neural networks. In: Proceedings of International Joint Conference on Neural Networks (IJCNN2004), pp. 985–990. Budapest, Hungary, 25–29 July 2004
2. Huang, G.B., Zhu, Q.Y., Siew, C.K.: Extreme learning machine: theory and applications. Neurocomputing **70**, 489–501 (2006)
3. Huang, G.B., Zhou, H., Ding, X., Zhang, R.: Extreme learning machine for regression and multiclass classification. IEEE Trans. Syst. Man Cybern. Part B Cybern. **42**(2), 513–529 (2012)
4. Lin, J., Yin, J., Cai, Z., Liu, Q., Li, K., Leung, V.C.: A secure and practical mechanism for outsourcing elms in cloud computing. IEEE Intel. Syst. **28**(6), 35–38 (2013)
5. He, Q., Shang, T., Zhuang, F., Shi, Z.: Parallel extreme learning machine for regression based on MapReduce. Neurocomputing (2012)
6. Wang, C., Ren, K., Wang, J.: Secure and practical outsourcing of linear programming in cloud computing. In: INFOCOM2011, pp. 820–828. (2011)
7. Yu, S., Wang, C., Ren, K., Lou, W.: Achieving secure, scalable, and fine-grained data access control in cloud computing. In: 2010 Proceedings IEEE INFOCOM, IEEE, pp. 1–9. (2010)
8. Cheng, Y., Ren, J.C., Mei, S., Zhou, J.: Keys distributing optimization of cp-abe based access control in cryptographic cloud storage. In: IEICE Transactions on Information Systems, vol. 95-D, pp. 3088–3091. (2012)
9. Krizhevsky, A., Hinton, G.: Learning multiple layers of features from tiny images. Master's thesis, Department of Computer Science, University of Toronto (2009)

H-MRST: A Novel Framework for Support Uncertain Data Range Query Using ELM

Bin Wang, Rui Zhu and Guoren Wang

Abstract Probabilistic range query is a typical problem in the domain of probabilistic database management system. There exist many efforts for supporting such query. However, the state of arts approaches can not efficiently index uncertain data when their probability density function are discrete. In this paper, we propose a general framework to construct summary for uncertain data with any type of PDF. Especially, if the PDF of uncertain data is discrete, we employ a novel machine learning technique named ELM to learn its distribution type and fit the specific function. If this method does not work, we propose a hybrid algorithm to construct its summary. Besides the hybrid summary construction algorithm, we propose a bitwise-based accessing algorithm to speed up the query. Theoretical analysis and extensive experimental results demonstrate the effectiveness of the proposed algorithms.

1 Introduction

Query processing over uncertain data has become increasingly important due to the ubiquitous data uncertainty in many real-world applications such as sensor data monitoring, location-based services (LBS), biometric databases, moving object search, to name but a few.

For instance, an intelligent transportation system monitors the traffic volume and the average vehicle speed on every road section. As a tradeoff between communication cost and the data accuracy, the statistics are transmitted to a central database

B. Wang (✉) · R. Zhu · G. Wang
College of Information Science and Engineering, Northeastern University,
Shenyang, China
e-mail: binwang@ise.neu.edu.cn

R. Zhu
e-mail: neuruizhu@gmail.com

G. Wang
e-mail: wanggr@ise.neu.edu.cn

© Springer International Publishing Switzerland 2016
J. Cao et al. (eds.), *Proceedings of ELM-2015 Volume 2*,
Proceedings in Adaptation, Learning and Optimization 7,
DOI 10.1007/978-3-319-28373-9_7

periodically (e.g., every 10 min). The database content may not exactly reflect the current value, since the actual value may have changed since it was last measured. A similar query in the intelligent transportation system may "identify" the road section whose traffic volume in [30, 50 V/h], average vehicle speed in [40, 55 km/h] with a with at least 0.8 likelihood.

Many researchers employ probabilistic model to manage such data. They often combine a set of data into a tuple $< o.r, o.pdf >$ (named uncertain data), where $o.r$ denotes the region these objects could appear, and $o.pdf$ denotes their distribution. In order to manage and retrieve uncertain data, many probability queries are proposed. In this paper, we study the problem of probability range query over uncertain data. Corresponding to uncertain data, it is expressed by a tuple $q < r, \theta >$, where q returns the objects that are located in region with the probability at least θ.

A straightforward approach for evaluating probabilistic range query is to examine the appearance probability of each object that lies in the query region by calculus. Obviously, the cost of such calculus approach might be expensive because of the calculation of the complex function. Therefore, the filter-refinement framework is more prevailing, which tries to filter those probabilistic objects that are (or not) able to become the query results. Thus, the key of optimizing a prob-range query is to provide, as tighter as possible, a bound for pruning/vaildating with a relatively small cost.

Several indexes have been proposed to answer the queries over uncertain data. Their key idea is to pre-compute the summary [1] of each object's PDF. Among these works, Tao et al. [2] introduces PCR (short for probabilistically constrained regions) as the summary of an object. Zhang et al. [1] develops a U-Quardtree, which employ the partition technique to summary the PDF of uncertain data. Aiming to their defects, Zhu et al. [3]. propose R-MRST to approximately capture the PDF of uncertain data, with strong pruning power and lower space cost. However, R-MRST only works when the PDF of uncertain data is a continuous function. When the PDF is discrete,R-MRST can not be used. Unfortunately, in most real applications, the PDF of uncertain data are usually represented by a group of sampling points. Therefore, although R-MRST could provides uncertain data with tight boundary, its applicability is limited.

In this paper, we propose the H-MRST (short for hybrid MRST) framework to extend R-MRST. It could construct summary for uncertain data with any types of PDF (e.g., continuous or discrete). Achieving this goal, we employ a classic learning method named ELM(short for Extreme Learning Machine) to solve this problem. As far as we know, ELM is an outstanding machine learning method which is widely used in the domain of text classification [4], multimedia recognition [5–7], bioinformatics [8], mobile objects [9], etc. It could both fast and accurate employ the classification. In this way, it helps us to identify every type of PDF (e.g., normal, uniform, binomial distribution, and so on). And then, we further study the characteristics of MRST and propose a bitwise-based algorithm for computing the probabilistic lower-bound(and upper-bound) of objects lying in the query region. Above all, the contributions are as follows:

A PDF-aware summary construction framework The PDF-aware could be reflected from the following two facets: (i) type-aware. According to the type of PDF, we develop a group of algorithms for constructing the summary of uncertain data. Specifically, if the PDF of uncertain data is continuous, we employ the algorithm proposed in [3] for summary construction. If the PDF of uncertain data is discreet, we propose a ELM-based algorithm for summary construction. (ii) distribution-aware. If the type of PDF is discreet, we employ ELM for identifying the distribution type of PDF. In this way, we could find a suitable function for fitting.

A Self-verification algorithm Through using ELM, we could obtain the distribution type of uncertain data. However, it could not achieve 100 % accuracy rate for one thing. For another, we could not obtain the coefficients of PDF. Aiming to this problem, we propose ELM-LSE. It combines the ELM and *least squares estimation*. It has two functions:(i)verifying whether the classification is accurate; (ii) compute the coefficients of PDF. Accordingly, the above issues could be effectively solved.

A Bitwise-based Summary accessing algorithm In order to speed up the query, we propose a bitwise-based algorithm for summary accessing. Specifically, via deeply study the characteristic of summary, we find that we could use a few bit operations to compute the topological relationship between query region and uncertain data region. It leads the following benefits. (i) fully usage the superiority of CPU; (ii)reducing the computing times(e.g., from 4 times to 2 times).

The rest of this paper is organized as follows: Sect. 2 gives related work and the problem definition. Section 3 proposes the H-MRST. Section 4 evaluates the proposed methods with extensive experiments. Section 5 is the conclusion and the future work.

2   Background

In Sect. 2.1, we explain the state of arts approaches. In Sect. 2.2, we briefly gives an overview of ELM. Last of this section, we formally define the problem of probabilistic range query on uncertain data.

2.1   Related Work

The MRST Zhu et al. [3] propose MRST to capture approximately the PDF of an uncertain object, with strong pruning power and lower space cost. The MRST is a data structure that can be regarded as an extension of gird file. It could capture the PDF of uncertain data via self-adaptively adjusting the resolution the partition. Given an uncertain data o, after constructing, the region of uncertain data is partitioned into a group of subregions. For each subregion $o(i)$ of o, if o_{pdf} in $o(i)$ changes dramatically, we use a fine resolution to partition $o(i)$. On the contrary, we use a coarse resolution to partition it.

At the same time, we associate each subregion $o(i)$ with plentiful probabilistic information. They are $app(o, i)$, $lb(o, i)$, and $(ub(o, i))$ respectively. They represent the likelihood of o falling in $o(i)$, the maximal (or minimal) probability density in $o(i)$. Based on these information, we could compute the lower-bound and upper-bound of o lying in $q_r \cap o(i)_r$ via Eqs. 1 and 2.

$$lb_{app}(q, i) = lb(o, i) \times (max(0, S(q, i) - ZS(o, i))) \tag{1}$$

$$ub_{app}(q, i) = min(ub(o, i) \times S(q, i), app(o, i)) \tag{2}$$

Recalling the summary construction strategy, $\forall i$, MRST could guarantee $ub_{app}(q, i) - lb_{app}(q, i)$ may be small enough. Accordingly, as is explained in Lemma 1, the MRST of o could provide it with a tighter probabilistic boundary. Its issue is it only works when the PDF of uncertain data is continuous.

Lemma 1 *Given an object o and a query q, when q_r overlaps with o's subregion $\bigcup_{i=1}^{i=n_1} o(i)$, the $lb_{app}(o, q) = \sum_{i=1}^{i=n_1} lb_{app}(q, i)$ and $ub_{app}(o, q)$ is $\sum_{i=1}^{i=n_1} ub_{app}(q, i)$.*

The Other Indexes In recent years, many effective indexes have been proposed to answer prob-range query over the uncertain data. Given a set $\mathcal{O}$ of uncertain data, a range region $\mathcal{R}$, and a probabilistic threshold θ, a probabilistic range query returns uncertain data $o \in \mathcal{O}$ with the probability θ. Among all these works, Cheng et al. [10] first address probabilistic range queries on one-dimensional uncertain data. And then, Tao et al. [2] proposed a U-tree which derives from R-Tree. It uses a finite set of probabilistically constrained regions(short for PCR) as the summary of uncertain data. Accordingly, the pruning rules of U-tree lie on the topological relationship between PCR and query region. However, as is depicted in [11], its pruning ability is weak. Zhang et al. proposed UI-Tree and UD-Tree. It is based on the partition technique. Specifically, given an object o, it partitions o_r into a group of subregions, pre-computes the probability of o lying in each subregion and store them. In this way, they can employ the probabilistic pruning between topological relationship between these subregion and query region. Its issue is the space cost is high,which leads a high I/O cost. Kalashnikov et al. [12] proposed grid-based index named U-grid for indexing probabilistic objects. U-grid is a 2-layers index. The 1st-layer index provides the query with spatial information, probability summarizations and a pointer to an entry with detailed probability information. The summary probability information stored in 2nd-layer is base on a "virtual grid" and the 2nd-layer index uses these aggregate information for pruning. However, the U-grid does not provide a lower-bound for the query and its storage cost is also very high. The others, Aggarwal and Yu [13] show how to construct an effective index structure in order to handle uncertain range queries in high dimensionality.

2.2 Extreme Learning Machine

Extreme learning machine (ELM) [14, 15] has been originally developed based on single-hidden layer feed-forward neural networks (SLFNs). Also, it has many variants [10, 13, 16–18]. As is discussed in [19], it provides a unified learning platform with a widespread type of feature mappings. Also, it get better performance due to its *universal approximation capability* [13, 17, 20] and *classification capability* [19].

It randomly assigns the input weights and hidden layer biases of SLFNs and analytically determines the output weights of SLFNs. In ELM, the input weight and the bias of the hidden node are randomly generated, and the output weight could avoid iterative tuning. The output function of SLFNs with L hidden nodes can be represented by

$$f_L(x) = \sum_{i=1}^{L} \beta_i g_i(x) = \sum_{i=1}^{L} \beta_i G(a_i, b_i, x), x \in R^d, \beta_i \in R^m \tag{3}$$

where g_i denotes the output function $G(a_i, b_i, x)$ of the i-th hidden node, $\beta_i = [\beta_{i1}, \ldots, \beta_{iL}]^T$ denotes the vector of the output weights between the hidden layer of L nodes, $h(x) = [G(a_1, b_1, x), \ldots, G(a_L, b_L, x)]^T$ is the output vector of the hidden layer with respect to the input x.

To approximate these samples with zero errors means that $\sum_{j=1}^{I_i} ||o_j - t_j|| = 0$, where β_i, w_i, and b_i satisfy Eq. 4.

$$\sum_{i=1}^{L} \beta_i g(w_i x_j + b_i) = t_j, j = 1, \ldots, N \tag{4}$$

Equation 4 is equivalent to $H\beta = T$, where $T = [t_1^T, \ldots, t_N^T]_{M \times N}^T$, $\beta = [\beta_1^T, \ldots, \beta_L^T]_{m \times L}^L$.

2.3 Problem Definition

Given a multidimensional probabilistic object o in the d-dimension space, it is described either continuously or discretely. In the continuous case, an object has two attributes: o_r and $o.PDF(x)$. The o_r is a d-dimension uncertainty region, where

---

Algorithm 1: ELM

---

1 **for** *i to 1 to* $\widetilde{N}$ **do**
2 $\quad\lvert\quad$ randomly assign input weight w_i;
3 $\quad\lvert\quad$ randomly assign bias b_i;
4 compute H;
5 compute $\beta = H^\dagger T$;
6 return ;

---

o may appear at any locations with certain probabilities. The o.PDF(x) is the probability of o appearing at location x. In the discrete case, o is represented by a set of sampled points $x_1, x_2, \ldots, x_m$, and o occurs at location x_i with probability $x_i.p$. Given a query region q_r, we use $app(o,q)$ to represent the likelihood of o falling in the query region q_r. $app(o,q)$ is also calculated by two cases. In the continuous case:

$$app(o, q) = \int_{o_r \cap q_r} o \cdot \text{PDF}(x)dx \tag{5}$$

where $o_r \cap q_r$ denotes the intersection of o_r and q_r, and o is a result if $p_{app}(o,q) \geq \theta$ (query probability threshold). In the discrete case:

$$app(o, q) = \sum_{i=1}^{n2} o.\text{PDF}(x_i) / \sum_{i=1}^{n1} o.\text{PDF}(x_i) \tag{6}$$

where n_1 is amount of the sampled points in o_r, and n_2 is the amount of the sampled points falling into $o_r \cap q_r$.

Definition 1 (*Probabilistic Range Query*). Given a set of probabilistic objects O and a range query q, the probabilistic range query retrieves all probabilistic objects $o \in O$ with $app(o, q) \geq \theta$, where θ is the probabilistic threshold and $0 \leq \theta \leq 1$.

3   Effectively Summarizing Uncertain Data

In this section, we propose a novel framework named H-MRST to capture the PDF of uncertain data. It could handle the uncertain data with different types of PDF (e.g., discrete or not).

Fig. 1 The H-MRST
Framework

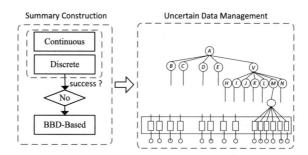

3.1 The H-MRST *Overview*

As is shown in Fig. 1, the framework could be divided into two parts. The first part is
to construct summary for uncertain data. The second part is to index uncertain data
and its summary.

For a set of N uncertain objects, the H-MRST is constructed as follows. Firstly, for
every probabilistic object, we firstly checks the PDF type of uncertain data. If the PDF
is continuous, we employ the algorithm discussed in [3] for summary construction.
On the contrary, we firstly employ a novel machine learning algorithm called ELM
for PDF classifying. And then we propose a novel algorithm named ELM-LSE for
fitting the PDF of uncertain data and verifying whether the classification is right.
After fitting all the PDF of uncertain data, we employ the algorithm discussed in [3]
for index construction and accessing.

In the following part, we firstly discuss the summary construction under different
kind of uncertain data. And then, we propose a novel bitwise algorithm for speeding
up the summary accessing.

3.2 The Summary Construction Algorithms

As is discussed before, if the PDF type of uncertain data is discrete, the superior
properties of R-MRST could not be used. To address this issue, we employ ELM and
least squares estimation for fitting the PDF if it is discrete.

The ELM-Based Classification Algorithm ELM is an outstanding machine learning method that is suitable for classification. In this paper, we employ ELM for identifying the distribution type of PDF(normal distribution, geometric distribution, and
etc.). In order to make ELM works, we build a vector for each uncertain data, where
the elements in the vector are the coordinates of sampling points. Based on this vector, we apply ELM for classification. In this way, we can determine the PDF's type of
uncertain data. As shown in Algorithm 2, given an uncertain data o and its sampling
point $\mathcal{O}$, we firstly generate the vector v. And then, we input the vector to the classifier. Note that, the classifier is generated by the training set, for limitation of space,
we skip the details of training.

Algorithm 2: PDF classification Using ELM

    Input: Uncertain Data $\mathcal{O}$ $\{o_1, o_2, \ldots, o_n\}$
    Output: PDF Type t
1 Vector $v \leftarrow$ build-Vector($\mathcal{O}$);
2 Type $t \leftarrow$ ELM-Classifier(v);
3 Function $f \leftarrow$ fitting-Function($\mathcal{O}, t$);
4 **if** *Verification(f, $\mathcal{O}$)$> \lambda$* **then**
5 | Return;

6 **else**
7 | BDTree($\mathcal{O}$) ;

8 **return** ;

The Self-Verification Algorithm A natural question is ELM only could obtain the distribution type of uncertain data. In addition, if the classification is error, the summary construction may also be not suitable. Aiming to these problems, we propose the ELM-LSE algorithm. It could both compute the accurate coefficient of PDF. For another, it could verify whether the classification is right.

$$
\begin{cases}
\dfrac{\partial S}{\partial a_0} = -2 \sum_{i=1}^{i=m} (s_i - a_j \sum_{j=0}^{j=n} x_i^j) x \\[2mm]
\dfrac{\partial S}{\partial a_1} = -2 \sum_{i=1}^{i=m} x_i (s_i - a_j \sum_{j=0}^{j=n} x_i^j) \\[1mm]
\ldots \\[1mm]
\dfrac{\partial S}{\partial a_n} = -2 \sum_{i=1}^{i=m} x_i^n (s_i - a_j \sum_{j=0}^{j=n} x_i^j) x^k
\end{cases}
\tag{7}
$$

Specifically, we employ *least squares estimation* for computing its coefficients. For simplicity, we use the n degree polynomial as the example to show our solution. Let the function $F_{fit}(x)$ be $\sum_{i=1}^{i=n} a_i x^i$ and the sampling set be the candidate set. According to *least squares estimation*, our aim is to make $\sum_{j=0}^{j=|C|} (\sum_{i=1}^{i=n} a_i x^i - s(x_j))^2$ as small as possible. Accordingly, we could obtain a_0 to a_n through solving Eq. 7.

After fitting, we verify whether the fitting function could be used. Specially, let $\mathcal{O}$ be the sampling points, F be the actual PDF of $\mathcal{O}$, and F_{fit} be the fitting function. If non-equation 8 is satisfied, we alert that the current fitting function is suitable. On the contrary, it indicates the predicted result may be wrong. In this case, we use the key idea of BD-Tree [21] for summary construction. After constructing, the sampling points in each subregion are roughly the same. For the limitation of space, we skip the details.

As shown in Algorithm 2, after classifying, we firstly apply the function fitting. Under this step, we could decide whether the classification is suitable. If so, we do nothing. Otherwise, we apply BD-Tree for summary construction.

$$
\sum_{x \in |\mathcal{O}|} F_{fit}(x) - F(x) < \lambda
\tag{8}
$$

3.3   Accessing the Summary of Uncertain Data

After extending the summary construction algorithms, in this section, we develop a novel Algorithm 1 named priority-based to access the summary of uncertain data. Algorithm 1 shows the details. It employs the key idea of greedy algorithm for accessing. Specifically, it uses a field called $d(q, i)$ to determine the accessing order of the nodes in MRST so as to early terminating the accessing of MRST as much as possible.

Given a query q, an object o and a subregion $o(i)$, if $q.r$ overlaps with $o.MBR$, we access the MRST of o to check whether o is a result of q. The $d(q, i)$ is computed through Eq. 9. Obviously, the larger the $d(q, i)$ is, the greater it contributes to $ub_{app}(o, i)$-$lb_{app}(o, i)$, and the corresponding $o(i)$ should be prior accessed. Compared with the traditional accessing method such as preorder traversal and inorder traversal, introducing this field to control the nodes accessing order is more efficiently to compute the bound. After accessing the MRST of an object, o is validated if the lower-bound of $app(o, q)$ is more than q_p. Also, o is pruned if the upper-bound of $app(o, q)$ is less than q_p.

$$d(i, q) = min(u(i, o) \times S(q, i), app(o, i)) - lb(i, o) \times (max(0, S(q, i) - ZS(o, i))) \quad (9)$$

4   Experimental Evaluation

4.1   Experimental Setting

This section experimentally evaluates the efficiency of the proposed techniques. We compare H-MRST with both U-Tree and UD-Tree. They are the classic technique and an advanced index respectively.

Data set. In this paper, we use two real spatial data sets that are *LB* and *CA*, where they contain 62 and 53 kb respectively. In these two sets, the elements are used as the center of probabilistic regions. Then, we generate sampling points for each object according to a given probability density function. Similarly, we generate another three synthetic data sets. They contain 128 k/256 k/512 k two-dimension points.

Sampling Points Generation. In this paper, we use rectangle to bound sampling points of objects. The side-length varies from 100 to 500 with the default with 500. There exist four types of PBF that are poisson distribution, normal distribution, geometric distribution, Heavy-tailed distribution. For each object, we randomly select a distribution and generate samplings for it.

Experimental methods. Our experiments mainly evaluate the following two aspects: (i) the accuracy of classification, where we compare our proposed algorithms with varying size of the training data in different datasets. (ii) the efficiency of H-MRST. In this setting, a workload contains 100 queries in our experiment. The

Fig. 2 Training time
evaluation. **a** CA. **b**
Synthesis

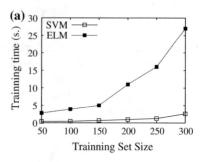

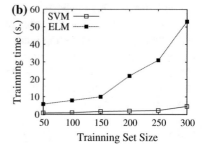

region of the queries are a rectangular with r_q varying from 500 to 1500. In our experiments, we randomly choose the probabilistic threshold $\theta \in (0,1]$ for each query.

4.2 Classification Evaluation

In this subsection, we are going to evaluate the effect of the classifiers based on the ELM and SVM under different data set. We firstly evaluate the training time against different training set.

From Fig. 2a, b, with the increasing of the training set size, the training time under ELM and SVM are all increasing. However, compared with SVM, the training time of ELM increases slowly. The reason behind is ELM has better performance than SVM. Next, we evaluate the classifying accuracy rate. We also compare ELM with SVM. The difference is we add another two algorithms for comparison. Recalling Sect. 3.2, we use *least squares estimation* for computing its coefficients. Its another function is to do the verification. We call this algorithm as LSE-ELM. Also, under SVM, we use *least squares estimation* for verification. We call this algorithm as LSE-SVM.

From Fig. 3a, b, the accuracy ratio of LSE-ELM is highest of all. For one thing, using LSE-SVM, we could find some mis-claiming classification. For another, ELM has better classifying ability.

Fig. 3 Accuracy ratio evaluation. **a** CA. **b** Synthesis

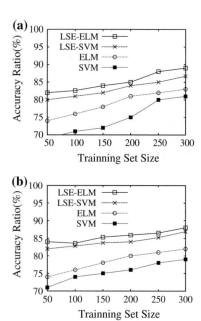

Last of this subsection, we evaluate the classifying time. We also compare LSE-ELM with LSE-ELM. From Table 1, the classifying time of LSE-ELM is shortest of all. The main reason is the classifying efficiency is higher than the SVM algorithms.

4.3 Query Performance

In this section, we evaluate the query performance. First of all, we compare H-MRST with the other state of arts indexes(e.g., UD-Tree and UD-Tree). In this group, we set the probabilistic threshold to the default value(e.g., 0.5), and vary the query region from 500×500 to 1500×1500. Our aim is to evaluate the impaction of query region to the algorithm performance. We show two results that are CPU time and candidate set. They evaluate the ability of pruning/validating and algorithm performance respectively. From Fig. 4a, the candidate size of H-MRST is smaller than the other indexes. The reason behind is MRST could more effectively reflect the feature of PDF. From Fig. 4b, the CPU time of H-MRST is shortest of all (Fig. 5).

Secondly, we set the query region to 1000×1000, and vary probabilistic threshold from 0.1 to 0.9. Our aim is to evaluate the impaction of probabilistic threshold to the algorithm performance. Similarly, we still compare H-MRST with UD-Tree and U-Tree. In addition, we still evaluate the ability of pruning/validating and algorithm

Table 1 Classification time

Data set	LSE-SVM		LSE-ELM	
	Classifying time (s)	Class	Classifying time (s)	Class
CA	38.2	4	6.3	4
LB	36.7	4	5.5	4
64 kb	40.7	4	6.7	4
128 kb	83.1	4	11.3	4
256 kb	126.2	4	23.9	4

Fig. 4 Cost versus diff R_u.
a CPU time. **b** Candidate size

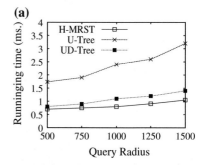

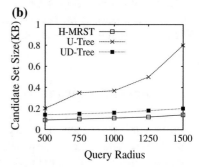

performance. As the same as the first group of evaluation, H-MRST performs best of all (Table 2).

In the last experiments, we compare the performance of H-MRST, UD-Tree, and U-Tree by different data sets. Five data sets (LB and three synthesize) are employed. The number of data points of each data set is 53 k, 64 k, 128 kb and 256 kb. We use default parameters in these experiments. As expected, H-MRST performs best of all.

Fig. 5 Cost versus diff θ.
a Candidate size. **b** CPU
time

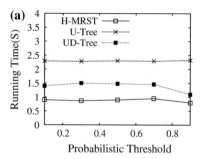

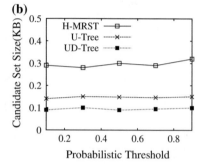

Table 2 Algorithm performance

Data set	H-MRST		UD-Tree		U-Tree	
	Running time	Candidate set	Running time	Candidate set	Running time	Candidate set
LB	1	0.92	1.2	1.08	2.21	2.36
64 kb	1.3	1.11	1.7	1.39	2.26	2.99
128 kb	2.1	1.23	3.3		3.9	5.74
256 kb	4.5	1.59	6.9	3.03	4.8	8.28

5 Conclusions

In this paper, we studied the problem of range query on probabilistic data. Through deep analysis, we proposed an effective indexing technique named E-MRST to manage uncertain data. E-MRST could provided a very tight bound for pruning/validating the objects that overlap(or non-overlap) with the query region in a lower cost. Our experiments convincingly demonstrated the efficiency of our indexing techniques. In the future, we will further study other indexes which are suitable for high-dimensional uncertain data and support probabilistic data update frequently.

Acknowledgments The work is partially supported by the National Natural Science Foundation of China for Outstanding Young Scholars (No. 61322208), the National Basic Research Program of China (973 Program) (No. 2012CB316201), the Joint Research Fund for Overseas Natural Science of China (No. 61129002), the National Natural Science Foundation of China for Key Program (No. 61572122), the National Natural Science Foundation of China (Nos. 61272178, 61572122).

References

1. Zhang, Y., Zhang, W., Lin, Q., Lin, X.: Effectively indexing the multi-dimensional uncertain objects for range searching. In: EDBT, pp. 504–515 (2012)
2. Tao, Y., Cheng, R., Xiao, X., Ngai, W.K., Kao, B., Prabhakar, S.: Indexing multi-dimensional uncertain data with arbitrary probability density functions. In: VLDB. pp. 922–933 (2005)
3. Zhu, R., Wang, B., Wang, G.: Indexing uncertain data for supporting range queries. In: Web-Age Information Management—15th International Conference, WAIM 2014, Macau, China, Proceedings, 16–18 June 2014, pp. 72–83 (2014)
4. Zhao, X.-G., Wang, G., Bi, X., Gong, P., Zhao, Y.: XML document classification based on ELM. Neurocomputing **74**, 2444–2451 (2011)
5. Lan, Y., Hu, Z., Soh, Y., Huang, G.B.: An extreme learning machine approach for speaker recognition. Neural Comput. Appl. **22**(3–4), 417–425 (2013)
6. Lu, B., Wang, G., Yuan, Y., Han, D.: Semantic concept detection for video based on extreme learning machine. Neurocomputing **102**, 176–183 (2013)
7. Zong, W., Huang, G.B.: Face recognition based on extreme learning machine. Neurocomputing **74**, 2541–2551 (2011)
8. Wang, G., Zhao, Y., Wang, D.: A protein secondary structure prediction framework based on the Extreme Learning Machine. Neurocomputing **72**, 262–268 (2008)
9. Wang, B., Wang, G., Li, J., Wang, B.: Update strategy based on region classification using elm for mobile object index. Soft Comput. **16**(9), 1607–1615 (2012)
10. Rong, H.J., Huang, G.B., Sundararajan, N., Saratchandran, P.: Online sequential fuzzy extreme learning machine for function approximation and classification problems. IEEE Trans. Syst. Man Cybern. **39**, 1067–1072 (2009)
11. Zhang, Y., Lin, X., Zhang, W., Wang, J., Lin, Q.: Effectively indexing the uncertain space. IEEE Trans. Knowl. Data Eng. **22**(9), 1247–1261 (2010)
12. Kalashnikov, D.V., Ma, Y., Mehrotra, S., Hariharan, R.: Index for fast retrieval of uncertain spatial point data. In: GIS, pp. 195–202 (2006)
13. Huang, G.B., Chen, L.: Enhanced random search based incremental extreme learning machine. Neurocomputing **71**, 3460–3468 (2008)
14. Huang, G.B., Zhu, Q.Y., Siew, C.K.: Extreme learning machine: a new learning scheme of feedforward neural networks. In: International Symposium on Neural Networks. vol. 2 (2004)
15. Huang, G.B., Zhu, Q.Y., Siew, C.K.: Extreme learning machine: theory and applications. Neurocomputing **70**, 489–501 (2006)
16. Feng, G., Huang, G.B., Lin, Q., Gay, R.K.L.: Error minimized extreme learning machine with growth of hidden nodes and incremental learning. IEEE Trans. Neural Netw. **20**, 1352–1357 (2009)
17. Huang, G.B., Chen, L.: Convex incremental extreme learning machine. Neurocomputing **70**, 3056–3062 (2007)
18. Huang, G.B., Zhu, Q.Y., Mao, K.Z., Siew, C.K., Saratchandran, P., Sundararajan, N.: Can threshold networks be trained directly? IEEE Trans. Circuits Syst. Ii: Analog Digital Signal Process. **53**, 187–191 (2006)
19. Huang, G.B., Zhou, H., Ding, X., Zhang, R.: Extreme learning machine for regression and multiclass classification. IEEE Trans. Syst. Man Cybern. **42**, 513–529 (2012)

20. Huang, G.B., Chen, L., Siew, C.K.: Universal approximation using incremental constructive feedforward networks with random hidden nodes. IEEE Trans. Neural Netw. **17**, 879–892 (2006)
21. Ohsawa, Y., Sakauchi, M.: The bd-tree—a new n-dimensional data structure with highly efficient dynamic characteristics. In: IFIP Congress, pp. 539–544 (1983)

The SVM-ELM Model Based on Particle Swarm Optimization

Miao-miao Wang and Shi-fei Ding

Abstract Extreme learning machine (ELM) is a simple and effective SLFNs single hidden layer feedforward neural network learning algorithm, in recent years, it has become one of the hot areas in machine learning research. But single hidden layer node lacks of judgement ability, to some extent the classification accuracy depends on the number of hidden layer nodes. In order to improve the judgement ability of single hidden layer node, Support Vector Machine (SVM) is combined with ELM, and a simplified SVM-ELM model is established. At the same time, in order to avoid the subjectivity of human to choose parameters, the SVM-ELM model uses Particle Swarm Optimization (PSO) algorithm to automatically select the parameters, finally PSO-SVM-ELM model is proposed. Experiments show that classification accuracy of the model is higher than the SVM-ELM and ELM, and it also has good robustness and adaptive generation ability.

Keywords Particle swarm optimization (PSO) · Support Vector Machine (SVM) · Extreme learning machine (ELM) · SVM-ELM

1 Introduction

Extreme learning machine (ELM) is a simple and effective SLFNs single hidden layer feedforward neural network learning algorithm [1]. In the process of algorithm execution, only need to set the number of hidden layer nodes of networks, input

M. Wang—Project supported by the National Natural Science Foundation of China (No.61379101), Postgraduate Cultivation Innovation project of Jiangsu Province in 2014 (No. No.SJLX_0636).

M. Wang · S. Ding (✉)
School of Computer Science and Technology, China University of Mining and Technology, Xuzhou 221116, China
e-mail: dingsf@cumt.edu.cn

M. Wang · S. Ding
Key Laboratory of Intelligent Information Processing, Institute of Computing Technology, Chinese Academy of Sciences, Beijing 100190, China

© Springer International Publishing Switzerland 2016
J. Cao et al. (eds.), *Proceedings of ELM-2015 Volume 2*,
Proceedings in Adaptation, Learning and Optimization 7,
DOI 10.1007/978-3-319-28373-9_8

weight and the bias of hidden layer take random assignment. Output layer weight is obtained the unique solution by the least squares method. Therefore classification accuracy of ELM depends on the number of hidden layer nodes. To some extent, there are more hidden layer nodes, and the classification accuracy of ELM will be higher. Since ELM is based on empirical risk minimization principle, random section of input weight and hidden layer bias are likely to lead to unnecessary input weight and hidden layer bias. Compared with the traditional learning algorithm, which is based on gradient descent algorithm, ELM may require more hidden layer neurons, but more hidden layer neurons will reduce the operating rate and training effect. So many algorithms are put forward to improve ELM.

Huang et al. put forward online sequential extreme learning machine (OS-ELM) algorithm based on the recursive least square (RLS) algorithm [2]. After that, online sequential fuzzy extreme learning machine (OS-fuzzy-ELM) model is proposed to solve the problem of function approximation and classification [3]. Rong et al. put forward pruned extreme learning machine (pruned-ELM) as systematic and auto-mated approach of ELM classification network, due to structural design of original ELM's classification network results in under or overfitting problems [4]. Group search optimization strategy is proposed by Silva et al. to select the input weight and hidden layer bias of ELM algorithm, shorten by GSO-ELM. Silva et al. used this method to improve ELM [5]. Compared with ELM, these optimization methods have gotten very good classification effect [6].

Support Vector Machine (SVM) is a new machine learning method and has devel-oped in recent years [7], it is based on Statistic Learning Theory, and its theoretical basis is structural risk minimization principle. Compare with neural network, SVM is more suitable for small sample to classify, and its generation performance is excel-lent [8]. SVM has achieved remarkable advances in pattern recognition, regression analysis, function estimation, time series prediction, etc. It also has been widely used in handwriting recognition, face detection, gene classification and many other fields [9]. SVM transforms linearly non-separable problems of low dimensional space into linearly separable problems of high dimensional by kernel function, and constructs optimal separating hyperplane to realize classification of data samples.

Classification accuracy of ELM depends on the number of hidden layer nodes, but large numbers of hidden layer nodes cause the structure of ELM overstaffing, also easily lead to insufficient memory. In order to improve the learning and judge-ment ability of single hidden layer node, this paper optimizes ELM by SVM, then establishes SVM-ELM model. In the SVM-ELM model, the leaning and judgement ability of single node are highly improved. Traditionally, the selection of parame-ter relies on experience value, and requires repeated trials, so the selection is very subjective. If the parameters are chosen by cross validation, it will consume a lot of time. Particle swarm optimization (PSO) is a new evolutionary algorithm developed in recent years, it is easily to implement, rapid convergence rate, high precision, and has been successfully applied to function optimization, pattern recognition, etc. In order to make the SVM-ELM model obtain better classification accuracy, this paper optimizes the SVM-ELM model by PSO, and realizes the classification target [10].

2 SVM-ELM Model

Extreme Learning Machine (ELM) was proposed by Guangbin Huang in 2004, a professor at Nanyang Technological University, and it is a kind of single hidden layer feedforward neural network (SLFN) [11]. Traditional ELM algorithm randomly determines weight and bias of hidden layer nodes, according to the experimental results, this method greatly saves the learning time of the system. But if higher classification accuracy is required, the network needs a lot of hidden layer nodes, while fewer nodes lead to poorer classification results. This indicates that the learning and judgement ability of single hidden layer node are insufficient and need more nodes to compensate. For a single node, insufficient learning ability is due to weight and bias of linear decision function randomly selected, and this causes the node under-fitting [12]. If the learning ability of the single hidden layer node in ELM network is greatly improved, then fewer hidden layer nodes are needed and ELM network can obtain better learning ability. Compared with original ELM, the network structure will be optimized.

In this paper, SVM is used to improve the judgement ability of the single node, to simplify the structure of ELM network, and to improve the generalization performance of ELM. Specific optimization method is as follows. According to the categories of data, the number of hidden layer nodes is determined. Changing the traditional way to determine the number of hidden layer nodes which adopts random value and requires repeated trials to find the best value, SVM is used to determine weight and bias of each hidden layer node [13], improve the learning ability and generalization ability of each node. In the case of k kinds of data to classify, the number of hidden layer nodes in the SVM-ELM model is k, the task of the ith node is to use SVM to separate the ith class data from the other k − 1 classes. On the basis of upper training result, weight and bias of each hidden layer node will be obtained.

If the data set is $S = \bigcup_{i=1}^{k} S_i$, S_i is the set of the ith class input vector x_j^i, $x_j^i = [x_{j1}^i, x_{j2}^i, \ldots, x_{jn}^i]^T \epsilon R^n$, the class label of x_j^i is $y^i = [y_{j1}^i, y_{j2}^i, \ldots, y_{jm}^i]^T \epsilon R^m$. The linear decision function of the ith node is $f_i\left(x_j^i\right) = w_i \cdot \phi\left(x_j^i\right) + b_i$, w_i is the pending weight and b_i is the pending bias, $x_j^i \epsilon S_i$ or $x_j^i \epsilon S - S_i$, and $\phi(\cdot)$ is the implicit function to mapping the sample in the low dimensional space to high dimension space. For the ith node, the weight and bias of the linear decision function $\phi(\cdot)$ can be identified by solving the following SVM optimization problem.

$$minL_p = \frac{1}{2}\left\|w_i\right\|^2 + C_i \sum_{j=1}^{|S|} \xi_j^i \tag{1}$$

$$s.t. y_j^i\left(w_i \cdot f\left(x_j^i\right) + b_i\right) \geqslant 1 - \xi_j^i, (j = 1, 2, \ldots, |S|), \xi_j^i \geqslant 0 \, (j = 1, 2, \ldots, |S|)$$

$$y_j^i = 1, \left(x_j^i \epsilon S_i\right), y_j^i = -1, \left(x_j^i \epsilon S - S_i\right)$$

In formula (1), x_j^i is the jth input vector of the ith class, y^i is its label to classify, C^i is the penalty parameter of the ith class, ξ_j^i is the relaxation factor of x_j^i. Lagrange objective function of the formula (1) as follows:

$$L\left(w_i, b_i, \alpha^i\right) = \frac{1}{2}\|w_i\|^2 - \sum_{p=1}^{|S|} \alpha_p^i [y_p^i \left(w_i^T x_p^i + b_i\right) - 1] \tag{2}$$

The dual problem of the formula (1) is:

$$maxL_D = \sum_{p=1}^{|S|} \alpha_p^i - \frac{1}{2} \sum_{p=1}^{|S|} \sum_{q=1}^{|S|} y_p^i y_q^i \alpha_p^i \alpha_q^i \cdot \phi\left(x_p^i\right) \cdot \phi\left(x_q^i\right) \tag{3}$$

$$s.t. \sum_{p=1}^{|S|} y_p^i \alpha_p^i = 0. y_p^i = 1, \left(x_p^i \epsilon S_i\right), y_p^i = -1, \left(x_p^i \epsilon S - S_i\right)$$

$$0 \leqslant \alpha_p^i \leqslant C_i, (p = 1, 2, \dots, |S|)$$

In formula (3), the support vector is x_p^i that is corresponding to the nonzero Lagrange multiplier α_p^i, y_p^i is the label of x_p^i. If $y_p^i = 1$, $x_p^i \epsilon S_i$, otherwise $x_p^i \epsilon S - S_i$. The weight w_p^i can be expressed as:

$$w_p^i = \frac{1}{N_j^i} \sum_{j=1}^{N_j^i} \alpha_j^i y_j^i \phi\left(x_j^i\right) \tag{4}$$

In formula (4), N_j^i is the number of the support vector x_j^i, y_j^i is the label of the support vector. According to formula (4), bring the value of the weight w_p^i into linear decision function $f(\cdot)$, then

$$f_i(x) = \frac{1}{N_j^i} \sum_{j=1}^{N_j^i} \alpha_j^i y_j^i \phi\left(x_j^i\right) \phi(x) + b_i \tag{5}$$

The input matrix of hidden layer can be expressed as:

$$H_k\left(w_1, w_2, \dots, w_k; b_1, b_2, \dots, b_k; x_1, x_2, \dots, x_{|S|}\right) = [g_1, g_2, \dots, g_{|S|}]_{|S|*k}^T$$

and

$$\begin{aligned}
g_1 &= g\left(w_1 \cdot \phi\left(x_1\right) + b_1\right) & \cdots & \quad g\left(w_k \cdot \phi\left(x_1\right) + b_k\right) \\
g_2 &= g\left(w_1 \cdot \phi\left(x_2\right) + b_1\right) & & \quad g\left(w_k \cdot \phi\left(x_2\right) + b_k\right) \\
&\vdots & \ddots & \quad \vdots \\
g_{|S|} &= g\left(w_1 \cdot \phi\left(x_{|S|}\right) + b_1\right) & \cdots & \quad g\left(w_k \cdot \phi\left(x_{|S|}\right) + b_k\right)
\end{aligned}$$

so

$$H_k = \begin{bmatrix} g\left(f_1\left(x_1\right)\right) & g\left(f_2\left(x_1\right)\right) & \cdots & g\left(f_k\left(x_1\right)\right) \\ g\left(f_1\left(x_2\right)\right) & g\left(f_2\left(x_2\right)\right) & \cdots & g\left(f_k\left(x_2\right)\right) \\ \vdots & \vdots & \ddots & \vdots \\ g\left(f_1\left(x_{|S|}\right)\right) & g\left(f_2\left(x_{|S|}\right)\right) & \cdots & g\left(f_k\left(x_{|S|}\right)\right) \end{bmatrix}_{|S|*k}$$

$g(x)$ is the activation function of ELM. The commonly used activation functions are Sigmoid function, Sine function, RBF function, etc. In this paper, we choose the Sigmoid function as the activation function because it has better effect. From these considerations, the number of hidden layer nodes is k, the activation function is $g(x)$, so the SLFN model is:

$$\sum_{i=1}^{k} \beta_i g_i\left(x_j^i\right) = \sum_{i=1}^{k} \beta_i g\left(w_i \cdot x_j^i + b_i\right) = y_j^i, (j = 1, 2, \ldots, |S|) \tag{6}$$

The learning system is $H_k \beta_k = Y$, learning parameter β_k can be expressed as follows:

$$\beta_k = \begin{bmatrix} \beta_1^T & \beta_2^T & \cdots & \beta_k^T \end{bmatrix}_{k*m}^T = \begin{bmatrix} \beta_{11} & \beta_{12} & \cdots & \beta_{1m} \\ \beta_{21} & \beta_{22} & \cdots & \beta_{2m} \\ \vdots & \vdots & \ddots & \vdots \\ \beta_{k1} & \beta_{k2} & \cdots & \beta_{km} \end{bmatrix}$$

The goal of the SVM-ELM model is to obtain the learning parameter $\hat{\beta}_k$ in the following formula:

$$\left\| H_k \hat{\beta}_k - Y \right\| = \min_{\beta_k} \left\| H_k \beta_k - Y \right\| \tag{7}$$

The solution of the formula (7) is the minimum norm least-squares solution $\hat{\beta}_k$, $\hat{\beta}_k = H_K^+ Y$, H_k^+ is Moore-Penrose generalized inverse of H_k.

3   Particle Swarm Optimization

Particle Swarm Optimization (PSO) algorithm is a kind of evolutionary computation technology, and is derived from the behavior of the birds and proposed by Eberhart et al. in 1995 [14]. The algorithm was originally inspired by the regularity of birds cluster activity, and then a simplified model is established through the swarm intelligence [15]. On the basis of observing the animals' cluster activity behavior, particle swarm optimization algorithm uses the information that the individual sharing in the group to make the movement of the population from disorder to order in the solution space of the problem, so as to obtain the optimal solution. Particle Swarm Optimization algorithm images every potential solution of the optimizing problem as a point of the D-dimensional search space, and the point is called 'particle'. Particle flies at a certain speed in the search space. According to its own flight experience and companion's flight experience, the speed is adjusted dynamically. All the particles have a fitness value that determined by the optimized objective function, the particles also records the best position (particle best, shorten as pbest) that found as far and current position of the particle, and these are seen as the particle's flying experience. At the same time, each particle also records the best position of the entire group (global best, shorten as gbest). Apparently gbest is the best value of pbest, and this is companion's flight experience.

The size of population is m, and the position of the ith particle in D-dimension space can be expressed as: $x_i = (x_{i1}, x_{i2}, \ldots, x_{id})$, $(i = 1, 2, \ldots, m)$, the speed $v_i = (v_{i1}, v_{i2}, \ldots, v_{id})$, $(i = 1, 2, \ldots, m)$ determines the displacement of the particle's iteration times in the search space, d represents the number of the independent variables in the problem that to be solved, the fitness function is determined by the optimized function in the practical problems. In each iteration, the particle updates its own speed and position by tracking two 'extremum'. One is the best solution that the particle itself has found so far, namely particle best, shorten as pbest, $pbest_i = (p_{i1}, p_{i2}, \ldots, p_{id})$, the other is the best solution that the entire population have found so far, namely global best, shorten as gbest, $gbest_i = (g_1, g_2, \ldots, g_d)$ [16].

$$v_{ij}(t+1) = v_{ij}(t) + c_1 \cdot r_1 \cdot \left(p_j(t) - x_{ij}(t)\right) + c_2 \cdot r_2 \cdot \left(g_j(t) - x_{ij}(t)\right) \qquad (8)$$

$$x_{ij}(t+1) = x_{ij}(t) + v_{ij}(t+1), (i = 1, 2, \ldots, m)(j = 1, 2, \therefore \ldots, d) \qquad (9)$$

In formulas (8) and (9), t is the iteration times, r_1 and r_2 is a random number in the (0,1) interval, c_1 and c_2 is acceleration factor, and commonly $c_1 = c_2 = 2$.

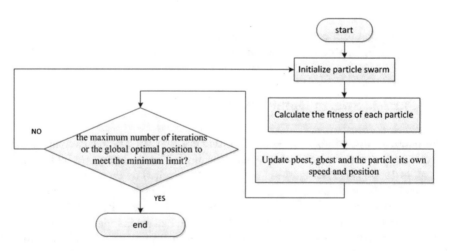

Fig. 1 The flow chart of PSO algorithm

Particle's speed of each dimension can't exceed the maximum speed v_{max} that set by the algorithm. If v_{max} is big, the global search ability is stronger, otherwise partial searching ability is stronger. In PSO algorithm, the termination condition for iteration is generally the maximum number of iterations or the global optimal position to meet the minimum limit [17, 18]. The flow chart of PSO algorithm is shown in Fig. 1.

In this paper, particle swam optimization algorithm is used to automatically select the parameters of SVM-ELM model [19]. It overcomes the subjectivity of artificial selection and avoids consuming too much time on artificial attempts. At the same time, the aim to control the direction of parameter selection can be realized through setting the fitness function of particles [20, 21].

4 Experimental Analysis

In order to verify the validity of the PSO-SVM-ELM model, four data sets are selected from UCI machine learning repository. These data sets are used for classification, and the description of these data sets is shown in Table 1.

In the SVM-ELM model, when using the result of SVM to optimize the weight and bias of hidden layer nodes, the kernel function of SVM is multiple, such as radial basis function which is namely RBF kernel, polynomial kernel function. According to the experimental results, select RBF kernel function which has better effect. Expression of RBF kernel function is $K(x, y) = \exp\left[-\|x - y\|^2 / 2\sigma^2\right]$. Sigmoid function is chosen as the activation function of ELM which has better performance, and its expression is $g(x) = 1 / (1 + e^{-x})$. In the SVM-ELM model based on particle swarm optimization algorithm, the parameters that to be optimized are the width parameter σ of RBF kernel function and penalty parameter C. The population size and the search range of the parameters are shown in Table 2.

Table 1 The description of data sets

Data sets	Number of samples	Number of categories	Number of attributes	Training data (%)	Testing data (%)
Wine	178	3	13	70	30
Seeds	210	3	7	60	40
Balance Scale	625	3	4	70	30
CNAE-9	1080	9	856	70	30

Table 2 The parameters search range of PSO-SVM-ELM model

Data sets	σ search range	C search range	Population size
Wine	[100, 10000]	[1, 10]	20
Seeds	[1, 1000]	[1, 10]	20
Balance Scale	[1, 100]	[1, 10]	20
CNAE-9	[1, 1000]	[1, 10]	20

Table 3 The parameters value of PSO-SVM-ELM model

	Wine	Seeds	Balance Scale	CNAE-9
σ	5247.37	37.1443	9.81278	724.485
C	8.190191	5.79651	2.65891	2.19863

Fig. 2 Traces flight of Wine's global best

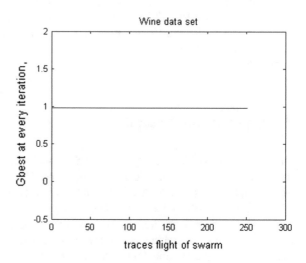

The accelerated factors c_1 and c_1 of the particle swarm are equal to 2, namely $c_1 = c_2 = 2$, the population size m = 20, inertia weight w = 0.9, and the maximum iteration number is 2000. Experimental environment: the processor is Intel(R) Xeon(R) CPU E3-1225 V2 @ 3.20 GHz, RAM is 4 GB, the operating system is 32-bit, and the main software is MATLAB R2012b. After several iterations, the parameters σ and C of each data set are shown in Table 3.

The flight path of gbest in Wine, Seeds, Balance Scale, CNAE-9 data sets are respectively shown in Figs. 2, 3, 4 and 5.

The classification results of the SVM-ELM model based on particle swarm optimization algorithm are shown in Table 4.

If PSO algorithm isn't used to optimize the SVM-ELM model, the parameters of the SVM-ELM model are determined by continuous attempts and experiments. Wine, Seeds and Balance Scale data sets have 3 kinds of data, so in the SVM-ELM model, the number of hidden layer nodes is 3. However CNAE-9 data set has 9 kinds of data, so its number of hidden layer nodes is 9. The classification results of SVM-ELM model without PSO algorithm are shown in Table 5. In the table, the parameters are the best of all the continuous attempts and experiments.

In the case of setting different number of hidden layer nodes, the classification results of ELM are shown in Table 6. In Table 6, the testing accuracy is the maximum of 10 times experiments, when the number of hidden layer nodes is the same.

Fig. 3 Traces flight of
Seeds's global best

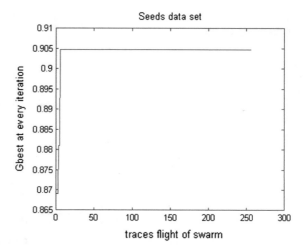

Fig. 4 Traces flight of
Balance Scale's global best

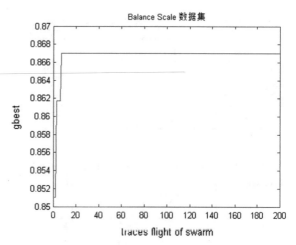

From Table 6, the testing accuracy of ELM is continuously improved with the number of hidden layer nodes increasing on Wine data set. On Seeds and CNAE-9 data sets, with the number of hidden layer nodes increasing, even though the testing accuracy of ELM appears partial fluctuations, the testing accuracy is still increasing as a whole. This phenomenon suggests that the learning effect of ELM has a great relationship with the number of hidden layer nodes. If a good learning effect is required, there must be enough hidden layer nodes. But large numbers of hidden layer nodes lead to the network bloated. On Balance scale data set, with the number of hidden layer nodes increasing, the testing accuracy of ELM is increasing at first and then decreasing. This suggests that the stability of the ELM is poorer and ELM easily appears overfitting phenomenon.

Fig. 5 Traces flight of
CNAE-9's global best

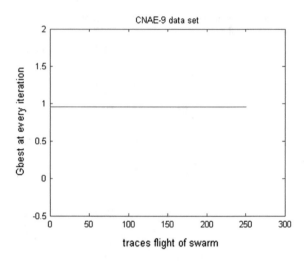

Table 4 The classification results of PSO-SVM-ELM model

Data sets	Number of hidden layer nodes	Training accuracy (%)	Testing accuracy (%)
Wine	3	92.00	98.11
Seeds	3	86.51	90.48
Balance Scale	3	86.51	90.48
CNAE-9	9	97.49	95.68

Table 5 The classification results of SVM-ELM model

Data sets	Number of hidden layer nodes	σ	C	Training accuracy (%)	Testing accuracy (%)
Wine	3	2000	2	89.60	96.00
Seeds	3	1000	5	90.48	86.90
Balance Scale	3	25	1	87.87	87.23
CNAE-9	9	500	2	97.35	94.75

Comparing Table 4 with Tables 5 and 6, we can find that the SVM-ELM model based on particle swarm optimization algorithm has a higher testing accuracy on the 4 data sets than SVM-ELM model and ELM. On Wine data set, the testing accuracy of the SVM-ELM model based on particle swarm optimization algorithm is 2.11 % higher than SVM-ELM model, and 5.66 % higher than ELM. On Seeds data set, the testing accuracy of the SVM-ELM model based on particle swarm optimization algorithm is 3.58 % higher than SVM-ELM model, and 5.96 % higher than ELM. The comparison about the classification accuracy of the SVM-ELM model based on particle swarm optimization algorithm, SVM-ELM and ELM on the selected data sets is shown in Table 7.

Table 6 The classification results of ELM

Data sets	Number of hidden layer nodes	Training accuracy (%)	Testing accuracy (%)
Wine	3	40.00	39.62
	100	69.60	71.70
	1000	96.80	83.02
	5000	100.00	92.45
Seeds	3	65.87	63.10
	10	90.48	83.33
	100	100.00	78.57
	1000	100.00	84.52
	5000	100.00	82.14
Balance Scale	3	76.20	72.87
	20	91.76	89.36
	50	94.28	81.91
	100	95.88	73.94
	1000	100.00	27.66
CNAE-9	9	51.46	54.96
	200	98.28	92.28
	300	100.00	92.59
	10000	100.00	91.98
	20000	100.00	91.36

Table 7 The comparison about the classification accuracy (%)

	Wine	Seeds	Balance Scale	CNAE-9
ELM	92.45	84.52	89.36	92.59
SVM-ELM	96.00	86.90	87.23	94.75
PSO-SVM-ELM	98.11	90.48	90.43	95.68

After many attempts, the parameters of SVM-ELM model are determined, but it consumes a lot of time and fails to find the optimal value. And on these data sets, when ELM achieves the highest testing accuracy through countless attempts, the number of hidden layer nodes respectively are 5000, 1000, 20 and 300, also the classification result of ELM is very unstable and prone to overfitting phenomenon. The number of hidden layer nodes in the SVM-ELM model based on particle swarm optimization algorithm is only equal to the number of categories. Its network is stability and there is no overfitting phenomenon, more important it is easily to find the optimal value.

5 Conclusion

This paper proposes the SVM-ELM model based on particle swarm optimization algorithm, namely PSO-SVM-ELM. The weight and bias of hidden layer nodes are determined by SVM, and then the parameters of the SVM-ELM model are optimized by particle swarm optimization algorithm. PSO-SVM-ELM model not only improves the generation ability of ELM, but also simplifies ELM. So when the number of hidden layer nodes is equal to the number of categories, the classification result is good. The experiments are enough to show that the SVM-ELM model based on particle swarm optimization algorithm is ideal in data classification.

References

1. Huang, G.B., Zhou, H., Ding, X.: Extreme learning machine for regression and multiclass classification. IEEE Trans. Syst. Man Cybern. Part B Cybern. **42**(2), 513–529 (2012)
2. Huang, G.B., Zhu, Q.Y., Siew, C.K.: Extreme learning machine: a new learning scheme of feedforward neural networks. In: IEEE International Joint Conference on Neural Networks, vols. 1–4, pp. 985–990 (2004)
3. Huang, G.B., Liang, N.Y., Rong, H.J., et al.: On-line sequential extreme learning machine. In: The IASTED International Conference on Computational Intelligence, pp. 232–237 (2005)
4. Rong, H.J., Ong, Y.S., Tan, A.H., et al.: A fast pruned-extreme learning machine for classification problem. Neurocomputing **72**(1–3), 359–366 (2008)
5. Silva, D.N.G., Pacifico, L.D.S., Ludermir, T.B.: An evolutionary extreme learning machine based on group search optimization. In: IEEE Congress on Evolutionary Computation, pp. 574–580 (2011)
6. Zhang, Y., Ding, S.: Research on Extreme Learning Machines Optimization Methods. China University of Mining Technology, Xu Zhou (2014)
7. Pradhan, B.: A comparative study on the predictive ability of the decision tree, support vector machine and neuro-fuzzy models in landslide susceptibility mapping using GIS. Comput. Geosci. **51**, 350–365 (2013)
8. Cristianini, N., Shawe-Taylor, J.: An Introduction to Support Vector Machines and Other Kernel-Based Learning Methods. Cambridge University Press, Cambridge (2000)
9. Ding, S., Qi, B., Tan, H.: An overview on theory and algorithm of support vector machines. J. Univ. Electron. Sci. Technol. China **40**(1), 2–8 (2011)
10. Liu, B., Hou, D., Huang, P., et al.: An improved PSO-SVM model for online recognition defects in eddy current testing. Nondestr. Test. Eval. **28**(4), 367–385 (2013)
11. Deng, W., Zheng, Q., Chen, L., et al.: Research on extreme learning of neural networks. Chin. J. Comput. **33**(2), 279–287 (2010)
12. Gang, M.A., Ding, S., Shi, Z.: Rough RBF neural network based on extreme learning. Microelectron. Comput. **29**(8), 9–14 (2012)
13. Shen, F., Wang, L., Zhang, J.: Reduced extreme learning machine employing SVM technique. J. Huazhong Univ. Sci. Technol. **42**(6), 107–110 (2014) (Nature Science Edition)
14. Rini, D.P., Shamsuddin, S.M., Yuhaniz, S.S.: Particle swarm optimization: technique, system and challenges. Int. J. Comput. Appl. **14**(1), 19–26 (2011)
15. Rabinovich, M., Kainga, P., Johnson, D., et al.: Particle swarm optimization on a GPU. In: 2012 IEEE International Conference on Electro/Information Technology (EIT), pp. 1–6. IEEE (2012)
16. Subasi, A.: Classification of EMG signals using PSO optimized SVM for diagnosis of neuromuscular disorders. Comput. Biol. Med. **43**(5), 576–586 (2013)

17. Huang, C.L., Dun, J.F.: A distributed PSOSVM hybrid system with feature selection and parameter optimization. Appl. Soft Comput. **8**(4), 1381–1391 (2008)
18. Bazi, Y., Melgani, F.: Semisupervised PSO-SVM regression for biophysical parameter estimation. IEEE Trans. Geosci. Remote Sens. **45**(6), 1887–1895 (2007)
19. Xu, X., Ding, S., Shi, Z.-Z., et al.: A self-adaptive method for optimizing the parameters of pulse coupled neural network based on QPSO algorithm. Pattern Recogn. Artif. Intell. **25**(6), 909–915 (2012)
20. Chen, W.N., Zhang, J., Lin, Y., et al.: Particle swarm optimization with an aging leader and challengers. IEEE Trans. Evol. Comput. **17**(2), 241–258 (2013)
21. Shawe-Taylor, J., Sun, S.: A review of optimization methodologies in support vector machines. Neurocomputing **74**(17), 3609–3618 (2011)

ELM-ML: Study on Multi-label Classification Using Extreme Learning Machine

Xia Sun, Jiarong Wang, Changmeng Jiang, Jingting Xu, Jun Feng, Su-Shing Chen and Feijuan He

Abstract Extreme learning machine (ELM) techniques have received considerable attention in computational intelligence and machine learning communities, because of the significantly low computational time. ELM provides solutions to regression, clustering, binary classification, multiclass classifications and so on, but not to multi-label learning. A thresholding method based ELM is proposed in this paper to adapted ELM for multi-label classification, called extreme learning machine for multi-label classification (ELM-ML). In comparison with other multi-label classification methods, ELM-ML outperforms them in several standard data sets in most cases, especially for applications which only have small labeled data set.

X. Sun (✉) · C. Jiang · J. Xu · J. Feng
School of Information Science and Technology, Northwest University, Xi'an 710069, China
e-mail: raindy@nwu.edu.cn

C. Jiang
e-mail: 526107208@qq.com

J. Xu
e-mail: 1476622348@qq.com

J. Feng
e-mail: fengjun@nwu.edu.cn

J. Wang
Institute of Information Engineering, Chinese Academy of Sciences, Beijing 100190, China
e-mail: jiarongrongg@gmail.com

S.-S. Chen
Systems Biology Lab, University of Florida, Gainesville, FL 32608, USA
e-mail: suchen@cise.ufl.edu

S.-S. Chen
Computer Information Science and Engineering, University of Florida, Gainesville, Florida 32608, USA

F. He
Department of Computer Science, Xi'an Jiaotong University City College, Xi'an 710069, China
e-mail: hfj@mail.xjtu.edu.cn

© Springer International Publishing Switzerland 2016
J. Cao et al. (eds.), *Proceedings of ELM-2015 Volume 2*,
Proceedings in Adaptation, Learning and Optimization 7,
DOI 10.1007/978-3-319-28373-9_9

Keywords Extreme learning machine · Multi-label classification · Thresholding strategy

1 Introduction

Multi-label classification deals with one object which possibly belongs to multiple labels simultaneously, which widely exist in real-world applications, such as text categorization, scene and video annotation, bioinformatics, and music emotion classification [1]. Multi-label classification has attracted a lot of attention in the past few years [2–5]. Nowadays, there mainly exist two ways to construct various discriminative multi-label classification algorithms: problem transformation and algorithm adaptation. The key philosophy of problem transformation methods is to fit data to algorithm, while the key philosophy of algorithm adaptation methods is to fit algorithm to data [6].

Problem transformation strategy tackles multi-label learning problem by transforming it into multiple independent binary or multi-class sub-problems, constructs a sub-classifier for each sub-problem using an existing technique, and then assembles all sub-classifiers into an entire multi-label classifier. It is convenient and fast to implement a problem transformation method due to lots of existing techniques and their free software. Representative algorithms include Binary Relevance [7], AdaBoost.MH [8], Calibrated Label Ranking [2], Random k-labelsets [9], and etc.

Algorithm adaptation strategy tackles multi-label learning problem by adapting popular learning techniques to deal with multi-label data. Representative algorithms include ML-kNN [10], ML-DT [11], Rank-SVM [12], BP-MLL [13] and etc. The basic idea of ML-kNN is to adapt k-nearest neighbor techniques to deal with multi-label data, where maximum a posteriori (MAP) rule is utilized to make prediction by reasoning with the labeling information embodied in the neighbors. The basic idea of BP-MLL is to adapt feed-forward neural networks to deal with multi-label data, where the error back propagation strategy is employed to minimize a global error function capturing label correlations.

In the multi-labeled setting, classes belonging to one instance are often related to each other. The performance of the multi-label learning system is poor if it ignores the relationships between the different labels of each instance. Therefore, the famous Rank-SVM defines the margin over hyperplanes for relevant-irrelevant label pairs, which explicitly characterizes label correlations of individual instance. Rank-SVM achieves great accurateness. Unfortunately, Rank-SVM has high computational cost. It is necessary to build novel efficient multi-label algorithms.

Recently, Huang et al. [14, 15] proposed a novel learning algorithm for single-hidden layer feedforward neural networks called extreme learning machine (ELM). ELM represents one of the recent successful approaches in machine learning. Compared with those traditional computational intelligence techniques, ELM is better generalization performance at a much faster learning speed and with least human intervenes. ELM techniques have received considerable attention in computational intelligence and machine learning communities, in both theoretic study and applications [16–21]. ELM provides solutions to regression, clustering, feature learning, binary classification and multiclass classifications, but not to multi-label learning. Therefore, a thresholding method based ELM is proposed in this paper to adapted ELM for multi-label classification, called ELM-ML (Extreme Learning Machine for Multi-Label classification). Experiments on 3 multi-label datasets show that the performance of ELM-ML is superior to some other well-established multi-label learning algorithms including Rank-SVM and BP-MLL in most cases, especially for applications which only have small labeled data set.

2 ELM-ML

In this section, we will describe our multi-label classification algorithm, called ex-treme learning machine for multi-label classification (ELM-ML).

From the standard optimization method point of view [15], ELM with multi-output nodes can be formulated as

$$\text{Minimize: } \frac{1}{2}\left\|\beta^2\right\| + C\frac{1}{2}\sum_{i=1}^{N}\left\|\xi_i\right\|^2$$
$$\text{s.t.} \quad h(x_i)\beta = t_i^T - \xi_i^T \tag{1}$$

Formula (1) tends to reach not only the smallest training error but also the smallest norm of output weights. Where, $1 \leq i \leq N$. N is the number of training samples. $\xi_i = [\xi_{i,1}, \ldots, \xi_{i,m}]^T$ is the training error vector of the m output nodes with respect to the training sample x_i. C is a user-specified parameter and provides a tradeoff between the distance of the separating margin and the training error. The predicted class label of a given testing sample is the index number of the output node which has the highest output value. Formula (1) provides a solution to multi-class classifications.

Multi-label learning is a harder task than traditional multi-class problems, which is a special case of multi-label classification. One sample belongs to several related labels simultaneously, so we cannot simply regard the index number of the highest output value as predicted class for a given testing sample. A proper thresholding function $th(x)$ should be set. Naturally, the predicted class labels for a given testing sample are those index numbers of output nodes which have higher output value than the predefined thresholding.

We believe that thresholding function $th(x)$ should be learned from instances. That is to say, different instances should correspond to different thresholdings in multi-label learning model. A naive method would be to consider the thresholding

---

Inputs:
$\mathcal{D} = (x_1, T_1),\dots,(x_i, T_i),\dots,(x_n, T_n)$
$\mathcal{Y} = \{y_1, y_2, \dots, y_m\}$;
Where, $\mathcal{Y}$ is label space with m possible class labels. x_i is a d-dimensional feature vector$(x_{i1}, x_{i2}, \dots, x_{id})^T$. T_i is a m-dimensional binary vector $(t_{i1}, \dots, t_{im})^T$. If t_{ij} is relevant with sample x_i then $t_{ij} = 1$, otherwise, $t_{ij} = 0$.

Output:
$h(x) = ELM\text{-}ML(\mathcal{D}, \mathcal{Y})$

Process:

Phase I: Train a multi-class classifier $f(x)$ based ELM with multi-output nodes

Step 1: Randomly assign input weight w_i and bias b_i, $i = 1, \dots, N$.
Step 2: Given activation function $g(x)$, and hidden node number $\tilde{N}$, calculate the hidden layer output matrix **H**.
Step 3: According to formula 7 solve the output weight $\boldsymbol{\beta}$.
Step 4: Given every sample x_i, ELM learning model $f(x_i) = \{f_1(x_i), \dots, f_m(x_i)\}^T$. Let $f_j(x_i)$ denote the output value of the jth output node.

Phase II: thresholding function learning based ELM

Step 1: Comparing vector Y_i with $f(x_i)$, if $t_{ij} = 1$ then $f_j(x_i) \subseteq Y_i$, other wise, $f_j(x_i) \subseteq \overline{Y_i}$. Y_i is relevant label set and $\overline{Y_i}$ is irrelevant label set of sample x_i respectively.
Step 2: Assign $th1(x_i) = max\ (\overline{Y_i})$, $th2(x_i) = min\ (Y_i)$. Given a sample x_i, set the threshold value $th(x_i) = (th1 + th2)/2$.
Step 3: Train a thresholding function learning model based ELM with single output nodes.
    a)   Randomly assign input weight w_i and bias b_i, $i = 1, \dots, q$.
    b)   Given data set $(x_1, th(x_1)),\dots, (x_q, th(x_q))$, activation function $g(x)$, and hidden node number $\tilde{N}$ calculate the hidden layer output matrix H_{th}.
    c)   calculate $\boldsymbol{\beta}_{th} = H_{th}{}^{\dagger} T$
Step 4: Given a unseen sample x, the predicted class label set
$$h(x) = \{y_j | f_j(x) \geq H_{th}\boldsymbol{\beta}_{th}, j = 1, \dots m\}$$

Fig. 1 The pseudo-code of ELM-ML

function $th(x)$ as a regression problem on the training data. In this paper, we use ELM algorithm with single output node to solve this regression problem. Overall, the proposed ELM-ML algorithm has two phases: multi-class classifier based ELM with multi-outputs and thresholding function learning based ELM. The pseudo-code of ELM-ML is summarized in Fig. 1.

3 Experiments

In order to compare the proposed thresholding strategy in this paper with strategy in Rank-SVM [12], we performed several experiments. We also compare the performance of different multi-label classification algorithms, including our algorithm ELM-ML, Rank-SVM, BP-MLL [13] on 3 multi-label classification data sets.

3.1 Datasets

In order to verify the performance of thresholding strategy and different multi-label classification algorithms, 3 data sets have been tested in our simulations (http://computer.njnu.edu.cn/Lab/LABIC/LABIC_Software.html).

Table 1 describes these 3 benchmark data sets, in which, LC and LD denote the label cardinality and density respectively. LC measures the degree of multi-labeledness. Accordingly, LD normalizes label cardinality by the number of possible labels in the label space.

As shown in Table 1, the data sets cover different range of cases whose characteristics are diversified. #Training and #Test means the numbers of training examples and test examples respectively.

3.2 Evaluation Measures

In this paper, we choose four evaluation criterias suitable for classification: Subset Accuracy, Hamming loss, Precision and Recall. These measures are defined as follows.

Assume a test data set of size n to be $S = \{(x_1, Y_1)\}, \ldots, (x_i, Y_i), \ldots, (x_n, Y_n)\}$ and $h(\cdot)$ be the learned multi-label classifier. A common practice in multi-label learning is to return a real-valued function $f(x, y)$. For a unseen instance x, the real-valued output $f(x, y)$ on each label should be calibrated against the thresholding function output $th(x)$.

Table 1 Information of 3 benchmark data sets

Dataset	Domain	#Training	#Test	Attributes	Labels	LC	LD
Genbase	Biology	463	191	1185	27	1.35	0.05
Emotions	Music	391	202	72	6	1.87	0.31
Enron	Text	1123	579	1001	53	3.38	0.06

- Hamming Loss
  The hamming loss evaluates the fraction of misclassified instance-label pairs,
  i.e. a relevant label is missed or an irrelevant is predicted.

$$hammingloss(h) = \frac{1}{n} \sum_{i=1}^{N} |h(x_i) \Delta Y_i| \tag{2}$$

  here, Δ stands for the symmetric difference between two sets.
- Subset Accuracy
  The subset accuracy evaluates the fraction of correctly classified examples, i.e. the
  predicted label set is identical to the ground-truth label set. Intuitively, subset
  accuracy can be regarded as a multi-label counterpart of the traditional accuracy
  metric, and tends to be overly strict especially when the size of label space is large.

$$subsetacc(h) = \frac{1}{n} \sum_{i=1}^{N} |h(x_i) = Y_i| \tag{3}$$

- Precision, Recall

$$Precison(h) = \frac{1}{n} \sum_{i=1}^{N} \frac{|Y_i \cap h(x_i)|}{|h(x_i)|} \tag{4}$$

Here, Y_i and $h(x_i)$ correspond to the ground-truth and predicted label set for x_i
respectively.

Obviously, except for the first metrics, the larger the metric value the better
the system's performance is.

3.3   Results

3.3.1   Thresholding Function

We compare two thresholding strategies: the proposed thresholding strategy in
ELM-ML and thresholding strategy in Rank-SVM. Rank-SVM employs the
stacking-style procedure to set the thresholding function th(x) [12]. We apply
multi-class ELM algorithm providing scores for each sample then use two
thresholding strategies, called ELM-ML and ELM-Rank-SVM to predict labels for
test data. All results are detailed in Tables 2, 3 and 4. In experiments, our computational platform is a HP workstation with Intel 2.67 GHz CPU.

As seen in Tables 2, 3 and 4, the proposed thresholding strategy in ELM-ML
achieves the highest performance among the other, whatever hamming loss or
subset accuracy. Training time and testing time are list in Table 4. We compare
running time of thresholding strategy in ELM-ML and thresholding strategy in
Rank-SVM. Obviously, ELM-ML achieves overwhelming performance. In conclusion, the proposed thresholding strategy in ELM-ML is effective and efficient.

Table 2 Hamming Loss for two thresholding strategies on 3 data sets (hamming loss ↓)

Datasets	Algorithms	
	ELM-ML	ELM-Rank-SVM
Genbase	**9.3058e-04**	0.9544
Emotions	**0.2087**	0.2145
Enron	**0.0851**	0.9290

Bold value denotes the best results

Table 3 Subset accuracy for three thresholding strategies on 3 data sets (subset accuracy ↑)

Datasets	Algorithms	
	ELM-ML	ELM-Rank-SVM
Genbase	**0.9749**	0
Emtions	**0.2673**	0.2426
Enron	**0.0708**	0

Bold value denotes the best results

Table 4 Computation time of thresholding strategies

Datasets	ELM-ML		ELM-Rank-SVM	
	Train time(s)	Test time(s)	Train time(s)	Test time(s)
Genbase	**1.0739**	**0.0212**	4.6656	0.0673
Emotions	**0.5016**	**0.0064**	0.7916	0.0190
Enron	**3.3922**	**0.0504**	19.6976	0.2460

Bold value denotes the best results

3.3.2 Multi-label Algorithms

We also compare the performances of different multi-label classification algorithms, including ELM-ML, Rank-SVM, BP-MLL on 3 multi-label classification data sets. We downloaded Matlab code of BP-MLL (http://cse.seu.edu.cn/people/zhangml/ Publication.htm). We accept their recommended parameter settings. The best parameters of BP-ML reported in the literatures [13] is used. For BP-MLL, the learning rate is fixed at 0.05, the number of hidden neurons is 20 % of the number of input neurons, the training epochs is set to be 100 and the regularization constant is fixed to be 0.1. For Rank-SVM developed in Matlab, Gaussian kernel is tested, which kernel parameter γ and cost parameter C need to be chosen appropriately for each data set. In our experiments, the Hamming Loss measure is regarded as a criterion to tune these parameters. To achieve an optimal parameter combination (γ, C), we use similar tuning procedure as in [9]. The optimal parameters on each data set are shown in Tables 5, 6 and 7. We develop ELM-ML algorithm in Matlab and choose sigmoid activation function. Set hidden nodes $\tilde{N} = 1000$. The detailed experimental results are shown in Tables 5, 6, 7 and 8.

From Table 5, 6, 7 and 8, our ELM-ML obtains the best performances in all three criterions on Genbase data set and Emotions data set. However, ELM-ML is

Table 5 Hamming loss for 3 algorithms on 3 data sets (hamming loss ↓)

Datasets	Algorithms		
	ELM-ML	BP-MLL	Rank-SVM
Genbase	**9.3058e-04**	0.0037	0.0865 (γ, C $= -3, 0.25$)
Emotions	**0.2087**	0.2252	0.3317 (γ, C $= -3, 0.125$)
Enron	0.0851	**0.0532**	0.0560 (γ, C $= -2, 8$)

Bold value denotes the best results

Table 6 Subset accuracy for 3 algorithms on 3 sata sets (subset accuracy ↑)

Datasets	Algorithms		
	ELM-ML	BP-MLL	Rank-SVM
Genbase	**0.9749**	0.9045	0.0000(γ, C $= -3, 0.125$)
Emotions	**0.2673**	0.2327	0.0000(γ, C $= -3, 0.125$)
Enron	0.0708	**0.1002**	0.0397(γ, C $= -2, 8$)

Bold value denotes the best results

Table 7 Precision for 3 algorithms on 3 data sets (precision ↑)

Datasets	Algorithms		
	ELM-ML	BP-MLL	Rank-SVM
Genbase	**0.9965**	0.9724	0.9875(γ, C $= -3, 0.125$)
Emotions	**0.8083**	0.6625	0.7005(γ, C $= -3, 0.125$)
Enron	0.5762	**0.6933**	0.5949(γ, C $= -2, 8$)

Bold value denotes the best results

Table 8 Recall for 3 algorithms on 3 data set (recall ↑)

Datasets	Algorithms		
	ELM-ML	BP-MLL	Rank-SVM
Genbase	**0.9918**	0.9761	0.9749(γ, C $= -3, 0.125$)
Emotions	**0.6491**	0.6370	0.6436(γ, C $= -3, 0.125$)
Enron	0.5187	**0.6422**	0.5664(γ, C $= -2, 8$)

Bold value denotes the best results

inferior to others on Enron data set. Genbase data set and Emotions data are all small size of training data whatever size of labels and feature dimensions. That is to say, the proposed ELM-ML is more suitable to solve those applications that a large among of labeled data is difficult to obtain. ELM-ML achieves better results in a small among of labeled data. Whereas BP-MLL works well when a large number of labeled data are obtained easily. Training time and testing time are list in Table 8. ELM-ML achieves the best in testing time (Table 9).

Table 9 Computation time of multi-label classification algorithms

	ELM-ML		BP-MLL	
	Train(s)	Test(s)	Train(s)	Test(s)
Genbase	**1.0739**	**0.0212**	1.0138e + 04	5.9486
Emotions	**0.5016**	**0.0064**	2.6062e + 03	1.6248
Enron	**3.3922**	**0.0504**	2.0979e + 04	22.513

Bold value denotes the best results

4　Conclusion

In this paper, we present ELM-ML algorithm to solve multi-label classification. ELM has been regarded as one of the recent successful approaches in machine learning, because ELM is the significantly low computational time required for training a learning model and provides better generalization performance with least human intervenes. However, ELM does not provide a solution to multi-label classifications. A post-processing step, threshold calibration strategies, should be used to predict label set of a given sample. A naive method would be to consider the thresholding function th(x) as a regression problem on the training data with class labels. In this paper, we first use ELM algorithm with multi-output nodes to train a learning model returning real-valued function, then use ELM algorithm with single output node to learn thresholding function. Experiments on 3 diverse benchmark multi-label datasets show that the performance of ELM-ML is effective and efficient.

Acknowledgements The authors wish to thank the anonymous reviewers for their helpful comments and suggestions. The author also thanks Prof. Zhihua Zhou, Mingling Zhang and Jianhua Xu, whose software and data have been used in our experiments. The authors also thank Changmeng Jiang and Jingting Xu for doing some related experiments. This work was supported by NSFc 61202184 and Scientific research plan projects 2015JQ6240 and 2013JK1152.

References

1. Xu, J.: Multi-label core vector machine with a zero label. Pattern Recogn. **47**(7), 2542–2557 (2014)
2. Fürnkranz, J., Hüllermeier, E., Mencía, E.L., Brinker, K.: Multilabel classification via calibrated label ranking. Mach. Learn. **73**(2), 133–153 (2008)
3. Ji, S., Sun, L., Jin, R., Ye, J.: Multi-label multiple kernel learning. In: Koller D., Schuurmans D., Bengio Y., Bott L., Schuurmans D., Bengio Y., Bottou L. (eds.) Advances in Neural Information Processing Systems 21, pp. 777–784. MIT Press, Cambridge (2009)
4. Guo, Y., Schuurmans, D.: Adaptive large margin training for multilabel classification. In: Proceedings of the 25th AAAI Conference on Artificial Intelligence, pp. 374–379. San Francico, CA (2011)
5. Quevedo, J.R., Luaces, O., Bahamonde, A.: Multilabel classifiers with a probabilistic thresholding strategy. Pattern Recogn. **45**(2), 876–883 (2012)

6. Zhang, M.-L., Zhou, Z.-H.: A review on multi-label learning algorithms. IEEE Trans. Knowl. Data Eng. **26**(8), 1819–1837 (2014)
7. Boutell, M.R., Luo, J., Shen, X., Brown, C.M.: Learning multi-label scene classification. Pattern Recogn. **37**(9), 1757–1771 (2004)
8. Schapire, R.E., Singer, Y.: Boostexter: a boosting-based system for text categorization. Mach. Learn. **39**(2/3), 135–168 (2000)
9. Xu, J.: An efficient multi-label support vector machine with a zero label. Expert Syst. Appl. **39**, 2894–4796 (2012)
10. Zhang, Min-Ling, Zhou, Zhi-Hua: ML-KNN: a lazy learning approach to multi-label learning. Pattern Recogn. **40**(7), 2038–2048 (2007)
11. Clare, A., King, R.D.: Knowledge discovery in multi-label phenotype data. In: De Raedt L., Siebes A. (eds.) Lecture Notes in Computer Science, pp. 42–53. Springer, Berlin (2001)
12. Elisseeff, A., Weston, J.: A kernel method for multi-labelled classification. In: Dietterich T.G., Becker S., Ghahramani Z. (eds.) Advances in Neural Information Processing Systems 14, pp. 681–687. MIT Press, Cambridge (2002)
13. Zhang, M.-L., Zhou, Z.-H.: Multilabel neural networks with applications to functional genomics and text categorization. IEEE Trans. Knowl. Data Eng. **18**(10), 1338–1351 (2006)
14. Huang, G.B., Zhu, Q.Y., Siew, C.K.: Extreme learning machine: theory and applications. Neurocomputing **70**(1–3), 489–501 (2006)
15. Huang, G.-B., et al.: Extreme learning machine for regression and multiclass classification. IEEE Trans. Syst. Man Cybern. Part B: Cybern. **42**(2), 513–529 (2012)
16. Rong, Hai-Jun, Ong, Yew-Soon, Tan, Ah-Hwee, Zhu, Zexuan: A fast pruned-extreme learning machine for classification problem. Neurocomputing **72**(1–3), 359–366 (2008)
17. Mohammed, A.A., Minhas, R., Jonathan Wu, Q.M., Sid-Ahmed, M.A: Human face recognition based on multidimensional PCA and extreme learning machine. Pattern Recogn. **44**(10–11), 2588–2597 (2011)
18. Wang, Yuguang, Cao, Feilong, Yuan, Yubo: A study on effectiveness of extreme learning machine. Neurocomputing **74**(16), 2483–2490 (2011)
19. Xia, Min, Zhang, Yingchao, Weng, Liguo, Ye, Xiaoling: Fashion retailing forecasting based on extreme learning machine with adaptive metrics of inputs. Knowl. Based Syst. **36**, 253–259 (2012)
20. Mishra, A., Goel, A., Singh, R., Chetty, G., Singh, L.: A novel image watermarking scheme using extreme learning machine. In: The 2012 International Joint Conference on IEEE Neural Networks (IJCNN), pp. 1–6 (2012)
21. Horata, Punyaphol, Chiewchanwattana, Sirapat, Sunat, Khamron: Robust extreme learning machine. Neurocomputing **102**, 31–44 (2013)

Sentiment Analysis of Chinese Micro Blog Based on DNN and ELM and Vector Space Model

Huilin Liu, Shan Li, Chunfeng Jiang and He Liu

Abstract Analysis of Chinese micro blog has great commercial value and social value. Based on the depth analysis of the language style of the Chinese micro blog, this paper makes a deep research on the sentiment analysis of Chinese microblog based on and DNN ELM and vector space model. First of all, the micro blog Abstract sentiment feature extraction technology is studied in depth. Based on the traditional text representation model, DNN algorithm is used to extract the feature of the abstract emotion. Combined with the characteristics of the short text of micro blog, this paper uses SAE to construct the DNN. In the construction process of vector space, in order to fully and effectively said microblogging text emotional information, in this paper, we introduce the emotional factor and structure factor of information gain feature selection method is improved and introduced the location information of the feature words of TF-IDF weighting calculation method to improve. Then, the sentiment classification of micro blog is deeply studied. In this paper, we use the concept model to express the emotion category of micro blog, and propose the spatial expansion algorithm based on the concept model (ESA). The experimental results show that the presented in this paper, based on DNN microblogging Abstract emotional feature extraction algorithm and the algorithm of conceptual model of spatial development based on the microblogging text emotional other identification *is* effective.

Keywords Sparse coding · Vector space model · ELM · Emotion factor

1 Introduction

Text sentiment analysis and opinion mining is the process of analyzing and processing the text with emotion color. Text sentiment analysis method [1] can be divided into machine learning, dictionary based and language based text sentiment analysis

H. Liu (✉) · S. Li · C. Jiang · H. Liu
College of Information Science and Engineering, Northeastern University,
Shenyang 110819, Liaoning, China
e-mail: liuhuilin@mail.neu.edu.cn

© Springer International Publishing Switzerland 2016   117
J. Cao et al. (eds.), *Proceedings of ELM-2015 Volume 2*,
Proceedings in Adaptation, Learning and Optimization 7,
DOI 10.1007/978-3-319-28373-9_10

method. Pand B [2] the first to use machine learning methods to text for sentiment analysis and reviews data as the evaluation corpus, respectively, using three kinds of machine learning naive Bayes method, maximum entropy and SVM to the evaluation corpus emotion category forecast. [3–6] is based on the SVM classifier for text classification, can be used in different areas of the classification of the emotional analysis.

When we analyze the sentiment of the micro blog, we need to use the spatial vector model to express the text. Because the micro blog text information has the word limit, the use of the low level of the emotional characteristics of the construction of the microblogging vector space model has a sparse problem. Deep learning can use low-level features to extract high-level abstraction features, which can be used to express the text of a higher level of abstraction. The main research contents are as follows:

(a) based on the traditional text representation model, DNN algorithm is used to extract the feature of the abstract emotion.For the emotional feature vector appear sparseness problem, in microblogging emotional features automatically, the SAE of DNN is constructed, reducing the due to the sparsity of the microblogging feature vector of the training error.

(b) According to the characteristics of micro blog, this paper carries on the transformation to the feature selection and weight calculation method of the traditional text processing, so that the vector representation of the text is more effective and effective.

(c) using the relationship between the spatial vector and the concept model, the spatial expansion algorithm based on the concept model is proposed.

2 Related Work

2.1 ELM

ELM [7] is a kind of single hidden layer feedforward neural network proposed by Professor Huang Guangbin in 2006, which is composed of three layers of neurons in the input layer, hidden layer and output layer. When the training classifier, for a given sample vector set $T = \{(x_i, t_i)|x_i \in R^n, t_i \in R^m, i = 1, 2, \ldots, N\}$, t_i sample vectors belonging to the category, ELM and traditional neural network is different, just determine the implicit bias of hidden layer node number and the input layer weights were randomly assigned. Standard ELM expression as shown in formula (1).

$$f_L(x) = \sum_{i=1}^{L} \beta_i G\left(w_i, b_i, x\right) \tag{1}$$

where l denotes the number of neuron node, w_i denotes hidden layer nodes I and input layer nodes of the weight vector, b_i represents the hidden layer node i corresponds to the offset value, $G\left(w_i, b_i, x\right)$ represents the hidden layer neuron node i output, its activation value, β_i indicates the weight vector of the hidden layer node i to the output layer neuron node. The activation function of the hidden layer node is g(x), then the output I expression of the hidden layer node I is shown in type (2).

$$G\left(w_i, b_i, x\right) = g\left(w_i \cdot x + b_i\right) b_i \subseteq R^+ \tag{2}$$

$$H\left(w_1, \ldots, w_L, b_1, \ldots, b_L, x_1, \ldots, x_N\right) = \begin{bmatrix} G\left(w_1, b_1, x_1\right) & G\left(w_2, b_2, x_1\right) & \cdots & G\left(w_L, b_L, x_1\right) \\ \vdots & & \cdots & \vdots \\ G\left(w_1, b_1, x_N\right) & G\left(w_2, b_2, x_N\right) & \cdots & G\left(w_L, b_L, x_N\right) \end{bmatrix}$$

$$= \begin{bmatrix} G\left(w_1 \cdot x_1 + b_1\right) & G\left(w_2 \cdot x_1 + b_2\right) & \cdots & G\left(w_L \cdot x_1 + b_L\right) \\ \vdots & & \cdots & \vdots \\ G\left(w_1 \cdot x_N + b_1\right) & G\left(w_2 \cdot x_N + b_2\right) & \cdots & G\left(w_L \cdot x_N + b_L\right) \end{bmatrix} \tag{3}$$

The weight matrix of the hidden layer and the output layer is solved according to the output matrix of the hidden layer $\hat{\beta}$.

$$\hat{\beta} = \left(H^T H\right)^{-1} H^T T \tag{4}$$

where T represents the microblog emotion vector category, according to the error minimization and structural risk minimization principle, solving the hidden layer and the output layer between the weight matrix $\hat{\beta}$ need to satisfy such as (5) shown in the goal programming:

$$\min \left(\|H\beta - T\| + \lambda \|\beta\|\right) \tag{5}$$

Among them,

$$\beta = \begin{bmatrix} \beta_1^T \\ \vdots \\ \beta_L^T \end{bmatrix}_{L \times M}, T = \begin{bmatrix} t_1^T \\ \vdots \\ t_N^T \end{bmatrix}_{N \times M} \tag{6}$$

λ represents the proportion of structured risk.

2.2 Vector Space Model

Vector space model [8] is a text representation model proposed by Salton in 1975. The main idea is: the text as a collection of feature items, for each feature item, according to the degree of its representation of the text to give a weight. Thus the text is represented as a collection of weighted feature items.

2.3 Information Gain

Information gain [9] is an important concept in information theory, the text classification commonly used feature selection algorithms, the feature of the difference in before and after the emergence of the information entropy in the text to represent the information gain feature. Information entropy H represents the average amount of information contained in the received message.

2.4 Weight Calculation Method TF-IDF

Frequency distribution of TF-IDF [10] inverse document frequency is proposed by McGill and Gerald Salton a used to represent text feature method, text mining in common a weighting technique for computing text in each characteristic of the text indicates that the degree of, namely feature of the text weights. According to TF (term frequency) that for a given feature item T, it appears in the text D frequency, the main purpose is to normalize the words, to prevent it in favor of the length of the text.

3 Research Content

3.1 Feature Selection

We introduce one of the two factors, one is the emotional factor, the other is the structural factor. In order to obtain the emotional words and its affective factors, it is needed to calculate the concentration of emotion, that is, the intensity of emotional words. Specific feature selection process, such as algorithm 3.1 description.

Algorithm 1 Feature selection based on improved information gain

Require: a micro blog corpus, a collection of candidate feature words, S, and structure candidate
feature set C;
Ensure: emotional feature word set S_1 and structure feature word set C_1;
 1: the character of the word t to meet the $t \in S$ or $t \in C$, computing the traditional information
gain value of the word t;
 2: e existence of T is not ergodic, $t \in S$ or $t \in C$, if $t \in C$ the implementation of step 6, otherwise
the implementation of step 3;
 3: the emotional factors of the T;
 4: combined with the emotional factors and the corresponding information gain to calculate the
emotional value of the emotional characteristics;
 5: according to the intensity of the emotional characteristics of the words of the intensity of the
sort, select the first N_s emotional words in the collection S_1, the implementation of 2;
 6: structural factors of the structure of T;
 7: combined with the emotional factors and the corresponding information gain to calculate the
emotional value of the structural characteristics of the structure;
 8: according to the structure of the strength of the strength of the structure of the word order,
select the first N_c feature words into the collection C_1, the implementation of 2;

1. Emotion concentration calculation

The basic idea is: first artificial selection for a number of positive and negative emotional words as benchmark words, then calculate the similarity between emotional words and benchmark words as the degree of emotion. According to the related strategies to obtain the value of the concentration.

2. Based on the logic function of the emotional factor structure (Fig. 1)

In sentiment analysis on Weibo, emotional words fierce early and weak late effects of text sentiment Co., emotional words of the intense emotion for interim distribution, influence of the weak change of sentiment is intense. Due to the minimal emotional concentration greater than 0, so the need of curve of translational processing, considering the late processing, also need gentle handling on the dotted line trend, by image analysis, this paper take $\alpha = 0.5$, $\beta = 6$, X in the range of [0, 24]. After

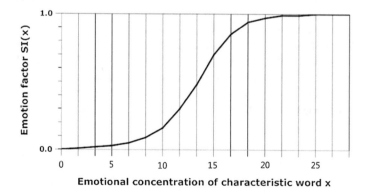

Fig. 1 Relation between Emotional Concentration x and Emotional factors SI

Fig. 2 Relation between
Emotional Concentration x
and Emotional factors SI

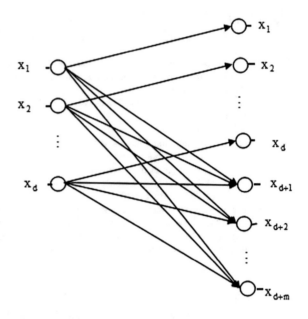

translation, gentle treatment after, emotional factors and emotional concentration trend curve as shown in Fig. 2.

Through the above analysis, the relationship between the emotional factors and the concentration of emotion is shown in the formula (23).

$$x = \begin{cases} \frac{1}{1+e^{-\alpha x+\beta}} & \theta_1 \geqslant x \geqslant \theta_2 \\ 1 & \theta_1 < \chi \end{cases} \tag{7}$$

θ_1 indicates that the threshold limit, when x is greater than θ_1, SI (x) value is 1, when x expressed emotion concentration, its value range is w, SI (x) value range is (0,1].

3. Structure factor based on position information

Structure factor is used to measure the information of text structure. In order to measure the value of continuous string sequence structure information, this paper on the basis of traditional information gain increased factor structure, to measure the size of the information structure. In view of the characteristics of the structure character sequences, the expression of the structure factor is shown in the formula (24).

$$CI(x) = e^{-|x-0.5|} \quad 1 \geqslant x \geqslant 0 \tag{8}$$

Where X represents the relative position of the structural word, the structure factor of the relative position of $CI(x)$ is x, the value of X is [0,1]. Can be seen from the expression of the emotion factor, when the relative position is 0.5, $CI(x)$ takes the maximum value 1. The minimum value of $CI(x)$ is more than 0.

4. Micro blog feature selection algorithm

In this paper, based on the traditional information gain feature selection method, the effect of emotion and structure is enhanced by the introduction of emotion factor and structure factor.

Emotional words feature selection expression, as shown in formula (25).

$$IGS(t) = SI(SO(t))IG(t) = SI(SO(t))[H(C) - H(C|t)] \tag{9}$$

$$= SI(SO(t))p(t)\left\{\sum_{i=1}^{i=n} p(C_i|t) \log\left[p(C_i|t)\right]\right\}$$

$$+ SI(SO(t))p(\bar{t})\left\{\sum_{i=1}^{i=n} p(C_i|\bar{t}) \log\left[p(C_i|\bar{t})\right]\right\}$$

$$- SI(SO(t))\sum_{i=1}^{i=n} p(C_i) \log\left[p(C_i)\right]$$

where T represents a feature word character sequence, $SO(t)$ expresses the emotional concentration of characteristic words, $SI(SO(t))$ is a feature word emotion factor, $IGS(t)$ representation of feature words t, $IG[t]$ represents the information gain of the feature word t.

From the formula (26) can be seen in emotional feature selection, considers not only the emotional words t distribution, also contains the emotional words t itself the emotion factor, $IGS(t)$ value is greater, t is chosen as emotional feature words the possibility of more.

Structural word feature selection expression, as shown in formula (26)

$$IGC(t) = CI(t)IG(t) = CI(t)[H(C) - H(C|t)] \tag{10}$$

$$= CI(x)p(t)\left\{\sum_{i=1}^{i=n} p(C_i|t) \log\left[p(C_i|t)\right]\right\}$$

$$+ CI(x)p(\bar{t})\left\{\sum_{i=1}^{i=n} p(C_i|\bar{t}) \log\left[p(C_i|\bar{t})\right]\right\}$$

$$- CI(x)\sum_{i=1}^{i=n} p(C_i) \log\left[p(C_i)\right]$$

3.2   Weight Calculation

Traditional TF-IDF computation influence of feature weight method which only considers the frequency distribution of its value and neglects factors of feature words in position effects to the weights. In order to describe the influence of position factor on the weight of its weight, the position coefficient is introduced in this paper.

The position coefficient expression is shown in the formula (27).

$$
x = \begin{cases} \frac{1}{1+e^{-\alpha x + \beta}} & \theta \geqslant x \geqslant 0 \\ 1 & \theta < x \end{cases}
$$
(11)

The θ indicates that the threshold limit, when the x is greater than θ, the structure coefficient $SP(x)$ value is 1, when x is the relative position of the feature words, its value range is $(0,1)$, the $SI(x)$ value range is $(0,1]$. α β for the corresponding coefficient, can be adjusted according to the actual situation. Improved TFIDF expression as shown in formula (28).

$$
tfidfSP(t,d) = SP(x)\, tfidf(t,d) = SP(x)\frac{n_t}{n_d} \log\left(\frac{N}{N_t}\right)
$$
(12)

Where t denotes the feature words, d denotes a text message, $tfidfSP(t,d)$ denotes feature words in text D weights x says that features t in text D in the relative position, $SP(x)$ t in d in the position of the coefficients and n_t the number of feature item t appear in the text d and n_d to denote the d text in a word frequency count, N_t contains the number of text feature item t, n denotes the total number of text.

3.3   Space Expansion Algorithm Based on Concept Model

This paper presents an efficient spatial expansion algorithm Expand (ESA) Spatial Algorithm, extended by the concept of class and its calculation model the distance of space.

In the vector space model, the similarity between samples is usually measured by Euclidean distance or angle cosine. For any two text and the sample space, the similarity of the two text vector is given by the angle cosine.

$$
sim(d_i, d_j) = \cos\theta = \frac{d_i \cdot d_j}{|d_i| \cdot |d_j|} = \frac{\sum_{k=1}^{n} w_{ik} \cdot w_{jk}}{\sqrt{\sum_{k=1}^{n} w_{ik}^2 \cdot \sum_{k=1}^{n} w_{jk}^2}}.
$$
(13)

where $sim\left(d_i, d_j\right)$ expresses the similarity between d_j and d_i, and θ indicates that the angle between d_i and d_j, $d_i = \left(w_{1i}, w_{2i}, \dots, w_{ni}\right)$ n is the vector dimension of text d.

The spatial expansion algorithm based on concept model ESA is shown in 2.

Algorithm 2 Expand Spatial Algorithm Based on Conceptual Model

---

Require: feature vector space x, $\left\{X_1, X_2, ..., X_n\right\}$, $X = \left\{x_1, x_2, ..., x_d\right\} \in \left\{X_1, X_2, ..., X_n\right\}$ representation of the feature vector space in any vector, d representation of the dimension of the feature vector, n vector space in the number of feature vector;

Ensure: $\left\{X'_1, X'_2, ..., X'_n\right\}$, $X' = \left\{x_1, x_2, ..., x_d, x_{d+1}, x_{d+2}, ..., x_{d+m}\right\}$, m is a number of vector categories, which is extended after the expansion;

1: from the concept set $M = \left\{M_1, M_2, ..., M_m\right\}$, select the m of $\left\{X_{m1}, X_{m2}, ..., X_{mm}\right\}$ as the basis of the category. This point, the category of basic point X_{mi} and the concept of the model M_i meet the conditions $X_{mi} \in M_i$;

2: traversal feature vector space $\left\{X_1, X_2, ..., X_n\right\}$, $X = \left\{x_1, x_2, ..., x_d\right\} \in \left\{X_1, X_2, ..., X_n\right\}$, executive line step 3, after the completion of the implementation steps 6;

3: calculate the distance between x and $\left\{X_{m1}, X_{m2}, ..., X_{mm}\right\}$, the value of i, $dis\left(X, X_{mi}\right)$ Surround i=1,2,3... m, executive step 4;

4: combined with the emotional factors and the corresponding information gain to calculate the emotional value of the emotional characteristics;

5: the $dis\left(X, X_{mi}\right)$ is normalized to calculate the basic points corresponding to the X_{mi} eigenvalue, the implementation of step 5;

6: output expansion feature vector, $X' = \left\{x_1, x_2, ..., x_d, x_{d+1}, x_{d+2}, ..., x_{d+m}\right\}$ the implementation of step 2;

7: output extended feature vector space $\left\{X'_1, X'_2, ..., X'_n\right\}$;

The space vector Q mapping relation is shown in Fig. 2.

4 Experiments and Results

4.1 Space Expansion Algorithm Based on Concept Model

Experiment 1 Verify the effectiveness of the algorithm based on DNN for the automatic extraction of micro blog sentiment

(A) using the traditional information gain feature selection algorithm and the traditional TF-IDF feature weight calculation method to generate the feature vector to train the classifier to predict the sample.

(B) the feature vector is used to train the classifier to predict the sample using the DNN method.

Table 1 by numerical form the emotional feature automatic extraction of sentiment classification effect, Fig. 3a through the form of histogram, and gives the emotional feature automatic extraction of the emotion categories estimate the influence degree. In the sentiment analysis, negative to the recognition of emotional information is the most important, through a combination of Table 1 and Fig. 3a, from an

Table 1 Effect of emotional features of automatic extraction for emotional classification

a	Positive (%)	Negative (%)	Objective (%)	Average (%)
Recall b–a	−1.76	12.94	−8.24	0.98
Accurate b–a	2.29	−2.64	9.27	2.98
F-value b–a	0.25	3.25	−2.47	0.34

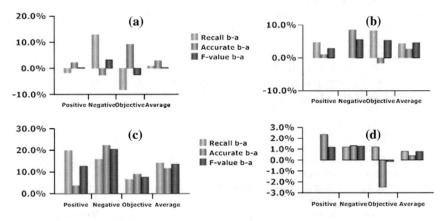

Fig. 3 Recognition performance of each emotional category under Test ab

average value of see the recall rate of promotion is most obvious, enhance the 2.98 %, F value of promotion effect is limited only 0.34 % of ascension.

Experiment 2 Verify the effectiveness of the algorithm based on the improved information gain feature selection algorithm and the improved TF-IDF algorithm to construct the original emotion feature space.

(A) using the traditional information gain feature selection algorithm and the traditional TF-IDF feature weight calculation method to generate the original feature vector.

(B) the improved information gain feature selection algorithm and the improved TF-IDF feature weight calculation method is generated feature vector.

2 the numerical form of the improved method of feature selection and weighting calculation method of sentiment classification effect, Fig. 3b through the form of histogram, are given to improve the characteristics of the choice of weighting method and calculation method of the emotion categories estimate the influence degree.

Combined with Table 2 and Fig. 3b, the effect of the method of improving feature selection and weight calculation is analyzed. From Fig. 3b it can be seen that the recall rate, accuracy rate and F-value of the difference in the emotional category are significantly improved, which is worth F Enhance the effect of the most obvious, improved by 4.60 %.

Table 2 Effect of improved feature selection and weight calculation method for emotional classification

a	Positive (%)	Negative (%)	Objective (%)	Average (%)
Recall b–a	4.71	0.00	8.24	4.31
Average b–a	1.05	8.507	−1.58	2.66
F-value b–a	2.90	5.55	5.35	4.60

Table 3 Effects of different classifier for emotional classification

a	Positive (%)	Negative (%)	Objective (%)	Average (%)
Recall b–a	19.91	15.82	6.53	14.09
Accurate b–a	3.65	22.23	9.01	11.63
F-value b–a	12.65	20.43	7.57	13.55

4.2 Micro Blog Sentiment Classification Experiment

Experiment 3 Verify the validity of ELM classifier

(A) select ELM as the micro blog text sentiment classification

(B) select SVM as the micro blog text sentiment classification

Table 3 by numerical are given in the form of the effect of different classifiers to classify emotions, Fig. 3c by the form of histogram, given the different classifiers for each emotion category forecast the influence degree. Analysis of the prediction effect of different classifiers based on the combination of Table 3 and Fig. 3c. From Fig. 3c it can be seen that the recall rate, accuracy rate and F-value of the difference in the different emotional categories are significantly improved, the recall rate of the most obvious effect, improve 11.63 %.

Experiment 4 To verify the effectiveness of the EFA based on the concept model of spatial expansion algorithm.

(A) the ELM model is trained and predicted by using the feature vector generated by the above steps a.

(B) the feature vector generated by the above steps is expanded by EFA algorithm, and the new features are generated. Table 4 through numerical are given in the form of conceptual model based spatial expansion algorithm to classify emotions influence, Fig. 3d through the form of histogram is given based on the conceptual model of spatial expansion algorithm of the emotion categories estimate the influence degree.

Based on the combination of Table 4 and Fig. 3d, the prediction effect of spatial expansion algorithm based on conceptual model is analyzed. From Fig. 3d it can be seen that the recall rate, accuracy rate and F-value of the difference in the different emotional categories of the average value have been improved to a certain extent.

5 Summary

Research on the automatic extraction technology of emotion feature of micro blog. Based on the traditional text representation model, DNN algorithm is used to extract the feature of the abstract emotion. Combined with the characteristics of micro blog text, this paper uses SAE to construct the DNN, and reduces the training error due to the sparsity of the feature vectors. In the word feature selection, the traditional information gain method only consider frequency and distribution characteristics, the lack of distinguishing the difference between emotional words and word structure.In this paper, we introduce the affective factors and structural factors from two different perspectives to measure the emotional words and the structure of words, so that the selection of the word features is effective.The in addition to considering the distribution of word frequency, will also consider the location information of the feature words, will use the traditional TF-IDF computation of weights and word feature location information acquired by the weight coefficient is weighted, to adjust the traditional TF-IDF calculation method of the.Neglect the influence of the feature words on the emotional value. The in addition to considering the distribution of word frequency, will also consider the location information of the feature words, will use the traditional TF-IDF computation of weights and word feature location information acquired by the weight coefficient is weighted, to adjust the traditional TF-IDF calculation method of the.

Research on the emotion classification of micro blog. As the relationship between the micro blog text and the specific emotion category plays a key role in the process of micro blog emotion classification. First, we give a conceptual model for representing text categories, and propose a conceptual model selection method based on the spatial density and mutual exclusion. Secondly, based on the relationship between the space vector and the concept model, the spatial expansion algorithm based on the concept model is proposed. In this paper, ELM is used as a classifier to prove the effectiveness of the spatial expansion algorithm.

Table 4 Effects of Spatial Expansion Algorithm (ESA) for emotional classification

a	Positive (%)	Negative (%)	Objective (%)	Average (%)
Recall b–a	0.00	1.18	1.18	0.78
Accurate b–a	2.35	1.31	−2.48	0.39
F-value b–a	1.17	1.27	−0.14	0.77

References

1. Thelwall, M., Buckley, K., Paltoglou, G.: Sentiment in Twitter events. J. Am. Soc. Inf. Sci. Technol. **62**(2), 406C418 (2011)
2. Pang, B., Lee, L., Vaithyanathan, S.: Thumbs up?: sentiment classification using machine learning techniques. Proc. EMNLP, 79–86 (2002)
3. Haddi, E., Liu, X., Shi, Y.: The role of text pre-processing in sentiment analysis. Procedia Comput. Sci., 26C32 (2013)
4. Wu, D.D., Zheng, L., Olson, D.L.: A decision support approach for online stock forum sentiment analysis. IEEE Trans. Syst. Man Cybern. Syst. **44**(8), 1077–1087 (2014)
5. Kranjc, J., Smailovi, J., Podpean, V., et al.: Active learning for sentiment analysis on data streams: Methodology and workflow implementation in the ClowdFlows platform. Inf. Process. Manag. **51**(2), 187C203 (2014)
6. Balahur, A., Turchi, M.: Comparative experiments using supervised learning and machine translation for multilingual sentiment analysis. Comput. Speech Lang. **28**(1), 56C75 (2014)
7. Huang, G.B., Zhu, Q.Y., Siew, C.K.: Extreme learning machine: theory and applications. Neurocomputing **70**, 489C501 (2006)
8. Salton, G., Wong, A., Yang, C.S.: A vector space model for automatic indexing. Commun. ACM **18**(10), 613–620 (1974)
9. Ouyang, C., Yang, X., Lei, L., et al.: Multi-strategy approach for fine-grained sentiment analysis of Chinese microblogy. Acta Scientiarum Naturalium Universitatis Pekinensis (2014)
10. Zhang, W., Yoshida, T., Tang, X.: A comparative study of TF*IDF, LSI and multi-words for text classification. Expert Syst. Appl. **38**(3), 2758–2765 (2011)

Self Forward and Information Dissemination Prediction Research in SINA Microblog Using ELM

Huilin Liu, Yao Li and He Liu

Abstract With the popularity of social network, information propagation prediction based on social network is also becoming popular. As far as we know, people do not concern the user who forwards its own microblog in information propagation prediction. In our investigation the self forward behavior can cause the further spreading of the information. Thus in this paper we propose a self forward prediction model to predict the self forward behavior. We use ELM to train and predict self forward behavior. Based on this model we proposed an algorithm to predict the information dissemination. The experiment results show that our algorithm is real and effective and it significantly improves the forecast accuracy. It also can be seen in the experimental results that the results of ELM has a better performance than SVM.

Keywords Self forward · Information dissemination prediction · ELM · SINA microblog

1 Introduction

The development of social network not only brought more convenient communication pattern, but also brought more efficient information transfer mode. For example, the earthquake information will be shown in the SINA microblog by the user the China earthquake networks studies after it happened 1 min. There is no doubt that researches on social network information dissemination have vital significance.

The information dissemination in social networks is based on the user behavior, especially the forward behavior. There are many studies about the information dissemination in social network. For example, Ye et al. [1] measured the message propagation in Twitter and evaluated the different social influences. Cao et al. [2]

H. Liu (✉) · Y. Li · H. Liu
College of Information Science and Engineering, Northeastern University,
Shenyang 110819, Liaoning, China
e-mail: liuhuilin@mail.neu.edu.cn

© Springer International Publishing Switzerland 2016
J. Cao et al. (eds.), *Proceedings of ELM-2015 Volume 2*,
Proceedings in Adaptation, Learning and Optimization 7,
DOI 10.1007/978-3-319-28373-9_11

Fig. 1 A simple
information propagation
prediction tree

found some features that influence the user behavior, and based on the user behavior predict model they studied the information propagation prediction. However, all these investigations are based on the follower-friends relationship without concerning the condition that users forward the microblogs published by the users themselves.

There are not much forward limits in social networks. Thus, in the individual perspective, people may want to add some information and forward the microblogs published by themselves. This behavior may cause the increasing in information dissemination.

We use Peking University PKUVIS Microblog visual analysis tools [3] to analyze those two conditions. Figure 1 is the visualization of the condition. In Fig. 1 the yellow points represent the publisher, and we can see directly that the self forward behavior causes the diffusing of the microblog.

To our best knowledge, we do not find any work about the condition that users forward the microblogs published by themselves. Thus based on the existing work, we study the influence of self forward behavior to the information dissemination in social network. We choose SINA microblog as the research object and using ELM as training and predicting algorithm. The main contributions of this paper are shown as below:

(1) We extracted the user self forward behavior features and build a model to predict the self forward behavior based on the features.

(2) We change the microblog propagation prediction algorithm based on the self forward behavior. The experiment results shows that the change improves the prediction accuracy of the algorithm.

(3) We use ELM to train and predict the data. The experiment results shows that using ELM has a better performance.

The rest of this paper is organized as follows. Section 2 briefly introduces the related work about information dissemination in social networks and ELM In Sect. 3, we give the features we extracted. The information dissemination model is also introduced in this section. In Sect. 4 we described the experiments and their results. The conclusions and future work will be shown in Sect. 5.

2 Related Work

2.1 Information Dissemination in Social Networks

There are many researches about the internet social networks. In the early time, people studied the marketing function [4], the popularity degree [5], the response of users to the news [6] and many other things in online social networks. In the further studies, researchers focus on the user features and analyzed the influence of users. They promoted the PageRank algorithm [7], TwitterRank algorithm [8], TURank algorithm [9] and many other variants of these algorithm. In the user behavior analysis, the influence of users was seen as a feature, and was used in user behavior prediction combining with other features. For example, the studies of Song et al. [10], Luo et al. [11], Cha et al. [12] and Suh et al. [13]. All these models are different. Beside the above achievements, some people investigate information propagation from a statistical standpoint and give some conclusion [1].

Along with the deepening of the information propagation in online social networks, some publishers want to know the approximate transmission range of the microblog before they publish it. Cao et al. [2] considered the user behavior and predicted the information propagation based on it. Petrovic et al. [15] did the information propagation prediction in the same way with [2], but used different model. In addition to the user behavior prediction model used by [2] and [15], other model in [10 13] can also be used in this pattern. Bandari [14] did not pay attention to the user behavior and think the content feature is more important. In [14], he proposed a new algorithm to predict the forward quantity of news which were shared to Twitter. In all the research above, whether the information dissemination is based on user behavior or content, they did not concern the influence of the condition in which people forward the microblog published by themselves.

Considering the above situation, we put forward an improved information propagation prediction model in online social networks based on user behavior prediction. This model can enhance information dissemination forecasting accuracy in online social networks.

2.2 ELM

Extreme Learning Machine (ELM) is a more simple and effective algorithm of single hidden layer feed forward network (SLFNs) algorithm [16]. It provides the best generalization ability and very fast learning speed. At the first time ELM is the full supervision machine learning algorithm, but now it has semi-supervised and unsupervised version [17]. According to studies [18, 19], the principle of ELM has been summarized below.

For different N samples (x_i, t_i), $x_i = \left[x_{i1}, x_{i2}, \ldots, x_{im}\right]^T \in R^n$, $t_i = \left[t_{i1}, t_{i2}, \ldots, t_{im}\right]^T \in R^m$. The mathematical model of standard single hidden layer feed forward neural network (SLFN) which has L hidden node and activation function g(x) is shown as function (1).

$$\sum_{i=1}^{L} \beta_i g_i (x_i) = \sum_{i=1}^{L} \beta_i g (w_i x_j + b_i) = o_j (j = 1, 2, \ldots, N) \tag{1}$$

The SLFNs which have L hidden node and activation function g(x) can infinitely close to the samples with the 0 error. That is $\sum_{j=1}^{L} \left\| o_j - t_j \right\| = 0$. Thus there are \$$\beta_i$, w_i and b_i that makes function (2) tenable.

$$\sum_{i=1}^{L} \beta_i g (w_i x_j + b_i) = t_j (j = 1, 2, \ldots, N) \tag{2}$$

The above function can use the function $H\beta = T$ to express in brief. H is the single hidden layer output matrix of the neural network. The jth list of H is the output of the jth hidden node whose input is $x_{i1}, x_{i2}, \ldots, x_{in}$. The least-square solutions to the optimal solution of the above linear system is $\hat{\beta} = H^\dagger T$. $H^\dagger$ is the Moore-Penrose generalized inverse matrix of H. Thus output sample function of ELM is the function (3).

$$f (x) = h (x) \beta = h (x) H^\dagger T \tag{3}$$

The machine learning-based algorithm without iterative tuning can be divided into three steps. The specific process of ELM is summarized as follows:

Step 1: Randomly assign input weight w_i and bias b_i, $i = 1; 2; \ldots; N'$.
Step 2: Calculate the hidden layer output matrix **H**.
Step 3: Calculate the output weight β, where $\beta = H^\dagger T$.

In professor Huang Guangbin Extreme Learning Machine for Regression and Multiclass Classification study, he has proved that the SVM obtains sub-optimal solution and needs higher computational complexity [20]. Thus in this paper ELM is a better choice and we only need full supervision ELM in this paper.

3 Self Forward Model and Information Dissemination Prediction

Before we start the investigation, we first introduce the definition we use. According to the Sect. 1, for the publishers who want to make some supplement, or want to expand the influence of microblog through further forwarding. We give the following definition.

Definition 1 If user u forward the microblog m which he has published or forwarded before, we call the u self forward m. The forward behavior is called self forward.

Then, we give the dataset description.

3.1 Dataset Description

We use SINA microblog API to get the data, and finally we got 89377 users in the dataset. Then we crawl all Microblogs of these users which published between May 1, 2015 and June 30, 2015 and get 857647 Microblogs.

Then we counted the quantity of the microblogs generated by direct self forward behavior. In all these microblogs, only 5.36 % of them are caused by direct self forward behavior. These microblogs are considered forward samples and put into forward dataset. The rest microblogs are ignore samples. In the actual experiments, due to the user relationship graph, we selected 35 % of ignore samples as ignore dataset.

Our study is based on the above data. In the next section, we will introduce the features we use and the corresponding evaluation index.

3.2 Direct Self Forward Prediction

Direct self forward behavior belongs to the forward behavior, so the feature in forward behavior prediction can also be used in direct self forward behavior prediction. The direct self forward behavior is in the individual perspective. Thus, the features about the interpersonal things are irrelevant to the direct self forward behavior.

According to the researcher in Sect. 2.1, we choose the following features.

(1) Microblog content importance

The microblog content importance is widely used in the existing work. Based computing weight of TF-IDF (term frequency inversed document frequency) algorithm on the text classification field, we calculate the importance of microblog [21]. We can use Formula 4 to calculate the importance.

$$f(d) = n_\omega \times \log \frac{N}{n_d} \tag{4}$$

In this formula, d represents the word d in the microblog ω, n_ω represents the number of d appears in ω, N represents the number of microblog that microbolg set W contains, n_d represents the number of microblogs contain d in the microblog set W. The TFIDF of microblog ω can be computed by adding the TF-IDF of all the word in ω.

$$tf(\omega) = \sum_j tf(d_j)$$ (5)

(2) User activity
We use the following formula to compute the activity of user.

$$PA = \frac{n}{t}$$ (6)

The PA in the Formula 6 represents the microblog number published over a period of time, n is the total number of microblogs (including the original microblog an the forwarding microblog), t is the unit time. In general, we set t to 1 day.

(3) User forward activity
The forward activity is percentage of users forwarding microblog account for all published microblogs in 1 day. We use Formula 7 to compute it.

$$RA = \frac{\sum\limits_{i \in t} r_i}{\sum\limits_{i \in t} p_i}$$ (7)

The t in this formula is represent the days. r_i is the number of users forwarding microblog in ith day, p_i is the number of users releasing microblog in ith day and RA represents the forward activity.

(4) The influence of user
People always use PageRank to compute the influence of user [7]. The PageRank formula they use is shown as Formula 8.

$$pr_i = \frac{1-q}{N} + q \sum_{j \in Follower(i)} \frac{pr_j}{|Friend(j)|}$$ (8)

In this formula, pr_i represent the PageRank value of user i, *Follower* (i) represents the fans list of user i, *Friend* (j) represents the collection of users that user j pays attention to, q is the damping coefficient, N is the total number of users.

(5) The status of user
In SINA microblog user has two states, one is certified and the other is noncertified. The status of user influence the behavior of user. We use status to represent it.

(6) The follower number and the friend number
These two feature were used in the works before which study the user behavior. We think they may also influence the self forward behavior.

Table 1 Features to predict self forward

Symbol	Description
$tf(\omega)$	The microblog content importance
PA	The publish activeness of user
RA	The forward activeness of user
FA	The self forward activeness of user
pr_i	The influence of user
Status	The status of user
Follower_num	The follower number of user
Friend_num	The friends number of user
FN	The direct self forward times

From the statistical conclusion we found that some people always do the direct self forward while some of them never. We think the direct self forward behavior related to this user habit. Thus, in addition to the above features, we proposed following feature.

(7) User self forward activity

The self forward activity is the percentage of users self forwarding microblog account for all published microblogs in 1 day. We use Formula 9 to compute it.

$$FA = \frac{\sum_{i \in t} f_i}{\sum_{i \in t} p_i} \tag{9}$$

The t in this formula is represent the days. f_i is the number of users self forwarding microblog in ith day, p_i is the number of users releasing microblog in ith day and FA represents the self forward activity.

(8) Forwarding numbers

Direct self forward behavior is limited. Although people can do direct self forward discretionarily, hardly anyone do it forever. People always end it in finitely time. Thus we consider direct self forward numbers as an important feature.

Thus in this paper we use 9 feature to predict the self forward behavior. The features we used in this paper are summarized in Table 1.

3.3　Information Dissemination Prediction

Based on the above, we use ELM to predict the self forward behavior. Because the user forward behavior prediction is not the emphasis of this paper, we use the model in the existing work to predict forward behavior. According to the self forward prediction and the forward prediction, we determine the spread scope of the microblog eventually.

Fig. 2 A simple
information propagation
prediction tree

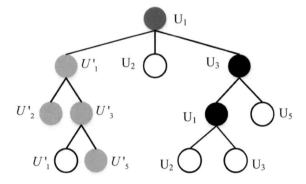

Fig. 3 The percentage of
each self forward numbers

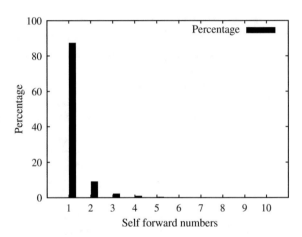

When we forecast the microblog transmission range, first of all, we predict the direct self forward behavior of one user, and then we predict the forwarding state of the user. In order to introduce our model in details, we give a simple model in Fig. 2.

In figure the dark grey point U_1 is the publisher of the microblog. It has the direct self forward behavior and the U'_1 sub tree is the predicting tree after self forwarding. U_2 and U_3 are the follower of U_1 (Fig. 2). The black points represent the forward behavior. In figure U_3 is predicted forwarding. Then we calculate the self forward state of U_3. U_3 do not self forward, so we do not add the subtree.

There is an end for forwarding behavior. From the existing work [23] we can know, after 5 jump, the percent of forward microblogs approach to 0. So when we calculate the information propagation, we only concern the first five jump. For the self forward behavior, Fig. 3 shows the average percent of self forward times.

From the Fig. 3 we can see, the percentage of self forward behavior close to 0 after 4. So when we calculate the information propagation, the self forward behavior we predict 3 times.

Thus, based on the above, the information propagation prediction algorithm is shown in Algorithm 1.

Algorithm 1 Forward_quantity_forecast(U)

Require: Microblog publisher U_0.
Ensure: Forwarded number C.
 1: Predict direct self forward behavior;
 2: **if** (U direct self forward and direct self forward number < 3) **then**
 3:     Add U_0 into forward queue;
 4:     $C = C + 1$;
 5: **end if**
 6: **for** (i = 0; i < N; i++) **do**
 7:     // N is the number of fans, $N \geqslant 0$
 8:     Predict the forward behavior;
 9:     **if** user forward the microbblog **then**
10:        $C = C + 1$;
11:        Add user into forward queue;
12:     **end if**
13: **end for**
14: Get a user U' from forward queue;
15: Forward_quantity_forecast(U');

---

4 Experiments and Results

In this section we introduce the experiments and results. We compared the results which are got by using ELM and SVM. We predicted the amount ofinformation dissemination, and tested the both accuracy between we concerned self forward behavior and not.

4.1 Self Forward Prediction

We use ELM to train and predict the data. The source code of ELM can be obtained from the website.[1] To prove the results are effective, we also use SVM to train and predict the data. We can get the lib-SVM tool from the website.[2] In order to guarantee the stability and effectiveness of the experiment, we use 10 times of cross validation method validation algorithm.

In order to further compare the results of ELM and SVM, we choose the evaluation index of information retrieval, including accuracy, recall and the value of F1. Table 2 shows the comparison of the self forward behavior prediction results.

---

[1]ELM Source Codes: ELM Source Codes: http://www.ntu.edu.sg/home/egbhuang/.
[2]Data set: http://www.csie.ntu.edu.tw/cjlin/libsvm/.

Table 2 The results comparison of ELM and SVM

	Recall	Accuracy	F_scall	Training time (s)	Testing time (s)
ELM	1	0.8349	0.91	2.8704	3.5256
SVM	0.9735	0.8241	0.8926	9.6565	66.9556

Because ELM has a better performance in processing time, we also compare the processing time of ELM and SVM (including training time and testing time).

We can see in Table 2 all the results get by using ELM is better than by using SVM. The training time and the testing time of ELM are much shorter than SVM. Especially in the testing time aspect, the testing time of ELM is 3.5256 s and the testing time of SVM is 66.9556 s. This is because we use 10 times of cross validation method validation algorithm, and each time we test the data by using SVM costs much. All the data proves that using ELM algorithm is better than using SVM algorithm. ELM algorithm has good performance.

4.2 *Information Dissemination Prediction*

In this section we predict the information dissemination based on the user behavior. The user behavior is predict by ELM. We use the model in our previous study [23].

In order to determine the scale of information dissemination, we divide the scale according to the $\frac{1}{2} \times 10^n$ order of magnitudes. If the information dissemination scale we predicted is in the same order of magnitude which is the actual information dissemination scale, we can say the prediction is right. We calculated the average predict information dissemination scale accuracy of 20000 microblogs.

We calculated the average predict information dissemination scale accuracy of 20000 microblogs. The figure shows the comparison of considering self forward behavior and not considering it. In figure, each column represents the average accuracy of different microblogs in different condition. Before represents that the experiments have nothing concerned. DSF represents that the experiments consider self forward behavior. For example, the first black column represents the average accuracy of 100 microblogs without considering self forward behavior. The first slash column represents the average accuracy of 100 microblogs considering direct self forward behavior.

In figure we can see considering self forward behavior can significantly enhance prediction accuracy (Fig. 4). The bigger the number of microblogs is, the more stable the accuracy is. When the microblogs over 10000, the accuracy stabilize at about 80 %. In contrast, the results that nothing has been concern approach to DSF only at 100. In other conditions it is much lower than DSF. This is because in 100, the sample is too small and cannot represent all of the situation.

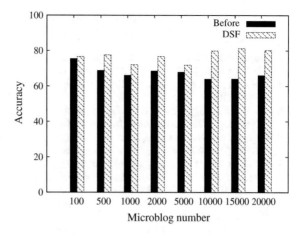

Fig. 4 Average prediction accuracy comparison between considering self forward behavior and not

5 Conclusions

Information propagation prediction based on social network is becoming popular in nowadays. In this paper, we study the condition that user forward the microblog of itself and based on this we change the information propagation prediction algorithm.

We extracted 9 features to predict the direct self forward behavior. The features are microblog content importance, user activity, user forward activity, user self forward activity, the influence of user, the status of user, the follower number, the friend number and the direct self forward times. We use ELM to train and predict the SINA microblog data to predict whether user self forward a microblog. Our experiment results show ELM has a better performance than SVM. Based on self forward behavior we proposed an algorithm to predict the information dissemination. The experiment results show that our algorithm has a good accuracy in information dissemination prediction.

Our model and algorithm in this paper can help the businesses and government to spread information in the social network. There is still something we need to improve in this paper. For example, the features we concerned in this paper is not integrated and some features may be repeated. We will take it into consideration in the future.

Acknowledgments This research was partially supported by the National Natural Science Foundation of China under Grant Nos. 61332006 and 61100022; the National BasicResearch Program of China under Grant No. 2011CB302200-G; the 863 Program under Grant No. 2012AA011004. The researcher claims noconflicts of interests.

References

1. Ye, S., Wu, S.F.: Measuring Message Propagation and Social Influence on Twitter.com[C]. SocInfo, pp. 223–228 (2010)
2. Cao, J., Wu, J., Wei, S., et al.: SINA microblog information diffusion analyse and prediction [J]. Chin. J. Comput. **37**(4), 779–790 (2014)

3. Ren, D., Zhang, X., Wang, Z., et al.: WeiboEvents: a crowd sourcing weibo visual analytic system [C]. In: Pacific Visualization Symposium (PacificVis), 2014 IEEE, Yokohama, pp. 330–334 (2014)
4. Jansen, B.J., Zhang, M., Sobel, K., Chowdhury, A.: [J] Twitter power: tweets as electronic word of mouth. J. Am. Soc. Inf. Sci. Technol. (2009)
5. Szabo, G., Huberman, B.A.: Predicting the popularity of online content [C]. Commun. ACM 80–88 (2010)
6. Lerman, K., Hogg, T.: Using a model of social dynamics to predict popularity of news [J]. In: Proceedings of International Conference on World Wide Web, pp. 621–630 (2010)
7. Page, L., Brin, S., Motwani, R. et al.: The PageRank citation ranking: bringing order to the Web [J]. Stanford University: Technical Report SIDL-WP, pp. 99–120 (1999)
8. Weng, J., Lim, E.P., Jiang, J. et al.: TwitterRank: finding topic-sensitive influential twitter-ers[C]. In: Proceedings of the Third ACM International Conference on Web search and Data Mining. ACM, pp. 261–270 (2010)
9. Yamaguchi, Y. et al.: TURank: Twitter user ranking based on user-tweet graph analysis [C]. WISE 243–246 (2010)
10. Song, G., Li, Z., Tu, H.: Forward or ignore: user behavior analysis and prediction on microblogging [C]. In: 2012 IEEE 16th International Conference on Computer Supported Cooperative Work in Design (CSCWD). IEEE, pp. 678–684 (2012)
11. Luo, Z., Osborne, M., Tang, J. et al.: Who will retweet me? Finding retweeters in Twitter [C]. In: Proceedings of the 36th International ACM SIGIR Conference on Research and Development in Information Retrieval. ACM (2013)
12. Cha, M., Haddadi, H., Benevenuto, F. et al.: Measuring user influence in Twitter: the million follower fallacy. In: Proceedings of the 4th International AAAI Conference on Weblogs and Social Media, Washington, pp. 10–17 (2010)
13. Suh, B., Hong, L., Pirolli, P. et al.: Want to be retweeted? Large scale analytics on factors impacting retweet in Twitter network [C]. In: IEEE International Conference on Social Computing/IEEE International Conference on Privacy, Security, Risk and Trust. IEEE, pp. 177–184 (2010)
14. Bandari, R., Asur, S., Huberman, B.: The pulse of news in social media: forecasting popularity [A]. In: Proceedings of Association of the Advancement of Artificial Intelligence (AAAI-12) [C], Toronto, pp. 26–33 (2012)
15. Petrovic, S., Osborne, M., Lavrenko, V.: RT to Win! Predicting message propagation in Twitter. ICWSM 586–589 (2011)
16. Huang, G.B., Zhu, Q.Y., Siew, C.K.: Extreme learning machine: a new learning scheme of feedforward neural networks [C]. In: International Joint Conference on Neural Networks (IJCNN2004), Budapest, Hungary, pp. 985–990 (2004)
17. Huang, G.B., Zhu, Q.Y., Siew, C.K.: Extreme learning machine: theory and applications [J]. Neurocomputing **70**(1), 489–501 (2006)
18. Huang, G., Song, S. Wu, C., You, K.: Semi-supervised and unsupervised extreme learning machines. IEEE Trans. Syst. Man Cybern. **44**, 2405–2417 (2014)
19. Huang, G.B., Wang, D.H., Lan, Y.: Extreme learning machines: a survey [J]. Int. J. Mach. Learn. Cybern. **2**(2), 107–122 (2011)
20. Huang, G.B., Zhou, H., Ding, X., et al.: Extreme learning machine for regression and multiclass classification [J]. IEEE Trans. Syst. Man Cybern. Part B: Cybern. **42**(2), 513–529 (2012)
21. Shi, C., Xu, C., Yang, X.: Study of TFIDF algorithm [J]. J. Comput. Appl. **6**(29), 167–170 (2009)
22. Lin, X., Wang, W.: Set and string similarity queries: a survey [J]. Chin. J. Comput. **34**(10), 1853–1862 (2011)
23. Liu, H., Li, Y.: Weibo information propagation dissemination based on user behavior using ELM. Math. Probl. Eng. (2015)

Sparse Coding Extreme Learning Machine for Classification

Zhenzhen Sun and Yuanlong Yu

Abstract As one of supervised learning algorithms, extreme learning machine (ELM) has been proposed for single-hidden-layer feedforward neural networks (SLFN) and shown great generalization performance. ELM randomly assigns the weights and biases between the input and hidden layers and trains the weights between hidden and output layers. Physiological research has shown that neurons at the same layer are laterally inhibited to each other such that the output of each layer is a type of sparse codings. However, it is difficult to accommodate the lateral inhibition by directly using random feature mapping in ELM. Therefore, this paper proposes a sparse coding ELM (ScELM) algorithm, which can map the input feature vector into a sparse representation such that the mapped feature is sparse. In this proposed ScELM algorithm, an unsupervised way is used for sparse coding in the sense that dictionary is randomly assigned rather than learned. Gradient projection (GP) based method is used for the sparse coding. The output weights are trained in the same supervised way which ELM presents. Experimental results on benchmark databases have shown that this proposed ScELM algorithm can outperform other state-of-the art methods in terms of classification accuracy.

Keywords Sparse coding · Extreme learning machine · Gradient projection

1 Introduction

During the past decades, neural network is widely studied in the areas of machine learning, pattern recognition and robotics since it is able to approximate complex nonlinear functions so as to provide much higher classification accuracy. Many

This work is supported by National Natural Science Foundation of China (NSFC) under grant 61473089.

Z. Sun · Y. Yu (✉)
College of Mathematics and Computer Science, Fuzhou University, Fuzhou 350116, Fujian, China
e-mail: yu.yuanlong@fzu.edu.cn

© Springer International Publishing Switzerland 2016
J. Cao et al. (eds.), *Proceedings of ELM-2015 Volume 2*,
Proceedings in Adaptation, Learning and Optimization 7,
DOI 10.1007/978-3-319-28373-9_12

learning algorithms have been proposed for training neural networks, for example, support vector machine (SVM) [1, 2] for single-hidden-layer neural networks (SLNN), back-propagation (BP) algorithm and deep learning algorithms [3–5] for multiple-hidden-layer neural networks (MLNN).

SVM can be seen as a training method for SLNN based on standard optimization method by maximizing the margin between two classes. However, it is difficult for SVM to deal with large-scale data since the quadratic programming required to obtain the optimal solution is computationally expensive when the number of training samples is too large.

Further efforts have also been put on training MLNNs. BP algorithm is a pioneer for this type of efforts. It minimizes the training errors based on gradient descent method and the errors are back-propagated from the output layer to previous hidden layers. However, in real applications, BP algorithm has not shown great performance for neural networks with much more hidden layers. This is because that the gradients become smaller and smaller with the back-propagation process from the top to lower layers such that the updates are weak at lower layers. Recently, several deep learning algorithms have been proposed, e.g., deep Boltzmann machine (DBM) [5–7], deep belief network (DBN) [4], convolutional neural network (CNN) [3], stacked denoise autoencoder (SDAE) [8–10] and stacked sparse autoencoder (SSAE) [11, 12]. The underlying idea of deep learning is that feature extraction and classification are combined together in a unified MLNN architecture. In these algorithms, learning of connection weights is basically divided into two processes. The first one is bottom-up layer-wise pre-training through unsupervised ways with a common objective function that output and input are as close as possible between two neighboring layers. For example, DBM performs Gibbs sampling to maximize the log-likelihood of training data and SSAE performs self-taught sparse coding. The second one is top-down fine-tuning of connection weight through a supervised way mainly based on gradient descent strategy. However, the gradient descent based pre-training and fine-tuning is likely to converge to a local optimum.

Recently, extreme learning machine (ELM) was proposed for training SLNNs [13]. One contribution of ELM is that the weights and bias between input and hidden layers are randomly generated such that only the weights between hidden and output layers require training. The other contribution of ELM is that it obtains an optimal output weights by minimizing not only the training errors but also the norm of output weights such that better generalization performance is achieved [14]. This objective function is solved by using Lagrange multiplier method. Theoretically, ELM can obtain a global optimum [15] and therefore it is unlikely to fall into a local optimum. In terms of computation, the training cost of ELM is much lower than other state-of-the-art learning methods.

However, it is difficult to accommodate the lateral inhibition between neurons by directly using random feature mapping in ELM. Physiological research has shown that neurons at the same layer are laterally inhibited to each other such that the output of each layer is a type of sparse codings [16]. Therefore, this paper proposes a sparse coding ELM (ScELM) algorithm which uses sparse coding technique to map the inputs to the hidden layer instead of the random mapping used in ELM. The gradient projection (GP) based method [17] is used in the encoding stage and the output weights between hidden and output layers are learned using ELM algorithm. The mapped feature representation captures more salient properties such that it can contribute more for classification.

The remainder of this paper is organized as follows. Section 2 reviews some related work on sparse coding. Section 3 presents details of this proposed ScELM algorithm. The experiment results are shown in Sect. 4.

1.1 Related Work on Sparse Coding

In 1959, Hubel and Wiesel studied the receptive fields of simple neurons in the visual stripe cortex of cats and posited that the receptive field of primary visual cortex V1 neurons can produce a sparse representation for visual perception signals [18]. Then the electrophysiological experiments on primate and cats shown that the expression of complex stimuli in the visual cortex adopt the sparse coding principle [19].

Sparse coding technique has been used for feature extraction in recent years. A face recognition algorithm using sparse representation coding (SRC) [20]. The SRC algorithm doesn't need to do dictionary learning but sets all training samples as atoms of the dictionary. Any new (test) sample will approximately lie in the linear span of the training samples from the same class. So the coefficient is sparse (most of the coefficients are zero). The paper [20] shown that, ever the face images have as much as 80 % of random noise, the SRC algorithm can also get a high recognition accuracy. Another advantage of SRC algorithm is that for the occlusion and corruption cases, it still be able to get a higher recognition performance. What's more, the SRC is more robust than other face recognition methods that the precise choice of feature space is no longer critical. The training time of SRC is much faster than other face recognition methods.

Another feature extraction method for image classification called sparse coding spatial pyramid matching (ScSPM) [21] was also proposed. It is an extension of the spatial pyramid matching (SPM) algorithm by generalizing vector quantization to sparse coding. Then it uses a linear SVM classifier to implement classification. The ScSPM method remarkably reduces the complexity of SVMs to O(n) in training and a constant in testing. The experimental results of the ScSPM have shown that the sparse coding of features always significantly outperforms other types of feature mapping.

Sparse coding is also integrated into the deep neural networks, e.g., SSAE [11]. SSAE uses an unsupervised learning algorithm for feature extraction by imposing a sparsity constraint on the hidden units so that it can discover interesting structure in the data.

2   The Proposed ScELM Algorithm

As Fig. 1 shows, this proposed ScELM is a method for learning single hidden neural network which contains one input layer, one hidden layer and one output layer. Between the input layer and hidden layer, it uses sparse coding technique to do feature mapping for the input data, the outputs of the hidden layer are the sparse representations of input data. In the encoding stage, it uses the GP algorithm [17] to calculated the sparse representations of input data. The output weights between the hidden layer and output layer are solved by a optimization problem which aim to minimize the training error as well as the output weights norm.

Fig. 1 The framework of ScELM

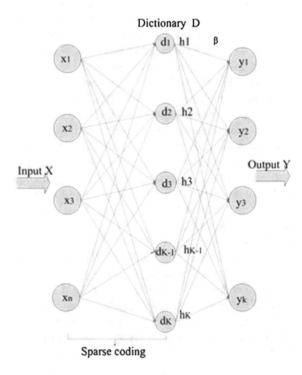

2.1 Encoding Stage

In this stage, the ScELM uses the GP algorithm to compute the sparse representations of input data since the GP algorithm is much faster than other state-of-the-art approaches, e.g., IST [22, 23] and the recent $l_1\_l_s$ package [24]. The GP aims to minimize the reconstruction error as well as the l^1-*norm* of the sparse representation:

$$\textbf{Minimize: } \frac{1}{2}||x - Dh||^2 + \lambda||h||_1 \tag{1}$$

where $x \in R^n$ is the input signal, $D = [d_1, d_2, \dots, d_m] \in R^{n \times m}(m >> n)$ is the overcomplete dictionary, $h \in R^m$ is the sparse representation of x.

The first key of GP algorithm is to express (1) as a quadratic program, this is done by splitting the variable h into its positive and negative parts.

$$h = u - v, u \geq 0, v \geq 0 \tag{2}$$

These relationship are satisfied by $u_i = (h_i)_+$ and $v_i = (-h_i)_+$ for all $i = 1, 2, \dots, n$, where $(h)_+ = max\{0, h\}$. Thus $||h||_1 = 1_n^T u + 1_n^T v$, where $1_n = [1, 1, \dots, 1]^T$ is the vector consisting of n ones, so (1) can be rewritten as the following bound-constrained quadratic program (BCQP):

$$\begin{array}{c} \min: \\ \textbf{u,v} \end{array} \frac{1}{2}||x - D(u-v)||_2^2 + \lambda 1_n^T u + \lambda 1_n^T v, \tag{3}$$
$$\text{s.t. } u \geq 0, v \geq 0.$$

Problem (3) can be written in more standard BCQP form

$$\begin{array}{c} \min: \\ \textbf{z} \end{array} c^T z + \frac{1}{2}z^T Bz \equiv F(z) \tag{4}$$
$$\text{s.t. } z \geq 0.$$

where

$$z = \begin{bmatrix} u \\ v \end{bmatrix}, \quad b = D^T x, \quad c = \lambda 1_{2n} + \begin{bmatrix} -b \\ b \end{bmatrix}$$
$$and \tag{5}$$
$$B = \begin{bmatrix} D^T D & -D^T D \\ -D^T D & D^T D \end{bmatrix}.$$

In the GP algorithm, z^k is iteratively updated to z^{k+1} as follows. First, it chooses some scalar parameter $\alpha^k > 0$ and set

$$w_k = (z^k - \alpha^k \nabla F(z^k))_+ \tag{6}$$

Then chooses a second scalar $\gamma^k > 0$ and set

$$z^{k+1} = z^k + \gamma^k(w^k - z^k). \tag{7}$$

In the basic approach, it searches from each iterate z^k along the negative gradient $-\nabla F(z^k)$, projecting onto the nonnegative orthant, and performing a backtracking line search until a sufficient decrease is attained in F. It uses an initial guess for α^k that would yield the exact minimizer of F along the direction if no new bounds were to be encountered. Specifically, it defines the vector g^k by

$$g_i^k = \left\{ \begin{array}{ll} (\nabla F(z^k))_i, & \text{if } z_i^k > 0 \text{ or } (\nabla F(z^k))_i < 0 \\ 0, & \text{otherwise.} \end{array} \right. \tag{8}$$

Then the initial guess can be computed explicitly as

$$\alpha_0 = \frac{(g^k)^T g^k}{(g^k)^T B g^k}. \tag{9}$$

The complete routine of gradient projection is shown in Algorithm 1.

Algorithm 1 Gradient projection for sparse reconstruction

1: **(Initialization)** Given z^0, choose parameters $\rho \in (0, 1)$ and $\mu \in (0, \frac{1}{2})$; set $k = 0$;
2: Compute α_0 from (9), and replace α_0 by $mid(\alpha_{min}, \alpha_0, \alpha_{max})$;
3: **(Backtracking Line Search)** Choose α^k to be the first number in the sequence
   $\alpha_0, \rho\alpha, \rho^2\alpha, \ldots$ such that $F((z^k - \alpha^k \nabla F(z^k))_+) \leq F(z^k) - \mu \nabla F(z^k)^T(z^k - (z^k - \alpha^k \nabla F(z^k))_+)$, and set $z^{k+1} = (z^k - \alpha^k \nabla F(z^k))_+$;
4: Perform convergence test and terminate with approximate solution z^{k+1} if it is satisfied; otherwise set $k = k + 1$ and return to Step 2.

The parameter λ in express (1) is chosen as suggested in [24]

$$\lambda = 0.1||D^T x||_\infty \tag{10}$$

Note that the zero vector is the unique optimal solution of (1) for $\lambda \geq 0.1||D^T x||_\infty$ [24, 25].

2.2 Calculation of Output Weights β

For an input sample (x, t), the output function of ScELM is

$$y = \sum_{i=1}^{L} \beta_i h_i(x) = h(x)\beta \tag{11}$$

where $\beta = [\beta_1, \ldots, \beta_L]^T$ is the output weight from the hidden layer to the output layer. $\mathbf{h}(x) = [h_1(x), \ldots, h_L(x)]$ means the sparse representation (output of hidden layer) with the respect to the input x.

The ScELM algorithm aims to not only minimize the training error but also the output weights which is learned from the ELM because the smaller norm of weights are, the better generalization performance the neuron network tends to has. The target of ScELM is

$$\textbf{Minimize: } L_{P_{ELM}} = \frac{1}{2}||\beta||^2 + C\frac{1}{2}\sum_{i=1}^{N}||\xi_i||^2 \tag{12}$$

$$\textbf{Subject to: } h(x_i)\beta = t_i^T - \xi_i^T, \quad i = 1, \ldots, N$$

The result of β can be divided into two cases:

(1) For the case where the number of training samples is NOT huge:

$$\beta = H(\frac{I}{C} + H^T H)^{-1} T^T. \tag{13}$$

(2) For the case where the number of training samples is huge:

$$\beta = (\frac{I}{C} + HH^T)^{-1} HT^T. \tag{14}$$

3   Experiments

3.1   Experimental Setup

A standard PC is used in our experiments and its hardware configuration is as follows:

(1) CPU: Intel(R) Pentium(R) CPU G2030 @3.00 GHz;
(2) Memory: 8.00 GB;
(3) Graphics Processing Unit (GPU): None.

The data sets used in our experiments are taken from *UCI Machine Learning Repository* [26]. There are 15 data sets including 7 binary-classification cases and 8 multi-classification cases. The details of these data sets are shown in Table 1.

Table 1 Data sets used in our experiments

Datasets	# train	# test	# features	# classes
Musk	6598	476	166	2
Madelon	2000	600	500	2
Diabetes	512	256	8	2
Isolet	6238	1559	617	26
Gesture_phase	1743	1260	50	5
Glass	142	72	9	6
Wine	118	60	13	3
Satimage	4435	2000	36	6
Image segmentation	1660	650	19	7
Australian credit	460	230	14	2
Vehicle	564	282	18	4
Breast cancer	380	189	30	2
Ecoli	224	112	7	8
Diabetic	786	365	19	2
Blood transfusion	500	248	4	2

This proposed ScELM algorithm is compared against ELM and SVM algorithms. In the ScELM and ELM, the number of hidden neurons and parameter C are tuned to find the optimal accuracy. The 'sigmoid' function is chosen as active function in ELM. Because of the randomness of overcomplete dictionary in ScELM and the randomness of input weights in ELM, the ScELM and ELM procedure were run 50 times with a fixed L (the number of hidden neurons) and a fixed parameter C each time to compare their average accuracy.

3.2   Evaluation

As shown in Tables 2 and 3, for most of the data sets, ScELM achieves higher accuracy than SVM and ELM. For example, ScELM outperforms ELM by more than 11 % and SVM by more than 60 % in the wine data. It can be seen that ScELM outperforms SVM and ELM in terms of classification accuracy.

Table 2 The maximal accuracy of each algorithm

Datasets	SVM (%)	ELM (%)	ScELM (%)
Musk	56.51	**92.86**	92.56
Madelon	50.00	61.50	**62.50**
Diabetes	67.58	70.70	**71.88**
Isolet	**95.70**	95.32	95.25
Gesture_phase	38.81	**67.22**	67.06
Glass	43.06	54.17	**63.89**
Wine	35.00	85.00	**96.67**
Satimage	23.85	**82.30**	79.10
Image segmentation	57.23	**94.92**	92.92
Australian credit	53.48	75.22	**79.57**
Vehicle	23.76	75.89	**78.72**
Breast cancer	57.14	94.71	**95.24**
Ecoli	67.85	85.71	**86.61**
Diabetic	56.16	**78.90**	74.79
Blood transfusion	65.32	72.98	**73.79**

Bold value denotes the best results

Table 3 The average accuracy of ELM and ScELM

Datasets	ELM (%)	ScELM (%)
Musk	**90.13 ± 1.6202**	89.52 ± 2.5394
Madelon	**59.76 ± 4.4355**	59.48 ± 3.0098
Diabetes	63.68 ± 6.0692	**66.84 ± 4.9217**
Isolet	**94.44 ± 1.7506**	94.31 ± 1.5622
Gesture_phase	63.12 ± 8.2969	**63.44 ± 5.2899**
Glass	48.06 ± 13.5425	**51.19 ± 27.7943**
Wine	67.97 ± 31.4388	**87.13 ± 18.0317**
Satimage	**81.36 ± 0.1802**	77.81 ± 0.4674
Image segmentation	**93.39 ± 0.4594**	91.39 ± 0.9767
Australian credit	72.30 ± 1.5975	**75.00 ± 3.6515**
Vehicle	71.62 ± 5.5272	**73.98 ± 3.8030**
Breast cancer	93.02 ± 0.7931	**93.10 ± 0.5713**
Ecoli	79.98 ± 6.8360	**81.36 ± 6.8865**
Diabetic	**74.82 ± 1.7961**	71.07 ± 4.1483
Blood transfusion	68.90 ± 4.6455	**70.33 ± 2.5755**

Bold value denotes the best results

4   Conclusions

This paper proposes a new method for learning single hidden layer feedforward neural network, called ScELM. It uses sparse coding technique to connect the input layer and hidden layer as the feature mapping of input signals so that it can decouple the signals as well as reduce the redundancy of signals. This paper conduct extensive experiments on publicly available databases to verify the efficacy of the proposed algorithm and the result shows that the ScELM gets a better performance than ELM and SVM in terms of classification. This proposed algorithm provides new insights into future research in classification algorithms, and promotes the development of machine learning.

Although the ScELM gets higher accuracy than ELM and SVM, there are still a lot of work should to be done. There are a lot of new sparse coding algorithms been proposed, such as l_1-homotopy [27], Predictive Sparse Decomposition(PSD) [28], and Locally Competitive Algorithm (LCAs) [29]. It is worth trying the new sparse coding algorithms to see wether the accuracy of ScELM will be improved.

Future work also includes how to find an over-complete dictionary that can significantly filter out the same characteristics between different classes and how to design the number of atoms of the dictionary. Doing that in a principle manner remains an important direction for future work.

References

1. Cortes, C., Vapnik, V.N.: Support vector networks. Mach. Learn. **20**, 273–297 (1995)
2. Hastie, T., Rosset, S., Tibshirani, R., Zhu, J.: The entire regularization path for the support vector machine. J. Mach. Learn. Res. **5**, 1391–1415 (2004)
3. Lecun, Y., Bottou, L., Bengio, Y., Haffner, P.: Gradient-based learning applied to document recognition. Proc. IEEE **86**(11), 2278–2324 (1998)
4. Hinton, G.E., Osindero, S., Teh, Y.W.: A fast learning algorithm for deep belief nets. Neural Comput. **22**, 781–796 (2006)
5. Salakhutdinov, R., Hinton, G.: An efficient learning procedure for deep boltzmann machines. Neural Comput. **24**(8), 1967–2006 (2012)
6. Hinton, G.E., Salakhutdinov, R.R.: Reducing the dimensionality of data with neural networks. Science **313**(5786), 504–507 (2006)
7. Salakhutdinov, R., Hinton, G.: Deep Boltzmann machine. J. Mach. Learn. Res. **5**, 448–455 (2009)
8. Bengio, Y., Lamblin, P., Popovici, D., Larochelle, H.: Greedy layer-wise training of deep networks. In: Proceedings of Advances in Neural Information Processing Systems, vol. 19, pp. 153–160 (2006)
9. Vincent, P., Larochelle, H., Bengio, Y., Manzagol, P.A.: International conference on machine learning. In: Proceedings of International Conference on Machine Learning, pp. 1096–1103 (2008)
10. Vincent, P., Larochelle, H., Lajoie, I., Bengio, Y., Manzagol, P.-A.: Stacked denoising autoencoders: Learning useful representations in a deep network with a local denoising criterion. J. Mach. Learn. Res. **11**, 3371–3408 (2010)

11. Coates, A., Andew, Y.N.: Search machine learning repository: the importance of encoding versus training with sparse coding and vector quantization. In: Proceedings of International Conference on Machine Learning, pp. 921–928 (2011)
12. Lee, H., Ekanadham, C., Ng, A.Y.: Sparse deep belief net model for visual area v2. In: Proceedings of Advances in Neural Information Processing Systems, vol. 20, pp. 1–8 (2008)
13. Huang, G.-B., Zhou, H.-M., Ding, X.-J., Zhang, R.: Extreme learning machine for regression and multiclass classification. IEEE Trans. Syst. Man Cybern. Part B Cybern. **42**(2), 513–529 (2012)
14. Bartlett, P.L.: The sample complexity of pattern classification with neural networks: the size of the weights is more important than the size of the network. IEEE Trans. Inf. Theory **44**(2), 525–536 (1998)
15. Huang, G.-B., Chen, L., Siew, C.-K.: Universal approximation using incremental constructive feedforward networks with random hidden nodes. IEEE Trans. Neural Netw. **17**(4), 879–892 (2006)
16. Olshausen, B.A., Field, D.J.: Sparse coding with an overcomplete basis set: a strategy employed by v1? Vis. Res. **37**(23), 3311–3325 (1997)
17. Figueiredo, M.A.T., Nowak, R.D., Wright, S.J.: Gradient projection for sparse representation: application to compressed sensing and other inverse problems. IEEE Trans. Select. Topics Signal Process. **1**(4), 586–597 (2007)
18. Hubel, D.H., Wiesel, T.N.: Receptive fields of signal neurons in the cat's striate cortex. J. Physiol. **148**, 574–591 (1959)
19. Roll, E.T., Tovee, M.J.: Sparseness of the neuronal representation of stmuli in the primate temporal visual cortex. J. Neurophysiol. **173**, 713–726 (1992)
20. Wright, J., Yang, A.Y., Ganesh, A.: Robust face recognition via sparse representation. IEEE Trans. Pattern Anal. Mach. Intell. **31**, 210–227 (2009)
21. Yang, J.-C., Yu, K., Gong, Y.H., Huang, T.: Linear spatial pyramid matching using sparse coding for image classification. In: IEEE Conference on Computer Vision and Pattern Recognition, vol. 1, pp. 1017–1022 (2005)
22. Daubechies, M.D.F.I., Mol, C.D.: An iterative thresholding algorithm for linear inverse problems with a sparsity constraint. Commun. Pure Appl. Math **57**, 1413–1457 (2004)
23. Figueiredo, M., Nowak, R.: An em algorithm for wavelet-bases image restoration. IEEE Trans. Image Process. **12**, 906–916 (2003)
24. Kim, S.J., Koh, K., Boyd, S.: An interior-point method for large-scale l_1-regularized least squares. Neural Comput. **24**(8), 1967–2006 (2012)
25. Fuchs, J.J.: More on sparse representations in arbitrary bases. IEEE Trans. Inf. Theory **50**, 1341–1344 (2004)
26. Blake, C.L., Merz, C.J.: Uci repository of machine learning databases. Department of Information and Computer Sciences, University of California, Irvine, CA (1998)
27. Asif, M.S., Romberg, J.: Sparse recovery of streamiing signals using l_1-homotopy. IEEE Trans. Signal Process. **62**(16), 4209–4233 (2014)
28. Kavukcuoglu, K., Ranzato, M., LeCun, Y.: Fast inference in sparse coding algorithms with applications to object recognition. In: Technical report. Computational and Biological Lerning Lab, NYU (2008)
29. Rozell, C.J., Johnson, D.H., Olshausen, B.A.: Sparse coding via thresholding and local competition in neural circuits. Neural Comput. (2008)

Continuous Top-K Remarkable Comments over Textual Streaming Data Using ELM

Rui Zhu, Bin Wang and Guoren Wang

Abstract The increasing popularity of location-based social networks encourages more and more users to share their experience. It deeply impact the decision of the other users. In this paper, we study the problem of top-K remarkable comments over textual streaming data. We first study how to efficiently identify the mendacious comments. Through using a novel machine learning technique named ELM, we could filter most of mendacious comments. We then study how to maintain these vital comments. For one thing, we propose a two-level index to maintain their position information. For another, we employ domination transitivity to remove meaningless comments. Theoretical analysis and extensive experimental results demonstrate the effectiveness of the proposed algorithms.

1 Introduction

The increasing popularity of location-based social networks encourages more and more users to share their experience for point-of-interest (POI) in a cyber world [1, 2]. When users want to visit a POI such as store, museum, hotel, they often retrieve the hot comments from the internet. Accordingly, some high-quality comments are important to the user decision. In this paper, we study the problem of user-defined top-k remarkable comments over location-based social networks.

Usually, comments associated with two important attributes that are textual content and geo-spatial content. They are generated by visitor check-in, buyer advertisement, etc. As a type of real-time based data, they are modeled as geo-textual data streams. Thus, we reduce this work to the problem of continuous top-k query

R. Zhu (✉) · B. Wang · G. Wang
College of Information Science and Engineering, Northeastern University, Shenyang, China
e-mail: neuruizhu@gmail.com

B. Wang
e-mail: binwang@ise.neu.edu.cn

G. Wang
e-mail: wanggr@ise.neu.edu.cn

© Springer International Publishing Switzerland 2016
J. Cao et al. (eds.), *Proceedings of ELM-2015 Volume 2*,
Proceedings in Adaptation, Learning and Optimization 7,
DOI 10.1007/978-3-319-28373-9_13

over location-based steaming textual data. Obviously, this problem is built up of two fundamental problems they are POI recommendations and continuous top-k query over streaming data.

For the first one, there exists a great deal of works answering POI recommendations. Ye et al. propose the classical user-based collaborative filtering techniques. Its key idea is to score POIs in terms of similarity between users' check-in activities, and return high score results to the users. However, this type of work lack of location information and face with data sparsity problem [1]. In order to overcome this problem, Yin et al. utilized local features (e.g., attractions and events) to improve the model learning and inference procedure for the recommendation purpose. However, users are usually more interested to a group of flavors that are exposed in a given range. Aim to this problem, Chen et al. proposed top-K location category based POI recommendation, by introducing information coverage to encode the location categories of POIs in a city. Because it is a NP-hard problem, they propose a greedy algorithm and an optimization algorithm solve this problem. However, these algorithms exists a common problem where they do not consider the truth of this comments. If a undesirable buyer send some fictitious but tempting information to the system (e.g., weibo facebook and etc.), the recommendation system may send some deception result to the users.

For the second one, these also exists a group of works. The representative work is the k-skyband based algorithm. As is reviewed in Sect. 2, an object o is a k-skyband object if there are less than k objects who come later than o and have scores higher than $F(o)$. As proved in [3], top-k results must be k-skyband objects. Consequently, this cluster of approaches could answer the query via scanning top-k skyband objects once. However, this kind of algorithms have to waste lots of computing cost in maintaining the dominate number of k-skyband, and reduce the overall performance. Especially, when the data are skewed distributed, this algorithms has to maintain multitudes of k-skybands. In this case, the computing cost of maintaining objects' dominate number is increased quite significantly.

This paper takes on the challenge of designing an efficient algorithm for solving the above problems. We propose the framework PBTI (short for personalized based top-k recommendation index). It improves the state of arts solutions by introducing the falsity comment identification and exploiting a family of filtering techniques. Furthermore, it employs the domination transmissibility to enhance the algorithm performances. The challenges and our contributions are as follows.

(i) How to efficiently identify falsity comments? As far as we know, this type of comments have two characteristics, that are mutable and volume. In order to fast identify them, we firstly study their behavior character. And then, we employ a novel machine learning algorithm called ELM for classification. From [4], since its classification speed is much faster than the other classical algorithms such as SVM, neural networks and so on, it is very suitable for identifying falsity comments over textual streaming data.

(ii) How to efficiently answer the top-k query? Firstly, we employ the multi-resolution grid technique to maintain the local information of these comments.

We then built up of a super top-K result set of all kinds of industries such as hotels, restaurants and so on. Since K is usually larger than that of user-defined parameter k, in order to reduce the computation cost, we employ the domination transmissibility. From the test evaluation, it could effectively reduce the computing cost.

The rest of this paper is organized as follows: Sect. 2 gives background. Section 3 proposes the PBTI. Section 4 evaluates the proposed methods with extensive experiments. Section 5 is the conclusion and the future work.

2   Background

In this section, we firstly discuss the related work. And then, we introduce an issue machine learning technique called ELM.

2.1   Related Work

The related work includes two parts. We firstly introduce the related work over location-based social networks. We then discuss the continuous top-k query over streaming data.

For the first one, there exists a large number of efforts studying the problem of LBSN [1, 5, 6]. Wang et al. proposed an efficiently algorithm to do selectivity estimation on Streaming spatio-textual data [2]. Hu et al. [7] study the problem of location-aware publish/subscribe for parameterized spatio-textual subscriptions. Ye et al. proposed an efficiently algorithm to tackle the problem of place semantic annotation in LBSNs [1]. They formulated as a multi-label classification problem and propose a two-phase algorithm to learn a binary SVM classifier for each tag in the entire tag space. They then develop a number of techniques for extracting population and temporal features, building a network of related places (NRP), and deriving label probability for each place in the system. Yin et al. [8] improve the model learning and inference procedure via utilizing local features. Chen et al. [1] studies the problem of point-of-interest recommendation for information coverage.

For the second one, there are also many efforts studying the problem of continuous top-k query over sliding window. Among all these efforts, [9] developed a novel algorithm named SMA to answer the top-k query [10]. However, its issue is when the data distribution is skewed, the window may be re-scanned many times. MinTopK algorithm [10] employed a popular technique named partition to handle this kind of query, where the partition size equals to the sidling length (e.g., s). In this way, it could avoid to maintain non-k skybands, and obtain a nice performance when $s \gg k$. However, when s is closing to k, its performance turn to worse.

2.2 *Extreme Learning Machine*

In this section, we present a brief overview of extreme learning machine (ELM) [4, 11], developed by Huang et al. ELM is based on a generalized single-hidden-layer feedforward network (SLFN). Also, it has many variants [12–16]. Compared with neural networks, its hidden-layer nodes are being randomly chosen instead of iteratively tuned. In this way, it provides good generalization performance at thousands of times faster speed than traditional popular learning algorithms (e.g., SVM, neural networks). Also, it get better performance due to its *universal approximation capability* [13, 14, 17] and *classification capability* [18].

As an improving version, the Online sequential extreme [16] learning machine (short for OS-ELM is proposed. Compared with the basic ELM, OS-ELM could learn data one-by-one or chunk-by-chunk with fixed or varying size. Thus, it is suitable for processing streaming data.

Specially, given a set of samples (x_i, t_i), where $x_i = [x_{i1}, x_{i2}, ..., x_{in}]^T \in \mathcal{R}^n$ and $t_i = [t_{i1}, t_{i2}, ..., t_{in}]^T \in \mathcal{R}^m$, OS-ELM select the type of nodes, the activation function, and the hidden node number and so on.

Then, OS-ELM is employed in a two-phase steps that are: (i) initialization phase and (ii) sequential learning phase. In the first phase, OS-ELM uses a small set of samples for training. Specially, given a set of training data $\mathcal{N}_0 = \{x_i, t_i\}_{i=1}^{N_0}$ is used for initializing.

From then on, the second phase employs the learning in a chunk-by-chunk way. In the k-th chunk of new training data, for N_{k+1} distinct arbitrary samples, OS-ELM firstly compute the partial hidden layer output matrix $\mathbf{H}_{k+1}$. And then, OS-ELM computes the output weight matrix β_{k+1}, where β_{k+1}, $\mathbf{H}_{k+1}$, and $\mathbf{T}_{k+1}$ are computed according to Eqs. 1–4.

$$\beta_{k+1} = \beta_k + \mathbf{P}_{k+1}\mathbf{H}_{k+1}^T \left(\mathbf{T}_{k+1} - \mathbf{H}_{k+1}\beta_k\right) \tag{1}$$

$$\mathbf{P}_{k+1} = \mathbf{P}_k - \mathbf{P}_k\mathbf{H}_{k+1}^T \left(I + \mathbf{H}_{k+1}\mathbf{H}_k\mathbf{H}_{k+1}^T\right)^{-1} \mathbf{H}_{k+1}\mathbf{P}_k \tag{2}$$

$$\mathbf{H}_{k+1} = \begin{bmatrix} \mathbf{G}\left(\mathbf{a}_1, b_1, \mathbf{x}_{\sum N_j + 1}\right) & \cdots & \mathbf{G}\left(\mathbf{a}_N, b_N, \mathbf{x}_{\sum N_j + 1}\right) \\ \vdots & \cdots & \vdots \\ \mathbf{G}\left(\mathbf{a}_1, b_1, \mathbf{x}_{\sum N_j + 1}\right) & \cdots & \mathbf{G}\left(\mathbf{a}_1, b_1, \mathbf{x}_{\sum N_j + 1}\right) \end{bmatrix}_{N_{k+1} \times l} \tag{3}$$

$$\mathbf{T}_{k+1} = \left[\mathbf{T}_{(\sum_{j=0}^{k} N_j)+1} \cdots \mathbf{T}_{(\sum_{j=0}^{k} N_j)}\right]_{N_{k+1} \times m}^T \tag{4}$$

3   The PBTI

In this section, we discuss how PBTI supporting continuous top-k query over textual streaming data. We firstly introduce the framework of this paper. We then discuss the comments management. Lastly, we discuss the efficiently continuous top-k algorithm.

3.1   *The ELM-Based Framework*

In this section, we introduce the framework of our solution. From Fig. 1a, it manages comments from the following three aspects. They are the spatial, the typological and contents.

The first part is a filter that is used to prune mendacious comments. In order to accurately them, we firstly find a serial of characteristic of the mendacious contents, extract them and build up of the characteristic vector according to them. And then, we employ the ELM technique to apply the training. After training, we use the classifier to do the pruning. Specially, given a newly arrival comment, if it is classified as a mendacious content, we discard it immediately. Otherwise, we manage it.

The second part is also a classifier. Compared with the first one, we classify them according to their associated key words, where the comments are classified into few types (e.g., accommodation, shopping and so on). Since there exists lots of works, we skip the details.

The third part is to manage their spatial information. Specially, we employ multi-resolution grid technique to manage the spatial information of the comments. We partition the management region into a group of subregions according to the density of the comments. If the density in a given region is high, we finely partition it. Otherwise, we coarsely partition it. After partition, the comments account in each cell are rough the same.

From Fig. 1b, c, these comments are managed in a two-level index. The first level is a multi-resolution grid. It manage the spatial information of the cells. For each cell, we use a group of inverted-list to manage these comments. Each list corresponds to a type of comments. They are sorted according to their arrival time. In order to

Fig. 1 The ELM-Based framework **a** Flowchart **b** Multi-resolution grid **c** Group-domination algorithm

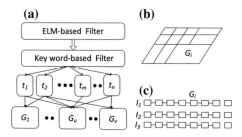

supporting continuous top-k query, we propose the group-domination algorithm to answer top-k queries. In this way, from the evaluation part, the computing cost could reduce a lot.

3.2 The ELM-Based Mendacious Comments Filtering

In this section, we discuss how ELM-Based identify mendacious comments. In order to achieve this goal, in Sect. 3.2.1, we firstly study the characteristics of the mendacious comments and construct the corresponding vector. We then discuss the ELM-based classification.

3.2.1 The Feature Vector Selection

In this section, we should points three mendacious comments characteristics. Considering a mendacious comment, it may be published by the third-party users. Thus, we need to analyze the characteristics of this kind of users. Through deeply study, the characteristics are abstracted as follows:

Friends Distribution. The amount of their friends $|I|$ may be relatively small, and they often concern lots of users (denoted by $|C|$). In this way, this kind of users is convenient to spread advertisement to the other users. Thus, the first characteristics is the value of $\frac{|I|}{|C|}$.

Comments Published Frequency. As far as we know, this kind of users often publish comments in a batch way. For example, they may publish a set of comments at time t_1. And then, they may publish another set of comments at time t_2. The interval between these two group are usually long. Thus, the second characteristics is constructed as follows: we firstly combine the comments $\{c_1, c_2, ...c_n\}$ published in a given interval into one comment set C_a. And then, we compute the comments published frequency according to Eq. 5, where c_u, c_v denote comments the same C_i, $c_u - c_v$ denotes the comments publishing time difference between two comments, $C_i - C_j$ denotes the comments publishing time difference between two set of comments. Obviously, if the user is an advertising publisher, $|\overline{c_u - c_v}|$ is usually small, and $|\overline{C_i - C_j}|$ usually high.

$$F_{fre} = \frac{|\overline{c_u - c_v}|}{|\overline{C_i - C_j}|} \tag{5}$$

Comments Similarity. Usually, this kind of comments are similarity. Therefore, we apply an effectively text similarity metric algorithm to evaluate the similarity of different comments, where it is computed by a score function. Thus, we use the score $|F|$ as the third attribute.

Algorithm 1: ELM-based Classification

Input: textual o_{in}, WhiteList $\mathcal{W}$, BlackList $\mathcal{B}$
Output: Classification b
1   User $Id \leftarrow$ findID(o_{in});
2   findList($Id, \mathcal{W}, \mathcal{B}$);
3   **if** $Id \in W$ **then**
4       $b=$ trustworthy;
5   **if** $Id \in B$ **then**
6       $b=$ mendacious;
7   **else**
8   vector $v \leftarrow$ buildVector(o_{in}, Id);
9      $b=$ELM-classifier(v);
10 **return** ;

3.2.2   Identifying Mendacious Comments Using ELM

According to the characteristics discussed before, we build a vector for each newly arrival comments c_{in}. Let u be the user that publishes c_{in}. Its corresponding vector v is $\left\langle \frac{|I_u|}{|C_u|}, F_{fre}, |F| \right\rangle$. Based on this vector, we apply ELM for classification. In general, the ELM-based classification can be summarized as two steps. The first step is training. In this phase, we select a set of comments for learning and obtain a binary classes ultimately. The second step is classifying. Using the trained classifier, we classify the newly arrived textual data. In this way, we can determine whether the comments are mendacious comments. If it is labeled as a mendacious comment, we discard it. Otherwise, we maintain it.

3.2.3   L-ELM

The above method could effectively find mendacious comments. However, it exists a key issue, where it has to timely process multitude of textual data. Thus, the classification speed is vital important. In order to speed up the classification, we propose a two-level filter. The first level is a user-based filter. This filter maintains two type of lists that are black-listing and white-listing. Specially, blacklisting maintains all the users that usually publish mendacious comments. Thus, if a comment is generated by a blacklisting user, we discard it immediately. Whitelisting maintains all the users that are trustworthy. Accordingly, if a comment is generated by a white-listing user, we maintain it.

Algorithm shows the process of newly arrival textual. Line2 to Line6 is the first-level, where it filter newly arrived textual o_{in} according to white-listing $\mathcal{W}$, black-listing $\mathcal{B}$. If o_{in} can not be filtered, we apply the second level filtering. In this phase, we firstly build a vector v according to the comment itself and the corresponding user's history comments. And then, we input the vector v into the classifier. Lastly, we output the result.

3.3 Domination Transitivity-Based Continuous Top-k Query over Textual Data

After filtering mendacious comments, we maintain the top-k representative comments of each type (e.g., accommodation, shopping and so on). Since the classify in this step is not our mainly work, we skip the details.

Given a comment c, we use a score function F to compute its representative. And then, we explain how to maintain candidates according to their score. We abstract a comment o to a tuple $\langle F(o), F(o) \rangle$, where $F(o)$ is value computed by the scoring function, $F(o)$ its arrived time. Also, after it generates, we insert it into the the candidate list. When it is not able to become the query result, we delete it. A natural question is how to identify an comment is able to become a query result.

We discuss our solution in the sliding window model. Specially, given a sidling window $\langle N, s \rangle$, we timely retrieve the k object with highest score in the window. This query window can be either time- or count-based. In this paper, we only focus on the count-based window. Accordingly, the parameter N denotes the object amount in the window, and s denotes number of objects that arrive whenever the window slides. We answer the query through using the domination of different tuples. In the following part, we propose the conception of dominance. We then propose the conception of domination number. As a basis, we discuss how to maintain candidates.

Definition 1 (*Dominance.*) Given two objects o and o', object o is dominated by o' if $F(o) > F(o')$ and $o.t < o't$, denoted as $o' \prec o$. Here, $o.t$ refers to the arrival order of object o. Given a sliding window W and a set of objects O_W in W, o's dominant number, denoted as $D(o, O_W, W)$, refers to the number of objects in W that can dominate o, i.e., $D(o, O_W, W) = |\{o' \in O_W | o' \prec o\}|$.

Definition 2 (*Dominance Number.*) Given two objects $o \in \mathcal{O}$, its dominate number (denoted by $D(o)$) records how many object in $\mathcal{O}$ dominating o.

Clearly, from [10], if an object is dominated by k objects, it is not able to become a query result. However, as is discussed in [10], they need $O(k)$ computing cost to delete an object. If k is relatively high, the cost of deleting meaningless object is still high. In this paper, we propose the domination transitivity to maintain candidates.

After discuss the above theorems, we discuss how to manage meaningful objects in the candidate set. Let C be the candidate set, C_{in} be the comments arrived in the system at the same time, and K be the account of retrieved objects. We firstly initialize the dominate number of the objects in C_{in}. Specially, for each object $o \in C_{in}$, if its score is the ith highest, we set its dominate number to i. Especially, if $i > k$, we discard it.

After initializing, we merge C_{in} with C. Since C_{in} and C are all ordered, we insert them based on the key of merge sort. Specially, when we insert $o_i \in C_{in}$ into C, its starting searching s_i position is the insertion position of the o_{i+1}. We insert it before the object whose score is minimal but higher than $F(o)$ (denoted by e_i). For the object o' between $[s_i, e_i]$, we add $D(o')$ to $D(o') + i$ (Fig. 2).

Fig. 2 Example of removable candidates disqualified by newly entered objects, where *grey cycles* refer to removable candidates and *dotted cycles* refer to outdated candidates

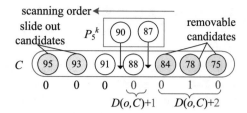

4 Indexing Local-Based Comments

In this section, we discuss the local-based comments indexing. As is discussed in Sect. 3.1, we use a two-level index named MGC (short for multi-resolution grid-comment) to manage these comments. The first level is to manage the local information of these comments. The second level is to maintain the textual contents. In the following part, we will discuss the comments management. Then, we explain the corresponding query algorithm.

The Updating Algorithm. Given a newly arrival comments c, we firstly prune c via ELM-based classifier discussed in Sect. 3.1. If c can not be pruned, we select a suitable comments list via keyword-based classifier. Then, we insert c into its corresponding cell and push c at the end of the suitable list.

After insertion, we delete the expire comments from the list. Noted that because the comments in each list are sort in descendent order by their arrival order, we only need to pop them from the head. Lastly, we update the corresponding candidate set which has been discussed in Sect. 3.3. We want to highlight that because s is usually larger than 1, we could maintain candidates in a batch way. We use a buffer $\mathcal{B}$ to maintain comments. When $|\mathcal{B}| = s$, we first sort the elements in $\mathcal{B}$. Then, we merge the k comments with highest score into candidate set. Lastly, we clear $\mathcal{B}$.

The Query Algorithm. Let $q\langle \mathcal{R}, \mathcal{K}, k \rangle$ be a query, where $\mathcal{R}$ denotes the query region, $\mathcal{K}$ denotes the key word set, and k denotes the answers amount. When handling q, we firstly find the cells that are overlapping with $\mathcal{R}$. Then, we access the corresponding list to find the k comments with highest score from these lists. For limitation of space, we skip the details.

5 Experimental Evaluation

5.1 Experimental Setting

This section experimentally evaluates the efficiency of the proposed techniques.

Data set. We used two real datasets: D-COMM and M-COMM. The D-COMM dataset was collected from dianping.com, which had 10 million comments with locations.

The M-COMM dataset was collected from meituan.com, which had 5 million comments with locations. In addition, the location of all the comments are all from shenyang.

Data clear. Since the comments contain many information, we extract a few useful information from the each comment. It could be expressed by the tuple $< uId, time, text, location, seller >$. They represent the user Id, comment publishing time, comment content, position coordinates and seller Id.

The baseline algorithm. In this section, we propose a baseline algorithm for comparison. Comparing with PBTI, the baseline algorithm exists two difference. Firstly, recalling Sect. 3.1, we use ELM for classifying. For comparison, we use SVM as the classifier of the baseline algorithm. Secondly, when maintaining the domination number of candidates, we introduce the concept of domination transitivity. For comparison, the baseline algorithm employ the basic k-skyband algorithm.

Experimental methods. Our experiments mainly evaluate the following two aspects: (i) the accuracy of classification, where we compare our proposed algorithms with varying size of the training data in different datasets. (ii) the efficiency of PBTI.

5.2 Classification Evaluation

In this subsection, we are going to evaluate the effect of the classifiers based on the ELM and SVM under different data set. We firstly evaluate the training time against different training set.

From Fig. 3a, b, with the increasing of the training set size, the training time under ELM and SVM are all increasing. However, compared with SVM, the training time of ELM increases slowly. The reason behind is ELM has better performance than SVM.

Next, we evaluate the accuracy rate of ELM. Compared with the first group, we add two algorithms for comparison. They are the L-ELM and L-SVM. L-ELM combines ELM and write-list (also black-list) for classifying; L-SVM combines SVM and write-list (also black-list) for classifying. From Fig. 4a, the testing time of L-ELM is fastest of all. The reason behind is it could filter many comments via white-list (or black-list). From Fig. 4a, b, the accuracy ratio of L-ELM is also highest of all.

Last of this subsection, we evaluate the classifying time. We also compare L-ELM with the baseline. From Table 1, the classifying time of L-ELM is shortest of all. The main reason is the classifying efficiency of ELM is higher than the SVM algorithms.

5.3 Query Performance

In this section, we evaluate the query performance. We compare PBTI with the baseline algorithm. Firstly, we evaluate the impaction of window length to the algorithm

Fig. 3 Training time
evaluation. **a** D-COMM,
b M-COMM

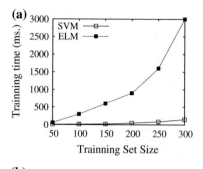

Fig. 4 Accuracy ratio
evaluation. **a** D-COMM,
b M-COMM

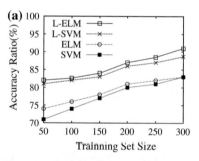

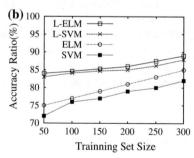

performance. We set the parameter k to 100, the amount of comments that have
the same timestamp to 1000, and vary the window size from 100 KB to 10 MB.
From Fig. 5a, b, we find that the PBTI has a better performance than that of baseline

Table 1 Classification time

Data set	L-SLM		L-SVM	
	Classifying time(S)	Class	Classifying time(S)	Class
D-COMM	15.6	2	325.6	2
M-COMM	7.3	2	147.1	2

Fig. 5 Running time under different window length. **a** D-COMM, **b** M-COMM

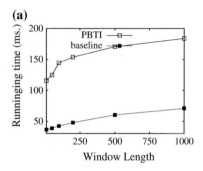

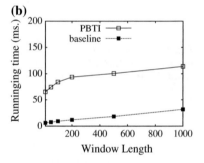

algorithm. The reason is, for one thing, the classify efficiency of PBTI is much higher than that of the baseline algorithm. For another, because we employ the domination transitivity to maintain candidates, the computing cost is further reduced.

Next, we evaluate the impaction of window length to the algorithm performance. From Fig. 6a, b, we find that the PBTI has a better performance than that of baseline algorithm. Another obversion is, the running time of PBTI increases sightly slow.

6 Conclusions

In this paper, we propose a novel and general framework namely PBTI, for supporting continuous top-k comments over textual streaming data. Different from all the existing works, PBTI adopts the effectively machine learning technique named ELM for identifying mendacious comments. And then, we classify the reality comments

Fig. 6 Running time under
different K. **a** D-COMM,
b M-COMM

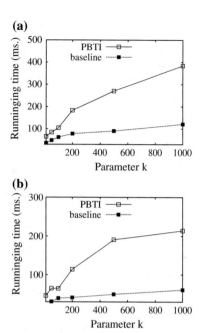

according to their associated key words. Then, we propose a novel index named MGC to manage these comments according their type and location. We have conducted extensive experiments to evaluate the performance of PBTI on several datasets. The results demonstrate the superior performance of PBTI.

Acknowledgments The work is partially supported by the National Natural Science Foundation of China for Outstanding Young Scholars (No. 61322208), the National Basic Research Program of China (973 Program) (No. 2012CB316201), the Joint Research Fund for Overseas Natural Science of China (No. 61129002), the National Natural Science Foundation of China for Key Program (No. 61572122), the National Natural Science Foundation of China (Nos. 61272178, 61572122, 61173029).

References

1. Chen, X., Zeng, Y., Cong, G., Qin, S., Xiang, Y., Dai, Y.: On information coverage for location category based point-of-interest recommendation. In: Proceedings of the Twenty-Ninth AAAI Conference on Artificial Intelligence, Austin, Texas, USA, pp. 37–43, 25–30 Jan 2015. http://www.aaai.org/ocs/index.php/AAAI/AAAI15/paper/view/9703
2. Wang, X., Zhang, Y., Zhang, W., Lin, X., Wang, W.: Selectivity estimation on streaming spatio-textual data using local correlations. PVLDB **8**(2), 101–112 (2014). http://www.vldb.org/pvldb/vol8/p101-wang.pdf
3. Shen, Z., Cheema, M.A., Lin, X., Zhang, W., Wang, H.: Efficiently monitoring top-k pairs over sliding windows. In: ICDE, pp. 798–809 (2012)

4. Huang, G.B., Zhu, Q.Y., Siew, C.K.: Extreme learning machine: a new learning scheme of feedforward neural networks. In: International Symposium on Neural Networks, vol. 2 (2004)
5. Bao, J., Zheng, Y., Mokbel, M.F.: Location-based and preference-aware recommendation using sparse geo-social networking data. In: SIGSPATIAL 2012 International Conference on Advances in Geographic Information Systems (formerly known as GIS), SIGSPATIAL'12, Redondo Beach, CA, USA, pp. 199–208, 7–9 Nov 2012. http://doi.acm.org/10.1145/2424321. 2424348
6. Liu, B., Xiong, H.: Point-of-interest recommendation in location based social networks with topic and location awareness. In: Proceedings of the 13th SIAM International Conference on Data Mining, Austin, Texas, USA, pp. 396–404, 2–4 May 2013. http://dx.doi.org/10.1137/1. 9781611972832.44
7. Hu, H., Liu, Y., Li, G., Feng, J., Tan, K.: A location-aware publish/subscribe framework for parameterized spatio-textual subscriptions. In: 31st IEEE International Conference on Data Engineering, ICDE 2015, Seoul, South Korea, pp. 711–722, 13–17 April 2015. http://dx.doi. org/10.1109/ICDE.2015.7113327
8. Yin, H., Sun, Y., Cui, B., Hu, Z., Chen, L.: LCARS: a location-content-aware recommender system. In: The 19th ACM SIGKDD International Conference on Knowledge Discovery and Data Mining, KDD 2013, Chicago, IL, USA, pp. 221–229, 11–14 Aug 2013. http://doi.acm. org/10.1145/2487575.2487608
9. Mouratidis, K., Bakiras, S., Papadias, D.: Continuous monitoring of top-k queries over sliding windows. In: SIGMOD Conference, pp. 635–646 (2006)
10. Yang, D., Shastri, A., Rundensteiner, E.A., Ward, M.O.: An optimal strategy for monitoring top-k queries in streaming windows. In: EDBT, pp. 57–68 (2011)
11. Huang, G.B., Zhu, Q.Y., Siew, C.K.: Extreme learning machine: theory and applications. Neurocomputing **70**, 489–501 (2006)
12. Feng, G., Huang, G.B., Lin, Q., Gay, R.K.L.: Error minimized extreme learning machine with growth of hidden nodes and incremental learning. IEEE Trans. Neural Networks **20**, 1352–1357 (2009)
13. Huang, G.B., Chen, L.: Convex incremental extreme learning machine. Neurocomputing **70**, 3056–3062 (2007)
14. Huang, G.B., Chen, L.: Enhanced random search based incremental extreme learning machine. Neurocomputing **71**, 3460–3468 (2008)
15. Huang, G.B., Zhu, Q.Y., Mao, K.Z., Siew, C.K., Saratchandran, P., Sundararajan, N.: Can threshold networks be trained directly? IEEE Trans. Circ. Syst. II Analog Digital Sig. Process. **53**, 187–191 (2006)
16. Rong, H.J., Huang, G.B., Sundararajan, N., Saratchandran, P.: Online sequential fuzzy extreme learning machine for function approximation and classification problems. IEEE Trans. Syst. Man Cybern. **39**, 1067–1072 (2009)
17. Huang, G.B., Chen, L., Siew, C.K.: Universal approximation using incremental constructive feedforward networks with random hidden nodes. IEEE Trans. Neural Networks **17**, 879–892 (2006)
18. Huang, G.B., Zhou, H., Ding, X., Zhang, R.: Extreme learning machine for regression and multiclass classification. IEEE Trans. Syst. Man Cybern. **42**, 513–529 (2012)

ELM Based Representational Learning for Fault Diagnosis of Wind Turbine Equipment

Zhixin Yang, Xianbo Wang, Pak Kin Wong and Jianhua Zhong

Abstract The data preprocessing, feature extraction, classifier training and testing play as the key components in a typical fault diagnosis system. This paper proposes a new application of extreme learning machines (ELM) in an integrated manner, where multiple ELM layers play correspondingly different roles in the fault diagnosis framework. The ELM based representational learning framework integrates functions including data preprocessing, feature extraction and dimension reduction. In the novel framework, an ELM based autoencoder is trained to get a hidden layer output weight matrix, which is then used to transform the input data into a new feature representation. Finally, a single layered ELM is applied for fault classification. Compared with existing feature extraction methods, the output weight matrix is treated as the mapping result with weighted distribution of input vector. It avoids wiping off "insignificant" feature information that may convey some undiscovered knowledge. The proposed representational learning framework does not need parameters fine-tuning with iterations. Therefore, the training speed is much faster than the traditional back propagation-based DL or support vector machine method. The experimental tests are carried out on a wind turbine generator simulator, which demonstrates the advantages of this method in both speed and accuracy.

Keywords Fault diagnosis · Wind turbine · Representational learning · Classification · Extreme learning machines · Autoencoder

Z. Yang (✉) · X. Wang · P.K. Wong · J. Zhong
Department of Electromechanical Engineering, Faculty of Science
and Technology, University of Macau, Macau SAR, China
e-mail: zxyang@umac.mo

X. Wang
e-mail: yb47410@umac.mo

P.K. Wong
e-mail: fstpkw@umac.mo

J. Zhong
e-mail: yb17416@umac.mo

© Springer International Publishing Switzerland 2016
J. Cao et al. (eds.), *Proceedings of ELM-2015 Volume 2*,
Proceedings in Adaptation, Learning and Optimization 7,
DOI 10.1007/978-3-319-28373-9_14

1 Introduction

As a new technique in a trend to supplement the traditional power generation methods, the reliability of wind turbine generation system (WTGS) becomes a new issue need to be concerned. Continuously condition monitoring and fault diagnosis technologies are necessary so as to reduce maintenance cost and work stably. In recent years, a large body of research suggests fault detection using machine learning-based approach for WTGS is feasible. Intelligent methods for fault diagnosis in WTGS and rotating machinery usually depend on the procedures of the vibration signal processing and fault pattern recognition. Generally, the raw vibration signals contains high-dimensional information and abundant noise (includes irrelevant and redundant signals), which cannot be fed into the fault diagnostic system directly [1]. Therefore, it's necessary to extract the useful information from the raw signals. There are several typical feature extraction methods available, such as wavelet packet transform (WPT) [2, 3], time-domain statistical features (TDSF) [4] and independent component analysis [5, 6]. The demerit of these methods is that these linear methods cannot extract nonlinear characteristics of input variables effectively, which may result in the weak performance of recognition. Obtained from WTGS, the vibration signals characterizes with high dimension and nonlinear. The aforementioned methods haven't capability to wipe off the "insignificant information" from the raw signals. This paper introduces the concept of autoencoder for feature extraction and explores its application. Regarding autoencoder, each layer in the stack architecture can be treated as an independent module [7]. The procedure shows briefly as follows, each layer is firstly trained to produce a new hidden representation of the observed patterns (input data), based on the representation it receives as input from the layer below, by optimizing a local supervised criterion. Each level produces a representation of the input pattern that is more abstract than the previous levels [8]. After representational learning for a mapping that produces a high level intermediate representations (intermediate matrix) of the input pattern, whereas, it is still complex and hard to calculate. Therefore, we need to decode it into low dimension and simple representations. Regarding algorithms for classification, neural network (NN) is widely used for rotating machinery fault diagnosis [1, 9]. However, NN has many inevitable drawbacks, such as local minima, time-consuming for determination of optimal network structure, and risk of over-fitting. Recent studies show that extreme learning machine (ELM) tends to have better scalability and achieves much better generalization performance at much faster learning speed than SVM [10, 11]. The ELM algorithm is easier to be implemented for multiclass problem. With the aforesaid advantages of ELM, a new application utilized ELM is introduced and by which building a fault diagnosis model for the WTGS. The rest of this paper is organized as follows. Section 2 presents the structure of fault diagnostic framework and the algorithms involved. Experimental rig setup and signals sample data acquisition with a simulated WTGS are discussed in Sect. 3. Section 4 discusses the experimental results of ELM and its comparisons with SVM, ML-ELM. Finally, a conclusion is given in the last Section.

2   Proposed Fault Diagnostic Framework

The framework consists of three parts, namely, autoencoder, matrix compressor and classifier. The roles of these parts are feature extraction, dimension reduction and fault classification. Figure 1 presents the structure of the proposed framework, which consists of three components: (a) ELM based autoencoder, (b) dimension compression transform, (c) supervised feature classification. The autoencoder has ability to reconstruct three types of representation (compressed, equal and sparse dimension). For classification, the original ELM classifier is applied for the final decision making.

2.1   ELM Based Autoencoder

ELM is a recently available learning method with SLFNs [12]. The character of ELM is that the model only has single hidden layer, of which the parameters need not to be tuned and can be initialized randomly. The parameters of the hidden layer are independent upon the target function and the training data [13]. Afterwards, the output weights which link hidden layer to output layer are determined analytically by a Moore-Penrose generalized inverse [12, 14]. Different from ANN, ELM tends to provide good generalization capability at learning speed benefits from its simple structure and efficient learning algorithm. The main idea of ELM algorithm is summarized as follows:

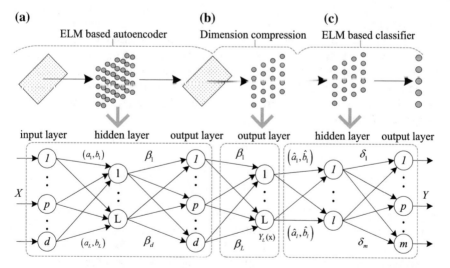

Fig. 1   Structure of the proposed method. **a** Framework of the proposed ELM based autoencoder. **b** Dimension compression. **c** The ELM based classifier

$$f_L(\mathbf{x}) = \sum_{i=1}^{L} \beta_i h_i(\mathbf{x}) = \mathbf{h}(\mathbf{x})\beta$$

$$\begin{cases} \beta = [\beta_1, \beta_2, ..., \beta_L] \\ h(\mathbf{x}) = [g_1(\mathbf{x}), g_2(\mathbf{x}), ..., g_L(\mathbf{x})] \end{cases} \tag{1}$$

where β_i is the output weight matrix between the hidden layer and the output layer. $h(\mathbf{x})$ is the hidden nodes outputs (random hidden features) for the input $\mathbf{x}$ and $g_i(\mathbf{x})$ is the output of the ith hidden node. Given N training samples $\{(x_i, t_i)\}_{i=1}^{N}$, the ELM aims to resolve the follow learning problems:

$$\mathbf{H}\beta = \mathbf{T} \tag{2}$$

where $\mathbf{T} = [t_1, ..., t_N]^T$ is the target labels and the matrix $\mathbf{H} = \left[\mathbf{h}^T(x_1), ..., \mathbf{h}^T(x_N)\right]^T$ is hidden nodes output. The output weights β can be calculated by Eq. (3)

$$\beta = \mathbf{H}^\dagger \mathbf{T} \tag{3}$$

where $\mathbf{H}^\dagger$ is the Moore-Penrose generalized inverse of matrix $\mathbf{H}$.

In order to have better generalization performance and to make the solution more robust, one can add a constrained parameter as shown in Eq. (4)

$$\beta = \left(\frac{1}{\mathbf{C}} + \mathbf{H}^T\mathbf{H}\right)^{-1} \mathbf{H}^T\mathbf{T} \tag{4}$$

The ELM algorithm in this paper is modified as follows: the input data is equal to the output data, namely $\mathbf{t} = \mathbf{x}$. The objective of ELM based autoencoder is to represent the input features meaningfully in three different representation: (1) Compressed representation, represent features from a higher dimensional input data space to a lower dimensional feature space. (2) Sparse representation, represent features from a lower dimensional input data space to a higher dimensional feature space. (3) Equal dimension representation, represent features from an input data space dimension equal to feature space dimension.

2.2 Dimension Compression

Regarding autoencoder, this paper adopts the regression method to training the parameters. By using these parameters, we can realize the output data equal to the input data. When it comes to dimension reduction, the above transform is not enough for the data compression, because the dimension of input data doesn't decrease. After all the parameters of autoencoder are identified, we apply a matrix transform to reconstruct the input data.

$$Y_L(\mathbf{x}) = \left(\beta f_L(\mathbf{x})\right)^T = (\beta \mathbf{X})^T \tag{5}$$

where $Y_L(\mathbf{x})$ is the final output of autoencoder. The dimension of $Y_L(\mathbf{x})$ is shown as Eq. (9). The subscripts N and L represent the number of input sample and hidden layer nodes respectively.

It is easy to find that the original dimension has been converted into a new low-dimensioned matrix. The procedure can be explained that each element in sample data $\mathbf{x}_{i(i\in N)}$ has relationship with β, in other words, β can be seen as the weight vector of $\mathbf{x}_{i(i\in N)}$. The procedure from $\mathbf{x}_{i(i\in N)}$ to $Y_L(\mathbf{x})$ is an unsupervised learning as the parameters have been identified in the second part as shown in Fig. 1.

Compared with the initial concept of autoencoder in deep learning, the autoencoder introduced in this paper has some differences as follows: (1) The autoencoder in this paper is a single hidden layer network, whereas, autoencoder in deep learning tends to adopt multi-hidden layers networks. (2) Deep learning tends to adopt back propagation (BP) algorithm to train all parameters of autoencoder, differently, this paper proposes the ELM to configure the networks with supervised learning (i.e. Let the output data is equal to input data). We can get the final output weight β so as to transform input data into a new representation through $Y_L(\mathbf{x}) = (\beta\mathbf{X})^T$. The dimension of converted data is much smaller than the raw data. (3) Autoencoder in deep learning tends to represent sparse feature for input data. However, this paper proposes the compression for the input data.

2.3 ELM Based Classifier

For a binary classification application, the decision function of ELM shows as:

$$f_L(\mathbf{x}) = \mathbf{sign}(\mathbf{h}(\mathbf{x})\delta) \tag{6}$$

ELM aims to reach not only the smallest training error but also the smallest norm of output weights. According to Bartletts theory [15], in order to reach small training error, the smaller the norms of weights are, the better generalization performance the networks tend to have.

$$Minimize : \ ||\mathbf{H}\delta - T||^2 \ and \ ||\delta|| \tag{7}$$

where $\delta = [\delta_1, \delta_2, ..., \delta_l]^T$ is the vector of the output weights between the hidden layer of l nodes and the output nodes.

$$\mathbf{H} = \begin{bmatrix} \mathbf{h}(\mathbf{x}_1) \\ \vdots \\ \mathbf{h}(\mathbf{x}_N) \end{bmatrix} = \begin{bmatrix} h_1(\mathbf{x}_1) & \cdots & h_l(\mathbf{x}_1) \\ \vdots & \vdots & \vdots \\ h_1(\mathbf{x}_N) & \cdots & h_l(\mathbf{x}_N) \end{bmatrix} \tag{8}$$

where $\mathbf{h}(\mathbf{x}) = [h_1(\mathbf{x}), h_2(\mathbf{x}), ..., h_l(\mathbf{x})]^T$ is the output vector of the hidden layer which maps the data from the d dimensional input space to the l dimensional hidden-layer space $\mathbf{H}$, $\mathbf{T}$ is the training data target matrix.

$$\mathbf{T} = \begin{bmatrix} \mathbf{t}_1^T \\ \vdots \\ \mathbf{t}_N^T \end{bmatrix} = \begin{bmatrix} t_{11} & \cdots & t_{1m} \\ \vdots & \vdots & \vdots \\ t_{N1} & \cdots & t_{Nm} \end{bmatrix} \tag{9}$$

Based on the KKT theorem, to training ELM is equivalent to solving the following dual optimization problem:

$$L_{D_{ELM}} = \frac{1}{2}||\delta||^2 + C\frac{1}{2}\sum_{i=1}^{N}||\xi_i||^2 - \sum_{i=1}^{N}\sum_{j=1}^{M}\alpha_{i,j}\left(\mathbf{h}(\mathbf{x}_i)\delta_j - t_{i,j} + \xi_{i,j}\right) \tag{10}$$

We can have the KKT corresponding optimality conditions as follows:

$$\frac{\partial L_{D_{ELM}}}{\partial \beta_j} = 0 \rightarrow \delta_j = \sum_{i=1}^{N}\alpha_{i,j}(\mathbf{h}(\mathbf{x}_i)^T \rightarrow \delta = \mathbf{H}^T a \tag{11}$$

$$\frac{\partial L_{D_{ELM}}}{\partial \xi_i} = 0 \rightarrow \alpha_i = C\xi_i, \quad i = 1, ..., N \tag{12}$$

$$\frac{\partial L_{D_{ELM}}}{\partial \alpha_i} = 0 \rightarrow \mathbf{h}(\mathbf{x}_i)\delta - t_i^T + \xi_i^T, \quad i = 1, ..., N \tag{13}$$

where $\mathbf{a}_i = [\alpha_{i,1}, \alpha_{i,2}, ..., \alpha_{i,M}]^T$. In this case, by substituting Eqs. (16) and (17) into Eq. (18), the aforementioned equations can be equivalently written as

$$\left(\frac{1}{C} + \mathbf{H}\mathbf{H}^T\right)\alpha = \mathbf{T} \tag{14}$$

From Eqs. (14)–(16), we have

$$\delta = \mathbf{H}^T\left(\frac{1}{C} + \mathbf{H}\mathbf{H}^T\right)^{-1}\mathbf{X} \tag{15}$$

The output function of ELM classifier is

$$\mathbf{f}(\mathbf{x}) = \mathbf{h}(\mathbf{x})\delta = \mathbf{h}(\mathbf{x})\mathbf{H}^T\left(\frac{1}{C} + \mathbf{H}\mathbf{H}^T\right)^{-1}\mathbf{T} \tag{16}$$

3   Experimental Results and Discussion

The test rig includes a computer for data acquisition, an electric load simulator, a prime mover, a gearbox, a flywheel and an asynchronous generator. The test rig can simulate many common periodic faults and irregular faults in a gearbox such as broken tooth, chipped tooth, wear of outer race of the bearing and so on. Table 1 presents a total of thirteen cases (including one normal case, eight single-fault and four simultaneous-fault cases) can be simulated in the gearbox, while which can generate the test and training dataset. It is necessary to note that some cases can be realized by specific tools or methods. (e.g., the mechanical misalignment of the gearbox was simulated by adjusting height of the gearbox with shims, and the mechanical unbalance case was simulated by adding one eccentric mass on the output shaft.)

In the diagnostic model, each simulated single fault was repeated two hundred times and one hundred times for each simultaneous-fault under various random electric loads. Each time, vibration signals in two seconds window was recorded with a sampling frequency of 2048 Hz. From a feasible data requisition point of view, the sample frequency must be much higher than the gear meshing frequency, which can ensure no missing signals during the process of sampling. In other words, each sampling dataset will record 8192 points (two accelerometers $*$ two seconds $*$ 2048) in each two seconds time window. There are 1800 sample dataset (i.e. (1 normal care + 8 kinds of single-fault cases) $*$ 200 samples) and 280 simultaneous-fault sample data (i.e. 4 kinds of simultaneous-fault data 100 samples). The procedure by using autoencoder for features extraction and dimension reduction has been shown in Fig. 1. The structure of autoencoder in this paper is set as $8192 * L * 8192$ and $8192 * L$. The output of the first part is an equal dimension representation of the input matrix, it is a supervised learning.

In order to verify the effectiveness of the proposed framework, this paper applies various combinations of methods to realize the contrast experiments. Testing accuracy and testing time are introduced to evaluate the prediction performance of the

Table 1 Sample single-faults and possible simultaneous-fault

Case no.	Condition	Fault description
C1	Normal	Normal
C2	Single fault	Unbalance
C3		Looseness
C4		Mechanical misalignment
C5		Wear of cage and rolling elements of bearing
C6		Wear of outer race of bearing
C7		Gear tooth broken
C8		Gear crack
C9		Chipped tooth
C10	Simultaneous fault	Gear tooth broken & chipped tooth
C11		Chipped tooth & wear of outer race of bearing
C12		Gear tooth broken & wear of cage & rolling elements of bearing
C13		Gear tooth broken & wear of cage and rolling elements of bearing & wear of outer race of bearing

classifier. In this paper, we choose the equal dimensional representation and use the ELM learning method to train the parameters. The function of autoencoder is to get an optimal matrix β, and the function of matrix transform is to reduce the dimension of input $\mathbf{X}$. Before the experiments, it is not clear that how many dimensions to be cut down is appropriate, in other words, the model need proper values of L and β to improve the testing accuracies. In order to get a set of optimal parameters (e.g., hidden layer nodes L in autoencoder, hidden layer nodes l in classifier), $D_{train}(D_{train}$ includes dataset D_{train_l} and D_{train_s}) is applied to train the networks. According to the experimental results not listed here, when $L = 80$, $l = 680$, and $C = 600$, the classifier has the best testing accuracy.

According to the feature extraction, this paper takes two kinds of methods as references. The two kinds of method are WPT+TDSF+KPCA combinations and EMD+SVD combination respectively. This paper takes the Db4 (Daubechies) wavelet as the mother wavelet and set the level of decomposition at the range from 3 to 5. The radial basis function (RBF) acts as kernel function for KPCA. To reduce the number of trials, the hyperparameter R of RBF based on 2^v is tried for v ranged from -3 to 3. In the KPCA processing, this paper selects the polynomial kernel with $d = 4$ and the RBF kernel with $R = 2$. After dimension reduction, a total of 80 principal components are obtained. After feature extraction, the next step is to optimize parameters of classifiers. This paper takes 4 kinds of methods, namely PNN, RVM, SVM and ELM. As mentioned previously, probabilistic based classifiers have their own hyperparameters for tuning. PNN uses spread s and RVM employs width ω. In this case study, the value of s is set from 1 to 3 at an interval of 0.5, and the values of ω is selected from 1 to 8 at an interval of 0.5. In order to find the optimal decision threshold, this paper sets the search region at the range from 0 to 1 at an interval of 0.01. For the configuration of ELM, this paper takes the sigmoid function as the activation function and sets the number of hidden modes l as 680 for a trial. According to the experimental results in Table 2, a total of 80 components are obtained from the feature extractor. It is clear that the accuracies with autoencoder are higher than those with WPT+TDSF+KPCA. The results can be explained that ELM based autoencoder holds all information of the input data during the representational learning. However, KPCA tends to hold the important information and inevitably lose some unimportant information. In order to compare the performances of classifiers, this paper sets the contrast experiments with the same ELM based autoencoder and different classifiers. As shown in Table 3, the number of hidden nodes L in autoencoder is 80, the last dimensions of training data D_{train} and testing data D_{test} are $1800 * 80$ and $280 * 80$ respectively. For parameters setting of ELM, this paper tried a wide range of l. For each dataset, we used 50 different values, namely set $l \in \{50, 75, ..., 1000\}$ respectively. As suggested in Table 2, this paper sets l is 680. According to the experimental results not listed here, SVM employed polynomial kernel with $C = 10$ and $d^* = 4$ show the best accuracy. Table 2 shows that the fault detection accuracy of ELM is similar to that of SVM, while the fault identification time of ELM and SVM take 20 and 157 ms respectively. The performance of ELM is much faster than that of SVM. Quick recognition is necessary for real-time fault diagnosis system. In actual WTGS application, the real-time fault diagnostic system is required to analyze sig-

Table 2 Evaluation of different combinations of methods using the optimal model parameters

Feature extraction	Classifier	Accuracies for test case (%)		
		Single-fault	Simultaneous-fault	Overall fault
WPT+TDSF+KPCA	PNN	83.64	83.64	83.76
	RVM	82.99	74.64	81.21
	SVM	**92.88**	**89.73**	90.78
	ELM	91.29	87.62	**90.89**
ELM AE	PNN	85.64	84.64	84.52
	RVM	83.99	77.64	83.21
	SVM	95.83	92.87	93.27
	ELM	**96.25**	**96.64**	**95.33**

Table 3 Evaluation of methods using ELM or SVM. ELM based autoencoder

Feature extraction	Fault type	Accuracies for test case (%)		Time for test case (ms)	
		SVM	ELM	SVM	ELM
ELM-AE	Single-fault	97.58 ± 2.25	96.58 ± 2.25	156 ± 0.9	18 ± 0.8
	Simultaneous-fault	95.33 ± 1.25	95.23 ± 3.25	158 ± 0.8	20 ± 0.5
	Overall fault	95.62 ± 3.15	94.53 ± 2.25	157 ± 0.4	20 ± 0.5

nals for 24 h per day. In terms of fault identification time, ELM is faster than SVM by 88.46 %. Although the absolute diagnostic time difference between SVM and ELM is not very significant in this case study, the time difference will be very significant in real situation because a practical real-time WTGS diagnostic system will analyze more sensor signals than the two sensor signals used in this case study.

4   Conclusions

This paper proposes a new application of ELM to the real-time fault diagnostic system for rotating machinery. At the stage of data preprocessing, this paper applies an ELM based autoencoder to train the network. During representational learning, the network generates a new representation with dimensional reduction which is put into the ELM based classifier. Compared with the widely-applied classifiers (e.g., SVM and RVM), ELM algorithm can search optimal solution from the feature space without any other constraints. Therefore, ELM is superior to SVM at producing lightly higher diagnostic accuracy. Besides, ELM tends to generate a smaller classification model and takes less execution time than SVM. This paper makes contributions in the following three aspects: (1) it is the first research to analyze the ELM based autoencoder as a tool of compressed representation. (2) it is the first application of ELM based autoencoder to the fault diagnosis for rotating machinery. (3) it is the original application of the proposed framework to the problem of WTGS diagnosis. As the proposed framework for fault diagnosis is general, it could be applied to other industrial problems.

Acknowledgments The authors would like to thank the University of Macau for funding support under Grants MYRG2015-00077-FST.

References

1. Wong, P.K., Yang, Z., Vong, C.M., Zhong, J.: Real-time fault diagnosis for gas turbine generator systems using extreme learning machine. Neurocomputing **128**, 249–257 (2014)
2. Bianchi, D., Mayrhofer, E., Grschl, M., Betz, G., Vernes, A.: Wavelet packet transform for detection of single events in acoustic emission signals. In: *Mechanical Systems and Signal Processing* (2015)
3. Keskes, H., Braham, A., Lachiri, Z.: Broken rotor bar diagnosis in induction machines through stationary wavelet packet transform and multiclass wavelet svm. Electr. Power Syst. Res. **97**, 151–157 (2013)
4. Ebrahimi, F., Setarehdan, S.-K., Ayala-Moyeda, J., Nazeran, H.: Automatic sleep staging using empirical mode decomposition, discrete wavelet transform, time-domain, and nonlinear dynamics features of heart rate variability signals. Comput. Methods Programs Biomed. **112**(1), 47–57 (2013)
5. Allen, E.A., Erhardt, E.B., Wei, Y., Eichele, T., Calhoun, V.D.: Capturing inter-subject variability with group independent component analysis of fmri data: a simulation study. Neuroimage **59**(4), 4141–4159 (2012)
6. Du, K.-L., Swamy, M.: *Independent Component Analysis*, pp. 419–450. Springer, London (2014)
7. Tang, J., Deng, C., Huang, G.-B.: Extreme learning machine for multilayer perceptron (2015)
8. Vincent, P., Larochelle, H., Bengio, Y., Manzagol, P.-A.: Extracting and composing robust features with denoising autoencoders. In: *Proceedings of the 25th International Conference on Machine Learning*, pp. 1096–1103. ACM (2008)
9. Yang, Z., Wong, P.K., Vong, C.M., Zhong, J., Liang, J.: Simultaneous-fault diagnosis of gas turbine generator systems using a pairwise-coupled probabilistic classifier. Math. Prob. Eng. **2013** (2013)
10. Huang, G.-B., Zhou, H., Ding, X., Zhang, R.: Extreme learning machine for regression and multiclass classification. IEEE Trans. Syst. Man Cybern. Part B Cybern. **42**(2), 513–529 (2012)
11. Huang, G.-B., Ding, X., Zhou, H.: Optimization method based extreme learning machine for classification. Neurocomputing **74**(1), 155–163 (2010)
12. Huang, G.-B., Zhu, Q.-Y., Siew, C.-K.: Extreme learning machine: theory and applications. Neurocomputing **70**(1), 489–501 (2006)
13. Luo, J., Vong, C.-M., Wong, P.-K.: Sparse bayesian extreme learning machine for multiclassification. IEEE Trans. Neural Networks Learn. Syst. **25**(4), 836–843 (2014)
14. Cambria, E., Huang, G.-B., Kasun, L.L.C., Zhou, H., Vong, C.M., Lin, J., Yin, J., Cai, Z., Liu, Q., Li, K., et al.: Extreme learning machines [trends & controversies]. IEEE Intell. Syst. **28**(6), 30–59 (2013)
15. Bartlett, P.L.: The sample complexity of pattern classification with neural networks: the size of the weights is more important than the size of the network. IEEE Trans. Inf. Theory **44**(2), 525–536 (1998)

Prediction of Pulp Concentration Using Extreme Learning Machine

Changwei Jiang, Xiong Luo, Xiaona Yang, Huan Wang
and Dezheng Zhang

Abstract Pulp concentration is one of the most important production parameters during ore dressing process. Generally, pulp concentration not only affects concentrate grade and recovery rate, but also has a major influence on the chemical and power consumptions during the flotation process. Recently, there has been a growing interest in the study of prediction for pulp concentration to improve the productivity and reduce consumption of various resources. Since the pulp concentration and other production parameters are nonlinearly related, it imposes very challenging obstacles to the prediction for this parameter. Because extreme learning machine (ELM) has the advantages of extremely fast learning speed, good generalization performance, and the smallest training errors, we employ ELM to predict pulp concentration in this paper. Pulp concentration data is first preprocessed using phase space reconstruction method. Then time series prediction model is adjusted from one dimension to multiple dimensions and thus it is established by several improved ELM algorithms, including traditional ELM, kernel-based ELM (Kernel-ELM), regularized ELM (R-ELM), and L_2-norm based ELM (ELM-L2). The experiments are conducted with a real-world production data set from a mine. The experimental results show the effectiveness of ELM-based prediction approaches, and we can also find that ELM-

This work was jointly supported by the National Natural Science Foundation of China under Grants 61174103, 61272357, and 61300074, and the National Key Technologies R&D Program of China under Grant 2015BAK38B01.

C. Jiang · X. Luo · X. Yang · D. Zhang (✉)
School of Computer and Communication Engineering, University of Science
and Technology Beijing, Beijing 100083, China
e-mail: zdzchina@126.com

C. Jiang · X. Luo · X. Yang · D. Zhang
Beijing Key Laboratory of Knowledge Engineering for Materials Science,
Beijing 100083, China

H. Wang
ANSTEEL MINING, Anshan 114001, China

H. Wang
School of Civil and Environmental Engineering, University of Science
and Technology Beijing, Beijing 100083, China

© Springer International Publishing Switzerland 2016
J. Cao et al. (eds.), *Proceedings of ELM-2015 Volume 2*,
Proceedings in Adaptation, Learning and Optimization 7,
DOI 10.1007/978-3-319-28373-9_15

179

L2 has better prediction effects than other algorithms with the increase of sample size. Both training speed and prediction accuracy are improved by employing ELM-L2 to the prediction of pulp concentration.

Keywords Prediction · Extreme learning machine (ELM) · Kernel-based elm · Phase space reconstruction · Regularized elm · Pulp concentration

1   Introduction

Pulp concentration is an important production index in flotation process [1]. But, it is difficult to detect pulp concentration by off-line testing. Then, we can predict pulp concentration with on-line learning method and some production problems may be inspected by comparing prediction result and actual production data [2].

Generally, the pulp concentration and other production parameters are nonlinearly related [3, 4]. And it imposes very challenging obstacles to the prediction for this value with traditional learning methods. Recently, extreme learning machine (ELM) for single-hidden layer feedforward network (SLFN) has attracted much attention because of its unique features, such as extremely fast learning speed, good generalization performance, and the smallest training errors [5, 6]. In this paper, we employ ELM related learning algorithms to train the production data and predict pulp concentration. Furthermore, several ELM algorithms are analyzed in this prediction applications. In addition to the traditional ELM [7], three improved ELM algorithms are also discussed. In view of the fact that the hidden layer output matrix is generated by random assignment and then the output of ELM model may be unsatisfactory with regard to stability and generalization in some cases, kernel-based ELM (Kernel-ELM) was proposed through the use of kernel function, and its computational performance in regression forecast is better than traditional ELM [8]. Meanwhile, considering the fact that traditional ELM does not address heteroskedasticity in practical applications and its performance will be affected seriously when outliers exist in the data set, regularized ELM (R-ELM) was developed in accordance with structural risk minimization principle and weighted least square [9]. The generalization performance of R-ELM is improved. However, the output weights in the implementation of R-ELM are also determined by Moore-Penrose generalized inverse as traditional ELM. Thus, using Kernel-ELM and R-ELM may cause memory overflow problem in the prediction applications while establishing time series prediction model on a data set with large sample size. In order to avoid such limitations and achieve better stability, L_2-norm based ELM (ELM-L2) was proposed since L_2 norm can shrink coefficients and is more stable [8]. It ensures a unique solution even when the number of hidden node is bigger than the number of samples, and leads to a grouping effect with higher robustness and higher learning accuracy compared with traditional ELM. Here, to ensure fast training speed and high prediction accuracy in the pulp concentration prediction, we can employ ELM-L2 to address this issue with the increase of sample size while dealing with pulp concentration data.

This paper is organized as follows. Section 2 analyzes related works about data preprocessing method and several ELM algorithms. Section 3 presents an ELM-based pulp concentration prediction method while providing an analysis for the advantages and limitations of those ELM related learning approaches. Then experimental results and discussions are provided in Sect. 4. Finally, a conclusion is given in Sect. 5.

2   Related Works

2.1   *Phase Space Reconstruction Theory*

Phase space reconstruction as a part of nonlinear time series analysis technique is an important step of chaotic time series processing. And reconstruction quality directly affects the establishment of system model. Phase space reconstruction theory has been used in time series prediction [10]. A core design strategy is that the evolution of any component is determined by other components of interaction. Therefore, the information of relative components is implicit in the development process of any component. In order to reconstruct an equivalent state space, we only need to consider a component and address it by processing measurement information obtained in some fixed time delay points. They identify a point in a multidimensional state space. Then some points can be obtained through repeating this process. In this way, it can reconstruct original dynamic system model and initially determine true information of the original system.

Phase space reconstruction plays an important role in time series prediction. In order to reconstruct a suitable phase space, the key is to select the appropriate embedding dimension r and time delay τ [11]. The selection of these two parameters is generally conducted using the following three criteria:

(1) Sequence dependent method, such as the autocorrelation method, the higher order correlation method, and many others.
(2) Phase space expansion method, such as fill factor method, swing method, and many others.
(3) Multiple autocorrelation and partial autocorrelation.

For a set of time series $\{z_i | i = 1, 2, \ldots, p\}$, the number of points in phase space is $T = p - (r - 1)\tau$ and phase space vector $\mathbf{Z}_i (i = 1, 2, \ldots, T)$ of reconstruction is:

$$
\begin{cases}
\mathbf{Z}_1 = [z_1, z_{1+\tau}, \ldots, z_{1+(r-1)\tau}], \\
\mathbf{Z}_2 = [z_2, z_{2+\tau}, \ldots, z_{2+(r-1)\tau}], \\
\vdots \\
\mathbf{Z}_T = [z_T, z_{T+\tau}, \ldots, z_{T+(r-1)\tau}].
\end{cases}
\tag{1}
$$

With regard to r and τ, there is a embedding time window Γ which makes phase space trajectory between z_i and $z_{i+\Gamma}$ relatively smooth in the measured time series $\{z_i|i = 1, 2, \ldots, p\}$. Here, the optimal embedding time window can be roughly determined through observing phase space trajectory.

2.2 Extreme Learning Machine (ELM)

For P arbitrary distinct training samples $\{(\mathbf{x}_i, \mathbf{t}_i)|i = 1, 2, \ldots, P\}$, where $\mathbf{x}_i = [x_{i1}, x_{i2}, \ldots, x_{im}]^T \in \mathbb{R}^m$ and $\mathbf{t}_i = [t_{i1}, t_{i2}, \ldots, t_{in}]^T \in \mathbb{R}^n$. The network structure of SLFN using ELM is in Fig. 1 and the output function of ELM for a input $\mathbf{x}$ can be defined as follows [8]:

$$f(\mathbf{x}) = h(\mathbf{x})\beta = h(\mathbf{x})\mathbf{H}^T(\frac{\mathbf{I}}{\gamma} + \mathbf{H}\mathbf{H}^T)^{-1}\mathbf{T}, \tag{2}$$

where $\beta = [\beta_1, \ldots, \beta_L]^T$ is the vector of the output weights between the hidden layer of L nodes and the output node, γ is the regularization parameter, $\mathbf{I}$ is the identity matrix, $\mathbf{T} = [\mathbf{t}_1^T, \mathbf{t}_2^T, \ldots, \mathbf{t}_P^T]^T$, $\mathbf{H} = [h(\mathbf{x}_1), h(\mathbf{x}_2), \ldots, h(\mathbf{x}_P)]^T$ is the feature mapping matrix of given P training samples, $h(\mathbf{x}_i)$ $(i = 1, 2, \ldots, P)$ and $h(\mathbf{x})$ are the output row vectors of the hidden layer with respect to the input $\mathbf{x}_i$ and $\mathbf{x}$.

Fig. 1 The network structure of SLFN using ELM

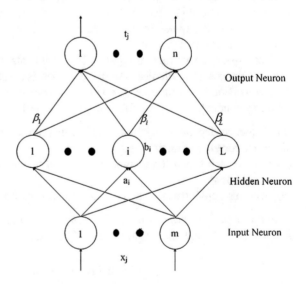

2.3 Kernel-Based Extreme Learning Machine (Kernel-ELM)

If the feature mapping $h(\mathbf{x})$ is unknown, kernel function is employed in Kernel-ELM and a kernel matrix can be defined as follows [8]:

$$\Omega_{\text{ELM}} = \mathbf{HH}^{\text{T}} = \begin{bmatrix} h(\mathbf{x}_1)h(\mathbf{x}_1) & \cdots & h(\mathbf{x}_1)h(\mathbf{x}_P) \\ \vdots & \vdots & \vdots \\ h(\mathbf{x}_P)h(\mathbf{x}_1) & \cdots & h(\mathbf{x}_P)h(\mathbf{x}_P) \end{bmatrix} = \begin{bmatrix} K(\mathbf{x}_1, \mathbf{x}_1) & \cdots & K(\mathbf{x}_1, \mathbf{x}_P) \\ \vdots & \vdots & \vdots \\ K(\mathbf{x}_P, \mathbf{x}_1) & \cdots & K(\mathbf{x}_P, \mathbf{x}_P) \end{bmatrix}, \quad (3)$$

where the kernel function $K(\mathbf{x}_i, \mathbf{x}_j) = h(\mathbf{x}_i) \cdot h(\mathbf{x}_j)$.

Here, the output function of Kernel-ELM can be redefined for (2) as follows:

$$f(\mathbf{x}) = h(\mathbf{x})\mathbf{H}^{\text{T}} \left(\frac{\mathbf{I}}{\gamma} + \mathbf{HH}^{\text{T}}\right)^{-1} \mathbf{T} = \begin{bmatrix} K(\mathbf{x}, \mathbf{x}_i) \\ \vdots \\ K(\mathbf{x}, \mathbf{x}_P) \end{bmatrix}^{\text{T}} \left(\frac{\mathbf{I}}{\gamma} + \Omega_{\text{ELM}}\right)^{-1} \mathbf{T}. \quad (4)$$

2.4 Regularized Extreme Learning Machine (R-ELM)

In view of the fact that traditional ELM works under empirical risk minimization theme and it tends to generate over-fitting model, R-ELM is designed in accordance with structural risk minimization, and it can be expected to provide better generalization ability than traditional ELM. And R-ELM is described as Algorithm 1 [9]. Here, a weight factor γ is introduced for empirical risk. By regulating γ, we can adjust the proportion of empirical risk and structural risk. Moreover, to obtain a robust estimate weakening outlier interference, the error variable can be weighted by using factor v_j, where v_j is a element of matrix $\mathbf{D} = \text{diag}(v_1, v_2, \ldots, v_P)$. And α is the Lagrangian multiplier. The inter quartile range **IQR** is the difference between the the 75th percentile and the 25th percentile. The constant c_1 and c_2 are set as $c_1 = 2.5$ and $c_2 = 3$, respectively.

2.5 L_2-Norm Based Extreme Learning Machine (ELM-L2)

In order to avoid some limitations of traditional ELM that is implemented normally under the empirical risk minimization scheme and may tend to generate a large-scale and over-fitting model, ELM-L2 is proposed to handle regression problem in a unified framework where L_2 norm can shrink coefficients and is more stable. The mathematic model of ELM-L2 can be described as [8]:

Algorithm 1: R-ELM

Input: Given a training set $\aleph = \{(\mathbf{x}_i, \mathbf{t}_i) | \mathbf{x}_i \in \mathbb{R}^m, \mathbf{t}_i \in \mathbb{R}^n, i = 1, \ldots, P\}$, a activation function
$g(x)$, and the hidden node number L;

1 Randomly assign input weights w_i and bias $b_i (i = 1, \ldots, L)$;

2 Calculate the hidden layer output matrix $\mathbf{H}$;

3 Calculate $\alpha = -\gamma(\mathbf{H}\beta - \mathbf{T})^{\mathrm{T}}$ and calculate $\varepsilon_i = \frac{\alpha_i}{\gamma}(i = 1, \ldots, P)$;

4 Calculate $\hat{s} = \frac{\mathbf{IQR}}{2 \times 0.6745}$ and calculate $v_j = \begin{cases} 1 & |\varepsilon_j/\hat{s}| \leq c_1 \\ \frac{c_2 - |\varepsilon_j/\hat{s}|}{c_2 - c_1} & c_1 \leq |\varepsilon_j/\hat{s}| \leq c_2 \\ 10^{-4} & \text{otherwise} \end{cases}$;

5 Update $\beta = (\frac{\mathbf{I}}{\gamma} + \mathbf{H}^{\mathrm{T}}\mathbf{D}^2\mathbf{H})^{\dagger}\mathbf{H}^{\mathrm{T}}\mathbf{D}^2\mathbf{T}$

$$\hat{\beta} = \arg\min_{\beta}\{||\mathbf{y} - \mathbf{H}\beta||_2^2 + \xi||\beta||_2^2\}, \tag{5}$$

where ξ is a ELM-L2 parameter. Then (5) can lead to a closed form solution as follows:

$$\hat{\beta} = (\mathbf{H}^{\mathrm{T}}\mathbf{H} + \xi\mathbf{I})^{-1}\mathbf{H}^{\mathrm{T}}\mathbf{y}. \tag{6}$$

3 Prediction of Pulp Concentration Using ELM Algorithms

We first reconstruct phase space of time series data of pulp concentration. Those reconstructed samples are trained by ELM algorithms, including traditional ELM, Kernel-ELM, R-ELM, and ELM-L2. Then the prediction results are obtained according to the evaluation criteria.

Specifically, we compare the prediction accuracy of those four algorithms to evaluate which kind of ELM algorithms perform best in addressing prediction of pulp concentration.

4 Experimental Results and Discussions

The actual pulp concentration data is a time series obtained every 5 s, and efficient data is collected within [40, 50] in terms of pulp concentration. We first reconstruct phase space of time series data where we set the time delay and the embedding dimension as 8 and 2, respectively.

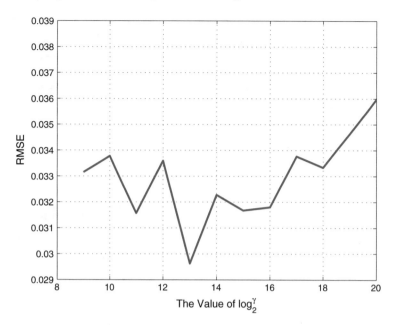

Fig. 2 The training of an optimal γ value

First, we use 7,000 data for the experiment while comparing prediction performance of traditional ELM algorithm and several improved ELM algorithms. For R-ELM, we need to select an optimal γ value. Then we conduct test for γ to evaluate root mean square error (RMSE). In Fig. 2, we find that the optimal γ value is between 10^9 and 10^{20} and we can also see that RMSE first comes down then goes up. And RMSE is minimum when the γ value is 10^{13}. Therefore, we choose 10^{13} as an optimal γ value.

The prediction results are shown in Figs. 3 and 4. In Fig. 3, we can see that the results of ELM-L2, R-ELM, and ELM are almost consistent with the actual value, and they are better than that of Kernel-ELM. Furthermore, we conduct the experiment using 15,000 data and the results are shown in Fig. 4. It should be pointed out that in our experiment Kernel-ELM, R-ELM, and ELM may cause memory overflow when sample size is relatively large. However, ELM-L2 can still work. In Fig. 4, we can see that the prediction result of ELM-L2 is with the trend of the actual curve. Therefore, ELM-L2 algorithm not only can be used in large sample size, but also has a fast training speed and high prediction accuracy.

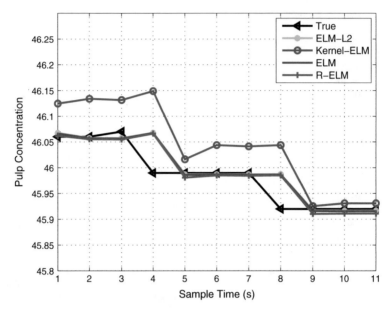

Fig. 3 Prediction results using ELM algorithms with time series of 7,000 pulp concentration data

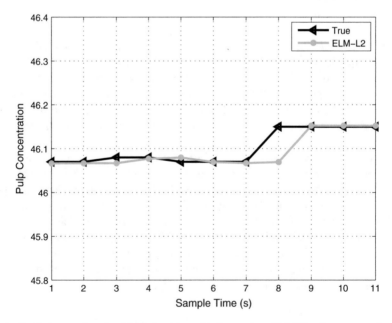

Fig. 4 Prediction results using ELM algorithm with time series of 15,000 pulp concentration data

5 Conclusion

In order to predict the trend of pulp concentration data, this paper uses ELM, Kernel-ELM, R-ELM, and ELM-L2 algorithms to establish time series prediction model of pulp concentration. Due to the unique features of ELM, the neural network learning prediction model used in this paper has few parameters to be adjusted, and it also retains fast convergence speed, strong generalization ability, and better prediction accuracy. However, ELM, R-ELM, and Kernel-ELM may be only suitable for small sample size and the prediction model may be memory overflow for large sample size. Meanwhile, ELM-L2 still has a better prediction effect in large sample size. Therefore, although pulp concentration is not easy to be predicted, we can apply ELM-L2 to do it well, so that we are able to prepare for the unknown changes in the future by revealing change rule of pulp concentration.

References

1. Subrahmanyam, T.V., Forssberg, K.S.: Grinding and flotation pulp chemistry of a low grade copper ore. Miner. Eng. **8**, 913–921 (1995)
2. Tipre, D.R., Dave, S.R.: Bioleaching process for Cu-Pb-Zn bulk concentrate at high pulp density. Hydrometallurgy **75**, 37–43 (2004)
3. Tucker, P.: The influence of pulp density on the selective grinding of ores. Int. J. Miner. Process. **12**, 273–284 (1984)
4. Morkun, V., Morkun, N., Pikilnyak, A.: Ultrasonic testing of pulp solid phase concentration and particle size distribution considering dispersion and dissipation influence. Metall. Min. Ind. **7**, 9–13 (2015)
5. Huang, G., Huang, G.B., Song, S., You, K.: Trends in extreme learning machines: a review. Neural Netw. **61**, 32–48 (2015)
6. Luo, X., Chang, X.H.: A novel data fusion scheme using grey model and extreme learning machine in wireless sensor networks. Int. J. Control Autom. Syst. **13**, 539–546 (2015)
7. Huang, G.B., Ding, X., Zhou, H.: Optimization method based extreme learning machine for classification. Neurocomputing **74**, 155–163 (2010)
8. Huang, G.B., Zhou, H., Ding, X., Zhang, R.: Extreme learning machine for regression and multiclass classification. IEEE Trans. Syst. Man Cybern. Part B Cybern. **42**, 513–529 (2012)
9. Deng, W.Y., Zheng, Q.H., Chen, L.: Regularized extreme learning machine. In: IEEE Symposium on Computational Intelligence and Data Mining, pp. 389–395. IEEE Press, New York (2009)
10. Zhang, H., Liang. J., Chai. Z.: Stock prediction based on phase space reconstruction and echo state networks. J. Algorithms Comput. Technol. **7**, 87–100 (2013)
11. Zhang, W., Ma, Y., Yang, G.: Study on parameter selection of phase space reconstruction for chaotic time series. Adv. Inf. Sci. Serv. Sci. **4**, 67–77 (2012)

Rational and Self-adaptive Evolutionary Extreme Learning Machine for Electricity Price Forecast

Chixin Xiao, Zhaoyang Dong, Yan Xu, Ke Meng, Xun Zhou
and Xin Zhang

Abstract Electricity price forecast is of great importance to electricity market participants. Given the sophisticated time-series of electricity price, various approaches of extreme learning machine (ELM) have been identified as effective prediction approaches. However, in high dimensional space, evolutionary extreme learning machine (E-ELM) is time-consuming and difficult to converge to optimal region when just relying on stochastic searching approaches. In the meanwhile, due to the complicated functional relationship, objective function of E-ELM seems difficult also to be mined directly for some useful mathematical information to guide the optimum exploring. This paper proposes a new differential evolution (DE) like algorithm to enhance E-ELM for more accurate and reliable prediction of electricity price. An approximation model for producing DE-like trail vector is the key mechanism, which can use simpler mathematical mapping to replace the original yet complicated functional relationship within a small region. Thus, the evolutionary procedure frequently dealt with some rational searching directions can make the E-ELM more robust and faster than supported only by the stochastic methods. Experimental results show that the new method can improve the performance of E-ELM more efficiently.

Keywords Approximation model · Differential evolution · Extreme learning machine · E-ELM · Electricity price prediction

C. Xiao (✉) · Y. Xu · K. Meng · X. Zhou
The Centre for Intelligent Electricity Networks, University of Newcastle,
Newcastle, NSW 2308, Australia
e-mail: chixinxiao@gmail.com

C. Xiao
The College of Information Engineering, Xiangtan University,
Xiangtan 411105, China

Z. Dong · X. Zhang
The School of Electrical and Information Engineering, University of Sydney,
Sydney, NSW 2006, Australia
e-mail: zydong@ieee.org

© Springer International Publishing Switzerland 2016
J. Cao et al. (eds.), *Proceedings of ELM-2015 Volume 2*,
Proceedings in Adaptation, Learning and Optimization 7,
DOI 10.1007/978-3-319-28373-9_16

1 Introduction

In past decades, a great of research efforts have been devoted to developing accurate and reliable methods for electricity price prediction [1–6]. Among, several state-of-the-art techniques, e.g., ARIMA [7], GARCH [8], and machine learning methods, the machine learning approaches have attracted the largest research attention due to its strong nonlinear modeling capacity [9, 10]. Although conventional machine learning methods, e.g., ANN [3], SVM [9], can extract the nonlinear relationships out of the input and output dataset, meanwhile, such kind of approaches are often adopted by users, they need time-consuming training, besides that, they mainly use gradient-descent information to direct the training, and they are often deemed as lacking efficient strategies to escape premature convergence or many local minima. Recently, another novel approach called Extreme learning machine (ELM) [11–15] as well as its variants have shown much better performance in respect both of training speed and training accuracy than transitional ones.

However, as for high dimensional problems as in the works [5, 6], ELM is still difficult to find out a satisfied regression or classification results by once calculation, because the inputs need to be considered with more characters than the ordinary ones, even though ELM can calculate output weights fast. E-ELM [16] treats one random matrix of hidden weights as a solution for the corresponding output vector for weights. Better solution leads to better output weights, then lower root mean square error, i.e., a better regression or classification result. The hidden weights can be combined into a single row solution of the objective function of ELM, the optimum can be obtained after an evolutionary procedure. Self-adaptive evolutionary extreme learning machine (SaE-ELM) [17] is a representative method, which can obtain output weights for a single hidden layer feed-forward network (SLFNs) with some promising features. However, in respect to training high dimensional data, SaE-ELM is also time-consuming in evolutionary iterations and seems a bit exhausted. For example, let the data is 100 dimensions and the number of the hidden layers is 10, the dimensionality of the solution individual will then reach 1000. Usually, dimensionality of the power market data or the data for electricity load forecasting is over 100. Thus, faster convergence and better quality of solution are two mandatory objectives should be considered in this paper.

As mentioned above, in conventional neural network, the gradient information provides some rapid exploring guides though often leading to local optima. This motivates us that gradient information perhaps can be properly used in E-ELM to provide some rational directions to accelerate the whole optimization procedure. But the complicated objective function of E-ELM is too difficult to mine the gradient information directly, furthermore, the basic framework of ELM seems to have pushed the gradient approaches out of date. Therefore, this paper proposes a new simple model composing an approximate mapping to simulate the old functional relationship of E-ELM within a comparative small region. Based on the new model, a hybrid DE algorithm is developed to ensure that the new E-ELM not only obtains global optima, the weights, more reliably than those dull gradient methods, but also inherits the rational searching features simultaneously, that is, the new algorithm can approach global optimal region faster than pure stochastic tools while keeping high level quality of the solutions. Thus it can be seen that the reliability of greedy means and their variants are no longer the patent of local optimum or premature convergence, on the contrary, their fast convergence becomes more attractive as long as with a well-design scheme.

The rest parts are organized as follows. Section 2 outlines some related backgrounds. The new approximation model is shown in Sect. 3. In Sect. 4, a new evolutionary algorithm for E-ELM learning high-dimensional data is proposed. The experimental results and discussions are placed in Sect. 5. Finally, a conclusion and future works are provided in Sect. 6.

2 Mathematical Background

This section gives brief reviews of aboriginal extreme learning machine (ELM) as well as some necessary methods used in the rest of this paper for completeness.

2.1 Extreme Learning Machine (ELM) [11]

The basic working mechanism of ELM is briefly generalized as follows.

Given N training samples $\{(x_i, t_i)\}_{i=1}^{N}$ which can be also described in matrix style $\{(P, T_{tar})\}$, where P is a $D \times N$ real matrix of input data and T_{tar} represents $N \times 1$ target vector. H is a $L \times D$ real matrix consisting of the hidden layer parameters generated randomly. β is a $L \times 1$ real vector of output weights. Their mathematical relationship can be expressed as Eq. (1)

$$f(H \cdot P + Bias)^{\mathrm{T}} \cdot \beta = T_{tar} \tag{1}$$

where $Bias$ is a $L \times N$ real matrix and function $f(\cdot)$ is a kind of activation functions [10], for instance, a log-sigmoid function,

$$\sigma(t) = \frac{1}{1 + e^{-c \cdot t}} \tag{2}$$

where c is a slop parameter. Usually, Eq. (1) can be presented in brief as Eq. (3)

$$H \cdot \beta = T_{tar} \tag{3}$$

where $H = f(H \cdot P + Bias)^{\mathrm{T}}$ is a $N \times L$ matrix. EML uses Moore-Penrose pseudoinverse $\widehat{H}^{\dagger}$ and target vector T_{tar} to obtain a least-square solution of such linear system as Eq. (3). That is, a least-square solution of output weight vector β can be analytically determined as Eq. (4)

$$\widehat{\beta} = \widehat{H}^{\dagger} \cdot T_{tar} \tag{4}$$

More details can be found in [11–14]. Instead of following traditional gradient descend approach, ELM minimizes training accuracy or the cost function Eq. (5) via the result gotten by the Eq. (4).

$$RSME = \sqrt{mse(H \cdot \widehat{\beta} - T_{tar})} \tag{5}$$

where $mse(\cdot)$ is the function to measure performance as the mean of absolute errors.

2.2   Basic Differential Evolution Framework

In classical differential evolution (DE) [18] framework, the remarkably simple trial vector generation scheme is a main character distinguished from other EAs. It processes a scaled difference of vectors originating from a fixed-size population of decision vectors. Usually, such three evolutionary operators as mutation, crossover and selection are included respectively. During the gth generation and in the basic DE mutation, a trial vector $u_{i,g}$ is produced by a crossover operation between old individual vector $x_{i,g}$ and a mutated vector $v_{i,g} = x_{r0,g} + F_i \cdot (x_{r1,g} - x_{r2,g})$, where F_i $(F_i > 0)$ is a scaling factor, $x_{r0,g}, x_{r1,g}, x_{r2,g}$ are three independent decision vectors selected randomly from the whole population $P = \{x_{1,g}, x_{2,g}, \ldots, x_{NP,g}\}$ in decision space. For each vector $x_{i,g} \in P$ in turn, there is a corresponding trial vector $u_{i,g}$ being generated. Each old vector $x_{i,g}$ in P will not be replaced unless its trial vector $u_{i,g}$ yields a better objective function value than itself. Consequently $x_{i,g}$ is also called a target vector in literature. More variants of DE can be checked in detail in [19–25].

2.3   SaE-ELM

Self-adaptive evolutionary extreme learning machine (SaE-ELM) [17] is upgraded from DE-LM [26] and E-ELM [16], which chooses trial vector generation strategies and some relative control parameters adaptively. Their common place is to explore the network input weights and hidden node biases of ELM aiming to get optimum of the network output weights. When training data set $X_{D \times N}$, L hidden layers and an activation function $f(\cdot)$ are given, the individuals to be evolved during the gth generation can be coded into as following vector [17],

$$\theta_{k,g} = (h_{11}^g, \ldots, h_{1D}^g, h_{21}^g, \ldots, h_{2D}^g, \ldots, h_{L1}^g, \ldots, h_{LD}^g, b_1^g, \ldots, b_L^g),$$

where $1 \leq k \leq NP$, NP is the population size, b_i^g, $1 \leq i \leq L$, represents the bias value for the ith hidden layer in g generations.

Based on the coding format, the parameters like $H, Bias$ are obtained as follows,

$$H = \begin{bmatrix} h_{11}^g, \ldots, h_{1D}^g \\ h_{21}^g, \ldots, h_{2D}^g \\ \vdots \\ h_{L1}^g, \ldots, h_{LD}^g \end{bmatrix}, P = X_{D \times N}, Bias = \begin{bmatrix} b_1^g \\ b_2^g \\ \vdots \\ b_L^g \end{bmatrix} \times J_{1 \times N} \qquad (6)$$

where $J_{1 \times N}$ is a one row and N columns matrix of ones. Then the corresponding fitness function is formulated as Eq. (7),

$$RSME = \sqrt{mse(f(H \cdot P_{test} + Bias)^{\mathrm{T}} \cdot \widehat{\beta} - T_{test})} \qquad (7)$$

where P_{test} and T_{test} are testing data set and testing target vector respectively.

The main aim of such kind of algorithms is to explore an optimum of H from population consisted of $\theta_{k,g}$ $(1 \leq k \leq NP)$ during g_{max} generations. The strategy for surviving can refer to [17].

3 Approximation Model

Although functional mapping based on the approximation model is not very accurate to replace the original functional relationship over whole hyper-plane, it can absolutely satisfy those practical demands within a limited region [27].

3.1 First-Order Approximation Model

Without loss of generality, a decision space can be formulated as a hyper-plane by one point attached with two vectors. Let $x \in \mathbb{R}^n$ is an arbitrary point in decision space $\mathbb{R}^n$ or the point can be denoted as a decision vector $(x_1, x_2, \ldots, x_n)^{\mathrm{T}}$, L is the hyper-plane, suppose $x^0 \neq x^1 \neq x^2$ are three distinct points selected randomly among $\mathbb{R}^n$, then any arbitrary point $x \in L$ can be formulated as such style as Eq. (8)

$$x = x^0 + t_1 \cdot (x^1 - x^0) + t_2 \cdot (x^2 - x^0) \tag{8}$$

where t_1, t_2 are two independent real variables.

According to Eq. (8), any $x \in L$ is linear corresponding to the variable vector (t_1, t_2) because rest parameters are constants, i.e., $x \Leftrightarrow (t_1, t_2)$, if and only if three arbitrary yet independent points $x^0 \neq x^1 \neq x^2$ have been fixed. In other words, if $x^0 \neq x^1 \neq x^2$ are located, any $x \in L$ can be evaluated based on variable vector (t_1, t_2) and Eq. (8). Therefore, when decision vector x approaches its optimum, x^*, there must exist a corresponding variable vector $(t_1^*, t_2^*) \Leftrightarrow x^*$, i.e.,

$$x^* = x^0 + t_1^* \cdot (x^1 - x^0) + t_2^* \cdot (x^2 - x^0) \tag{9}$$

Likewise, for any pair of fitness function $f(x)$ and its variable x, there has another pair of image $g(\cdot)$ and its variable vector (t_1, t_2). Their common place is $g(t_1, t_2) = f(x)$, while the difference is the functional relationship of $g(\cdot)$ is simpler than the one of $f(\cdot)$. The conversion relationship between $g(\cdot)$ and $f(\cdot)$ is defined as Eq. (10)

$$g(t_1, t_2) = f(x) = g^0 + t_1 \cdot (g^1 - g^0) + t_2 \cdot (g^2 - g^0) \tag{10}$$

where g^0, g^1, g^2, can be dealt with as constants if $x^0 \neq x^1 \neq x^2$ have been fixed as mentioned above. In order to obtain the constants, g^0, g^1, g^2 simply, some special points are considered here. Assume (t_1, t_2) is substituted by vectors, (0, 0), (1, 0), (0, 1) respectively, then $g^0 = f(x_0)$, $g^1 = f(x_1)$, $g^2 = f(x_2)$ can be easily extracted out via Eqs. (10) and (8). Equation (10) hereby provides an approximation equation as well to replace the original fitness function since $g(t_1, t_2) = f(x)$. Till this step, the complicated functional relationship between the decision variable $x \in \mathbb{R}^n$ and its original image $f(x)$ has been estimated via the new mapping between $g(\cdot)$ and (t_1, t_2).

3.2 Direction to Optimum

In fact, Eq. (10) also provides a linear functional relationship between variable vector (t_1, t_2) and its image $g(t_1, t_2)$. Through conventional optimization theories, $g(t_1, t_2)$ at point (t_1, t_2) has a vector of first partial derivatives, or gradient vector

$\nabla g(t_1, t_2) = ((g^1 - g^0), (g^2 - g^0))$. Hence, the local minimum optimum of (t_1^*, t_2^*) is most probably being placed in the opposite direction of $\nabla g(t_1, t_2)$.

$$(t_1^*, t_2^*) = (0, 0) - \alpha \cdot \nabla g(t_1, t_2) = -\alpha \cdot \nabla g(t_1, t_2) \tag{11}$$

where α is a step parameter. In one word, any three distinct decision variables, $x^0 \neq x^1 \neq x^2$, can deduce out the local optimum x^* via Eq. (9–11), which can be expressed as Eq. (12),

$$x^* = x^0 - \alpha \cdot [(g^1 - g^0) \cdot (x^1 - x^0) + (g^2 - g^0) \cdot (x^2 - x^0)] \tag{12}$$

4 Proposed Algorithm

In order to balance global exploration and local exploitation, another DE mutation strategies, 'DE/current-to-best/1' [18], is enrolled as well to construct a hybrid rational and self-adaptive mutation strategy named RSM mutation just as shown in Fig. 1.

```
function RSM-Trial()
input: x_{r0,g} ≠ x_{r1,g} ≠ x̂_{r2,g}, x_{i,g}, x^P_{best,g} (one of  the P
best individuals in current population, P=5 in
this paper)
output: two trial vectors u^1_{i,g}, u^2_{i,g}
g^0 = f(x_{r0,g}); g^1 = f(x_{r1,g}); g^2 = f(x̂_{r2,g}) ;
t_1 = -(g^1 - g^0); t_2 = -(g^2 - g^0);
s - Step_{i,g}/√(t_1^2 + t_2^2) ;
v^1_{i,g} = x_{r0,g} + s · [t_1 · (x_{r1,g} - x_{r0,g}) + t_2 · (x̂_{r2,g} - x_{r0,g})] ;
v^2_{i,g} = x_{r0,g} + F_i · (x^P_{best,g} - x_{i,g}) + F_i · (x_{r0,g} - x_{r1,g});
for j =1 to D
for k =1 to 2
if (j = k_{rand}) or rand(0,1) <CR^k_i
u^k_{j,i,g}=v^k_{j,i,g}
else
u^k_{j,i,g}=x^k_{j,i,g}
end if
end for
end for
end func
```

Fig. 1 Pseudo-code of producing hybrid trial vectors

4.1 Diversity Mechanism

Similar to JADE [20], RSM mutation applies a historical pool to temporarily reserve a part of individuals sifted out from the population. Each time, one of three distinct individual is picked out the union of current population and the historical pool, denoted by $\hat{x}_{r2,g}$, while the others $x_{r0,g}$, $x_{r1,g}$ are still selected from the current population. The size of the historical pool is set to a quarter of the population and the initial state is empty. After being full, the pool permits the individual perished from current population to replace the worst one if the perished one is better.

Motivated by [20, 23], many control parameters in the new algorithm are extended into solution individuals for controlling self-adaptively (see Table 1). The parameters are evolved simultaneously whilst the classical population of solutions is being processed in evolution procedure. Main parameters for self-adaptive control, such as $Step_{i,g} \in [0,2]$, $CR_i^1, CR_i^2 \in [0,0.9]$ and $F_{i,g} \in [0.1,1.0]$, are initialized within their definition domain. The successful parameters survive to the next generation, while the unsuccessful ones are replaced by a normal distribution of the mean $P_{m,g}$ and standard deviation $sigma$ as shown in Eq. (13).

$$P_{i,g} = P_{m,g} + sigma \cdot randn_{i,g} \tag{13}$$

where $P_{i,g}$ represents the variable of parameters for the ith individual in g generation. The sigma of each parameter equals to $\min(|P_{m,g} - P_{Ub,g}|, |P_{m,g} - P_{Lb,g}|)$. The mean values are initialized as follows, $Step_{m,1} = 1.1$, $F_{m,1} = 0.6$, $CR_i^k = 0.6$, $(k=1,2)$. Parameter $Step_{i,g}$ controls the incremental degree of the mutation. In Fig. 1, $v_{i,g}^1 = x_{r0,g} + s \cdot [t_1 \cdot (x_{r1,g} - x_{r0,g}) + t_2 \cdot (\hat{x}_{r2,g} - x_{r0,g})]$, where $s = Step_{i,g}/\sqrt{t_1^2 + t_2^2}$. At the beginning of whole evolving procedure, $Step_i \geq 1$ helps population converge to optimum fast, while $Step_i < 1$ is good at effective exploitation, especially for solutions approaching to the optimum.

4.2 Hybrid Strategy for Selection

In the procedure of selection, one in two new trail vectors is picked up

- Case 1: $f(x_{i,g}) < f(u_{i,g}^1) < f(u_{i,g}^2)$
  Both two trail vectors are successful trail ones, i.e., success(i,1) = 1, success (i,2) = 1, all their parameters can be kept to the next generation.

Table 1 Encoding format of self-adapting individuals

$x_{1,g}$	$Step_{1,g}$	$CR_{1,g}^1$	$F_{1,g}$	$CR_{1,g}^2$
...	...	...	...	...
$x_{NP,g}$	$Step_{NP,g}$	$CR_{NP,g}^1$	$F_{NP,g}$	$CR_{NP,g}^2$

– Case 2: $f(u^1_{i,g}) < f(x_{i,g}) < f(u^2_{i,g})$

$u^1_{i,g}$ is named as a successful trail vector and success(i,1) is set to 1.

– Case 3: $f(u^1_{i,g}) < f(u^2_{i,g}) < f(x_{i,g})$

This case means all the parameters need to be adjusted. As similar as above, the converse is also followed the rules.

At end of each generation, the mean of each parameter is adjusted by Eq. (14)

$$P_{i,g+1} = 0.85 \cdot P_{m,g} + 0.15 \cdot mean(P_{success,g}) \tag{14}$$

where mean(.) is a function of arithmetic mean.

4.3 RSM-DE Algorithm

The main body of RSM-DE algorithm:

Input   *NP*: the size of the population;
          Maxgen: the number of the maximum iteration;
          Fitness function;
          D: The dimension of decision space.

Output Optima of the fitness function

Step 1 Initialization
Create a random initial population $\{x_{i,0} | i = 1, \ldots, NP\}$. Initialize parameters within their definition regions.

  For $g = 1, \ldots, Maxgen,$ **do**
  Step 2 Evolution Items
  For $i = 1, \ldots, NP$ **do**
  Step 2.1 New Parameters Generating: Unsuccessful parameters are refreshed based on Eq. (14).
  Step 2.2 Mating: One of the P best individuals and other three independent individuals, $x^P_{best,g}, x_{r0,g} \neq x_{r1,g} \neq \hat{x}_{r2,g}$, are picked out. $\hat{x}_{r2,g}$ is from the union of current population plus historical pool and $x^P_{best,g}$ is one out of from current population, $P = 5$ in this paper.
  Step 2.3 Call Function RSM-Trial(): To produce two trail vectors by two strategies respectively.
  Step 2.4 Call Function Selection(): To select successful trail vectors and parameters into the next generation.
  Step 2.5 Renew Historical Pool: If the historical pool is not full then the eliminated individuals are pushed into the pool, otherwise the worst one in the pool is replaced when the eliminated on is better.
  Step 2.6 Summarize the Statistical Result of Successful Trail Vectors: To evaluate the arithmetical mean value of each parameter by Eq. (14).

Fig. 2 Pseudo-code of
RSM-DE-ELM

```
function Optimal_Layer()
input: [L₁,L₂],Train,Test
output: Optimum of Lbest,Fitnessmin
m₁ = RSM_DE_ELM(Train,Test,L₁) ;
m₂ = RSM_DE_ELM(Train,Test,L₂) ;
while(L₁!=round(((L₁+L₂)/2+0.1) or
L₁!=round(((L₁+L₂)/2-0.1))
   m₃ = RSM_DE_ELM(Train,Test,round((L₂
                           + L₁)/2 ));
if(m₁>m₂)
swap(m₁,m₂), swap(L₁,L₂);
endif
m₂ = m₃;   L₂ = round((L₁ + L₂)/2);
if(m₁>m₂)
swap(m₁,m₂); swap(L₁,L₂);
endif
endwhile
Lbest = L₁; Fitnessmin = m₁;
End func
```

Step 3 Stopping criteria

When stopping criterion is satisfied, the algorithm stops here and outputs corresponding results. Otherwise, goes to **Step 2**.

4.4 RSM-DE-ELM

Given a set of training data, a set of testing data, a candidate range for L hidden layers and an objective function $g(\cdot)$, RSM-DE-ELM algorithm is summarized as following Fig. 2. $RSM\_DE\_ELM(\cdot)$ represents a procedure to optimize ELM based on the RSM-DE algorithm. It returns the optimum of the net. $[L_1, L_2]$ is the candidate range and $Train, Test$ denote training set, testing set respectively.

5 Experimental Results

In this section, several sequential data series extracted from the Australian Energy Market Operator (AEMO) website [28] are used to test the performance of our new method.

For a convenient comparison, the first dataset in our case study and dataset format are a whole year's RRPs from QLD market just as [5]. It includes total of 17,520 observations and the period crosses over 01 June, 2006 to 31 May, 2007.

Parameter setting for RSM-DE-ELM:

The population size $NP = 100$ and maximum generation is 60. Since the iteration number of RSM-DE is not high, so the mean values of the parameters are initialized as, $Step_{m,1} = 1.1$, $F_{m,1} = 0.7$, $CR_{m,1}^k = 0.75$, $(k = 1, 2)$. The candidate range $[L_1, L_2]$ of hidden layers L is set to $[10,150]$.

Table 2 Comparison of four methods on RRP forecast

Season		Model	MAE	MAPE (%)	RMSE
Training	W	I	1.1566	5.1936	1.5225
		II	0.8345	3.8910	**1.1430**
		III	0.9458	4.3466	1.3544
		IV	**0.80145**	**2.9366**	1.1698
	S	I	1.3195	6.2802	1.7272
		II	**1.0048**	4.7648	**1.3311**
		III	1.2548	5.9897	1.6882
		IV	1.1896	**4.6761**	1.35257
	Su	I	4.4040	15.3203	6.0542
		II	**3.1721**	11.4565	**4.3307**
		III	3.7803	12.3331	5.5882
		IV	3.2682	**7.4954**	4.4953
	A	I	5.3081	9.6121	6.7923
		II	**3.9860**	7.6349	**5.1718**
		III	5.5116	10.0494	6.8658
		IV	5.1377	**7.43097**	5.7401
Testing	W	I	2.3611	9.9423	3.3470
		II	2.1046	8.5440	3.0537
		III	**2.0278**	8.3372	**2.9371**
		IV	2.1365	**6.0780**	2.9835
	S	I	2.2337	9.9291	3.1190
		II	2.6382	11.5712	3.6026
		III	2.3021	10.2642	3.1781
		IV	**1.9443**	**9.1967**	**3.0549**
	Su	I	10.9983	24.4636	17.2313
		II	10.8783	22.7230	17.7526
		III	**10.1656**	21.8798	16.5881
		IV	10.1797	**21.1376**	**16.2054**
	A	I	7.8198	13.7900	10.7256
		II	7.9618	13.5447	11.4401
		III	7.3193	12.7363	10.3818
		IV	**7.0116**	**11.9539**	**10.2671**

W Winter, *S* Spring, *Su* Summer, *A* Autumn
I BPNN, *II* RBFNN, *III* ELM, *IV* RSM-DE-ELM
Bold value represent the best value among *I* to *IV* methods by different criteria in different seasons

Results Analysis:

Table 2 shows the comparisons between the new approach with three existing methods [5]. All these results are the mean values collected by multiple trails which include 50 independent forecasts of each season model. From Table 2, the new algorithm wins most of the lowest testing criteria in four season dataset among all these four approaches. For testing in Spring and Autumn, the performances have been dramatically improved. For example, the testing MAE of Spring using RSM-DE-ELM is 1.9443, while the other three methods, the testing MAEs of this season dataset are all greater than 2.2000.

Figure 3. gives out the prediction results run by RSM-DE-ELM on first half of 7*48 observation points belong to testing dataset. The error curve shows the new algorithm can forecast with low and stable error rate in most points.

In terms of the training time, due to our approach falls into E- ELM category, the training procedure practically consists of several sub-trainings of basic ELM, thus it takes longer time in training than one single basic ELM. However, the proposed approach is definitely faster than SaDE-ELM [17] because only two basic DE strategies are included rather than four in SaDE-ELM. Secondly, rational DE model provides our method fast convergence in addition to promising experimental results, e.g., RSM-DE-ELM can get better results within 60 generations whilst SaDE-ELM need 100 more generations to reach the same magnitude. What's more, our approach is no longer running on the way mentioned in the previous literature [16, 17], in which the number of hidden layers is often gradually increased and the

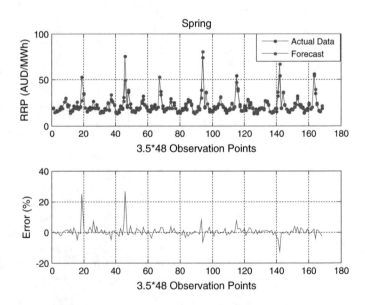

Fig. 3 Average RRP of forecast by RSM-DE-ELM in spring

one with the best generalization performance is adopted in final. In our proposed approach, the binary search frame helps the algorithm not only find the optimum at last, but also keep in less time complexity.

6 Conclusion

In this paper, a self-adaptive DE frame embedding with a rational approximation operator is proposed intending to optimize E-ELM fast and stably. Experimental results illustrate some mathematical auxiliary guides for evolving optima can bring optimization procedure more reliable performances than stochastic strategies do, as long as the design is a slight proper.

Acknowledgment This work is supported in part by the Australian Research Council (ARC) through a Linkage Project (grant no. 120100302), in part by the University of Newcastle through a Faculty Strategic Pilot Grant, in part by the Research Foundation of Education Bureau of Hunan Province, China (Grant No. 14A136). The author would like to thank Prof. Qingfu Zhang (UK and Hongkong) for fruitful discussions and patient tutoring. Thank Dr. Jingqiao Zhang for providing the source code of JADE.

References

1. Zhang, R., Dong, Z.Y., Xu, Y., Meng, K., Wong, K.P.: Short-term load forecasting of Australian National Electricity Market by an ensemble model of extreme learning machine. IET Gen. Trans. Dist. **7**(4), 391–397 (2013)
2. Wan, C., Xu, Z., Pinson, P., Dong, Z.Y., Wong, K.P.: Probabilistic forecasting of wind power generation using extreme learning machine. IEEE Trans. Power Syst. **29**(3), 1033–1044 (2014)
3. Meng, K., Dong, Z.Y., Wong, K.P.: Self-adaptive RBF neural network for short term electricity price forecasting. IET Gen. Trans. Dist. **3**(4), 325–335 (2009)
4. Pindoriya, N.M., Singh, S.N., Singh, S.K.: An adaptive wavelet neural network-based energy price forecasting in electricity markets. IEEE Trans. Power Syst. **23**(3), 1423–1432 (2008)
5. Chen, X., Dong, Z.Y., Meng, K., Xu, Y., Wong, K.P., Ngan, H.W.: Electricity price forecasting with extreme learning machine and bootstrapping. IEEE Trans. Power Syst. **27**(4), 2055–2062 (2012)
6. Wan, C., Xu, Z., Pinson, P., Dong, Z.Y., Wong, K.P.: A hybrid artificial neural network approach for probabilistic forecasting of electricity price. IEEE Trans. Smart Grid **5**(1), 463–470 (2014)
7. Conejo, A.J., Plazas, M.A., Espinola, R., Molina, A.B.: Day-ahead electricity price forecasting using the wavelet transform and ARIMA models. IEEE Trans. Power Syst. **20**(2), 1035–1042 (2005)
8. Garcia, R.C., Contreras, J., Akkeren, M.V., Garcia, J.B.C.: A GARCH forecasting model to predict day-ahead electricity prices. IEEE Trans. Power Syst. **20**(2), 867–874 (2005)
9. Bishop, C.M., et al.: Pattern Recognition and Machine Learning, vol. 1. Springer, New York (2006)
10. Goldberg, D.E., Holland, J.H.: Genetic algorithms and machine learning. Mach. Learn. **3**(2), 95–99 (1988)

11. Li, M.-B., Huang, G.-B., Saratchandran, P., Sundararajan, N.: Fully complex extreme learning machine. Neurocomputing **68**, 306–314 (2005)
12. Huang, G.B., Chen, L., Siew, C.K.: Universal approximation using incremental constructive feedforward networks with random hidden nodes. IEEE Trans. Neural Netw. **17**(4), 879–892 (2006)
13. Huang, G.B., Chen, L.: Convex incremental extreme learning machine. Neurocomputing **70** (16–18), 3056–3062 (2007)
14. Huang, G.B., Chen, L.: Enhanced random search based incremental extreme learning machine. Neurocomputing **71**(16–18), 3460–3468 (2008)
15. Feng, G., Huang, G.B., Lin, Q., Gay, R.: Error minimized extreme learning machine with growth of hidden nodes and incremental learning. IEEE Trans. Neural Netw. **20**(8), 1352–1357 (2009)
16. Zhu, Q.-Y., Qin, A.K., Suganthan, P.N., Huang, G.-B.: Evolutionary extreme learning machine. Pattern Recogn. **38**(10), 1759–1763 (2005)
17. Cao, J., Lin, Z., Huang, G.-B.: Self-adaptive evolutionary extreme learning machine. Neural Process. Lett. **36**, 285–305 (2012)
18. Storn, R., Price, K.: Differential evolution–a simple and efficient heuristic for global optimization over continuous spaces. J. Global Optim. **11**(4), 341–359 (1997)
19. Das, S., Suganthan, P.N.: Differential evolution: a survey of the state-of-the-art. IEEE Trans. Evol. Comput. **15**(1), 4–31 (2011)
20. Zhang, J., Sanderson, A.C.: JADE: adaptive differential evolution with optional external archive. IEEE Trans. Evol. Comput. **13**(5), 945–958 (2009)
21. Brest, J, Greiner, S., Boskovic, B., Mernik, M., Zumer, V.: Self-adapting control parameters in differential evolution: a comparative study on numerical benchmark problems. IEEE Trans. Evol. Comput. **10**(6), 646–657 (2006)
22. Abbass, H.A.: The self-adaptive pareto differential evolution algorithm. In: Proceedings of the 2002 Congress on Evolutionary Computation, 2002, CEC'02, vol. 1, pp. 831–836 (2002)
23. Qin, A.K., Suganthan, P.N.: Self-adaptive differential evolution algorithm for numerical optimization. In: The 2005 IEEE Congress on Evolutionary Computation, vol. 2, pp. 1785–1791 (2005)
24. Brest, J., Greiner, S., Boskovic, B., Mernik, M., Zumer, V.: Self-adapting control parameters in differential evolution: a comparative study on numerical benchmark problems. IEEE Trans. Evol. Comput. **10**(6), 646–657 (2006)
25. Das, S., Konar, A., Chakraborty, U.K.: Two improved differential evolution schemes for faster global search. In: Proceedings of the 2005 Conference on Genetic and Evolutionary Computation, 2005, pp. 991–998
26. Subudhi, B., Jena, D.: Differential evolution and Levenberg Marquardt trained neural network scheme for nonlinear system identification. Neural Process. Lett. **27**(3), 285–296 (2008)
27. Montgomery, D.C.: Design and Analysis of Experiments. Wiley.com, p. 405 (2006)
28. Australian Energy Market Operator (AEMO), www.aemo.com.au

Contractive ML-ELM for Invariance Robust Feature Extraction

Xibin Jia and Hua Du

Abstract Extreme Learning Machine (ELM), a single hidden layer feedforward neural networks works efficiently in many areas such as machine learning, pattern recognition, natural language processing, et al. due to its powerful universal approximation capability and classification capability. This paper uses multiply layer ELM (ML-ELM) which stacks many ELMs based on Auto Encoder (ELM-AE) as main framework. ELM-AE lets the input data as output data and chooses orthogonal random weights and random biases of the hidden nodes to perform unsupervised learning. To extract more invariance robust feature, we propose Contractive ML-ELM (C-ML-ELM referring to the work of Rifai et al.). Contractive ML-ELM applys a penalty term corresponding to the Frobenius norm of the Jacobian matrix of the encoder activations with respect to the input in each layer of ML-ELM. Experiment results show that Contractive ML-ELM achieves state of the art classification error on Mnist dataset.

Keywords Extreme Learning Machine · Contractive Auto Encoder · Multi-Layer ELM

1 Introduction

Extreme Learning Machine (ELM) is proposed by Huang et al. [1] for single-hidden layer feedforward neural network (SLFN) that has two powerful capability: universal approximation capability and classification capability [2]. ELM uses nonlinear

X. Jia (✉) · H. Du
Beijing Municipal Key Laboratory of Multimedia and Intelligent
Software Technology, Beijing Key Laboratory on Integration
and Analysis of Large-scale Stream Data, Beijing University of Technology,
Beijing 100124, China
e-mail: jiaxibin@bjut.edu.cn
URL: http://www.bjut.edu.cn

H. Du
e-mail: dhgacky@gmail.com

© Springer International Publishing Switzerland 2016
J. Cao et al. (eds.), *Proceedings of ELM-2015 Volume 2*,
Proceedings in Adaptation, Learning and Optimization 7,
DOI 10.1007/978-3-319-28373-9_17

piecewise continuous random hidden nodes such as Sigmoid nodes and RBF nodes to obtain universal approximation capability and classification capability. The input-hidden weights w and the hidden biases b of ELM are randomly generated independently from the training data and the output weights β are analytically computed. ELM is also a fast learning speed algorithm.

Multi-Layer Extreme Learning Machine (ML-ELM) [3] stacks many Extreme Learning Machines based on Auto Encoder [4] (ELM-AE) initializing the hidden layer weights. ELM-AE [3] performs layer-wise unsupervised training by applying the input data as output data and choosing orthogonal random weights and random biases of the hidden nodes. ML-ELM's performance is better than auto encoders based deep networks and Deep Belief Networks (DBN) for MNIST dataset. ML-ELM is yet significantly faster than any state-of-the-art deep networks due to not requiring fine tuning using BP.

However, either ELM or ML-ELM is based on optimize the model under the assumption with minimized prediction error and parameter energy. Neither of them consider of the variance of data. In practical problem, data is normally changing consecutively rather than extremely. Therefore, finding a solution of the trained model has good performance to fit this consecutive characteristic is benefit for improving the performance of algorithm.

In this paper, we propose Contractive ML-ELM (C-ML-ELM) to extract invariance robust features, referring to the work of Rifai et al. [5]. The method applies a penalty term corresponding to the Frobenius norm of the Jacobian matrix of the encoder activations with respect to the input in each layer of ML-ELM. Experiments show that Contractive ML-ELM achieves higher performance against other classification algorithms on Mnist and other datasets.

The paper is organized as follows. The first section of this paper introduces the Multi-Layer ELM, next section shows our proposed method Contractive ML-ELM, and last section gives the experiments results.

2   Background of Multi-Layer ELM

2.1   ELM

ELM [1] is a SLFN which randomly chooses the input weights and the hidden biases and analytically determines the output weights. This algorithm can provide the best generalization performance at extremely fast learning speed due to not requiring iterative training process.

ELM with L hidden nodes can be represented by the following equation:

$$f_L(\mathbf{x}) = \sum_{i=1}^{L} G_i(\mathbf{x}, \mathbf{w}_i, b_i) \cdot \beta_i, \mathbf{w}_i \in \mathbb{R}^d, b_i, \beta_i \in \mathbb{R} \tag{1}$$

where $G_i(\cdot)$ denotes the ith hidden node activation function, w_i is the input weight vector connecting the input layer to the ith hidden layer, b_i is the bias weight of the ith hidden layer, and β_i is the ith output weight vector. $G_i(\cdot)$ using sigmoid s as activation function is defined as follows:

$$G_i(\mathbf{x}, \mathbf{w}_i, b_i) = s(\mathbf{w}_i \cdot \mathbf{x} + b_i) \tag{2}$$

ELM theory aims to reach the smallest training error but also the smallest norm of output weights:

$$\textbf{Minimize}: \quad \lambda \|\mathbf{H}\beta - \mathbf{T}\|_2^2 + \|\beta\|_2^2 \tag{3}$$

where $\mathbf{H}$ is the hidden layer output matrix, $\mathbf{H} = [G_1(\mathbf{x}, \mathbf{w}_1, b_1), \ldots, G_L(\mathbf{x}, \mathbf{w}_L, b_L)]$. $\mathbf{T}$ is prelabeled target matrix $\mathbf{T} = [t_1^T, \ldots, t_N^T]^T$ of output in the training dataset.

The output weights are trained based on the smallest norm least squares solution using the Moore-Penrose generalized inverse [1] of the linear problem Eq. 3 as follows.

$$\hat{\beta} = \left(\frac{I}{C} + \mathbf{H}^T\mathbf{H}\right)^{-1} \mathbf{H}^T\mathbf{T} \tag{4}$$

2.2 Auto Encoder

Auto Encoder [4] has two parts: encoder and decoder. The encoder is a function f that maps an input $\mathbf{x} \in \mathbb{R}^{d_x}$ to hidden representation $h(\mathbf{x}) \in \mathbb{R}^{d_h}$

$$h(\mathbf{x}) = s_f(w\mathbf{x} + b_h) \tag{5}$$

where s_f typically use the logistic sigmoid $s(z) = \frac{1}{1+e^{-z}}$

The decoder function g represented by $\hat{\mathbf{x}}$ maps hidden representation h back to a reconstruction y

$$\hat{\mathbf{x}} = g(\mathbf{h}) = s_g(w^T\mathbf{h} + b_y) \tag{6}$$

where s_g typically either the identity or a logistic sigmoid.

Auto Encoder minimizes loss function

$$J_{AE}(\theta) = \sum_{x \in D} L(\mathbf{x}, \hat{\mathbf{x}}) \tag{7}$$

to fit the optimized model parameters $\theta = \{W, b_h, b_y\}$. $L(\mathbf{x}, \hat{\mathbf{x}})$ typically uses the second order norm: $\|\mathbf{x} - \hat{\mathbf{x}}\|_2^2$.

2.3   Multi-Layer ELM-AE

ELM-AE [3] is a technique by modified AE to perform unsupervised learning that input data is used as output data $t = x$, random weights and random biases of the hidden nodes are chosen to be orthogonal. The output weight is

$$\hat{\beta} = \left(\frac{I}{C} + \mathbf{H}^T \mathbf{H} \right)^{-1} \mathbf{H}^T \mathbf{x} \tag{8}$$

for sparse and compressed ELM-AE representations. The only difference of Eqs. 8 and 4 is last variable, $\mathbf{x}$ instead of $\mathbf{T}$. For equal dimension ELM-AE representations, output weights are calculated by the Singular Value Decomposition (SVD) [3].

In ML-ELM, each hidden layer weights are initialized using ELM-AE which performs layer wise unsupervised training. However in contrast to other deep networks such as DBN [6] and DBM [7], ML-ELM doesn's require fine tuning using BP.

3   Contractive ML-ELM

Contractive ML-ELM uses method Contractive Auto-Encoder (CAE) [5] to extract invariance robust feature.

3.1   Contractive Auto-Encoder

CAE is a deterministic Auto Encoder to find invariance feature by adding a penalty term, the Frobenius norm of the Jacobian matrix of the encoder activations. Rifai et al. [5] proof that CAE has the advantages for representations to be locally invariant in many directions of change of the raw input.
The loss function of CAE is

$$J_{CAE}(\theta) = \sum_{x \in D} \left(L(x, \hat{x}) + \lambda \| J_f(x) \|_F^2 \right) \tag{9}$$

where

$$\| J_f(x) \|_F^2 = \sum_{ij} \left(\frac{\partial h_j(x)}{\partial x_i} \right) \tag{10}$$

3.2   Contractive-ML-ELM

In Contractive-ML-ELM, we make use of Contractive Auto-Encoder instead of traditional Auto-Encoder in each ELM-AE layer to achieve both fast learning capability

Table 1 Test Accuracy % on Mnist

C-ML-ELM	ML-ELM	ELM	DBN	SAE
99.775 %	99.037 %	98.595 %	98.870 %	98.623 %

and robust generalization capability. According to Eqs. 3 and 9, Contractive-ML-ELM represents as:

$$\text{Minimize}: \quad \|\mathbf{H}\beta - \hat{\mathbf{x}}\|_2^2 + \lambda\|J_f(\mathbf{x})\|_F^2 \tag{11}$$

Using the same method of computing Eq. 4, the output weights of Contractive-ML-ELM is trained as

$$\hat{\beta} = \left(\frac{I}{C} + \mathbf{H}^T\mathbf{H} + \lambda\|\frac{\partial\mathbf{H}}{\partial\mathbf{x}}\|\right)^{-1}\mathbf{H}^T\mathbf{x} \tag{12}$$

Computing $\|\partial\mathbf{H}/\partial\mathbf{x}\|$ divides two steps. Firstly, we compute the norm of partial derivative of each columns of $\mathbf{H}$ with $\mathbf{x}$; Secondly, combine the result of first step to a diagonal matrix.

4 Experiments and Results

The MNIST [8] of handwritten digits is a commonly used dataset for testing performance of machine learning and pattern recognition algorithms in current papers, which has 60000 training samples and 10000 testing samples with 28×28 size of each image. We test the proposed algorithm C-ML-ELM on the original MNIST without any preprocessing comparing with ML-ELM, ELM, SAE (sparse auto-encoder) and DBN. The performance results are shown in Table 1. The experiments were performed on a workstation with a core Intel(R) Xeon(R) CPU E5-2687W 3.10 GHz processor and 32.0 GB RAM. The structure of deep network of C-ML-ELM is 784-700-700-15000-10 with ridge parameters $[10e-1, 0, 1e3, 1e8]$ same as the one of ML-ELM used by Kasun et al. [3] As we can see, C-ML-ELM has best performing against other algorithms.

5 Conclusion

This paper proposes a classification algorithm called contractive ML-ELM, which aims to provide a solution to make sure the consecutiveness of data by applying a penalty term corresponding to the Frobenius norm of the Jacobian matrix of input in each layer based on ML-ELM. In ML-ELM, many ELMs based on Auto Encoder(ELM-AE) stack a multi-layer network. In each layer, ELM-AE lets the input data as output data and chooses orthogonal random weights and random biases of the hidden nodes to perform unsupervised learning. By adding the consecutive constraint in the loss function, it is benefit for extract invariance robust feature to improving the classification results. Experiments show the method comparing with the ML-ELM and deep learning methods of DBN achieves state of the art classification error on Mnist dataset. The further work will be done in performance testing on problem of face recognition, expression recognition. It has prominent future in these application by adding the constraint of consecutive. The effectiveness performance will be evaluated in the future work.

Acknowledgments This research is partially sponsored by Natural Science Foundation of China (Nos. 61175115 and 61370113), Beijing Municipal Natural Science Foundation (4152005), Specialized Research Fund for the Doctoral Program of Higher Education (20121103110031), the Importation and Development of High-Caliber Talents Project of Beijing Municipal Institutions of 2014 (No. 067145301400), Jing-Hua Talents Project of Beijing University of Technology (2014-JH-L06), and the International Communication Ability Development Plan for Young Teachers of Beijing University of Technology (No. 2014-16).

References

1. Huang, G.B., Zhu, Q.Y., Siew, C.K.: Extreme learning machine: theory and applications. Neurocomputing **70**(1), 489–501 (2006)
2. Huang, G.B.: What are Extreme Learning Machines? Filling the gap between Frank Rosenblatt's Dream and John von Neumanns Puzzle. Cognitive Comput. **7**(3), 263–278 (2015)
3. Kasun, L.L.C, Zhou, H., Huang, G.B., Vong, C.M. Representational learning with extreme learning machine for big data. IEEE Intell Syst. **28**(6), 31C4 (2013)
4. Bengio, Y.: Learning deep architectures for AI. Found. trends Mach. Learn. **2**(1), 1–127 (2009)
5. Rifai, S., Vincent, P., Muller, X., Glorot, X., Bengio, Y.: Contractive auto-encoders: Explicit invariance during feature extraction. In: Proceedings of the 28th International Conference on Machine Learning (ICML-11), pp. 833–840 (2011)
6. Hinton, G.E., Salakhutdinov, R.R.: Reducing the dimensionality of data with neural networks. Science **313**(5786), 504–507 (2006)
7. Salakhutdinov, R., Larochelle, H.: Efficient learning of deep Boltzmann machines. In: International Conference on Artificial Intelligence and Statistics, pp. 693–700 (2010)
8. LeCun, Y., Bottou, L., Bengio, Y., et al.: Gradient-based learning applied to document recognition. Proc. IEEE **86**(11), 2278–2324 (1998)

Automated Human Facial Expression Recognition Using Extreme Learning Machines

Abhilasha Ravichander, Supriya Vijay, Varshini Ramaseshan and S. Natarajan

Abstract Facial expressions form a vital component of our daily interpersonal communication. The automation of the recognition of facial expressions has been studied in depth and experiments have been performed to recognize the six basic facial expressions as defined by Paul Ekman. The Facial Action Coding System (FACS) defines Action Units, which are movements in muscle groups on the face. Combinations of Action Units yield expressions. In this paper, we propose an approach to perform automated facial expression recognition involving two stages. Stage one involves training Extreme Learning Machines (ELMs) to recognize Action Units present in a face (one ELM per AU), using Local Binary Patterns as features. Stage two deduces the expression based on the set of Action Units present.

Keywords Action unit · Action unit recognition · Facial expression analysis · Extreme learning machines · Human-computer interface

1 Introduction

Facial expressions are inherent to social communication and are also one of the most universal forms of body language. A persons face, mainly the eyes and mouth give away the most obvious and important cues regarding an emotion. These cues drive our daily interpersonal interaction and present an additional information component for informed decision-making. There is a growing interest in improving the interactions between humans and machines and automatic recognition of facial expressions covers an aspect of it. It has several applications such as lie detection for criminals,

A. Ravichander (✉) · S. Vijay · V. Ramaseshan
Department of Computer Science and Engineering, P.E.S Institute of Technology,
Bangalore, India
e-mail: abhilashacs005@gmail.com

S. Natarajan
Department of Information Science and Engineering, P.E.S Institute of Technology,
Bangalore, India

© Springer International Publishing Switzerland 2016
J. Cao et al. (eds.), *Proceedings of ELM-2015 Volume 2*,
Proceedings in Adaptation, Learning and Optimization 7,
DOI 10.1007/978-3-319-28373-9_18

estimating how people react to a particular product/brand (brand perception). Also, in the field of Artificial Intelligence, such applications would enable the services that require a good appreciation of the emotional state of the user such as robots which can benefit from the ability to recognize expressions where they can gauge emotions and perform better based on the response they read from the user. Facial expression recognition thus has the potential to vastly improve the quality of human-computer interaction leading us to more natural interfaces and environments.

The automation of the detection of Action Units present on a face is a difficult problem, and more so is its study from static images. FACS is a human observer based coding system, requiring FACS experts to instruct and observe subjects to depict Action Units [1]. The six basic facial expressions (happiness, sadness, surprise, anger, fear and disgust) occur frequently [2]. However, the FACS system was developed to depict the multitude of subtle changes in human facial features that contribute to facial expressions that extend beyond the six basic expressions. Typically, Action Unit detection is done using feature point tracking as in [3] over a series of images depicting the transition into the expression from a neutral facial expression. Bartlett et al. [4] recognized six upper face Action Units using a combination of methods, but none occurred in combination with one another.

Several techniques (image processing and machine learning based) have been attempted for Action Unit detection. Esroy [5] examines a new method to detect expression. His paper discusses a method involving triangles of shadowing and warping surrounding the eyes, nose and mouth to reflect emotions. Schmidt [6] used Hidden Markov Models (HMM) as the learning technique for Action Unit detection. HMMs used for facial expression recognition because they perform well and are analogous to human performance (e.g. for speech and gesture recognition). Simon [7] examined the usage of segment based SVMs for Action Unit recognition. The average recognition rate was around 90 %. In the paper by Abidin [8], BPNN (Back Propagation Neural Networks) was used as the classifier to categorize facial expression images into seven-class of expressions. Feature extraction was done through Haar wavelet, Daubechies wavelet and Coiflet wavelet. The use of these wavelet techniques along with the right configuration of the neural network provided an average accuracy of 94 %. Ionescue [9] explores a method which uses a bag of words representation for facial expression recognition. It extracts dense SIFT descriptors either from the whole image or from a pyramid that divides the image into increasingly finer sub-regions. For machine learning, the method makes use of multi-class SVMs where the learning is local. This improves results from the global classification by around 5 %. Yadan et al. [10] used Deep Belief Networks for facial expression recognition. The DBNs are tuned through logistic regression. The detectors first detect face, and then detect nose, eyes and mouth A deep architecture pretrained with stacked autoencoder is applied to facial expression recognition with the concentrated features of detected components. The paper by Xiaoming Zhao [11] also examines the use of DBNs for this problem.

A number of systems have already dealt with exploring the use of neural networks for facial Action Unit recognition. Lisetti et al. [12] explored the use of neural networks for facial expression detection by isolating different areas of the face and training a neural network for each of these areas. Kanade [3] also examined the use of neural networks for Action Unit detection. A three layered neural network was used for both the upper and the lower face AUs. The average recognition rate for upper face was around 93 % and lower face was around 95 %. Gargesha et al. [13] used Multi Layer Perceptrons and Radial Basis Function Networks to achieve an expression classification accuracy of 73 %.

However traditional approaches as neural networks suffer from slow learning and poor scalability. Also, most of these systems are designed to work on image sequences and use the movement of facial muscles when transitioning from a neutral expression to the peak expression to make predictions. In our work, we propose a system that analyses static images in order to automate the detection of Action Units, and hence expressions, using Extreme Learning Machines [14, 15] and texture based feature extraction methods. The system automatically detects the face and recognises the Action Units. We observe that for the six basic facial expressions, the contributing Action Units occur in combination with one another. These analyses are done on static images and the usage of extreme learning machines significantly speeds up the time taken for the network to learn while offering comparable results. The results obtained using ELMs are compared to those obtained using Support Vector Machines which have been shown to perform well for the Action Unit recognition task. Differences in the experimental procedures results in variations which should be taken into account when comparing the recognition results between systems [3]. For this reason, the experimental conditions for comparing the performance of ELMs against SVMs were kept as close to each other as possible. The system has been trained and tested on the Cohn-Kanade dataset.

Organization of the paper. In Sect. 2 we introduce the architectural setup we used to perform facial expression analysis. Following that, in Sect. 3 we describe the experimental methodology and in Sect. 4 we present out experimental results including a comparison of the performances of Extreme Learning Machines against Support Vector Machines for Action Unit recognition. Finally, in Sect. 5 we elaborate on the results as well as present future directions this work should take.

2 System Architecture

Our system uses a two stage approach to recognize facial expressions—stage one generates classifiers to recognize Action Units in an image, and stage two predicts the expression based on the Action Unit combinations present. Hence, our system structure consists of a training phase (Fig. 1), and a testing phase (Fig. 2).

The analyses are done on static images. Each sample in the Cohn-Kanade database consists of a sequence of images starting from a neutral expression and graduating into a peak expression image. The peak image of every sequence is used, where the Action Units are at their fullest representation.

We make use of the Facial Action Coding System to recognize Action Units present in an image. There are over 44 AUs participating in facial expressions,

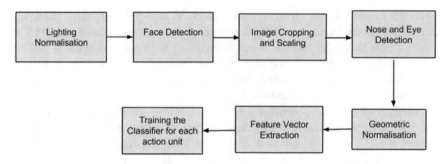

Fig. 1 Experimental flow for training

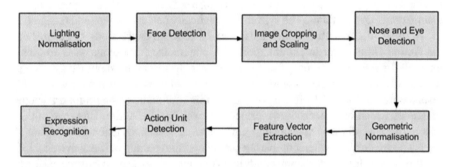

Fig. 2 Experimental flow for testing

categorized into upper and lower face AUs. One classifier per Action Unit is trained and used, for the AUs participating in the six facial expressions the system recognizes.

3 Experimental Approach

3.1 Dataset

A facial expression can be defined as one or more motions or positions of the muscles beneath the skin of the face. The Facial Action Coding System (FACS) [1] makes use of the face muscles to taxonomize human facial movements by their appearance on the face. FACS defines Action Units (AUs), which are a contraction or relaxation of one or more muscles.

We conducted our experiments on the Cohn-Kanade AU-Coded Expression Database [16]. The database consists of a series of subjects exhibiting various facial expressions. The subjects include representatives of various ethnic groups, thus accounting for diversity in facial features and ways of showing emotion. Each sample consists of a series of images starting from a neutral expression and transitioning

Fig. 3 Example of a shortened image sequence depicting transition to peak expression

into the peak expression (Fig. 3). The peak image is coded for the presence of Action Units as well as the expression. All images are in grayscale and full frontal view of the face of each subject. For our experiments, following the normalization and removal of non-viable samples, 552 images were used to train the Extreme Learning Machines (ELMs) to recognize Action Units. No other conditions were placed on the selected images.

3.2 Normalization

The images in the CK Database are in grayscale. However, lighting normalization [17] was performed in order to improve the contrast in the images using Contrast Limited Adaptive Histogram Equalization (CLAHE). This enables our system to work on images even when they are taken in poor lighting conditions or when the light source does not illuminate the entire face. Face detection using the Viola-Jones object detection framework [18] and cropping was performed (to remove background) and all cropped images were scaled to a standard size in order to ensure consistent size across all the training images. In the first training image used, the positions of the eyes and nose were noted after detection, and set as the standard positions to be maintained across all training images. Since a texture based feature extraction method such as LBP is used, which is sensitive to local changes in an image, it was imperative that the alignment and positions of the facial features remained consistent across the training set (and subsequently for each test image).

In instances when eye or nose detection failed, certain reconstructions were performed in order to approximate the positions. In our work, we propose a novel approach to overcome inaccurate nose and eye detection. It is useful to note here that after cropping the face from an image and scaling it, the image window being worked with consists of the detected face, with marginal background. To begin with, in an image window of dimension height × width, if more than two eyes were found, the eye windows found in the upper half of the face ($y < height/2$) are retained and stored (for selection of the best fits subsequently). Beyond this, the error handling mechanisms depend highly on the accuracy of nose detection. If the nose is correctly detected, four possible situations arise.

1. Two eyes detected successfully: this situation is ideal, and the nose and eyes are returned for the transform.
2. > Two eyes detected: in this case, the two eyes closest to the nose on the left and right side are chosen and returned.

Fig. 4 Triangulation
approach to overcome
inaccurate eye detection

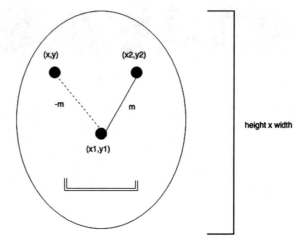

3. One eye detected: in this case, the position of the second eye is estimated using a triangulation method. In this method, the slope of the line connecting the nose and one detected eye is used to estimate the position of the second eye (as depicted in Fig. 4).

$$m = \frac{y_2 - y_1}{x_2 - x_1} \tag{1}$$

$$-m = \frac{y - y_1}{x - x_1} \qquad y = y_2 \qquad x = x_1 + \frac{y_1 - y_2}{m} \tag{2}$$

4. Zero eyes detected: In this case, the image cannot be used without risk of introducing erroneous data.

If zero noses are detected, one of two situations arise.

1. Two detected eyes, in which case the position of the nose is estimated using an approach similar to that described above.
2. For all other cases, the correct positions of the eye and nose could not be estimated from the given data.

If more than one nose is found, the nose closest to the middle of the image is selected. Following this, normal error handling procedures for one detected nose (as described above) is followed. Since the Cohn-Kanade database contains faces in the full frontal position, this method worked well and enabled us to retain images for training instead of discarding them.

Given the positions of the eyes and nose in an image and their expected positions in the transformed image, an affine transform was performed to align the eyes and nose to the same coordinates in every image. An affine transform [17] is a well-known geometric transform that maps points in one space to corresponding points in another space through a matrix multiplication followed by vector addition, and is often used for normalization tasks in image processing. This transform can be a combination of scaling (linear transformation), rotation (linear transformation) and translation (vector addition). The equations given below depicts how an affine transformation is performed. Vector A performs the linear transformation of rotation and scaling, while vector B is for translation, together leading to the transformation vector M. Using the positions of the eyes and nose in the source and destination images, the angle of rotation, scaling factor, and extent of translation can be derived, in order to yield the transformation vector M that will be applied to the source image.

$$A = \begin{bmatrix} a_{00} & a_{01} \\ a_{10} & a_{11} \end{bmatrix} \quad B = \begin{bmatrix} b_{00} \\ b_{10} \end{bmatrix} \tag{3}$$

$$M = \begin{bmatrix} A & B \end{bmatrix} \quad M = \begin{bmatrix} a_{00} & a_{01} & b_{00} \\ a_{10} & a_{11} & b_{01} \end{bmatrix} \tag{4}$$

$$X = \begin{bmatrix} x \\ y \end{bmatrix} \quad T = A \cdot \begin{bmatrix} x \\ y \end{bmatrix} + B \quad or \quad T = M \cdot \begin{bmatrix} x & y & 1 \end{bmatrix}^T \tag{5}$$

$$T = \begin{bmatrix} a_{00}x + a_{01}y + b_{00} \\ a_{10}x + a_{11}y + b_{10} \end{bmatrix} \tag{6}$$

3.3 Feature Extraction

Local Binary Patterns is a texture-based feature extraction technique, and is one of the best performing texture descriptors [19]. It overcomes challenges such as lighting, by being invariant to monotonic gray level changes [20]. LBP works on local grids within an image or window of study. The essence of LBP is to split an image into small grids, assign to each pixel in an image a value based on its relationship to its neighbours within its local grid and then construct a histogram of values over the grid. The histogram of values over each local grid is constructed, following which all the histograms are concatenated over the entire image to yield the resulting feature vector. The basic LBP operator works by calculating a code for each pixel in an image as shown in Fig. 5. For a pixel, the 8-pixel neighbourhood is thresholded against the center pixel, assigning 1 to a neighbour where the pixel value of the neighbour is greater than that of the center pixel and 0 otherwise. This yields an 8-bit binary string, the integer value of which (by multiplying each digit by its weight) yields the pixels code. 256 bins hence exist for the histogram, yielding a feature vector 256 components long.

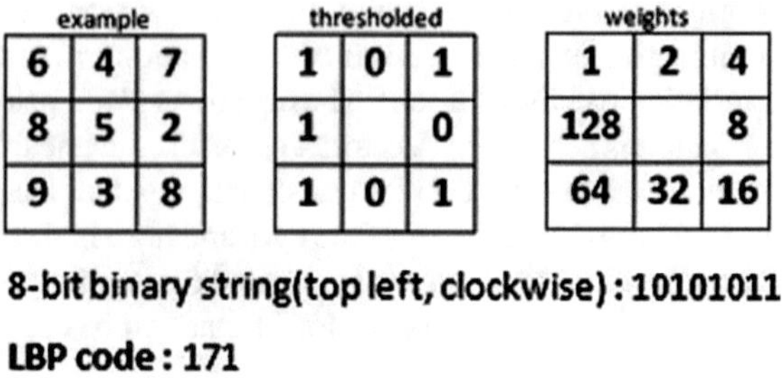

Fig. 5 Local binary pattern features

example		
6	4	7
8	5	2
9	3	8

thresholded		
1	0	1
1		0
1	0	1

weights		
1	2	4
128		8
64	32	16

8-bit binary string(top left, clockwise) : 10101011

LBP code : 171

3.4 Training

The proposed system, in order to be able to effectively recognize Action Units present in an image, uses one classifier, *i.e* one Extreme Learning Machine per Action Unit, that decides whether an Action Unit is present in the image or not (0 or 1 respectively). The entire database was split for training, testing and validation using the holdout sampling method (80-10-10). The Cohn-Kanade Database provides for each peak image a list of Action Units present. Using this, the training data was prepared by running LBP on each image (yielding a feature vector describing the image) and constructing a training set for each Action Unit. For each Action Unit, oversampling on the training set was performed in order to solve the class imbalance problem. If the number of negative samples exceeded the number of positive samples by a threshold value, oversampling was performed. The activation function used was the hyperbolic tangent function. For each AU, the configuration of the Extreme Learning Machine that performed best needed to be found, for which the classifier was trained multiple times using a different number of nodes in the hidden layer each time. Different parameters gave the best results for each AU.

3.5 Expression Prediction

The training phase yields one classifier per Action Unit of interest, trained using ELMs. On a test image, following normalization and feature vector extraction, the obtained feature vector is run through each classifier, at the end of which a list of Action Units present is obtained. A series of various combinations of AUs that yield an expression is defined, the satisfaction of any one of which indicates the possibility of the presence of the expression (Table 1).

However, the combinations of Action Units that result in a particular expression are not strictly defined. No complete evidence exists to show that these combinations result in the emotions shown with great universality. The table represents the central or most common actions for each emotion, but many minor variants of these combinations which also result in the same expression do exist. Our system proposes an approach where one classifier is used for each AU as opposed to assigning a classifier

Table 1 Action units for facial expressions

Expression	Action units
Happiness	6, 12
Surprise	1, 2, 5, 26, 27
Fear	1, 2, 4, 5, 20, 25
Sadness	1, 4, 6, 11, 15, 17
Disgust	9, 10, 15, 16, 17, 25, 26
Anger	4, 5, 7, 10, 16, 17, 22, 23, 24, 25, 26

for each Action Unit combination. For this reason, to improve expression prediction we also propose and implement a method to detect expressions which involves a maximal normalised subset intersection between the Action Units detected and the combination expected for each expression along with thresholding to determine if the expression is actually present.

4 Experimental Results

The testing of the trained extreme learning machines was performed on forty images which were randomly sampled and held out from the dataset. Each ELM was used to detect the presence of a particular Action Unit (AU) within the given image. We compare the accuracy of the Extreme Learning Machine to that of the Support Vector Machine for facial Action Unit detection under the same experimental conditions. Support Vector Machines are amongst the most common approaches used for this task. The results obtained for upper face and lower face Action Units [21, 22] are outlined in Tables 2 and 3 respectively. A comparison of the two classifiers is illustrated in Fig. 6.

The Extreme Learning Machines trained for this task used hyperbolic tangent as the activation function. Details of the configurations of the ELM's used for each facial Action Unit and the corresponding accuracies, sensitivities and specificities

Table 2 Upper Face AUs

Action unit	Description	Accuracy (ELM)	Accuracy (SVM)
1	Inner brow raiser	82	84
2	Outer brow raiser	86	98
4	Brow lowerer	72	88
5	Upper lid raiser	81	90
6	Cheek raiser	88	88
7	Lid tightener	78	81
9	Nose wrinkler	92	92

Table 3 Lower Face AUs

Action unit	Description	Accuracy (ELM)	Accuracy (SVM)
11	Nasolabial deepener	81	92
12	Lip corner puller	91	94
15	Lip corner depressor	79	88
16	Lower lip depressor	82	87
17	Chin raiser	72	92
20	Lip stretcher	82	88
22	Lip funneler	98	98
23	Lip tightener	78	92
24	Lip pressor	88	94
25	Lips part	73	92
26	Jaw drop	81	85
27	Mouth stretch	90	96

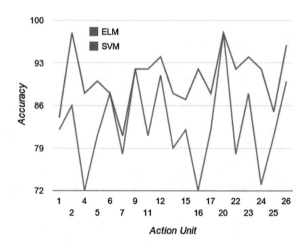

Fig. 6 A Comparitive study of ELM's versus SVM's for action unit recognition

are described in Tables 4 and 5 for upper and lower face Action Units respectively. The sensitivity measures the proportion of cases where the Action Unit is present and is correctly identified and the specificity measures the proportion of the cases where the absence of the Action Unit is correctly identified.

The Action Units obtained from each image were then used to predict the facial expression of the test subjects. Different combinations of Action Units detected resulted in being able to determine the facial expression of the test subject as demonstrated in Figs. 7 and 8 for a test subject from the CK+ database.

Table 4 ELM configuration for upper face AUs

Action unit	No. of hidden nodes	Accuracy (%)	Sensitivity (%)	Specificity (%)
1	150	82	82.3	82.0
2	75	86	84.6	86.6
4	100	72	72.2	71.0
5	100	81	90	78.2
6	75	88	90.9	86.7
7	120	78	66.7	80.4
9	150	92	85.7	92

Table 5 ELM Configurations for lower face AUs

Action unit	No. of hidden nodes	Accuracy (%)	Sensitivity (%)	Specificity (%)
11	120	81	25.0	85.0
12	220	91	77.0	95.3
15	220	79	60.0	83.0
16	120	82	66.7	83.0
17	100	72	60.0	78.9
20	150	82	75.0	83.3
22	150	98	00.0	100
23	150	78	57.1	80.3
24	100	88	33.3	94.0
25	120	73	83.8	60.0
26	75	81	66.6	82.3
27	120	90	77.7	91.6

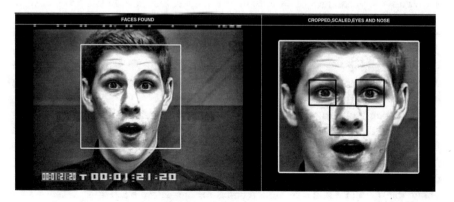

Fig. 7 Face detection, preprocessing, cropping, scaling

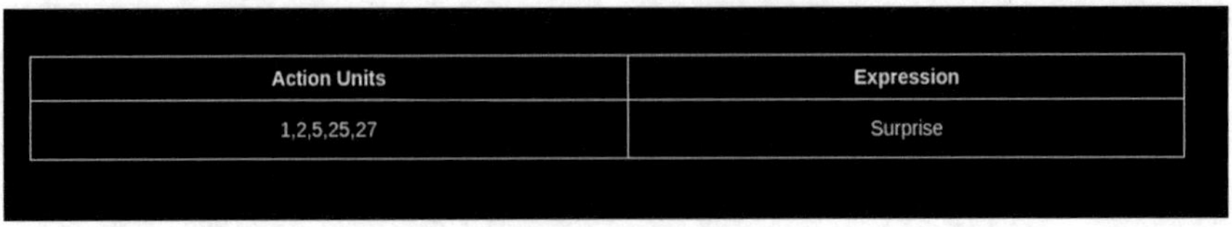

Action Units	Expression
1,2,5,25,27	Surprise

Fig. 8 Action unit detection and facial expression recognition

5   Conclusion and Future Work

Facial Expression Recognition is an important challenge in the domain of artificial intelligence and human-computer interaction. There is a need for robust and highly accurate systems which can be leveraged and used for many other unexplored problems in these areas.

We presented a systematic approach to build a fully automated facial expression recognition system using Extreme Learning Machines (ELMs) as the machine learning technique. The system we propose uses Local Binary Patterns for feature extraction from the test image and one classifier (ELM) per Action Unit is used to decide if an Action Unit is present within the image or not. The system runs the optimal configuration of the ELM for each Action Unit so as to be able to produce the best results. Lighting and geometric normalization are performed on each image before feature extraction for consistency. The system handles situations where the eye or nose detection fails initially and ensures the size of the training set is not reduced by such discrepancies. We present a series of results indicating the highest accuracy obtained for each Action Unit through ELMs. We also present results comparing the ELM accuracy to the highest accuracy obtained through Support Vector Machines under the same experimental conditions. SVMs have been amongst the most common machine learning techniques used for this problem. Support Vector Machines show an average accuracy of 88.71 % for seven upper face AUs and 91.5 % for twelve lower face AUs. In contrast, the proposed system using ELMs has an average accuracy of 82.7 % for upper face AUs 82.9 % for lower face AUs, the accuracies matching that of SVMs for AU1, AU6, AU9 and AU22. The performance of the system using Extreme Learning Machines against that of a system using Support Vector Machines under the same experimental conditions indicate that ELMs can offer comparable performances to that of SVMs for a few Action Units.

Our experiments suggest that better results can be achieved by applying better feature extraction techniques such as Gabor Filters and through the use of feature tracking models such as AAM (Active Appearance Models). In addition, better generalisation can be achieved by including training data from different FACS-coded datasets such as the MMI facial expression database. It would also be interesting to obtain the optimal number of hidden nodes for the Extreme Learning Machines by using techniques such as the algorithm suggested by Huang et al. [23] or by applying a pruning algorithm as described by Rong et al. [24]. For arriving at the optimal network architecture, incremental Extreme Learning Machines (I-ELMs) [25, 26] or Enhanced Incremental Extreme Learning Machines (EI-ELMs) [27] could also be employed. This experiment has been conducted with basic Extreme Learning Machines. For the Action Unit recognition task, the performance of ELM variants such as using an ELM ensemble [28] for each Action Unit with a bagging [29] or boosting [30] approach could also be investigated. In our work we have compared the performance of ELMs with that of SVMs. It would also be of interest to us to explore the performance of Support Vector Machines with an ELM kernel as suggested by Frenay and Verleysen [31].

References

1. Ekman, P., Friesen, W.V., Hager, J.C.: Facial Action Coding System (FACS): Manual. A Human Face. Salt Lake City, USA (2002)
2. Ekman, P., Friesen, W.V., Hager, J.C., Ellsworth, P.: Emotion in the Human Face. Salt Lake City, USA (1972)
3. Tian, Y., Kanade, T., Cohn, J.F.: Recognizing Action units for facial expression analysis. IEEE Trans. Pattern Anal. Mach. Intell. 97–115 (2001)
4. Bartlett, M.S., Viola, P.A., Sejnowski, T.J., Golomb, B.A., Hager, J.C., Ekman, P.: Classifying facial action. IEEE Trans. Pattern Anal. Mach. Intell. 974–989 (1996)
5. Ersoy, Y.: Express Recognition: Exploring Methods of Emotion Detection. Stanford CS229 Stanford University (2013)
6. Schmidt, M., Schels, M., Schwenker, F.: A Hidden Markov model based approach for facial expression recognition in image sequences. In: 4th IAPR TC3 Proceedings on Artificial Neural Networks in Pattern Recognition, pp. 149–160 (2010)
7. Simon, T., Nguyen, M.H., De la Torre, F., Cohn, J.F.: Action unit detection with segment-based SVMs. In: IEEE Conference on Computer Vision and Pattern Recognition (CVPR). IEEE Press, New York (2010)
8. Abidin, Z., Alamsyah, A.: Wavelet based approach for facial expression recognition. Int. J. Adv. Intell. Inform. 7–14 (2015)
9. Ionescu, R.T., Popescu, M., Grozea, C.: Local learning to improve bag of visual words model for facial expression recognition. In: Workshop on Challenges in Representation Learning, ICML (2013)
10. Lv, Y., Feng, Z., Xu, C.: Facial expression recognition via deep learning. In: International Conference on Smart Computing (SMARTCOMP), pp. 303–308. IEEE (2014)
11. Zhao, X., Shi, X., Zhang, S.: Facial expression recognition via deep learning. IETE Tech. Rev. 1–9 (2015)
12. Lisetti, C.L., Rumelhart, D.E.: Facial expression recognition using a neural network. In: Proceedings of the Eleventh International Florida Artificial Intelligence Research Society Conference, pp. 328–332 (1998)
13. Gargesha, M., Kuchi, P., Torkkola, I.D.K.: Facial expression recognition using artificial neural networks. EEE 511: Artificial Neural Computation Systems (2002)
14. Huang, G.B., Zhu, Q.Y., Siew, C.K.: Extreme learning machine: a new learning scheme of feedforward neural networks. In: Proceedings of International Joint Conference on Neural Networks, pp. 985–990 (2006)
15. Huang, G.B., Zhu, Q.Y., Siew, C.K.: Extreme learning machine: theory and applications. Neurocomputing. 489–501 (2006)
16. Lucey, P., Cohn, J.F., Kanade, T., Saragih, J., Ambadar, Z., Matthews, I.: The Extended Cohn-Kanade dataset (CK+): A complete dataset for action unit and emotion-specified expression. In: 2010 IEEE Computer Society Conference on Computer Vision and Pattern Recognition Workshops (CVPRW), pp. 94–101. IEEE (2010)

17. Bradski, G.: Dr. Dobb's Journal of Software Tools (2002)
18. Viola, P., Jones, M.: Rapid object detection using a boosted cascade of simple features. In: Proceedings of the 2001 IEEE Computer Society Conference on Computer Vision and Pattern Recognition (2001)
19. Ojala, T., Pietikäinen, M., Mäenpää, T.: Multiresolution gray-scale and rotation invariant texture classification with local binary patterns. IEEE Trans. Pattern Anal. Mach. Intell. 971–987 (2002) (IEEE Computer Society)
20. Ahonen, T., Hadid, A., Pietikainen, M.: face description with local binary patterns: application to face recognition. IEEE Trans. Pattern Anal. Mach. Intell. 2037–2041 (2006) (IEEE Computer Society)
21. Ekman, P., Friesen, W.V.: Facial Action Coding System. Consulting Psychologists Press, Palo Alto (1978)
22. Cohn, J.F., Ambadar, Z., Ekman, p.: Observer-based measurement of facial expression with the Facial Action Coding System, pp. 203–221. Oxford University Press, Oxford (2007)
23. Huang, Y., Lai, D.: Hidden node optimization for extreme learning machine. AASRI Procedia. 375–380 (2012)
24. Rong, H.J., Ong, Y.S., Tan, A.H., Zhu, Z.: A fast pruned-extreme learning machine for classification problem. Elsevier. 359–366 (2008)
25. Huang, G.B., Chen, L., Siew, C.K.: Universal approximation using incremental constructive feedforward networks with random hidden nodes. IEEE Trans. Neural Netw. 879–892 (2006)
26. Huang, G.B., Chen, L.: Convex incremental extreme learning machine. Neurocomputing. 3056–3062 (2007)
27. Huang, G.B., Chen, L.: Enhanced random search based incremental extreme learning machine. Neurocomputing. 3460–3468 (2008)
28. Hansen, L.K., Salamon, P.: Neural network ensembles. IEEE Trans. Pattern Anal. Mach. Intell. 993–1001 (1990)
29. Breiman, L.: Bagging predictors. Mach. Learn. 123–140 (1996) (Kluwer Academic Publishers)
30. Freund, Y.: boosting a weak learning algorithm by majority. Inf. Comput. 256–285 (1995)
31. Frénay, B., Verleysen, M.: Using SVMs with randomised feature spaces: an extreme learning approach. In: 18th European Symposium on Artificial Neural Networks Proceedings (2010)

Multi-modal Deep Extreme Learning Machine for Robotic Grasping Recognition

Jie Wei, Huaping Liu, Gaowei Yan and Fuchun Sun

Abstract Learning rich representations efficiently plays an important role in multi-modal recognition task, which is crucial to achieve high generalization performance. To address this problem, in this paper, we propose an effective Multi-Modal Deep Extreme Learning Machine (MM-DELM) structure, while maintaining ELM's advantages of training efficiency. In this structure, unsupervised hierarchical ELM is conducted for feature extraction for all modalities separately. Then, the shared layer is developed by combining these features from all of modalities. Finally, the Extreme Learning Machine (ELM) is used as supervised feature classifier for final decision. Experimental validation on Cornell grasping dataset illustrates that the proposed multiple modality fusion method achieves better grasp recognition performance.

Keywords Representation learning · Multi-Modal · Deep Extreme Learning Machine

1 Introduction

In the past several decades, due to their universal approximation on compact training samples and modeling capabilities for a large class of natural and artificial phenomena, feedforward neural networks have been widely popular in many fields. Many researchers have also explored the universal approximation capabilities of standard multi-layer feedforward neural networks [1–3]. However, there lack faster learning algorithms for neural networks.

J. Wei · G. Yan
Department of Electronic Information, Taiyuan University of Technology, Taiyuan, Shanxi, People's Republic of China

H. Liu (✉) · F. Sun
Department of Computer Science and Technology, State Key Laboratory of Intelligent Technology and Systems, Tsinghua University, Tnlist, Beijing, People's Republic of China
e-mail: hpliu@tsinghua.edu.cn

© Springer International Publishing Switzerland 2016
J. Cao et al. (eds.), *Proceedings of ELM-2015 Volume 2*,
Proceedings in Adaptation, Learning and Optimization 7,
DOI 10.1007/978-3-319-28373-9_19

Traditionally, gradient descent-based methods have mainly been used as learning algorithms of feedforward neural networks, in which all the parameters need to be tuned and may easily converge to local minima [4]. Thus these algorithms are usually far slower than required and always exist the dependency between different layers.

Recently, Huang et al. have proposed a new learning algorithm for single-hidden-layer feedforward neural network (SLFN) named the extreme learning machine (ELM) [5, 6]. Unlike traditional gradient descent-based methods, ELM avoids tuning control parameters and reaches good solutions analytically. That means the ELM learning algorithm not only tends to reach the smallest training error but also the smallest norm of weights [4]. The learning speed of ELM is extremely fast compared to other traditional methods [7].

Based on the fast learning speed and computational efficiency, ELM is more flexible and computationally attractive than traditional learning methods. However, its shallow architecture makes it difficult to capture relevant higher-level abstractions. In order to break this limitation, more and more deep ELM learning algorithm has been proposed. Reference [8] firstly introduces the ELM auto-encoder (ELM-AE), which represents features based on singular values. Reference [9] utilizes ELM as a base building block and incorporates random shift and kernelization as stacking elements. Reference [10] employs the ELM-AE as the learning unit to learn local receptive fields at each layer. Reference [11] proposes an efficient image set representation and the reconstruction error plays a role as a standard of classification. Reference [12] presents a new ELM sparse auto-encoder, which is utilized as the basic elements of H-ELM.

However, the aforementioned works do not refer to multi-modal problem. Thus, in this paper, we extend the deep ELM and propose a Multi-Modal Deep ELM-AE (MM-DELM) framework. The proposed MM-DELM is applied to multi-modal learning task, while maintaining its advantages of training efficiency. The contributions of this work are summarized as follows:

1. We propose a deep architecture—multi-modal ELM-AE framework, to construct the nonlinear representation from different aspects of information sources. An important merit of such a method is that the training efficacy is highly improved.
2. We perform the experimental validation on the recently developed publicly available Cornell grasping dataset. The obtained results show that the proposed fusion method obtains rather promising results.

The remainder of this paper is organized as follows: Sect. 2 introduces the related works, including the fundamental concepts and theories of ELM; Sect. 3 describes the proposed MM-DELM framework; Sect. 4 compares the performance of MM-DELM with single modality and one kind of concatenation framework on Cornell grasping dataset; while Sect. 5 concludes this paper.

2   Brief Introduction About Extreme Learning Machines

Different from traditional gradient-based algorithms, ELM's input weights and
single-hidden layer biases are arbitrarily chosen without iterative adjust, and the only
parameters to be learned in training are the output weights which can be calculated
by solving a single linear system [6].

Given N training samples, $\{\mathbf{X}, \mathbf{T}\} = \{\mathbf{x}_j, \mathbf{t}_j\}_{j=1}^{N}$, where $\mathbf{x}_j \in \mathbf{R}^p$ and $\mathbf{t}_j \in \mathbf{R}^q$ are the
j-th input and target vectors respectively. The parameters p and q are the dimension
of input and target vector respectively. To seek a regressor function from the input
to the target [4], the standard Single Hidden Layer Feed-forward network can be
mathematically modeled as (Fig. 1):

$$\mathbf{o}_j = \sum_{i=1}^{n_h} \beta_i g(\mathbf{w}_i^T \mathbf{x}_j + b_i) = \mathbf{t}_j, \tag{1}$$

where $\mathbf{o}_j \in \mathbf{R}^q$ is the output vector of the j-th training sample, $\mathbf{w}_i \in \mathbf{R}^p$ is the input
weight vector connecting the input nodes to the i-th hidden node, b_i is the bias of the i-
th hidden node, $g(\cdot)$ denotes hidden nodes nonlinear piecewise continuous activation
functions.

The above N equations can be written compactly as:

$$\mathbf{H}\beta = \mathbf{T}, \tag{2}$$

where the matrix $\mathbf{T}$ is target matrix,

Fig. 1   The model of basic
ELM

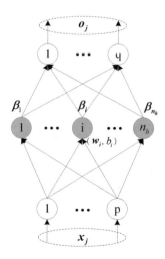

$$\mathbf{H} = \begin{bmatrix} g(\mathbf{w}_1^T \mathbf{x}_1 + b_1) & \cdots & g(\mathbf{w}_{n_h}^T \mathbf{x}_1 + b_{n_h}) \\ \vdots & \cdots & \vdots \\ g(\mathbf{w}_1^T \mathbf{x}_N + b_1) & \cdots & g(\mathbf{w}_{n_h}^T \mathbf{x}_N + b_{n_h}) \end{bmatrix}, \tag{3}$$

$$\boldsymbol{\beta} = \begin{bmatrix} \boldsymbol{\beta}_1^T \\ \vdots \\ \boldsymbol{\beta}_{n_h}^T \end{bmatrix}, \mathbf{T} = \begin{bmatrix} \mathbf{t}_1^T \\ \vdots \\ \mathbf{t}_N^T \end{bmatrix}. \tag{4}$$

The matrix $\mathbf{H}$ is the hidden layer output matrix, which can be randomly generated independent of the training data. $\boldsymbol{\beta} = [\boldsymbol{\beta}_1, \boldsymbol{\beta}_2, \ldots, \boldsymbol{\beta}_{n_h}]^T$ ($\boldsymbol{\beta}_i \in \mathbf{R}^q$) is the output weight matrix between the hidden nodes and the output nodes. Thus, training SLFNs simply amounts to getting the solution of a linear system (2) of output weights $\boldsymbol{\beta}$ [13].

A simple representation of the solution of the Eq. (2) is given explicitly by Huang et al. [4] as

$$\hat{\boldsymbol{\beta}} = \mathbf{H}^\dagger \mathbf{T}, \tag{5}$$

where $\mathbf{H}^\dagger$ is the Moore-Penrose generalized inverse of the hidden layer output matrix $\mathbf{H}$.

To improve generalization performance and make the solution more robust, we can add a regularization term [14], as shown in the Eqs. (6) and (7),

$$\hat{\boldsymbol{\beta}} = (\frac{\mathbf{I}}{\lambda} + \mathbf{H}^T \mathbf{H})^{-1} \mathbf{H}^T \mathbf{T}, \tag{6}$$

$$\hat{\boldsymbol{\beta}} = \mathbf{H}^T (\frac{\mathbf{I}}{\lambda} + \mathbf{H}^T \mathbf{H})^{-1} \mathbf{T}. \tag{7}$$

Thus, the ELM tends to reach the solutions straightforward without the issue of overfitting. These two features make ELM more flexible and attractive than traditional gradient-based algorithms.

3 Multi-modal Deep ELM-AE

However, due to its shallow architecture, feature learning using ELM cannot capture relevant higher-level abstractions, even with a large number of hidden nodes [9]. In order to learn rich representations efficiently, ELM-based Auto-Encoder(Deep ELM-AE) can be applied to extract the high level abstraction from different aspects of information sources.

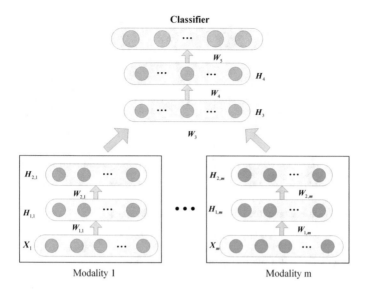

Fig. 2 The proposed Multi-modal architecture

3.1 Model Architecture

To address this issue, a hierarchical learning framework, Multi-Modal Deep ELM-AE (MM-DELM) is proposed . The multi-modal training architecture is structurally divided into three separate phases: unsupervised feature representation for each modality separately, feature fusion representation and supervised feature classification, as shown in Fig. 2.

As shown in Fig. 2, we perform feature learning to have high-level representations of each modality before they are mixed, in which features of m modalities are connected to two layers for constructing high level representation individually, where $\mathbf{X}_m \in \mathbf{R}^{n_m}$. The parameter n_m is the dimension of the m-th modal feature. Mathematically, the output of the two hidden layer in the m-th modality can be separately represented as:

$$\mathbf{H}_{1,m} = g(\mathbf{W}_{1,m}\mathbf{X}_m + \mathbf{B}_{1,m}), \tag{8}$$

$$\mathbf{H}_{2,m} = g(\mathbf{W}_{2,m}\mathbf{H}_{1,m} + \mathbf{B}_{2,m}), \tag{9}$$

where $g(\cdot)$ is activation function and we choose the sigmoid function. $\mathbf{H}_*$ (∗ can represent that the hidden nodes belong to which layer and modality) is hidden layer matrix representing non-linear representations extracted from features of all modalities. For example, $\mathbf{H}_{i,m}$ represents the i-th layer feature representation of m-th modality.

These high level representations of different information sources–$[\mathbf{H}_{2,1}, \ldots, \mathbf{H}_{2,m}]$ are mixed in a two-layer stacked structure to get well joint representation $\mathbf{H}_4$. At the

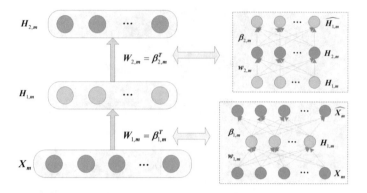

Fig. 3 Detailed illustration of the DELM representation learning

inference stage, the combination process is as follows:

$$\mathbf{H}_2 = \left[\mathbf{H}_{2,1}^T, \dots, \mathbf{H}_{2,m}^T \right]^T, \tag{10}$$

$$\mathbf{H}_j = g(\mathbf{W}_j \mathbf{H}_{j-1} + \mathbf{B}_j), \ for \ j = 3, 4. \tag{11}$$

Finally, the original ELM is performed to make final decision based on the joint representation:

$$\widehat{\mathbf{T}} = g(\mathbf{W}_5 \mathbf{H}_4). \tag{12}$$

Through the proposed approach, multi-modal system can be developed as one whole system rather than being developed as separate expert systems for each modality.

3.2 Unsupervised Feature Representation

Here, we consider a fully connected multi-modal multi-layer network with $h = 4$ hidden layers. Let $\mathbf{L} = \{\mathbf{W}_1, \mathbf{W}_2, \dots, \mathbf{W}_{h+1}\}(\mathbf{W}_i = \{\mathbf{W}_{i,1}, \dots, \mathbf{W}_{i,m}\}, i = 1, 2)$ denotes the parameters of the network that need to be learned. In our paper, the Deep ELM-AE is applied to learning the parameters $\mathbf{L}$, which is designed by using the encoded outputs to approximate the original inputs by minimizing the reconstruction errors [8, 12].

Figure 3 illustrates the process of learning representation from the features of m-th modality $\mathbf{X}_m$, which is similar to the separate learning of other modalities and fusion learning. To simplify training, each hidden layer of Deep ELM-AE is an independent ELM, and functions as a separated feature extractor, whose target is same as its input.

Fig. 4 The original ELM
makes the final decision

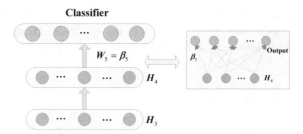

For instance, $\mathbf{W}_{1,m}$ is learned by considering a corresponding ELM with target vector $\mathbf{t} = \mathbf{X}_m$.

Thus, the system becomes a linear model and the output weights β can be analytically determined by the Eqs. (6) or (7) depending on the number of nodes in the hidden layer.

3.3 Supervised Feature Classification

Through the unsupervised feature representation, we can get a part of the parameters need to be learned:

$$\mathbf{W}_k = \hat{\beta}_k^T \ (k = 1, \dots, h). \tag{13}$$

Finally, the learned features $\mathbf{H}_h$ are transferred to the original ELM to model the mapping between feature representation and the label. The $\mathbf{H}_h$ can be regarded as the hidden layer of the original ELM, in which β_{h+1} can be obtained easily. Thus, we can get the parameter $\mathbf{W}_{h+1}$ as following equation (Fig. 4):

$$\mathbf{W}_{h+1} = \hat{\beta}_{h+1}. \tag{14}$$

In MM-DELM, the information of all modalities are combined in an effective way, and particularity and features for specific modalities have been learned. And in contrast to deep networks, MM-DELM also does not require expensive iterative fine tuning of the weights.

4 Experimental Result

Recognizing which part of an object is graspable or not is important for intelligent robot to perform some complicated tasks. In practice, the grasping performance not only depends on the pose and configuration of the robotic gripper, but also the shape and physical properties of the object to be grasped [15, 16]. The acquisition of depth information has made it easier to infer the optimal grasp for a given object beyond traditional RGB information. RGB-D image classification is a multi-modality learning problem [17, 18]. Therefore, the MM-DELM framework is applied to combining RGB information with depth information in RGB-D based image classification [19].

4.1 Dataset

We used the Cornell grasping dataset [20] for our experiments, which is available at http://pr.cs.cornell.edu/deepgrasping. This dataset contains 1035 images of 280 graspable objects. Each image is labeled with roughly equal numbers of ground-truth positive and negative grasping rectangles, representing these rectangles are graspable or non-graspable.

The color features are extracted from RGB's three 24×24 pixel channels, giving $24 \times 24 \times 3 = 1728$ input features. The three channels are the image in color space, used because it represents image intensity and color separately. The depth features simply contain the depth channel of the image, giving $24 \times 24 = 576$ input features. They are computed after the image is aligned to the gripper so that they are always relative to the gripper plates.

4.2 Result

We compare our algorithm in the Cornell grasping dataset with other single-modality networks trained in a similar manner where two separate sets of first layer features are learned for the depth channel, the combination of the RGB channels. A kind of naive method is also compared, which perform multi-modal fusion by combining color features and depth features as a concatenated vector that act as input of the framework.

Table 1 summarizes that compared with single-modality networks, our Multi-Modal Deep ELM-AE is able to learn rich representations efficiently which outperforms single-modal ones for recognition. Compared with RGB features, depth features can obtain the higher performance.

The proposed method and concatenation method are both integrating multi-modal information, but the proposed method outperforms the latter. Because in the concatenation method, the information sources with different statistical properties are

Table 1 Recognition results for different modalities

Modality	Accuracy (%)
RGB	86.34
Depth	89.71
Concatenation method	89.23
Proposed method	90.85

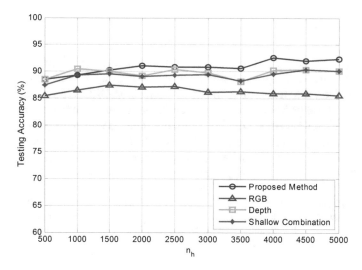

Fig. 5 The testing accuracies of different methods versus the number of hidden layer nodes

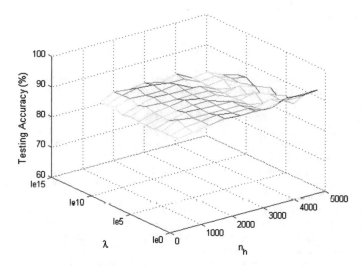

Fig. 6 Testing accuracy of MM-DELM in terms of n_h and λ

mixed in the first hidden layer ignoring the particularity of information about specific modalities. Therefore, the performance cannot be expected to be satisfactory (Fig. 5).

To analyse the roles of these parameters, we perform the sensitivity analysis. The most important two parameters in the proposed MM-DELM include the parameter λ for the regularized least mean square calculation, and the number of hidden nodes n_h. Therefore, we vary the value of λ within the set $\{10^0, 10^1, \ldots 10^9, 10^{10}\}$, and the value of n_h within the set $\{500, 1000, \ldots, 5000\}$ to analyze the performance variations. The results are shown in Fig. 6.

5 Conclusion

In this paper, we have proposed a novel multi-modal training scheme MM-DELM, in which information of all modalities has been learned and combined in an effective way without iterative fine-tuning. In this structure, MM-DELM takes full advantage of the hierarchical ELM to learn the high level representation from multi-modal data. Thus, the proposed method could obtain more robust and better performance. We also verified the generality and capability of MM-DELM on the Cornell grasping dataset. Compared with the single-modal and concatenation method, the training of MM-DELM is much faster and achieves higher learning accuracy.

Acknowledgments This work was supported in part by the National Key Project for Basic Research of China under Grant 2013CB329403; in part by the National Natural Science Foundation of China under Grant 61210013; and in part by the Tsinghua University Initiative Scientific Research Program under Grant 20131089295.

References

1. Hornik, K.: Approximation capabilities of multilayer feedforward networks. Neural Netw. **4**, 251C257 (1991)
2. Huang, G., Babri, H.A.:.Upper bounds on the number of hidden neurons in feedforward networks with arbitrary bounded nonlinear activation functions. IEEE Trans. Neural Netw. **9**(1), 224C229 (1998)
3. Leshno, M., Lin, V. Y., Pinkus, A., Schocken, S.: Multilayer feedforward networks with a non-polynomial activation function can approximate any function. Neural Netw. **6**, 861C867 (1993)
4. Huang, G., Zhu, Q., Siew, C.: Extreme learning machine: theory and applications. Neurocomputing **70**, 489–501 (2006)
5. Huang, G., Zhu, Q., Siew, C.: Extreme learning machine: a new learning scheme of feedforward neural networks. In: Proceedings of International Joint Conference on Neural Network(IJCNN), vol. 2, pp. 985–990 (2004)
6. Huang, G., Zhou, H., Ding, X., Zhang, R.: Extreme learning machine for regression and multiclass classification. IEEE Trans. Syst. Man Cybern.-Part B: Cybern. **42**(2), 513–529 (2012)
7. Li, M.B., Huang, G.B., Saratchandran, P., Sundararajan, N.: Fully complex extreme learning machine. Neurocomputing **68**, 306C314 (2005)
8. Cambria, E., Huang, G.: Extreme learning machines-representational learning with ELMs for big data. IEEE Intell. Syst. **28**(6), 30–59 (2013)
9. Yu, W., Zhuang, F., He, Q., Shi, Z.: Learning deep representations via extreme learning machines. Neurocomputing **149**, 308–315 (2015)
10. Zhu, W., Miao, J., Qing, L., Huang, G.: Hierarchical extreme learning machine for unsupervised representation learning. Neurocomputing (in press)
11. Uzair, M., Shafait, F., Ghanem, B., Mian, A.: Representation learning with deep extreme learning machines for efficient image set classification, pp. 1–10 (2015). arXiv:1503.02445
12. Tang, J., Deng, C., Huang, G.: Extreme learning machine for multilayer perceptron. IEEE Trans. Neural Netw. Learn. Syst., 1–13 (2015)
13. Feng, G., Huang, G., Lin, Q., Gay, R.: Error minimized extreme learning machine with growth of hidden nodes and incremental learning. IEEE Trans. Neural Netw. **20**(8), 1352–1357 (2009)
14. Ding, S., Zhang, N., Xu, X., Guo, L., Zhang, J.: Deep extreme learning machine and its application in EEG classification. Math. Probl. Eng., 1–12 (2014)

15. Sahbani, A., El-Khoury, S., Bidaud, P.: An overview of 3D object grasp synthesis algorithms. Robot. Auton. Syst. **60**, 326–336 (2012)
16. Bohg, J., Morales, A., Asfour, T., Kragic, D.: Data-driven Grasp SynthesisłA survey. IEEE Trans. Robot. **30**(2), 289–309 (2014)
17. Lai, K., Bo, L., Ren, X., Fox, D.: A large-scale hierarchical multi-view RGB-D object dataset. In: International Conference on Robotics and Automation(ICRA), pp. 1817–1824 (2011)
18. Bai, J., Wu, Y.: SAE-RNN deep learning for RGB-D based object recognition. Intell. Comput. Theory, 235–240 (2014)
19. Beksi, W.J., Papanikolopoulos, N.: Object classification using dictionary learning and RGB-D covariance descriptors. In: International Conference on Robotics and Automation (ICRA), pp. 1–6 (2015)
20. Lenz, I., Lee, H., Saxena, A.: Deep learning for detecting robotic grasps. Int. J. Robot. Res. **34**(4–5), 705–724 (2015)

Denoising Deep Extreme Learning Machines for Sparse Representation

Xiangyi Cheng, Huaping Liu, Xinying Xu and Fuchun Sun

Abstract In last decade, a large number of research has focused on the sparse representation for signal. As a dictionary learning algorithm, K-SVD, is introduced to efficiently learn an redundant dictionary from a set of training signals. In the mean time, there is an interesting technique named extreme learning machines (ELM), which is an single-layer feed-forward neural networks (SLFNs) with a fast learning speed, good generalization and universal classification capability. In this paper, we propose an denoising deep extreme learning machines based on autoencoder (DDELM-AE) for sparse representation. It makes the conventional K-SVD algorithm perform better. Finally, we show the experimental rusults on our optimized method and the typical K-SVD algorithm.

Keywords K-SVD · Extreme learning machines · Denoising · Deep ELM-AE

1 Introduction

Since Michael Elad and colleagues introduced the K-SVD algorithm [1], sparse signal reconstruction has gained considerable interests. Kinds of sparse representation have been applied to a variety of areas such as image denoising [2], image restoration [3], and image classification [4, 5]. In Ref. [5], using sparsity as a prior leads to state-of-art results.

K-SVD algorithm is an iterative method, which alternates between sparse coding of the examples based on the current dictionary, and a process of updating the dictionary atoms to better fit the examples. Under strict sparsity constrains, a signal $\mathbf{y} \in \mathbf{R}^n$

X. Cheng · X. Xu
Department of Electronic Information, Taiyuan University of Technology,
Taiyuan 030024, Shanxi, People's Republic of China

H. Liu (✉) · F. Sun
Department of Computer Science and Technology, State Key Laboratory of Intelligent
Technology and Systems, TNLIST, Tsinghua University, Beijing,
People's Republic of China
e-mail: hpliu@tsinghua.edu.cn

© Springer International Publishing Switzerland 2016
J. Cao et al. (eds.), *Proceedings of ELM-2015 Volume 2*,
Proceedings in Adaptation, Learning and Optimization 7,
DOI 10.1007/978-3-319-28373-9_20

can be represented by an redundant dictionary $\mathbf{D} \in \mathbf{R}^{n \times K}$ which includes Katoms, $\{\mathbf{d}_j\}_{j=1}^{K}$. Furthermore, the update of the dictionary columns is combined with an update of the sparse representations, thereby accelerating convergence.

During this period, a new theory named "deep learning" was put forward. References [6, 7] showed that a restricted Boltzmann machine (RBM) and auto-encoders could be used for feature engineering. The two types of auto-encoder-based deep networks are the stacked auto-encoder(SAE) [6] and the stacked denoising autoen-coder(SDAE) [8]. Both of them are constructed by stacking auto-encoders. The existing results show that deep networks outperform traditional multilayered neural networks.

In the meantime, as an emerging technology, extreme learning machines(ELM) has also achieved exceptional performance in large-scale settings, and is well suited to binary and multi-class classification, as well as regression tasks. Huang et al. [9] introduced ELM as a single-layer feed-forward neural networks with a fast learning speed and good generalization capability, whose hidden node parameters are randomly generated and the output weights are analytically computed. Like deep networks, Huang proposed multilayered ELM (ML-ELM) performs layer-by-layer unsupervised learning. And it also introduces the ELM based on autoencoder (ELM-AE), which represents features based on singular values. Similar to deep networks, ML-ELM stacks on top of ELM-AE to create a multilayered neural network. It learns significantly faster than existing deep networks, outperforming DBNs, SAEs, and SDAEs and performing on par with DBMs on the MNIST5 database.

The main contribution of our work is that we employ a denoising "input" of the raw data to the off-the-shelf K-SVD algorithm, which is generated by the denoising deep ELM-AE(DDELM-AE). Then, according to the restructure error, we solve the image classification problem. And our best results are much better than the simple K-SVD. Specially, we gain the test accuracies of 96.1 % on USPS [17] and 99.79 % on Coil-20.

This paper is organized as follows. In Sect. 2, we briefly review the K-SVD algorithm. Section 3 presents the deep ELM-AE. In Sect. 4, we state our optimization method on sparse representation. Section 5 demonstrates the experimental results and analyses the effect of parameters. Finally, Sect. 6 concludes the paper with the summary and demonstrates the superiority of our proposed method.

2   Sparse Representation and Dictionary Learning

Sparse and redundant signal representations have recently drawn much interest in computer vision, signal analysis and image processing [10, 11]. And several algorithms have been developed for the task of learning a dictionary. Two of the most well-known algorithms are the method of optimal directions (MOD) [12] and the K-SVD algorithm [1].

In this work, we adopt the K-SVD algorithm [1] for development. Given a set of N signals $\mathbf{Y} = [\mathbf{y}_1, \ldots, \mathbf{y}_N]$, the goal of K-SVD algorithms is to find a dictionary $\mathbf{D}$ and a sparse coding matrix $\mathbf{X}$ which solves the following optimization problem:

$$(\hat{\mathbf{D}}, \hat{\mathbf{X}}) = \arg\min_{\mathbf{D},\mathbf{X}} \|\mathbf{Y} - \mathbf{DX}\|_F^2,$$
$$s.t. \|\mathbf{x}_i\|_0 \leq T_0, \ \forall i = 1, \ldots, N, \tag{1}$$

where $\mathbf{x}_i$ represents the $i - th$ column of $\mathbf{X}$, $\|\mathbf{A}\|_F$ denotes the Frobenius norm of $\mathbf{A}$, and T_0 denotes the sparsity level. K-SVD is an iterative method that alternates between sparse-coding and dictionary update steps. First, a dictionary $\mathbf{D}$ with ℓ_2 normalized columns is initialized. Then, the main iteration is composed of the following two stages:

1. Sparse coding: In this step, we fix $\mathbf{D}$ and solve the following optimization problem over $\mathbf{x}_i$ for each example $\mathbf{y}_i$

$$\min_{\mathbf{x}_i} \|\mathbf{y}_i - \mathbf{Dx}_i\|_2^2,$$
$$s.t. \|\mathbf{x}_i\|_0 \leq T_0, \ \forall i = 1, \ldots, N. \tag{2}$$

2. Dictionary update: In this step, we fix the coding coefficient matrix and update the dictionary atom-by-atom in an efficient way.

With an update of dictionary columns and combining with an update of the sparse representations, traditional K-SVD algorithm achieves sparse signal representations from the raw signals. However, untreated data may be noisy, it is against this algorithm itself. According to the reconstruct error it inevitably leads to a poor classification result.

Motivated by the drawbacks of the current methods and the needs of many practical applications, we propose a preprocessing method used for the conventional K-SVD. In our paper, we use high level of representations as input rather than the original image, which is extracted by the denoising deep ELM based on autoencoder. Results demonstrate that high level of features can be better preserved and can reduce the reconstruction error effectively.

3 Principle of ELM based on Autoencoder

3.1 Extreme Learning Machines

Given N training samples $\{(\mathbf{x}_i, \mathbf{t}_i)\}_{i=1}^{N}$, the extreme learning machines can resolve the following learning problem:

$$\mathbf{H}\beta = \mathbf{T}, \tag{3}$$

where $\mathbf{T} = [\mathbf{t}_1, \ldots, \mathbf{t}_N]^T$ are target labels, and $\mathbf{H} = [\mathbf{h}^T(\mathbf{x}_1), \ldots, \mathbf{h}^T(\mathbf{x}_N)]^T$. We can calculate the output weights β from

$$\beta = \mathbf{H}^\dagger \mathbf{T}, \tag{4}$$

where $\mathbf{H}^\dagger$ is the Moore-Penrose generalized inverse of matrix $\mathbf{H}$.

To improve generalization performance and make the solution more robust, we can add a regularization term as shown [13]:

$$\beta = \left(\frac{\mathbf{I}}{C} + \mathbf{H}^T\mathbf{H} \right)^{-1} \mathbf{H}^T\mathbf{T}. \tag{5}$$

3.2 ELM Based on Autoencoder

In this section, we describe a common framework about deep ELM which is based on autoencoder [14] used for representation learning.

The ELM can be modified as follows: input data is used as output data $\mathbf{t} = \mathbf{x}$, and random weights and biases of the hidden nodes are chosen to be orthogonal. Widrow et al. [15] introduced a least mean square (LMS) implementation for the ELM and a corresponding ELM based on autoencoder that uses nonorthogonal random hidden parameters (weights and biases). Orthogonalization of these randomly generated hidden parameters tends to improve ELM-AE's generalization performance.

In ELM-AE, the orthogonal random weights and biases of the hidden nodes project the input data to a different or equal dimension space, as shown by the Johnson-Lindenstrauss lemma [16] and calculated as

$$\begin{aligned} \mathbf{h} &= g(\mathbf{a}\mathbf{x} + b), \\ \mathbf{a}^T\mathbf{a} &= \mathbf{I}, b^Tb = 1, \end{aligned} \tag{6}$$

where $\mathbf{a} = [\mathbf{a}_1, \ldots, \mathbf{a}_L]$ are the orthogonal random weights, and $\mathbf{b} = [b_1, \ldots, b_L]$ are the orthogonal random biases between the input and hidden nodes.

As stated above, ELM-AE's attractive property is that the output data is actually the input data, thus, we can calculate the output weights β as follows:

$$\beta = \left(\frac{\mathbf{I}}{C} + \mathbf{H}^T\mathbf{H} \right)^{-1} \mathbf{H}^T\mathbf{X}. \tag{7}$$

Finally, we learn representation in an unsupervised way using an ELM based on autoencoder. Experimental results also turn out that the learning procedure of ELM-AE is highly efficient and has good generalization capabilities.

4 Proposed Method

In this part, we intend to focus on our scheme for multilayered representation, and how this "deep" representation creates a meaningful learning used for the sparse representation.

4.1 Learning Representation with Deep ELM-AE

Learning high level of representations is vital for achieving better performance. We can often see stacked autoencoders (SAE) and stacked denoising autoencoders (SDA), whose outputs are equal the real input. Furthermore, many deep neural networks have yielded good performance in various tasks, they are generally very slow in training phase. Instead, our deep ELM-AE has obvious advantages in the calculation speed, even though the high dimensional image.

In our paper, we also set the output of an ELM network equal to the input, then, we will get the new representation $\widehat{X}$ from the whole deep ELM-AE(DDELM-AE). Figure 1 shows the process of learning a ELM-AE model from the training set X and what is the representation of X ultimately. We consider a fully connected deep network with L hidden layers and $\mathbf{W} = \{\mathbf{W}^1, \mathbf{W}^2, \dots, \mathbf{W}^{L+1}\}$ to denote the parameters of the deep ELM-AE that need to be learned, namely, $\beta = \{\beta_1, \beta_2, \dots, \beta_{L+1}\}$. To reduce the training cost, each layer is decoupled within the network and processed as an single ELM, of which targets are the same as its inputs. As shown in Fig. 1, $\mathbf{W}^1$, in other words, β_1^T is learned by considering a corresponding ELM with $\mathbf{T} = \mathbf{X}$.

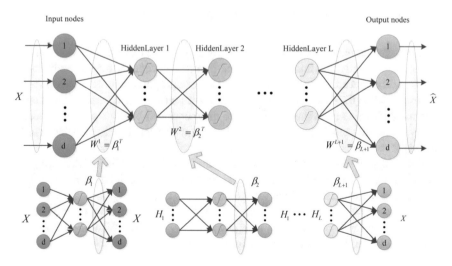

Fig. 1 A ELM-AE model from the samples of the training set X

The weight vectors connecting the input layer to each unit of the first hidden layer are orthonormal to each other, effectively leading to projection of the input data to a random subspace. Compared to initializing random weights independent of each other, orthogonalization of these random weights tends to better preserve pairwise distances in the random ELM feature space [16] and improves ELM based on autoencoder generalization performance. Next, β_1 is calculated by Eq. (7) depending on the number of nodes in the hidden layer. Therefore, this projection matrix is data-driven and hence used as the weights of the first layer ($\mathbf{W}^1 = \beta_1^T$).

$$\beta^* = \min \|\mathbf{H}\beta - \mathbf{X}\|_F^2,$$
$$s.t. \beta^T \beta = \mathbf{I}. \tag{8}$$

Similarly, the value of $\mathbf{W}^2$ is learned by forcing that the input and output of Hidden Layer 2 to $\mathbf{H}_1$ i.e. the output of Hidden Layer 1. In this way, all parameters of the multilayered ELM can be computed step by step. Using (7) does not ensure orthogonality of the computed weight matrix β. Imposing orthogonality in this case results in a more accurate solution since the data always lies in the same space. Therefore, the output weights β are calculated as the solution to the Orthogonal Procrustes problem.

In deep ELM-AE, the orthogonal random weights and biases of the hidden nodes project the input data to a different or equal dimension space. The deep ELM-AE models can automatically learn the non-linear structure of data in a very efficient manner. Compared with deep neural networks, deep ELM-AE does not require expensive iterations of fine tuning.

4.2 Using a Denoising Representation

Conventional autoencoder generates a simply copy of the input or similarly uninteresting ones trivially maximizes mutual information. A wide variety of modification of the traditional autoencoder framework have been proposed in order to learn sparse representations [6]. Pascal Vincent and colleagues introduced a very different strategy and defined a new representation into the mentioned below: "a good representation is one that can be obtained robustly from a corrupted input and that will be useful for recovering the corresponding clean input".

Here we propose a similar but different method. Using DDELM-AE, we get a clean input $\hat{\mathbf{x}} = f(\mathbf{x}) = g(\mathbf{W}\mathbf{x} + \mathbf{b})$ comparing with the initial input $\mathbf{x}$. See Fig. 2 for a denoising representation of the procedure.

Figure 3 shows two pairs of samples, the former is the initial input, the latter is representations applied the clever mapping f of DDELM-AE. What the representation DDELM-AE has learned demonstrates the features are denoised.

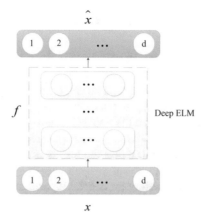

Fig. 2 The denoising deep ELM-AE architecture

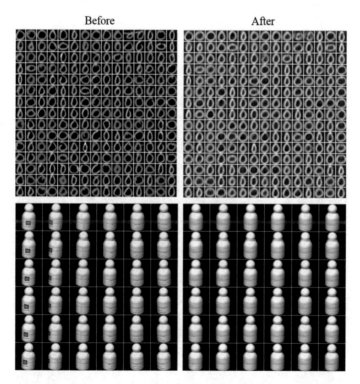

Fig. 3 Representation before and after performing the DDELM-AE

5 Experimental Results and Analysis

In the same case of K-SVD's parameters, such as $T_0 = 5$, maximum number of training iterations is set to 80, the above framework includes a number of parameters that can be changed: (i) the number of hidden layers, L, (ii) the number nodes of hidden layers of DDELM-AE, (iii) the ridge parameter $C = [C_1, C_2, C_3]$. In this section, we present our experimental results on the impact of these parameters on performance.

First, we will evaluate the effects of these parameters on the USPS dataset and Coil-20 dataset. Secondly, we will report the results achieved on these two dataset. Besides, the parameter settings that our analysis suggests is best overall(i.e., in our final results, we use the same setting for K-SVD algorithm.)

5.1 Digit Recognition

We apply our approach on the real-world handwritten digits classification problem. We use the USPS database [17] shown in Fig. 4, which contains ten classes of 256-dimensional handwritten digits. For each class, we select $N_{training} = 500$ samples for training and $N_{test} = 200$ samples for testing. Specifically, we choose the following parameters for learning the dictionaries for all classes: each class dictionary is learned with $K = 300$ atoms, $T_0 = 5$, maximum number of training iterations is set to 80.

Fig. 4 Random Samples on the USPS database [17]

We use $\mathbf{Y}_i = [\mathbf{y}_{i,1}, \dots, \mathbf{y}_{i,N}] \in \mathbf{R}^{256 \times 500}$ to represent the set of training samples of the ith class, where $i \in \{1, \dots, 10\}$. During our training procedure, there are two sequential stages: we first learn the stable and robust representations through a DDELM-AE. In the meanwhile, we also get 10 different kinds of deep ELM-AE model used to reconstruct testing samples later. The whole learning process is a very efficient and rapid manner in comparison to autoencoder.

In the second stage, K-SVD is applied to get the dictionary of the training set, $\mathbf{D}_i \in \mathbf{R}^{256 \times 300}$, where $i \in \{1, \dots, 10\}$. In other words, all above is to get the model of each class and the dictionary of every training class.

In the test phase, given a query image $\mathbf{z} \in \mathbf{R}^{256 \times 1}$, we first perform the DDELM-AE to get its denoising representation and then implement OMP algorithm(defined function s) separately for each $\mathbf{D}_i$, to get the sparse code $\mathbf{x}_i$. The sparse setting is the same as the training phase, namely $T_0 = 5$. Finally, the reconstruction error r_i is computed as:

$$r_{(i,N)} = \left\| \mathbf{z} - \mathbf{D}_i \mathbf{x}_i \right\|_F^2 = \left\| \mathbf{z} - (\mathbf{Y}_i \mathbf{X}) \mathbf{x}_i \right\|_F^2,$$
$$i \in \{1, \dots, 10\}, \ N = \{1, \dots, 2000\}, \tag{9}$$

where $\mathbf{X} \in \mathbf{R}^{256 \times 300}$ is sparse coefficient matrix of training samples, $\mathbf{x}_i \in \mathbf{R}^{300 \times 1}$ is sparse coefficient matrix of each testing sample and N is the total number of testing samples. The test sample is simply classified to the class that gives the smallest reconstruction error.

Before we present classification results, we first show the influence of the number of the hidden layers and the effect of different hidden nodes between hidden layers.

Number of Hidden Layers. Our experiments consider that how many hidden layers are favorable for the denoising DELM-AE. Through extensive experiments we find that 3 hidden layers based on our method is better than 2 layers. Furthermore, we also realize that increasing the number of hidden layer is not too good in surprise.

Table 1 shows that the effect of with different nodes in the case of two hidden layers. Extensive experiments turn out 3 hidden layers of DDELM-AE and 50 hidden nodes in each layer will perform better.

Table 1 Comparison of digit recognition accuracies for various hidden nodes in the case of two hidden layers

(a)

Structure	Accuracy (%)
100–50	85.55
100–100	95.4
100–200	84.35

(b)

Structure	Accuracy (%)
50–50	95.65
50–100	84.4
50–200	84.4

Ridge Parameters. Through the above experiments, we get a certain conclusion that we should adapt 3 hidden layers. And we speculate we may get the best result with 50 neurons in each layer. Next, we will further confirm the setting of parameter $C = [C_1, C_2, C_3]$. We set the parameter C in the range from $\{10^{-2}, 10^{10}\}$ from the first hidden layer to the last one.

Experiments are repeated with different nodes but all is 2 layers. There are 2 groups of figures which show the effect of C_1, C_2 to relevant layer with different nodes. In Figs. 5 and 6, we can see the results with 50 or 100 nodes in the first layers.

Based on experience, we set $C_3 = 10^8$, and then we obtain the best testing accuracy 96.1 % in the case of 50 nodes when $C_1 = 0.1$, $C_2 = 0.01$. Moreover, the result with 100 nodes is a little poor, only 95.7 % when $C_1 = 0.1$, $C_2 = 10^5$, $C_3 = 10^8$.

Final Classification Results. As we have mentioned, we employ the best and the most applicable number of neurons and hidden layers on the USPS database. In Table 2 we can see that our proposed method performs better obviously compared with the traditional K-SVD algorithm.

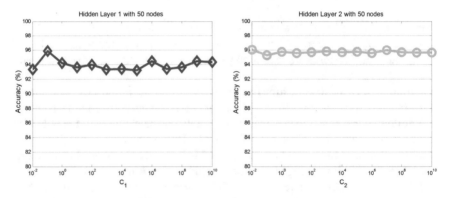

Fig. 5 Effect of different parameters C_1, C_2 to each layers with 50 nodes

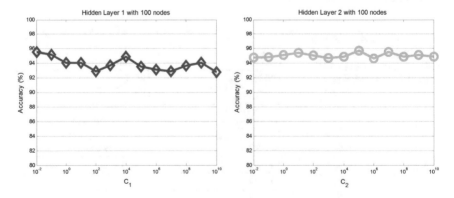

Fig. 6 Effect of different parameters C_1, C_2 to each layers with 100 nodes

Table 2 Test recognition accuracy on USPS

Algorithm	Accuracy (%)
K-SVD	93.1
Proposed method	96.1

5.2 Coil-20 Recognition

In this part, we will report the classification results on the Coil-20 database, which is established by Columbia Object Image Library and contains 20 classes objects.

The data set consists of gray-level images with $128 \times 128 = 16384$ pixels in 20 classes. Each class includes 72 images, we take 50 of them as training samples, the rest are chosen to test. So there are 1000 samples for training and 440 samples for testing in total.

As same as the setting of USPS, we set $T_0 = 5$, maximum number of training iterations is still set to 80, and input training samples are the vectorization of Coil-20 images with the dimension of $n = 16384$. Similarly, we employ $\mathbf{Y}_i = [\mathbf{y}_{i,1}, \ldots, \mathbf{y}_{i,N}] \in \mathbf{R}^{16384 \times 50}$ to represent the set of training samples of the ith class, where $i \in \{1, \ldots, 20\}$. And it should be emphasized that each class dictionary is learned with $K = 30$ atoms to use the K-SVD algorithm. There are still two sequential stages during our training procedure like the experiments on the USPS dataset. And the test phase can also follow the method of the USPS dataset.

Number of Hidden Layers. We use the same parameters setting and verify that 3 hidden layers in the DDELM-AE are better than 2 layers, and more layers are not useful to the result. Table 3 states the results with different nodes in 2 layers.

Therefore, we decide to use 3 hidden layers with 20 neurons in each layer.

Ridge Parameters. As same as the method of USPS, we still perform the same experiments on Coil-20 to verify the effect of various C for the results. We find $C_1 = 10$, $C_2 = 10^3$, $C_3 = 10^8$ generates the best testing accuracy.

Final Classification Results. Experimental results show that the proposed method outperforms the conventional K-SVD in the view of classification rate, so the proposed method has more practical value. Our best result is illustrated in Table 4.

Table 3 Comparison of Coil-20 recognition accuracies for various hidden nodes in the case of two hidden layers

Structure	Accuracy (%)
20–20	99.55
20–50	99.41

Table 4 Test recognition accuracy on Coil-20

Algorithm	Accuracy (%)
K-SVD	91.36
Proposed method	99.79

6 Conclusion

In this paper we have conducted extensive experiments on the USPS dataset and Coil-20 dataset using the representation learned by the denoising deep ELM-AE to characterize the effect of various parameters on classification performance. When combining our denoising representation with the K-SVD algorithm, we have shown more importantly that these elements such as ridge parameter C can be as significant as our proposed method itself. There are many classical and distinguished method about denoising such as SAE. However, for sparse representation, our denoising deep ELM-AE have greater representational power and fast speed. Compared with more complex algorithms, it can be highly competitive.

Acknowledgments This work was supported in part by the National Key Project for Basic Research of China under Grant 2013CB329403; in part by the National Natural Science Foundation of China under Grant 61210013; and in part by the Tsinghua University Initiative Scientific Research Program under Grant 20131089295.

References

1. Aharon, M., Elad, M., Bruckstein, A.: K-SVD: an algorithm for designing overcomplete dictionaries for sparse representation. IEEE Trans. Image Process. **54**(11), 4311–4322 (2006)
2. Elad, M., Aharon, M.: Image denoising via sparse and redundant representations over learned dictionaries. IEEE Trans. Image Process. **15**(12), 3736–3745 (2006)
3. Mairal, J., Elad, M., Sapiro, G.: Sparse representation for color image restoration. IEEE Trans. Image Process. **17**(1), 53–69 (2008)
4. Yang, J., Yu, K., Gong, Y., Huang, T.: Linear spatial pyramid matching using sparse coding for image classification. In: IEEE Conference on Computer Vision and Pattern Recognition (CVPR) (2009)
5. Wright, J., Yang, A., Ganesh, A., Sastry, S., Ma, Y.: Robust face recognition via sparse representation. IEEE Trans. Pattern Anal. Mach. Intell. **31**(2), 210–227 (2009)
6. Vincent, P., Larochelle, H., Lajoie, I., Bengio, Y., Manzagol, P.A.: Stacked denoising autoencoders: learning useful representations in a deep network with a local denoising criterion. Mach. Learn. Res. **11**, 3371–3408 (2010)
7. Hinton, G.E., Simon, O.: A fast learning algorithm for deep belief nets. Neural Comput. **18**, 1527–1554 (2006)
8. Salakhutdinov, R., Larochelle, H.: Efficient learning of deep boltzmann machines. Mach. Learn. Res. **9**, 693–700 (2010)
9. Huang, G., Zhu, Q., Siew, C.K.: Extreme learning machine: theory and applications. Neurocomputing **70**, 489–501 (2006)
10. Rubinstein, R., Bruckstein, A.M., Elad, M.: Dictionaries for sparse representation modeling. Proc. IEEE **98**(6), 1045–1057 (2010)
11. Elad, M., Mario, A.T., Figueiredo, M.A.T., Ma, Y.: On the role of sparse and redundant representations in image processing. Proc. IEEE **98**(6), 972–982 (2010)
12. Engan, K., Aase, S.O., Husoy, J.H.: Method of optimal directions for frame design. IEEE Trans. Signal Process. **5**, 2443–2446 (1999)
13. Huang, G., Zhou, H., Ding, X., Zhang, R.: Extreme learning machine for regression and multiclass classification. IEEE Trans. Syst. Man Cybern. **42**(2), 513–529 (2012)
14. Tang, J., Deng, C., Huang, G.: Extreme learning machine for multilayer perceptron. IEEE Trans. Neural Netw. Learn. Syst. 1–13 (2015)

15. Widrow, B., Greenblatt, A., Kim, Y., Park, D.: The no-prop algorithm: a new learning algorithm for multilayer neural networks. Neural Netw. **37**, 182–188 (2013)
16. Johnson, W., Lindenstrauss, J.: Extensions of Lipschitz mappings into a Hilbert space. Modern Anal. Probab. **26**, 189–206 (1984)
17. Hull, J.J.: A database for handwritten text recognition research. IEEE Trans. Pattern Anal. Mach. Intell. **16**(5), 550–554 (1994)

Extreme Learning Machine Based Point-of-Interest Recommendation in Location-Based Social Networks

Mo Chen, Feng Li, Ge Yu and Dan Yang

Abstract Researches on Point-of-Interests (POIs) have attracted a lot of attentions in Location-based Social Networks (LBSNs) in recent years. Existing studies on this topic most treat this kind of recommendation as just a type of point recommendation according to its similar properties for collaborative filtering. We argue that this recommending strategy could yield inaccuracy because these properties could not illustrate complete information of POIs for users. In this paper, we propose a novel Extreme Learning Machine (ELM) based approach named ELM Based POI Recommendation (EPR), which takes into account user preference, periodical movement and social relationship to discover the correlation of a user and a certain POI. Furthermore, we model recommendation in EPR as the problem of binary-class classification for each individual user and POI pair. To our best knowledge, this is the first work on POI recommendation in LBSNs by exploring the preference property, social property and periodicity property simultaneously. We show through comprehensive evaluation that the proposed approach delivers excellent performance and outperforms existing state-of-the-art POI recommendation methods, especially for cold start users.

Keywords Point-of-Interest recommendation · Extreme learning machine · Location-Based social networks · Classification · Cold start users

M. Chen (✉) · F. Li · G. Yu · D. Yang
Computing Center, Northeastern University, Liaoning, China
e-mail: chenmo@mail.neu.edu.cn

F. Li
e-mail: lifeng@mail.neu.edu.cn

© Springer International Publishing Switzerland 2016
J. Cao et al. (eds.), *Proceedings of ELM-2015 Volume 2*,
Proceedings in Adaptation, Learning and Optimization 7,
DOI 10.1007/978-3-319-28373-9_21

249

1   Introduction

With the rapid growth of wireless communication techniques and the popularity of mobile devices, a number of location-based social network (LBSN) services, e.g., Loopt, Foursqure, FindMe, Facebook Place and Wheel, have emerged in recent years. All these LBSN applications allow users to record their visits to point-of-interests (POIs) and share them, e.g., stores, cinema, restaurant, etc., to friends or other users. These services could provide people with more opportunities to experience life than ever before. Generally, users can share POIs to their friends on social network, meanwhile find new interesting spots to go according to other users' visiting records. For this reason, how to efficiently make a satisfying decision among the large number of POIs in LBSNs becomes a tough problem, which makes POI recommendation be a hot and interesting research area.

In the traditional recommendation approaches, collaborative filtering techniques are widely adopted, which provide better performance than content-based filtering techniques. User-item matrix is utilized in collaborative filtering approach to generate a prediction. However, in LBSNs, there are three aspects different form traditional recommendation, including social network, user movement and preference factors. For example, people prefer to visit the spots which have been visited by their friends, moreover, they are more likely to visit a spot which is nearer to users' POI that they are in favor of. Finally, it is less possible for a user to check in the spots that has big deviation from users' preference.

In this paper, we propose a novel POI recommendation approach, which integrates *user preference*, *periodical movement* and *social relationship* through extreme learning machine (ELM), named Extreme machine learning Based Point-of-interest Recommendation (EPR). We extract features of user and POI pair (named UP pair for short) in the above three aspects. The features extracted from user preference, corresponding to a given UP pair, can be derived by modeling topic associations between observable properties of users and spots. To consider the factor of users' periodical movement, we extract the feature from temporal and geographical distributions of users' check-ins by capture the relevance of two POIs. In addition, social relationships of users have significant influence on their check-in behavior, based on which we can extract the third feature.

The rest of the paper is organized as follows. We briefly review the related work in Sect. 2 and describe our POI recommendation approach EPR in Sect. 3. Results of experiments are shown in Sect. 4, and we conclude this paper in Sect. 5.

2   Related Work

In the POI recommendation researches, most existing approaches focus on dealing with similarity measurement for making recommendations. Huang et al. [1] build a Bayesian network model to recommend locations using business hours of a store, stay duration on a location and user occupation. Ye et al. [2] propose a unified POI recommendation framework, which fuses user preference to a POI with social influence and geographical influence. Hsieh et al. [3] argues a good route should consider the popularity of places, the visiting order of places, and the proper transit time from one place to another. In comparison with the traditional approaches in 1 and 2, the trip route recommendation in [3] is more suitable for tourists in traveling mode. Some works [4–6] proposed collaborative filtering approaches using social networks for POI recommendation. Memory-based collaborative filtering approaches are proposed in [4, 5], while a model-based collaborative filtering approach is proposed in [6]. All of the three papers search users' friends in the social network and recommend POIs visited by his/her friends. There are some other works [7–10] for top-k POI recommendation, which are not based on social networks. The most related study to our work so far is [11] which exploring the geographic, textual and social properties on followed recommendation in LBSNs. To our best knowledge, this is the first work on POI recommendation in LBSNs by exploring user preference, periodical movement and social friendship simultaneously.

3 ELM Based Point-of-Interest Recommendation

The problem of personalized POI recommendation is to recommend POIs to a user. As discussed earlier, the traditional POI recommendations could not work well. The reason is that traditional social-based POI recommendations always make recommendation by identifying the k most similar users to the active user, and then aggregate the similar users' choices to make recommendation. In our approach, POI recommending based on not only social factor but also POIs generated data and users' personalized check-in behaviors. Let $U = \{u_1, u_2, \ldots, u_n\}$ be a set of users and each user u_i has observable properties x_i, e.g., a user's profile. Let P be a set of POIs, and each u_i has a set of checked spots P_i, of which the locations are represented by <longitude, latitude>. Meanwhile, the check-in time of each user is also recorded by system. Observable properties y_k of each spot p_k is usually obtained by it's textual description. Moreover, users usually do not offer explicit ratings to a spot in most of LBSN applications, we regard a user's number of check-ins as the score of this spot.

We explore three different aspects which can affect POI recommendation, including *user preference*, *periodical movement* and *social relationship*. given a set of users U and point-of-interests P, the problem of POI recommendation can be formulated as classifying the relation of a given ordered UP pair, u_i and p_k, into binary class, 1 and 0. Here, class 1 means that recommends point p_k to user u_i, and class 0 means that does NOT recommends p_k to u_i.

$$f(u_i|p_k) \rightarrow \{0, 1\}, where\ u_i \in U\ and\ p_k \in P \tag{1}$$

The features are extracted based on prompted aspects above, corresponding to a given UP pair. We give the details of the approach in the following section.

4 Feature from User Preference

As mentioned above, user preference is an important factor to POI recommendation. There are two aspects contribute to user preference: one is how the user is interested in the spot and the other one is how popular the spot is. First, we calculate the user interesting score of a spot p_k, which plays a very important role in user preference. If a user prefers coffee to other drinks, the user is more interested in going to a coffee shop. Thus interesting score is extracted from textual feature description of users' profile and spots' tags. We adopt the aggregated LDA model in [12] to explore the observable features. The latent Dirichlet Allocation (LDA) model is a popular technique which can identify latent topic from a large document collection. Based on the model, we use proposed matrix factorization approach in [13] to obtain the interesting score:

$$in(u_i, p_k) = \mathbf{u_i^T p_k} + x_i^T W y_k \tag{2}$$

where $\mathbf{u_i}$ and $\mathbf{p_k}$ are user u_i and spot p_k factors, x_i and y_k are user and spot properties respectively, and W is a matrix used to transfer the observable prosperity space into the latent space.

As we know, popularity can affect the user check-in behavior deeply. Individual's decision to check in a POI is affected by its world wide opinions, which can be viewed as the popularity of POI. Thus, we calculate the popular score of each spots in P, which is involved in the later feature extraction. We normalize the popularity score for a spot p_k by following equation [12]:

$$po_k = \frac{1}{2} \left(\frac{Tpe(p_k) - 1}{\max(Tpe(p_k)) - 1} + \frac{Tch(p_k) - 1}{\max(Tch(p_k)) - 1} \right) \tag{3}$$

where $Tpe(p_k)$ and $Tch(p_k)$ are the total numbers of people have visited in p_k and the times they check in there respectively. Furthermore, max() is an operation of calculating the maximum value.

We extract a feature from user preference property based on the above calculations which can be represented as a linear combination of interesting score and popularity score as follows:

$$UPrefer(u_i, p_k) = in(u_i, p_k)po_k \tag{4}$$

5 Feature from Periodical Movement

A lot of studies have shown that temporal and geographic periodicity influences user check-ins. Intuitively, we expect some types of spots are visited regularly during the same times of the day. For example, some users go to café for afternoon tea, while some other users like go shopping after work. In order to explore periodicity of user movement, we separate a daily time into several time periods with different segment, i.e., $TP = \{tp_1, tp_2,...,tp_m\}$, of which an example is described in the following Table 1 ($m = 5$).

One simple but effective way to quantify the correlation of temporal periodicity of two spots (e.g., p_k and p_d) in tp_i is as follows:

$$sim\_tp_i(p_k, p_d) = \frac{ch\_t_i(p_k) + ch\_t_i(p_d)}{\sum_{i=1}^{m}(ch\_t_i(p_k) + ch\_t_i(p_d))} \quad (5)$$

where $ch\_t_i(p_k)$ and $ch\_t_i(p_d)$ represent total check-in number of p_k and p_d in period tp_i respectively. Then we normalized the similarity value into [0, 1]:

$$sim\_tp_i^{\diamond}(p_k, p_d) = \frac{sim\_tp_i(p_k, p_d) - \min_{n=1}^{m}(sim\_tp_i(p_k, p_d))}{\max_{n=1}^{m}(sim\_tp_i(p_k, p_d)) - \min_{n=1}^{m}(sim\_tp_i(p_k, p_d))} \quad (6)$$

Some works have shown that geographical proximities of spots have a significant influence on users' check-in behavior. Ye et al. [2] verifies the phenomenon that users are interested in exploring nearby spots of a POI that they are favor of, even it is far away from home. As a result, the spots visited by the same user have an implicit influence to POI recommendation. Thus in the following, we explore the geographical influence based on user existed check-in behavior at POIs, aiming to utilize it in feature extraction. We model the personalized distribution of the distance between any pair of spots visited by user u_i using kernel density estimation [14], since it is not limited to any distributions and without the assumption that the distance distribution is known. Given a spot p_d which belongs to user u_i's set of checked spots, we compute the Euclidean distance between the two spots as follows:

$$dis(p_k, p_d) = \sqrt{(x_k - x_d)^2 + (y_k - y_d)^2}, \quad \forall p_d \in P_i \quad (7)$$

Let SD_i be the sample for u_i that is drawn from some distribution with an unknown density f_i, its kernel density estimator $\hat{f}$ over distance $dis(p_k, p_d)$ using SD_i is given by:

$$\hat{f}_i(dis(p_k, p_d)) = \frac{1}{|SD_i|h} \sum_{sd_i' \in SD_i} K\left(\frac{dis(p_k, p_d) - sd_i'}{h}\right) \quad (8)$$

Table 1 Example of time period segmenting

Name	tp_1	tp_2	tp_3	tp_4	tp_5
Time period	7:00–9:00	11:00–13:00	15:00–16:00	17:00–20:00	20:00–24:00

where $K(.)$ is the kernel function and h is bandwidth. In this paper, we adopt the most popular normal kernel:

$$K(x) = \frac{1}{\sqrt{2\pi}} e^{-\frac{x^2}{2}} \tag{9}$$

and the optimal bandwidth [15] is as follows where $\hat{\sigma}$ is the standard deviation of the sample in SD_i:

$$h = \left(\frac{4\hat{\sigma}^5}{3n}\right)^{1/5} \approx 1.06\hat{\sigma}n^{-1/5} \tag{10}$$

We obtain the geographical influence measurement of P_i to p_k by taking the mean probability as follows:

$$p\_ge(p_k|P_i) = \frac{1}{|P_i|} \sum_{d=1}^{|P_i|} \hat{f}(dis(p_k, p_d)) \tag{11}$$

Finally, we can exploit the periodicity of user movement by fusing temporal and geographic periodicity measurement, based on which we can efficiently extract the feature:

$$UMove(u_i, p_k) = \eta \cdot sim\_tp_i(p_k, p_d) + (1 - \eta) \cdot p\_ge(p_k|P_i) \tag{12}$$

where parameter η is used to adjust the weight of temporal and geographical influence. For example, during the work week users check-in behaviors are more regular than weekend since they may go to the spots roughly at the same time period, while they prefer long-distance travel during weekend or holidays. Short-distance check-ins are more correlated with periodical movement, while long-distance check-ins is more influenced by geographical movement.

6   Feature from Social Relationship

Usually, friends have similar behavior because they might share a lot of common interests, which leads to similar check-in behaviors [16, 17]. For example, friends may hang out to have dinner together, or a user prefer to go to the restaurant recommended by his friends. In a word, users' check-in behaviors might be potentially affected by their friends. Thus we need to take into account of social influence based on social friendship for POI recommendation. Moreover, it is not possible that every friend of a user has the similar interest with him. Perhaps they are just only "social" friends. For example, a user is a foodie traveler meanwhile his colleague likes visiting museums. Therefore, user-user similarity is calculated which leads to more accurate recommendation. We combine the user similarity from friendship with similarity from check-in behaviors.

We use check-in matrix to describe users' check-in behavior, in which users' check-in records are viewed as feature vectors. Every user vector consists of n feature slots, one for each available spot. The value of each slot can be either the check-in number that counts the total number of a user u_i checks in spot p_k, or 0 if no such check-in record. We can compute the proximity between two users u_i and u_j by calculating the similarity between their vectors. We use Cosine Similarity as follows:

$$Sim\_ch(u_i, u_j) = \frac{Ch_i \cdot Ch_j}{\|Ch_i\| \|Ch_j\|} \tag{13}$$

where Ch_i and Ch_j are two vectors of check-ins from users u_i and u_j respectively. In order to avoid sparse problem when few spots are checked in, we limit the length of the vectors to parameter l. Interaction matrix is used to represent user-user similarity based on their social friendship. If user u_i give a comment on u_j's check-in information on the social network, we consider it as an interaction. If the number of the comments given by u_i to u_j is x, the value of cell co_{ij} is x. Otherwise, the value is 0. Considering individual user has different social interaction habit, for example some users may like to comment each updated status information of their friends while others may not like to comment frequently, we obtain user similarity by Adjusted Cosine Similarity based on this interaction matrix as follows:

$$Sim\_co(u_i, u_j) = \frac{\sum_{u_{i,j} \in U} (co_{ij} - \overline{co_i})(co_{ji} - \overline{co_j})}{\sqrt{\sum_{u_{i,j} \in U} (co_{ij} - \overline{co_i})^2} \sqrt{\sum_{u_{i,j} \in U} (co_{ji} - \overline{co_j})^2}} \tag{14}$$

where $\overline{co_i}$ is the average of the number of user u_i's total comments.

Fusing Similarity is calculated by combing above two similarities in a weighted way as follows:

$$Sim_{u_i, u_j} = \omega \cdot sim\_ch(u_i, u_j) + (1 - \omega) \cdot sim\_co(u_i, u_j) \tag{15}$$

Parameter ω is a tuning parameter ranging within [0, 1]. The bigger ω is, the more important role that behavior similarity plays. In other words, we can tune the influence of either check-in behavior or interaction to users' similarity calculating.

For a given UP, its inter correlation could be depicted by measuring the ratio of u's friends that have checked in p to all the friends of u. We extract a feature to formulate the social property of a UP pair named *UFriend*:

$$UFriend(u_i, p_k) = \frac{sim_{u_i, u_j \in U_i^{\dagger}}}{\sum_{u_j \in U_i^{\dagger}} sim_{u_i, u_j}} ch(u_j, p_k) \tag{16}$$

where $U_i^{\dagger}$ is the set of friends of u_i with top-N fusing similarity. Notice that the size of friends set U_i is usually much smaller than user set U. For a given user, only his friends are involved in the similarity calculation and contribute to feature extraction. Thus, we estimate the computation cost for the feature extraction is $|U_i||P| + N|P|$.

7 POI Recommendation

After the phase of feature extraction, features derived from all of user preference, social relationship and attraction of locations are used as inputs for the POI recommendation phase to learn a classification model for each UP pair. We choose ELM to classify the data, because compared with other traditional learning algorithms for recommendation tasks, ELM provides extremely faster learning speed, better generalization performance and with least human intervention.

The proposed ERP approach is designed a two-phase algorithm, as shown in Fig. 1, to address the problem of UP pair similarity mining for POI recommendation. The first phase processes the feature extraction (lines 1 to 4), while the second phase describes the POI recommendation (lines 5 to 7). Feature extraction explores three aspects that are discussed in introduction, such as user preference, social relationship and attraction of location. These features are used to learn an ELM model for each POI to classify whether the spot could be recommended to the user.

Input: Users Set U

POIs Set P

Check-in Datasets D

Output: Classification Result R

1. Feature Set $F \leftarrow \Phi$ // Feature Extraction
2. $F \leftarrow F \cup UPrefer(u,p)$
3. $F \leftarrow F \cup UMove(u,p)$
4. $F \leftarrow F \cup UFriend(u,p)$
5. Training Set $T \leftarrow F \cup D$ // POI Classification
6. Classifier $C \leftarrow ELM(T)$
7. $R \leftarrow C(U \times P)$

Fig. 1 EPR algorithm

8 Experiments

In this section, we conduct a series of experiments to evaluate the performance of our method. All the experiments are implemented on Intel Core i5-3320M CPU 2.60 GHz machine with 4 GB of memory running Microsoft Windows win7. We use the Foursquare dataset for our experiments, and we show the results of our experiments as the following discussions.

We extract the data from 04/2011 to 06/2011, which correspond to the city of New York and Chicago, denoted by NY and CH respectively. In New York, the number of the users is 6978, the number of the POIs is 49,021, the number of users' check-ins is 368,346. In Chicago, the number of the users is 5403, the number of the POIs is 44,132, the number of users' check-ins is 367,437. All of the data is divided into the training data and the testing data. About 80 % of the data is training data, and others are the testing data.

$$\text{Precision} = \frac{p^+}{p^+ + p^-} \qquad (17)$$

$$\text{Recall} = \frac{p^+}{R} \qquad (18)$$

Fig. 2 Comparison with various recommenders

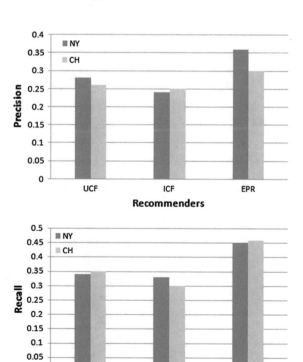

As the earlier study, The Precision, Recall are defined as followed, where p^+ and p^- indicate the number of correct recommendations and incorrect recommendations, and R indicates the total number of links in the testing data. Our experiment evaluates the effectiveness of our method (EPR) comparing with User-based Collaborative Filtering (UCF) and Item-based Collaborative Filtering (ICF). UCF and ICF are two main approaches in POI recommendation, of which the former aggregate k similar users to make recommendation and the latter computes the k most similar spots for each spot to make recommendation. The results are shown as below.

Figure 2 shows the performance of all approaches under evaluation. The experiments use both NY and CH datasets. The precision and recall for them are plotted in the two figures. In these figures, EPR always exhibits the best performance in terms of precision and recall, showing the strength of combines all three factors of user preference, periodical movement and social friendship influence. The results show that our approach outperforms UCF by about 22 and 32 % improvement in precision and recall performance averagely in dataset NY and CH, and

Fig. 3 Tuning parameter η

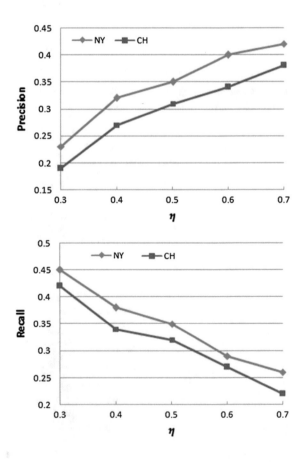

outperforms ICF by about 36 and 42 % improvement in precision and recall respectively.

As mentioned, parameter η can be controlled to tune the influence of either temporal or geographical factor to POI recommendation. Here we vary it in EPR to understand the roles of these two factors plays in our datasets.

Figure 3 shows the performance results of EPR under different η settings. The optimal value of η can be observed from the figures, which will lead an optimal tradeoff between precision and recall. For different datasets, the optimal value of is different since the proportion of users with short-distance travel habit varies, but usually η can be set in [0.5, 0.6].

In Fig. 4, we examine recommendations for cold start users whose numbers of friends less than 3 and less than 5 check-ins as the cold start users. As shown in the figures, the EPR method is more effective than the other two methods for the cold start users. Note that ICF and UCF perform worse for cold start users due to lack of user records because the two filtering methods cannot calculate similarity on little users' history data accurately. In addition, EPR for cold start users outperforms EPR for general users both in precision and recall.

Fig. 4 Comparison with various recommenders for cold start users

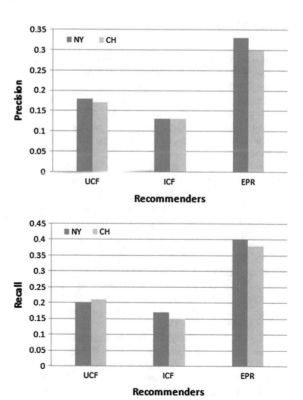

9   Conclusions

In this paper, we have proposed a novel approach named Extreme machine learning Based Point-of-interest Recommendation (EPR) for recommendation of POIs by mining the relationships of users and POIs. The core task of POI recommendation can be transformed to the problem of binary classification. We evaluate the possibility of each UP pair by learning ELM model. In EPR, we have explored user preference, social relationship and periodical movement by exploiting the check-in data in LBSN to extract descriptive features. To our best knowledge, this is the first work on POI recommendation that consider preference property, social property and mobility property for user and OPI pairs simultaneously. Through a series of experiments by on the real datasets obtained from Foursquare, we have validated our proposed EPR. As for the future work, we plan to design more advanced classification strategies to enhance the quality of POI recommendation for LBSNs.

Acknowledgments The research is supported by NSFC No. 61402093 and 61402213, Fundamental Research Funds for the Central Universities N141604001.

References

1. Huang, Y., Bian, L.: A Bayesian network and analytic hierarchy process based personalized recommendations for tourist attractions over the internet. In: Expert Systems with Applications, pp. 933–943 (2009)
2. Ye, M., Yin, P., Lee, W.C., Lee, D.L.: Exploiting geographical influence for collaborative point-of-interest recommendation. In SIGIR, pp. 325–334 (2011)
3. Hsieh, D.L., Li, C.T., Lin, S.D.: Exploiting large-scale check-in data to recommend time-sensitive routes. In: UrbComp, pp. 55–62 (2012)
4. Chen, X., Zeng, Y., Cong, G., Qin, S., Xiang, Y., Dai, Y.: On information coverage for location category based point-of-interest recommendation. In: AAAI, pp. 37–43 (2015)
5. Gao, H., Tang, J., Liu, H.: gSCorr. Modeling geo-social correlations for new check-ins on location-based social networks. In: CIKM, pp. 1582–1586 (2012)
6. Cheng, C., Yang, H., King, I., Lyu, M.R.: Fused matrix factorization with geographical and social influence in location-based social networks. In: AAAI (2012)
7. Liu, B., Fu, Y., Yao, Z., Xiong, H.: Learning geographical preferences for point-of-interest recommendation. In: KDD, pp. 1043–1051 (2013)
8. Yin, H., Sun, Y., Cui, B., Hu, Z., Chen, L.: Lcars: a location-content aware recommender system. In: KDD, pp. 221–229 (2013)
9. Hu, B., Jamali, M., Ester, M.: Spatio-temporal topic modeling in mobile social media for location recommendation. In: ICDM, pp. 1073–1078 (2013)
10. Hu, B., Ester, M.: Spatial topic modeling in online social media for location recommendation. In: RecSys, pp. 25–32 (2013)
11. Ying, J., lu, E., Tseng, V.: Followee recommendation in asymmetrical location-based social networks. In: UbiComp, pp. 988–995 (2012)
12. Liu, B., Xiong, H.: Point-of-interest recommendation in location based social networks with topic and location awareness. In: SDM, pp. 396–404 (2013)
13. Liu, B., Fu, Y., Yao, Z., Xiong, H.: Learning geographical preferences for POI recommendation. In: KDD, pp. 1043–1051 (2013)

14. Zhang, J., Chow, C.: iGSLR: personalized geo-social location recommendation-a kernel density estimation approach. In: GIS, pp. 324–333 (2013)
15. Silverman, B.: Density estimation for statistics and data analysis. Chapman and Hall, Boca Raton (1986)
16. Ma, H., King, I., Lyu, M.R.: Learning to recommend with social trust ensemble. In SIGIR, pp. 203–210 (2009)
17. Ma, H., Yang, H., Lyuand, M.R., King, I.: SoRec: social recommendation using probabilistic matrix factorization, pp. 931–940 (2008)

The Granule-Based Interval Forecast
for Wind Speed

Songjian Chai, Youwei Jia, Zhao Xu and Zhaoyang Dong

Abstract With the increasing penetration of wind power in modern power systems, sound challenges have emerged for system operators due to the uncertain nature of wind power. Deterministic point forecasting has become less effective to power system operators in terms of information accuracy and reliability. Unlike the conventional methods, a granule-based interval forecasting approach is proposed in this paper, which effectively considers the uncertainties involved in the original time series and regression models, other than only generating a plausible yet less reliable value. By incorporating Extreme Learning Machine (ELM) into the granular model construction, a specific interval can be simply obtained by granular outputs at extremely fast speed. Case studies based on 1-min wind speed time series demonstrate the feasibility of this approach.

Keywords Wind speed forecast · Information granule · Granular time series · Extreme learning machine (ELM) · Granular ELM

S. Chai (✉) · Y. Jia · Z. Xu
Department of Electrical Engineering, The Hong Kong Polytechnic University,
Hong Kong, China
e-mail: sj.chai@polyu.edu.hk

Y. Jia
e-mail: corey.jia@connect.polyu.hk

Z. Xu
e-mail: eezhaoxu@polyu.edu.hk

Z. Dong
School of Electrical and Information Engineering, The University of Sydney,
Sydney, NSW, Australia
e-mail: zydong@ieee.org

© Springer International Publishing Switzerland 2016
J. Cao et al. (eds.), *Proceedings of ELM-2015 Volume 2*,
Proceedings in Adaptation, Learning and Optimization 7,
DOI 10.1007/978-3-319-28373-9_22

1   Introduction

The surging growth in the wind energy sector has changed the electricity mix in many countries and brought significant environmental benefits for last decade. The energy structure has been changed with more and more fossil fuels being displaced by renewable energies, especially for wind energy, which turn out to be more environmental friendly. Over the last decade, the installed capacity of wind power has increased remarkably all around the world. In the Europe, electricity generated by wind shares 8 % of total electricity generation by the end of 2013 [1]. In Denmark, such amount even reached 33.2 % of national electricity generation in the same year.

In the modern power system, forecasting accuracy of wind power/wind speed has significant influences on reliable system operation and electricity market participants. Wind power producers are required to take part in the bidding and system operation by providing day-ahead generation profiles and undergoing penalties in case of deviations from the schedule. The grid operators could determine the reverse capacity and the strategy of generation commitment and dispatch based on the forecasting results. However, prediction errors always exist due to the variability and intermittence of the wind power. This always gives rise to potential risks to power system security and reliability, and economic loss to energy producers. Therefore, it is important to evaluate the uncertainties associated with the deterministic wind energy forecasts, and incorporate the knowledge into the operating practices.

Generally, the stochastic uncertainty and knowledge uncertainty [2] are regarded as the two main contributors to the prediction errors in a modeling system. The former is relevant to the inherent variability of the observed values due to the natural physical phenomenon, measured error, device failure and the like; the latter reveals the uncertainty in knowledge transfer, such as imperfect representation of processes in a model, the imperfect knowledge of the parameters associated with these processes, etc. Recently, information granule [3] has been attracting much attention as its underlying idea is intuitive and appeals to our commonsense reasoning [4]. By being abstract constructs, information granules and their ensuing processing provide a powerful vehicle to deal with an array of decision-making, control and prediction problems. In this paper, the information granule is utilized to quantify the underlying uncertainties in the forecasting system. By granulating model inputs and parameters through e.g. interval, rough set, and fuzzy set, a granular input-output mapping can be established; the results produced by the granular model are also granular. In this manner, they become reflective for various possible attributes in the process of knowledge transfer. In addition, ELM is employed as the basis model to construct granular ELM. Granular outputs are obtained by both coverage and specify criterion. That is, the optimal performance index makes sure the coverage criterion is satisfied to the nominal coverage rate while the average width of granular outputs maintains narrow.

The remainder of this paper is organized as follows. Section 2 describes four methods of granulating the original time series. Section 3 proposes the granular ELM and its underlying training strategy, where the granular parameters are selected by means of particle swarm optimization (PSO) algorithm. Case studies are reported in Sect. 4. 1-min wind speed dataset measured by Hong Kong Observatory are used to test the proposed granular model. Section 5 concludes this paper.

2 Information Granulation of Time Series

The granulation of time series aims to capture the variability and abstract the complexity of the time series with high volatility and intermittent, thus properly quantifying the uncertainty of inputs. The basic idea of granular time series is to build a collection of temporal windows with larger time scale by the approaches like intervals, rough sets, fuzzy sets and probabilistic density function (PDF) [3]. In this paper, the interval and fuzzy representation [2, 5] has been applied to the 1-min wind speed inputs using four different methods.

2.1 Interval Representation

The frequently used methods to represent interval information granule are min-max based granulation (min-max-Gr) and mean-std based granulation (mean-std-Gr). The min-max-Gr method aims to construct the intervals by taking the minimum and maximum value of certain larger time segment, which can be formulated as

$$X_i = \{x_{i,1}, x_{i,2}, \ldots, x_{i,k}\} \rightarrow G_i = [\min(X_i), \max(X_i)] \tag{1}$$

where X_i is the original time series belonging to the ith time window, k is the length of each time segment, which is a user-dependent variable.

Mean-std-Gr method characterizes the variability in each time window by two statistical parameters, the average value and standard deviation, as described by

$$X_i = \{x_{i,1}, x_{i,2}, \ldots, x_{i,k}\} \rightarrow G_i = [\text{mean}(X_i) - \text{std}(X_i), \text{mean}(X_i) + \text{std}(X_i)] \tag{2}$$

2.2 Fuzzy Sets Representation

The fuzziness of granules, their attributes and their values is characteristic of ways in which humans granulate and manipulate information [6]. In this paper, the membership-function-based granulation (MF-Gr) and fuzzy c-means clustering based granulation (FCM-Gr) are considered.

MF-Gr approach is mainly based on fuzzy membership function theory [7]. The commonly used membership functions include triangular membership function, trapezoidal membership function and Gaussian membership function. For simplify, only the triangular membership function is utilized in this paper.

The subseries X_i is firstly sorted and divided into $S_{low,i} = \{x'_{i,1}, \ldots x'_{i,[k/2]}\}$ and $S_{up,i} = \{x'_{i,[k/2]+b}, \ldots x'_{i,k}\}$, where $[k/2]$ represents the maximal integer not more than $[k/2]$, when k is even $b = 1$, otherwise, $b = 2$. Based on the triangle membership function theory, the low granule and up granule is produced by $S_{low,i}$ and $S_{up,i}$, respectively, and are expressed by [5].

$$
\begin{cases}
G_{low,i} = (2\sum_{j=1}^{[k/2]} x'_{i,j})/[k/2] - median\{x'_{i,1}, x'_{i,2}, \ldots, x'_{i,k}\} \\
G_{up,i} = (2\sum_{j=[k/2]+b}^{k} x'_{i,j})/(k - [k/2] - b + 1) - median\{x'_{i,1}, x'_{i,2}, \ldots, x'_{i,k}\}
\end{cases}
\tag{3}
$$

Hence, the $X_i = \{x_{i,1}, x_{i,2}, \ldots, x_{i,k}\}$ is granulated as $G_i = [G_{low,i}, G_{up,i}]$.

Similarly, FCM-Gr approach also takes advantages of fuzzy sets theory that each data point could belong to two or more clusters with different degrees of membership measured in [0,1], thus giving the flexibility to represent the membership relationship of each data. Further, the shape of membership function is not required to be pre-assumed in this method; instead, it depends on the clusters' centers to establish low and up granules [5]. The algorithm is presented as follows:

1. Initialize the partition matrix $U^{(0)} = [u_{mn}]$, $m = 1, 2, \ldots, k$.
2. Calculate the centers vector $C^{(t)} = [c_n]$ with $U^{(k)}$ at tth step, where $C_n = \sum_{m=1}^{k} u_{mn}^l x_{i,m} / \sum_{m=1}^{k} u_{mn}^l$.
3. Update $U^{(k)}$, $U^{(k+1)}$ by $u_{mn} = 1/\left(\sum_{t=1}^{C}\left(\|x_{i,m} - c_n\| / \|x_{i,m} - c_t\|^{2/(l-1)}\right)\right)$.
4. If $\|U^{(k+1)} - U^{(k)}\| < \xi$ then stop; otherwise, return to step 2. where ξ is termination criterion between 0 and 1.

Here, we take $C = 2$, that is two clusters in each time segment. Finally, the smaller center $c_{low,i}$ is set as low granule and the larger one $c_{up,i}$ is set as up granule, $G_i = [c_{low,i}, c_{up,i}]$.

3   Construction of Granular ELM

3.1   Extreme Learning Machine (ELM)

A novel learning algorithm termed as Extreme Learning Machine (ELM) is proposed in [8] to train single hidden-layer feedforward neural networks (SLFNs). In this algorithm, the input weights and biases of hidden nodes are randomly assigned and free to be tuned further. The output weights of SLFNs are analytically determined by a direct matrix calculation. According the experimental results reported in [8], ELM achieves better generalization performance with extremely fast learning speed. In this paper, ELM is utilized to efficiently generate Granular outputs. ELM algorithm is briefly introduced in the following.

Given N arbitrary distinct samples $\{(x_i, t_i) | x_i \in R^n, t_i \in R^m\}_{i=1}^N$, where x_i denotes the input vector and t_i denoted the target vector. ELM with a specific activation function $g(\cdot)$ and randomly assigned input weights and biases can efficiently approximate all sample data with zero error:

$$f(x_j) = \sum_{i=1}^{L} \beta_i g(w_i x_j + b_i) = t_j, \quad j = 1, 2, \ldots, N$$

$$\sum_{i=1}^{L} \beta_i g(w_i x_j + b_i) = o_j, \quad j = 1, 2, \ldots, N \tag{4}$$

$$\sum_{j=1}^{N} \|o_j - t_j\| = 0$$

where w_i is the weight vector associated with the ith hidden node and all input nodes; β_i is the weight vector associated with the ith hidden node and all output nodes; b_i is the threshold of the ith hidden node; and L is the number of hidden neurons. Equation (4) can be rewritten as the following matrix form:

$$\mathbf{H}\beta = \mathbf{T}$$

where $\mathbf{H}$ is expressed as:

$$H = \begin{bmatrix} g(w_1 x_1 + b_1) & \cdots & g(w_L x_1 + b_L) \\ & \vdots & \\ g(w_1 x_N + b_1) & \cdots & g(w_L x_N + b_L) \end{bmatrix}$$

β and $\mathbf{T}$ are respectively expressed as:

$$\beta = \begin{bmatrix} \beta_1^T \\ \vdots \\ \beta_L^T \end{bmatrix}_{L \times m} \quad \text{and} \quad T = \begin{bmatrix} t_1^T \\ \vdots \\ t_N^T \end{bmatrix}_{N \times m}$$

Since the input weights and biases are random assigned, training an SLFN based on ELM is equivalent to obtaining the smallest norm least-squares of the linear system in Eq. (4), which is expressed as:

$$\beta^* = \mathbf{H}^\dagger \mathbf{T}$$

where $\mathbf{H}^\dagger$ is the Moore-Penrose generalized inverse of $\mathbf{H}$ [9].

ELM effectively overcomes the limitation of high computational burdens of traditional gradient-based neural networks, which best suits the requirement in our application to achieve timely interval forecasting.

3.2 Granular ELM (GELM) Training Strategy

The granular neural network (GNN) was first introduced by Pedrycz [10, 11], where the parameters presented in the original neural network are regarded as information granules rather than numeric entities to account for the uncertainties in the process of knowledge transfer, the general architecture is shown in Fig. 1. Hence, the traditional numeric learning issue is transformed into a granular regression problem. Essentially, the granular parameters are trained using an optimization vehicle aiming at minimizing a cost function in terms of different performance index, such as interval error function [12, 13], coverage criterion [11] and alike. However, the quality and efficiency of the optimization process might be negatively affected by the large number of model parameters (double of the number of the original numeric parameters) to be optimized and the complex interval arithmetic. In order to tackle this issue, this paper proposes a novel ELM-based granular neural network (GELM), which could immensely take advantages of the wealth of well-established learning strategies of ELM, as described in Sect. 3.1.

The entire training process is outlined as follows. The low granules G_{low} and up granules G_{up} are extracted to train two numerical ELM networks with the same structure and same values of input weights and bias, respectively. Hence, two sets of numerical output weights $\underline{\beta}_{ini}$ and $\bar{\beta}_{ini}$ corresponding to up and low granules are determined. Afterwards, a spread percentage vector $s = \{[\underline{s}_i, \overline{s}_i], i = 1, 2, \ldots, m\}$ is introduced, where s_i ranges in $[-100, 100\ \%]$ and is assigned to each obtained output weight value, which can be formulated as,

$$\left[\underline{\beta}_i, \overline{\beta}_i\right] = \left[\underline{\beta}_{i, ini} + \underline{s}_i \left|\underline{\beta}_{i, ini}\right|, \overline{\beta}_{i, ini} + \overline{s}_i \left|\overline{\beta}_{i, ini}\right|\right] \tag{5}$$

Fig. 1 Architecture of granular neural network

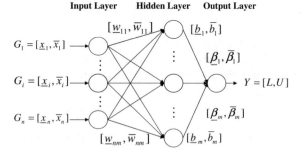

Finally, the whole learning process is well articulated and translated into the optimization stage only concerned with the design asset s_i. In this sense, the number of parameters to be optimized is significantly reduced and the searching space is narrowed. Furthermore, the topology of traditional neural network is augmented and simplified since the granular values of input weights and bias can be regarded as crisp values, which are randomly generated and as same as that in the well-trained numerical ELM networks.

3.3 Optimal Granular Prediction Outputs Construction

The granular outputs of GELM are evaluated by certain performance indices with regard to the numeric targets according to the required physical meaning of the underlying problem. Generally, the targets could be regarded as the means of each subseries, the real data points with original resolution, as well as the granular data [2].

In this paper, we evaluate the derived granular outputs in the light of both average coverage rate (ACR) and level of specificity with regard to the means of every granulation time window.

Average Coverage Rate (ACR): This criterion means the probability of the real targets lie in the granular outputs, defined as

$$ACR = \frac{1}{N_t} \sum_{i=1}^{N_t} c_i \tag{6}$$

where N_t is the number of targets, and t_i is the real target. If t_i is included in the PI, $c_i = 1$; otherwise, $c_i = 0$. Normally, a normalized coverage rate is assumed associated with the underlying granular outputs (e.g. 95, 90, 80 %). ACR is expected to be closet to the nominal probability. Hence, another evaluation index, Absolute Coverage Error (ACE), is introduced, indicating the difference between nominal coverage probability and estimated average coverage rate. Apparently, ACE should be diminished towards zero.

Specificity: It is focused at expressing a level of specificity of the granular outputs produced by the GELM, which is defined as the average width of the granular outputs, expressed by

$$Sp = \frac{\sum_{i=1}^{N_t} |\hat{G}_{up,i} - \hat{G}_{low,i}|}{N_t} \tag{7}$$

where $\hat{G}_i = [\hat{G}_{low,i}, \hat{G}_{up,i}]$ is the granular output with $\hat{G}_{low,i}$ and $\hat{G}_{up,i}$ as the lower and upper bound, respectively.

Normally, these two criterions are very likely in conflict, hence they should be combined together to construct the cost function in Eq. (8)

$$s_{opt} = argmin(ACE + Sp) \tag{8}$$

After obtaining the optimal spread percentage vector, the optimal GELM parameters are finally determined.

4 Case Study

In this section, the proposed GELM is tested on a very-short-term forecast (i.e. one time step) based on data measured by Hong Kong Observatory from January 1st, to December 30th, 2013, which comprises 525,600 pieces of data in total. By utilizing the granulation approaches introduced in Sect. 2, 17,520 and 8760 granules are respectively obtained in 30-min and hourly time window. Among the granular time series, the first 80 % of samples are used to train GELM, and the remaining is used for validation. All the measured data are pre-normalized into [0.1, 0.9] and 95 % nominal coverage rate is assumed. Experimental results based on different granulation methods are reported in Table 1.

As is shown in Table 1, there is no significant difference between the training results and testing results, and all testing ACR values are close to the nominal coverage rate 95 % with narrow prediction intervals, which demonstrates the effectiveness of GELM predictor in producing granule-based forecasts in different granulation level. Moreover, since the data size is highly reduced by time series granulation and the learning speed of ELM is extremely fast, the entire training process is computationally efficient. It is also observed that the interval-form granulation approaches, Min-max-Gr and Mean-std-Gr lead to a better coverage rate than fuzzy-granulation approaches perform, while the fuzzy-granulation methods could derive a slightly sharper granular output. Hence, one might comment that interval-based inputs can obtain higher quality granular outputs than

Table 1 The performance of granular outputs with different methods and different granulation level

Granulation time window	30 min				1 h			
	Training		Testing		Training		Testing	
Performance index	ACR (%)	Specificity	ACR (%)	Specificity	ACR (%)	Specificity	ACR (%)	Specificity
Min-max-Gr	94.94	0.1163	94.60	0.1165	94.63	0.1310	93.87	0.1298
Mean-std-Gr	94.99	0.1094	94.10	0.1133	95.00	0.1193	94.13	0.1198
Trimf-Gr	94.92	0.1099	93.90	0.1100	94.98	0.1133	93.66	0.1137
FCM-Gr	94.98	0.1133	93.83	0.1103	95.00	0.1181	92.60	0.1180

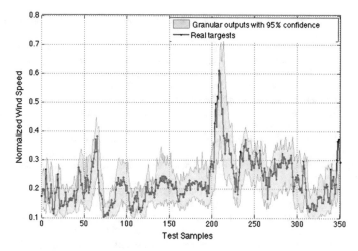

Fig. 2 Plot of granular outputs with 95 % confidence and 30-min granulation level

fuzzy sets based inputs. The 30 min min-max-Gr based granular outputs with 95 % confidence level is partially plotted in Fig. 2, where the real targets can be well enclosed in the granular prediction intervals.

5 Conclusion

This paper proposed a novel interval forecasting method based on GELM, which can quantitatively represent the uncertainties involved in time series and regression models. Interval and fuzzy sets based granulation approaches are employed to capture the variability in the original time series and reduce the data size. In addition, PSO is used to select granular parameters. Real wind speed time series are used in case study to test the proposed approach. Experimental results reveal that granular outputs can effectively reflect the stochastic uncertainty and knowledge uncertainty involved in the predictor, which demonstrates the effectiveness of the proposed approach. Future work is underway to quantify the uncertainty in other representation forms and to further enhance the performance of the granule-based predictor.

References

1. REN21: Renewables 2014 Global Status Report, REN21 Secr2014 (2014)
2. Ak, R., Vitelli, V., Zio, E.: An Interval-Valued Neural Network Approach for Uncertainty Quantification in Short-Term Wind Speed Prediction. IEEE, New York (2015)
3. Pedrycz, W.: Granular Computing: Analysis and Design of Intelligent Systems. CRC Press, Baco Raton (2013)

 4. Pedrycz, W., Skowron, A., Kreinovich, V.: Handbook of Granular Computing. Wiley-Interscience, New York (2008)
 5. Ruan, J., Wang, X., Shi, Y.: Developing fast predictors for large-scale time series using fuzzy granular support vector machines. Appl. Soft Comput. **13**(9), 3981–4000 (2013)
 6. Zadeh, L.A.: Toward a theory of fuzzy information granulation and its centrality in human reasoning and fuzzy logic. Fuzzy Sets Syst. **90**, 111–127 (1997)
 7. Zadeh, L.A.: Fuzzy sets. *Inf Control* **8**(6), 338–353 (1965)
 8. Huang, G.-B., Zhu, Q.-Y., Siew, C.-K.: Extreme learning machine: theory and applications. Neurocomputing **70**, 489–501 (2006)
 9. Rao, C.R., Mitra, S.K.: Generalized Inverse of Matrices and its Applications. Wiley, New York (1971)
10. Pedrycz, W., Vukovich, G.: Granular neural networks. Neurocomputing **36**(2), 205–224 (2001)
11. Song, M., Pedrycz, W.: Granular neural networks: concepts and development schemes. IEEE Trans. Neural Netw. Learn. Syst. **24**(4), 542–553, (2013) 2013
12. Roque, A., Maté, C., Arroyo, J., Sarabia, Á.: iMLP: applying multi-layer perceptrons to interval-valued data. Neural Process. Lett. **25**, 157–169, (2007)
13. Cimino, M.G.C.A., Lazzerini, B., Marcelloni, F., Pedrycz, W.: Genetic interval neural networks for granular data regression. Inf. Sci. **257**, 313–330 (2014)

KELMC: An Improved K-Means Clustering Method Using Extreme Learning Machine

Lijuan Duan, Bin Yuan, Song Cui, Jun Miao and Wentao Zhu

Abstract As a critical step for unsupervised learning, clustering is widely used in scientific data analysis and engineering systems. However, the shortage of categories information makes clustering an inconvenient issue. As an efficient and effective supervised learning algorithm, Extreme Learning Machines (ELMs) can be also adaptive for clustering tasks by constructing class labels properly. In this paper, we present a new clustering algorithm, K-means ELM Clustering (KELMC), which uses the output of an extreme learning machine instead of the similarity metrics in k-means. Extreme learning machine in KELMC is trained from potential cluster centers with its categories artificially labeled. For further improvement, we tried KELMC on an ELM-AE-PCA feature space and proposed another algorithm called EP-KELMC. Empirical study on UCI data sets demonstrates that the proposed algorithms are competitive with the state-of-the-art clustering algorithms.

Keywords Extreme learning machine · K-means · Clustering · Embedding

L. Duan · B. Yuan · S. Cui
Beijing Key Laboratory of Trusted Computing, College of Computer Science
and Technology, Beijing University of Technology, Beijing 100124, China
e-mail: ljduan@bjut.edu.cn

B. Yuan
e-mail: 19890926yb@emails.bjut.edu.cn

S. Cui
e-mail: cuisong@emails.bjut.edu.cn

L. Duan · B. Yuan · S. Cui
Beijing Key Laboratory on Integration and Analysis of Large-Scale Stream Data,
College of Computer Science and Technology, Beijing University of Technology,
Beijing 100124, China

J. Miao (✉) · W. Zhu
Key Lab of Intelligent Information Processing of Chinese Academy of Sciences (CAS),
Institute of Computing Technology, CAS, Beijing 100190, China
e-mail: jmiao@ict.ac.cn

W. Zhu
e-mail: wentaozhu1991@gmail.com

© Springer International Publishing Switzerland 2016
J. Cao et al. (eds.), *Proceedings of ELM-2015 Volume 2*,
Proceedings in Adaptation, Learning and Optimization 7,
DOI 10.1007/978-3-319-28373-9_23

1 Introduction

Data clustering is an inevitable step in most of scientific data analysis and in engineering systems. Generally speaking, clustering is a task of grouping similar objects together [1]. Since each algorithm has their own characters, researchers have proposed many principles and models for different kinds of problems. Therefore, a large family of clustering algorithms were developed by researchers to adapt different kinds of situations. Methods such as k-means clustering [2, 3], k-centers clustering [2], and the expectation maximization (EM) [4] were proposed in the early stage of machine learning, and keep a practical applicability until now. Laplacian eigenmaps (LE) [5] and spectral clustering (SC) [6] both use spectral techniques for clustering and get wonderful results in some data sets. Deep auto encoder (DAE) [7] processes the ability to capture multi-modal aspects of the input distribution.

The emergent machine learning technique, extreme learning machines (ELMs) [8–13], has become a hot area of research over the past years. ELM can be used for regression, classification, clustering and feature learning with its ability to approximate any objective function with fast learning speed and good generalization capability [8–10]. Recently, new variants of clustering methods based on ELM are emerging in an endless stream. He et al. [14] proposed ELM kMeans algorithm which transform the original data into the ELM feature space and use k-means to clustering. Huang et al. [15] proposed unsupervised ELM (US-ELM) based on manifold regularization framework. Kasun et al. [16] proposed Extreme Learning Machine Auto Encoder Clustering (ELM-AEC) which extends ELM for clustering using Extreme Learning Machine Auto Encoder (ELM-AE) [17]. Miche et al. [18] proposed a SOM-ELM clustering algorithm which projects the data samples to the cluster center space and using a heuristic method to optimization.

But above all, most of clustering algorithms, such as LE, SC, DA, ELM kMeans, US-ELM, ELM-AEC and so on, follow a unified framework that consist of two stages: (1) feature mapping or embedding and (2) clustering using k-means. They all received remarkable results by modifying the feature mapping strategy. In this paper, we present an improvement of k-means algorithm, K-means ELM Clustering (KELMC), which uses the outputs of an extreme learning machine instead of the similarity metrics in k-means. Extreme learning machine in KELMC is trained from potential cluster centers with its categories artificial labeled. This means a non-linear and evolvable similarity measurement was introduce in k-means. And inspired by the [16], which proposed ELM-AEC and shows that performing k-means clustering on the embedding ELM auto encoder feature space produces wonderful results, we proposed another improvement algorithms called K-ELMs Clustering on ELM-AE-PCA feature space (EP-KELMC).

The rest of the paper is organized as follows. In Sect. 2, we give a brief review of ELM and ELM-AEC. Section 3 introduces our two clustering algorithms. Extensive experimental result on clustering are presented in Sect. 4. Finally, some concluding remarks are provided in Sect. 5.

2 ELM

2.1 Brief Review of ELM

ELM, proposed by Huang et al. [12], has been widely adopted in pattern classification in recent years. It has many advantages, which not only avoids many problems encountered by traditional neural network learning algorithms based on gradient such as local minima, various training parameters, but also learns much faster with higher generalization performance than the established learning methods on the basis of gradient. ELM is a kind of generalized single hidden layer feed-forward networks (SLFNs). The essence of ELM is that the hidden layer of SLFNs need not to be tuned. The structure of the ELM is shown in Fig. 1.

Given N arbitrary distinct samples $(\mathbf{x}_j, t_j), j = 1, 2, \ldots, N$, where $\mathbf{x}_j = [x_{j1}, x_{j2}, \ldots, x_{jd}]^T \in \mathbb{R}^d$, and $t_j = [t_{j1}, t_{j2}, \ldots, t_{jm}]^T$, the output function of SLFNs with L hidden nodes in the output layer can be expressed by Eq. 1:

$$f(\mathbf{x}_j) = \sum_{i=1}^{L} \boldsymbol{\beta}_i G(\mathbf{a}_i, b_i, \mathbf{x}_j) = \mathbf{o}_j, \quad \mathbf{a}_i \in \mathbb{R}^d, \ b_i \in \mathbb{R}, \ \boldsymbol{\beta}_i \in \mathbb{R}^m \qquad (1)$$

where $\mathbf{a}_i = [a_{i1}, a_{i2}, \ldots, a_{id}]^T$ is the weight vector connecting the ith hidden neuron and input neurons, and b_i is the bias of the ith hidden neurons, $\boldsymbol{\beta}_i = [\beta_{i1}, \beta_{i2}, \ldots, \beta_{im}]^T$ is the weight vector connecting the ith hidden neuron and the output neurons.

For Extreme Learning Machine, all the equations above can be written compactly as $H\boldsymbol{\beta} = T$. In general, regularized ELM [19] is to solve the following learning problems:

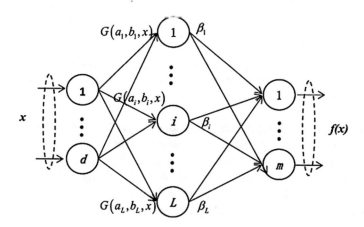

Fig. 1 Single hidden layer feed-forward networks [12]

$$\min_{\beta} \frac{1}{2} \|\beta\|^2 + \frac{C}{2} \|T - H\beta\|^2 \tag{2}$$

where C is a penalty coefficient on the training errors. We have the following closed form solution for (2):

$$\beta = \left(H^T H + \frac{I}{C} \right)^{-1} H^T T \tag{3}$$

$$\text{or } \beta = H^T \left(HH^T + \frac{I}{C} \right)^{-1} T \tag{4}$$

where I is an identity matrix.

2.2 Extreme Learning Machine Auto Encoder for Clustering

Extreme Learning Machine Auto Encoder for Clustering (ELM-AEC) [16] was proposed by Kasun L.L.C. et al. in 2014. It shows that k-means clustering in the embedding space of ELM-AE produces better results than clustering in the original space. They considered the Singular Value Decomposition (SVD) [20] of input data as:

$$D = U\Sigma V \tag{5}$$

where $U = [u_1, u_2, \ldots, u_N]$ is the eigenvectors of the gram matrix DD^T, $\Sigma = [\sigma_1, \sigma_2, \ldots, \sigma_n]$ is the singular value of D, $V = [v_1, v_2, \ldots, v_n]$ is the eigenvectors of the covariance matrix $D^T D$, N is the number of input data and n is the dimension of each data.

Theorem 1 *The embedding DV^T reduces the distances between the data points in the same cluster, while the distances between data points in different clusters are not changed* [21].

Theorem 1 shows that, projecting the input data D along the eigenvectors of the covariance matrix V^T is as similar as k-means clustering the data points. And it presents that k-means produces better results on the embedding DV^T than on original space. It has been shown that ELM-AE learns the variance information [17]. In [16], Kasun L.L.C. et al. shows that k-means clustering in the embedding $D\beta^T$ produces better results than clustering in the original space D.

3 Methodology

3.1 K-Means ELMs Clustering—KELMC

In k-means, clusters assignment always follow the nearest cluster center principle with Euclidean distance as its similarity measurement. In this paper, we proposed an improved clustering method for k-means based on ELM which changes the clusters assignment rules in k-means by using an extreme learning machine.

As we can see in the Fig. 2, the proposed method primarily involves the following processes. (a) Randomly choose k clustering centers. (b) Tag centers and build training data. (c) Train ELM and update β. (d) Cluster data with ELM. (e) Justify termination conditions. (f) Compute new cluster centers. Algorithm 1 describes KELMC algorithm. The following steps are introduced in detail.

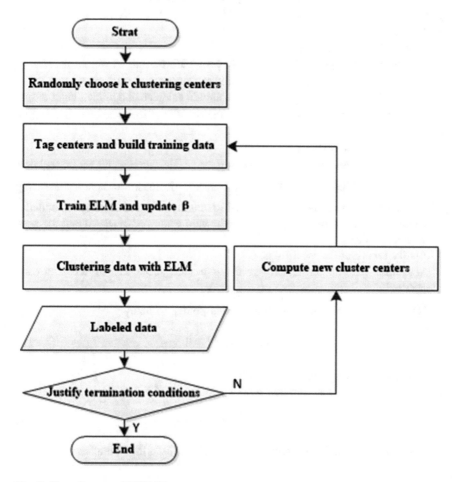

Fig. 2 Flow diagram of KELMC

Randomly choose k clustering centers.

Given a data set $D = \{x_i | x_i \in \mathbb{R}^n, i = 1, \ldots, N\}$, where n is the dimension of each data and the number of data is N. Like k-means, KELMC takes k initial clustering centers as exemplar set $E^{(0)} = \{e_i | e_i \epsilon D, i = 1, \ldots, k\}$, where k is the number of clusters. And clustering centers could be random values or random data points.

Tag centers and build train data.

After get the initial exemplar set $E^{(0)}$, we should tag centers with the train label which is typically given by a 1-in-all code. The configuration of target output, $T = \{t_i | t_i \in \mathbb{R}^k, i = 1, \ldots, k\}$, is set according to train label. Then, we build initial train data $TD^{(0)} = \{(e_i, t_i) | e_i \in D, t_i \in T, i = 1, \ldots, k\}$. If at least after a round of iteration, another form of exemplar set, $E^{(m)} = \{e_{ij} | e_{ij} \epsilon D, i = 1, \cdots, k, j = 1, \cdots, \rho\}$ where m is the number of iteration and ρ is the number of exemplars for each cluster, is used in proposed methods. In later iteration, we could use knowledge of data distribution learned by ELM to pick up more exemplars for each cluster. Therefore, we get training data $TD^{(m)} = \{(e_{ij}, t_i) | e_{ij} \in D, t_i \in T, i = 1, \ldots, k, j = 1, \ldots, \rho\}$.

Train ELM and update β.

Only in the initial phase, ELM needs to assign arbitrary input weight $\mathbf{w}_i$, bias b_i, $i = 1, \ldots, \tilde{N}$ where $\tilde{N}$ is the number of hidden neurons and set activation function $g(x)$, regularization factor γ. After that, we should calculate the hidden layer output matrix $H^{(m)}$ and get the target output matrix $T^{(m)}$. Then, update the output weight $\beta^{(m)}$.

Clustering data with ELM.

In this process, we use ELM to clustering the whole data set, and get labeled data $LD^{(m)} = \{(x_i, t_i^{(m)}) | x_i \in D, t_i^{(m)} \in T, i = 1, \ldots, N\}$. It must be noted that using ELM to clustering is equal to using ELM to classify. The reason why we called it "clustering" is that proposed methods use the auto generated label without the need for manual labeling.

Justify termination conditions.

Algorithm will be terminated, if the condition meet any of the following situations:

1. No change in the prediction label of data points, formally:

$$\left| LD^{(m)} - LD^{(m-1)} \right|_0 < \varepsilon \tag{6}$$

where ε is the threshold of label differences.
2. Reached the maximal iteration number limit, formally:

$$m > MaxIteration \tag{7}$$

where m is the number of iteration and $MaxIteration$ is the limitation of iteration.

Compute new cluster centers.

In order to minimize the within-cluster sum of squares (WCSSs), calculating the new mean value in each of the new clusters is a general measure. We make use of ELM in a reversed way. For each cluster, we pick up ρ data points as its new exemplar set (belong to this cluster) which is closest to the mean. After this step, we can get the new exemplar sets $E^{(m)} = \{e_{ij}|e_{ij}\epsilon D, i=1, \ldots, k, j=1, \ldots, \rho\}$.

Algorithm 1. k-ELMs clustering (KELMC)

Input: Data set $D = \{x_i|x_i \in \mathbb{R}^n, i=1, \ldots, N\}$, activation function $g(x)$, hidden neuron number $\tilde{N}$, number of clusters k, number of exemplar for each cluster ρ, and regularization factor γ.

Output: The labeled data $LD^{(m)}$.

Initialization: Randomly choose k clustering centers and build initial train data, then assign arbitrary input weight $\mathbf{w}_i$ and bias b_i and calculate $\boldsymbol{\beta}^{(0)}$ based on initial train data.

    For $m = 1: MaxItreation$

        1. Use ELM (with $\boldsymbol{\beta}^{(m-1)}$) cluster the data set, and get labeled data $LD^{(m)}$;

        2. Compute new cluster centers and get new exemplar set $E^{(m)}$;

        3. Tag centers and build train data $TD^{(m)}$;

        4. Calculate the hidden layer output matrix $H^{(m)}$ and get the target output matrix $T^{(m)}$;

        5. Update the output weight $\boldsymbol{\beta}^{(m)}$;

        6. If $\left|LD^{(m)} - LD^{(m-1)}\right|_0 < \varepsilon$, jumps to end

End

3.2 KELMC in ELM-AE-PCA Feature Space

In order to further improve the performance of the algorithm, we try KELMC on different kinds of feature spaces. In [16], the experiments show that performing k-means clustering on the ELM-AE feature space (embedding $D\boldsymbol{\beta}_{AE}^T$, where $\boldsymbol{\beta}_{AE}^T$ stands for the output weights of ELM-AE) get better results than performing k-means clustering in the original data space. And Ding C. et al. [21] have proved that using principal component analysis (PCA) [22] to get PCA feature space (embedding DV^T) can reduces the distances between the data points in the same cluster, while the distances between data points in different clusters are not changed. We made a combination with these two ideas to get an ELM-AE-PCA embedding and perform proposed KELMC in this embedding space.

Firstly, we transform the original data into the ELM-AE feature space. Then, projecting the current space to PCA feature space and get ELM-AE-PCA embedding. Finally, we operate KELMC on ELM-AE-PCA embedding space. Algorithm 2 describes EP-KELMC algorithm.

Algorithm 2. KELMC in ELM-AE-PCA embedding space

Input: Data set $D = \{x_i | x_i \in \mathbb{R}^n, \ i = 1, \ldots, N\}$, activation function $g(x)$, hidden neuron number $\tilde{N}$, number of clusters k, number of exemplar for each cluster ρ, regularization factor γ, regularization factor γ_{AE} for ELM-AE, number of embedding dimension n_e for PCA.

Output: The labeled data $LD^{(m)}$.

1. Calculate and normalize ELM-AE output weights β_{AE} for input data D.
2. Create and normalize the ELM-AE embedding $D\beta_{AE}^T$.
3. Calculate ELM-AE embedding's SVD $D\beta_{AE}^T = U\Sigma V$, which equals a PCA transform on ELM-AE embedding.
4. Create and normalize the ELM-AE-PCA embedding $D\beta_{AE}^T V^T$.
5. Operate KELMC on ELM-AE-PCA embedding $D\beta_{AE}^T V^T$.

---

4 Experiments

We evaluated our algorithms on a wide range of clustering tasks. Comparisons are made with the related state-of-the-art algorithms, e.g., k-means, ELM kMeans [14], DA [7], SC [6], LE [5], US-ELM [15] and ELM-AEC [16]. All algorithms were implemented using MATLAB R2014a on a 3.10 GHz machine with 4 GB of memory and Windows 8 (64 bits) operating system.

The data sets used for testing our two proposed algorithms include six UCI data sets, namely *Iris*, *Wine*, *Segment*, *Synthetic Control*, *Libras Movement* and *Seeds*. The characteristics of these data sets are presented in Table 1.

In our experiment, we run two proposed methods on all the data sets. For comparison purposes, we also present the results of seven other clustering algorithms, k-means, ELM kMeans, deep auto encoder (DA), spectral clustering (SC), Laplacian eginmaps (LE), unsupervised ELM (US-ELM), and Extreme Learning Machine Auto Encoder Clustering (ELM-AEC). Among these seven algorithms, the results of k-means and ELM kMeans are tested by ourselves on all the data sets. K-means algorithm is the original version in MATLAB R2014a and ELM kMeans

Table 1 Details of the data sets

Data sets	Cluster	Dimension	N
IRIS	3	4	150
Wine	3	13	178
Segment	7	19	2310
Synthetic control	6	60	600
Libras movement	15	90	360
Seeds	3	7	210

Table 2 Performance comparison in UCI data sets

Algorithms	Accuracy	IRIS	Wine	Segment	Synthetic control	Libras movement	Seeds
K-means	Average	85.37 ± 7.38	95.34 ± 0.49	58.78 ± 2.99	71.97 ± 5.12	46.76 ± 1.88	89.05 ± 0.01
	Best	88.67	96.63	72.21	83.50	51.67	89.05
LE [5]	Average	80.85 ± 13.82	96.63 ± 0	62.39 ± 5.03			
	Best	89.33	96.63	68.27			
SC [6]	Average	76.16 ± 8.92	93.32 ± 11.36	65.64 ± 4.72			
	Best	84.00	96.63	77.10			
DA [7]	Average	89.69 ± 14.01	95.24 ± 0.48	59.06 ± 4.03			
	Best	97.33	95.51	62.21			
US-ELM [15]	Average	86.06 ± 15.92	96.63 ± 0	64.22 ± 5.64			
	Best	97.33	96.63	74.50			
ELM-AEC [16]	Average	**96.00 ± 0**	97.26 ± 0.36	**72.20 ± 3.35**			
	Best	96.00	97.75	78.57			
ELM KMEANS [14]	Average	89.67 ± 1.92	94.94 ± 0.71	61.57 ± 1.24	75.72 ± 3.45	50.00 ± 2.07	92.62 ± 0.26
	Best	92.67	96.63	64.37	83.33	53.33	92.86
KELMC	Average	94.22 ± 1.24	97.19 ± 0	**69.40 ± 0.38**	67.92 ± 5.23	50.79 ± 1.70	95.24 ± 0
	Best	96.67	98.31	71.90	76.33	55.83	95.24
EP-KELMC	Average	**95.11 ± 0.54**	**98.50 ± 0.92**	67.18 ± 1.30	**81.56 ± 4.98**	**52.96 ± 2.24**	**95.81 ± 1.14**
	Best	96.67	98.88	71.65	87.67	59.17	96.67

Bold value denotes the best results

is implemented by ourselves according to the [14]. Our ELM-AEC's MATLAB version is from Kasun L.L.C. in [16], but we only have best configuration of parameters in first three data sets (*Iris*, *Wine* and *Segment*). The rest of experimental results for DA, SC, LE, and US-ELM are quoted from [15] (Table 2).

5 Conclusion

In this paper, we have proposed two algorithms, KELMC and EP-KELMC, to extend the traditional ELM for clustering task. The proposed KELMC is an improved k-means algorithm which changes the cluster assignment rules with an ELM instead of the nearest cluster center. An ELM is suitable for this task through its fast nonlinear cost function. Compared to the k-means algorithm, the proposed KELMC has better performance on most of six UCI data sets. In order to obtain further improvement in performance, we try KELMC in the ELM-AE-PCA feature space and propose EP-KELMC which receives comparable results in six UCI data sets. EP-KELMC also led to competitive results with several state-of-the-art clustering algorithms. In future works, we have two aspects to follow up. First, we need to prove the convergence of the proposed algorithm by theoretical and mathematical derivations. Second, during the experiments, we found KELMC has an ability to eliminate the interference of the noise characteristics. It will be used in some data sets which has useless information in feature expression.

Acknowledgements This research is partially sponsored by Natural Science Foundation of China (Nos. 61175115, 61370113 and 61272320), Beijing Municipal Natural Science Foundation (4152005 and 4152006), the Importation and Development of High-Caliber Talents Project of Beijing Municipal Institutions (CIT&TCD201304035), Jing-Hua Talents Project of Beijing University of Technology (2014-JH-L06), Ri-Xin Talents Project of Beijing University of Technology (2014-RX-L06), the Research Fund of Beijing Municipal Commission of Education (PXM2015_014204_500221) and the International Communication Ability Development Plan for Young Teachers of Beijing University of Technology (No. 2014-16).

References

1. Cattell, R.B.: The description of personality: basic traits resolved in clusters. J. Abnorm. Soc. Psychol. **38**(4), 476 (1943)
2. MacQueen, J. et al.: Some methods for classification and analysis of multivariate observations. In: Proceedings of the Fifth Berkeley Symposium on Mathematical Statistics and Probability, vol. 1, no. 14, pp. 281–297. Oakland, CA, USA (1967)
3. Hartigan, J.A., Wong, M.A.: Algorithm AS 136: a k-means clustering algorithm. Appl. Stat. **28**, 100–108 (1979)
4. Dempster, A.P., Laird, N.M., Rubin, D.B.: Maximum likelihood from incomplete data via the EM algorithm. J. Roy. Stat. Soc. B **39**(1), 1–38 (1977)
5. Belkin, M., Niyogi, P.: Laplacian eigenmaps for dimensionality reduction and data representation. Neural Comput. **15**(6), 1373–1396 (2003)

6. Ng, A.Y., Jordan, M.I., Weiss, Y.: On spectral clustering: analysis and an algorithm. Adv. Neural Inform. Process. Syst. vol. 2, pp. 849–856. MIT Press (2002)
7. Bengio, Y.: Learning deep architectures for AI. Found. Trends Mach. Learn. **2**(1), 1–127 (2009)
8. Huang, G.-B., Chen, L., Siew, C.-K.: Universal approximation using incremental constructive feedforward networks with random hidden nodes. IEEE Trans. Neural Netw. **17**(4), 879–892 (2006)
9. Huang, G.-B., Chen, L.: Convex incremental extreme learning machine. Neurocomputing **70**, 3056–3062 (2007)
10. Zhang, R., Lan, Y., Huang, G.-B., Xu, Z.-B.: Universal approximation of extreme learning machine with adaptive growth of hidden nodes. IEEE Trans. Neural Netw. Learn. Syst. **23**(2), 365–371 (2012)
11. Huang, G.-B., Chen, L.: Enhanced random search based incremental extreme learning machine. Neurocomputing **71**(16–18), 3460–3468 (2008)
12. Huang, G.-B., Zhu, Q.-Y., Siew, C.-K.: Extreme learning machine: theory and applications. Neurocomputing **70**, 489–501 (2006)
13. Huang, G.-B., Zhou, H., Ding, X., Zhang, R.: Extreme learning machine for regression and multiclass classification. IEEE Trans. Systems Man Cybern. Part B **42**(2), 513–529 (2012)
14. He, Q., et al.: Clustering in extreme learning machine feature space. Neurocomputing **128**, 88–95 (2014)
15. Huang, G., Song, S., Gupta, J.N.D., Wu, C.: Semi-supervised and unsupervised extreme learning machines. IEEE Trans. Cybern. **99**, 1 (2014)
16. Kasun, L.L.C., Liu, T.-C., Yang, Y., Lin, Z.-P., Huang, G.-B.: Extreme learning machine for clustering. Proc. ELM **2014**(1), 435–444 (2014)
17. Kasun, L.L.C., Zhou, H., Huang, G.-B., Vong, C.M.: Representational learning with extreme learning machines for big data. IEEE Intell. Syst. **28**(6), 31–34 (2013)
18. Miche, Y., Akusok, A., et al.: SOM-ELM-self-organized clustering using ELM. Neurocomputing **165**, 238–254 (2015)
19. Deng, W.-Y., Zheng, Q.-H., Chen, L.: Regularized extreme learning machine. In: Proceedings IEEE Symposium on Computational Intelligence and Data Mining, CIDM 2009, pp. 389–395 (2009)
20. Eckart, C., Young, G.: The approximation of one matrix by another of lower rank. Psychometrika **1**(3), 211–218 (1936)
21. Ding, C., He, X.: K-means clustering via principal component analysis. In: Proceedings of the Twenty-First International Conference on Machine Learning, ICML 2004, p. 29. ACM (2004)
22. Jolliffe, I.: Principal Component Analysis, 2nd edn. Springer, New York (2002)

Wind Power Ramp Events Classification Using Extreme Learning Machines

Sujay Choubey, Anubhav Barsaiyan, Nitin Anand Shrivastava, Bijaya Ketan Panigrahi and Meng-Hiot Lim

Abstract Wind power is becoming increasingly popular as a renewable source of energy. Being a non-dispatchable energy resource, wind power facilities entail efficient forecast mechanisms to estimate the production of various wind power utilities available. In an integrated grid system, a balance must be maintained between production and consumption. Given that wind power is directly affected by meteorological factors (wind speed etc.) accurately predicting such fluctuations becomes extremely important. These events of fluctuation are termed as ramp events. Forecast of wind power is important but accurate prediction of ramp events is much more crucial to the safety of the grid as well as the security and reliability of the grid. In this paper we employ the ELM (Extreme Learning Machine) technique on wind power data of 2012 Alberta, Canada market for different sampling times to predict wind power ramp events. We also try to compare it with respect to other existing standard algorithms of feed-forward Neural Networks to analyze the efficacy of the technique in the area. ELM is shown to outperform other techniques in terms of computation time whereas prediction performance is at par with other neural network algorithms.

Keywords Ramp events classification · Extreme learning machine · Neural networks

S. Choubey (✉) · A. Barsaiyan · N.A. Shrivastava · B.K. Panigrahi
Indian Institute of Technology Delhi, Hauz Khas 110016, New Delhi, India
e-mail: sujay.sprng@gmail.com

A. Barsaiyan
e-mail: anubhavgpt08@gmail.com

N.A. Shrivastava
e-mail: anandnitin26@gmail.com

B.K. Panigrahi
e-mail: bijayaketan.panigrahi@gmail.com

M.-H. Lim
School of Electrical and Electronic Engineering, Nanyang Technological University, Nanyang Avenue, Singapore 639798, Singapore
e-mail: emhlim@ntu.edu.sg

© Springer International Publishing Switzerland 2016
J. Cao et al. (eds.), *Proceedings of ELM-2015 Volume 2*,
Proceedings in Adaptation, Learning and Optimization 7,
DOI 10.1007/978-3-319-28373-9_24

1 Introduction

Wind power is a renewable eco-friendly source of energy that is spreading rapidly in popularity. The global share of wind power increased to 4 % in 2014 rising from 283 to 370 GW [1]. However wind energy comes with challenges of its own—the prime one being its intermittency. Variability of wind power is usually different on different timescales—hour, day, and season. When wind power generators are injected into the grid, their instantaneous production and consumption must remain in balance for maintenance of grid stability. While this calls for an accurate model that could reliably forecast power production in advance to plan loads and generators accordingly, it also entails specific functionality of these techniques to predict events of abrupt change (ramp events) in wind power of farms. Ramp events refer to relatively rapid movements in the power of the wind farm. Say the magnitude of wind power change over an interval of 10 min crosses a certain set threshold, then a ramp-up or ramp-down is said to occur. The concept of thresholds is introduced because of the behavior of power utilities in the grid system. If ramp events are small they are left to the generation control mechanisms but large ramp events require intervention such as re-dispatching which might even cause stress and load shedding. For example, ramp events of 15 MW/10 min and 25 MW/10 min may be dealt with by the mechanism but a ramp event of 150 MW/10 min will require different mechanism.

The above example establishes the importance of classification in comparison to point-forecasting. The thresholds vary as per the power system under study, and are system operator defined parameters. In modern grid applications involving self-scheduling by demand-side participants, certain ramp thresholds are more relevant than point forecasts because the behavior of machines changes across specific thresholds. However the selection of severity thresholds depends on the specific wind-farm under consideration. Hence wind forecast and classification is gaining popularity as a research area. Many methods have been proposed in the past to carry out the prediction. A survey of these methods can be found in [2]. A number of wind power models have been developed internationally, such as WPMS, WPPT, Prediktor, ARMINES, Previento, WPFS Ver1.0 etc. [3]. These include numeric weather prediction (NWP), statistical approaches, artificial neural network (ANN) and hybrid techniques over different time-scales.

This paper in particular tries to apply a novel method "Extreme Learning Machine" to the problem and document its performance compared to other common classification methods in the class of Neural Network. This includes gradient descent back-propagation (traingd), gradient-descent with adaptive learning rate back-propagation (traingdx) and gradient-descent with momentum back-propagation (traingdm). Historic data are used to predict to perform one-step ahead and multi-step ahead classification.

Classification problem is being widely worked upon in machine learning. However the available literature is still limited [4]. Neural network-based methods have been extensively used for classification purpose. Their good learning and generalizing ability with any kind of data has made them popular. However practical application is tedious since there are large numbers of parameters that must be iteratively tuned for training [5]. Their training algorithms are known to be slow and likely to get stuck into local minima. ELM is a novel learning algorithm for single feed-forward neural networks (SLFN). Unlike the traditional gradient-based learning algorithms, in ELM the input weights and hidden biases are randomly chosen, and the output weights are analytically determined by using the Moore–Penrose (MP) generalized inverse.

ELM has been applied to many contemporary research problems such as plain text classification [6, 7], Bioinformatics [8], XML document classification [9] and electricity price forecasting [10]. This paper is structured as follows. The methodology is introduced in Sect. 2. The underlying details of the algorithm are mentioned in Sect. 3. Section 4 discusses the approach while modeling input data and analyzing performance. Section 5 contains the experimental results. Conclusion from the study is summarized in Sect. 6.

2 Methodology

A ramp is basically "a large increase or decrease in energy output over a short period of time". As discussed earlier, ramp event is defined in a highly localized manner depending on local wind farm attributes. Thus no universal benchmarks or thresholds exist. The generic definition goes as follows:

A ramp event is considered to occur at the start of an interval if the magnitude of the increase or decrease in generation at a time ΔT ahead of the interval is greater than a predefined threshold, Tr:

$$|MW(T + \Delta T) - MW(T)| > Tr.$$

As discussed earlier, in our case, we have arbitrarily chosen the critical values of -300, -50, 50, 300 MW to define thresholds and classes. Hence

- Ramp values ranging from -500 to -30 MW are labeled as ramp class c1.
- Ramp values ranging from -30 to 30 MW are labeled as ramp class c2;
- Ramp values ranging from 30 to 500 MW are labeled as ramp class c3;

3   Extreme Learning Machine

ELM is a Single Layer Feed forward Network (SLFN) in which input weight matrix W is chosen randomly and the output weight matrix b of SLFN is analytically determined. For a data set with N arbitrary distinct samples (x_i, t_i) where $x_i = [x_{i1}, x_{i2}, \ldots, x_{in}]T \in Rn$ and $t_i = [t_{i1}, t_{i2}, \ldots, t_{im}]T \in Rm$.

The mathematical model of a standard SLFN with $\widetilde{N}$ hidden nodes and activation function g(x) for the given data can be formulated as see [11]

$$\sum_{i=1}^{\widetilde{N}} \beta_i g_i(x_j) = \sum_{i=1}^{\widetilde{N}} \beta_i g_i(w_i x_j + b_i) = y_j, \quad j = 1 \ldots, N \tag{1}$$

Here $w_i = [w_{i1}, w_{i2}, \ldots, w_{in}]T$ is the weight vector connecting the ith hidden node and the input nodes, $t_i = [t_{i1}, t_{i2}, \ldots, t_{im}]T$ is the weight vector connecting the ith hidden node and the output nodes, and b_i is the threshold of the ith hidden node. The operation $w_i . x_j$ in (1) denotes the inner product of w_i and x_j.

Let us consider that the standard SLFNs with $\widetilde{N}$ hidden nodes with activation function g(x) can approximate these N samples with zero error, then we have

$$\sum_{j=1}^{N} \|y_j - t_j\| \tag{2}$$

Here y is the actual output value of the SLFN. This indicates the existence of β_i, w_i and b_i such that

$$\sum_{i=1}^{\widetilde{N}} \beta_i g_i(w_i x_j + b_i) = t_j \quad j = 1 \ldots, N \tag{3}$$

The above N equations can be written compactly as

$$H\beta = T \tag{4}$$

Here H is the hidden layer output matrix:

$$\begin{bmatrix} h(x_1) \\ \ldots \\ h(x_N) \end{bmatrix} = \begin{bmatrix} h_1(x_1) & \ldots & h_{\widetilde{N}}(x_1) \\ \ldots & \ldots & \ldots \\ h_1(x_n) & \ldots & h_{\widetilde{N}}(x_n) \end{bmatrix} \tag{5}$$

$$\beta = \begin{bmatrix} \beta_1^T \\ \vdots \\ \beta_{\widetilde{N}}^T \end{bmatrix} \tag{6}$$

$$T = \begin{bmatrix} T_1^T \\ \vdots \\ T_N^T \end{bmatrix} \tag{7}$$

In ELM, the input weights and hidden biases are randomly generated and do not require any tuning as in the case of SLFN. The evaluation of the output weights linking the hidden layer to the output layer is equivalent to determining the least-square solution to the given linear system. The minimum norm least-square (LS) solution to the linear system (4) is

$$\widehat{\beta} = H^\dagger T \tag{8}$$

The $H^\dagger$ in the above equation is the Moore–Penrose generalized inverse of matrix H [12, 13]. The resultant solution is unique and has the minimum norm among all the LS solutions. ELM tends to obtain a good generalization performance with a radically increased learning speed by using the Moore–Penrose inverse method. In this work, we are using ELM for the classification problem and the ELM classifier handles the multi-class classification problem by using a network of multiple output nodes which are equal to the number of pattern classes. Therefore, for each training sample, the target t_j is m-dimensional vector $(t_1, t_2, \ldots t_m)T$. The final class of the given sample is determined by the output node having the highest output.

4 Datasets and Approach

The data that we used for our study is the wind power data of the year 2012 from Alberta, Canada. Alberta wind farms are mostly located in the southern part of the province, close to the foot of the Rocky Mountains where severe weather changes are experienced throughout the year [5]. Wind power production data for the year 2012 was available at 10-min resolution. This data was used to derive sample values at 30-min intervals and 60-min intervals. Ramp values were also derived from the data itself. For the sake of the classification problem, four thresholds were used uniformly based on the distribution of ramp values to encapsulate the severity of ramp event. Hence three classes were formed, c1 (−500 to −30 MW), c2 (−30 to 30 MW) and c3 (30 to 500 MW). Classification done was univariate using only historical wind power production data as features. A 16-member vector was used for the classification comprising historical wind power data at different lags spanning up to two weeks. One set of 24 features was also used for studying the effect of features on results.

To quantify the performance of each algorithm, simple percentage measure of correct predictions upon total number of predictions (testing data size) was used. Test data used (i.e. n-step ahead prediction) varied according to resolution used.

Thus the classification accuracy was assessed through Percentage Correct Classification (PCC) index.

$$PCC = \frac{Number\ of\ Correct\ Classifications}{Total\ cases} \times 100 \tag{9}$$

PCC indices were evaluated for representative months of different seasons (January, April, August, and November) to compare performance of different algorithms.

5 Experimental Results and Discussion

The ELM algorithm showed a decent performance on the data. Wind power is highly affected by weather variations so we treated samples from different seasons separately to analyze the effect on predictive capability of algorithms.

As can be seen in Figs. 2 and 3, occurrences of ramps does not follow any simple repetitive pattern that holds for each season. Clearly the occurrence of ramp is stronger and more frequent in May vis-a-vis February. Hence it is useful to analyze the problem in context of meteorological cycles. For example, there is greater occurrence of ramps in July than in February (Fig. 1). The results corresponding to different experiments are presented in Table 1.

The ELM-method was applied on two sets of the 10-min samples of wind power data with different number of features (basically lagged values i.e. historical wind power data). One selection was 16-feature based and the second was 24-feature based (Table 2).

We see that the addition of some extra features doesn't introduce any significant improvement in the prediction accuracy. Thus the dependence of the performance of the final algorithm depends highly on proper selection of significant features.

Fig. 1 Schematic representation of extreme learning machine [14]

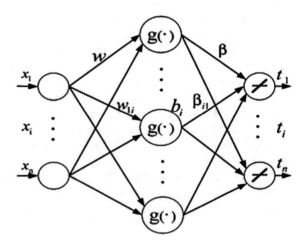

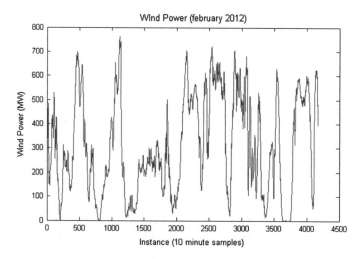

Fig. 2 Wind power versus instance (February 2012)

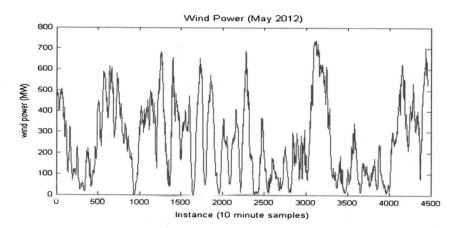

Fig. 3 Wind power versus instance (May 2012)

Timing is another parameter on the basis of which the resourcefulness of a particular algorithm is decided. In this respect, ELM shows explicit advantage over others. While Neural Network is fast, ELM with its unique constitution performs much faster than all neural network algorithms (traingdm, traingd, traingdx) (Tables 3 and 4).

Table 1 Comparison of accuracy through three classification methods (ELM, NN and multiSVM)

	ELM (10-min)	NN-traingdm (10-min)	NN-traingd (10-min)	NN-traingdx (10-min)
January	55.84	44.16	60.89	70.16
April	86.04	46.43	71	69.32
August	81.49	81.49	64.68	69
November	87.66	83.77	72.32	78
	ELM (30-min)	NN-traingdm (30-min)	NN-traingd (30-min)	NN-traingdx (30-min)
January	73	71.14	70	47.50
April	77	76.86	75.43	58.50
August	85.57	85.57	82.29	65.25
November	75	74.71	74.14	54.75
	ELM (60-min)	NN-traingdm (60-min)	NN-traingd (60-min)	NN-traingdx (60-min)
January	60.75	56.25	73	61
April	62.75	54.75	77	72.75
August	73	63.25	85.57	77.50
November	62	58	75	73.25

Table 2 Effect of adding more features to ELM algorithm on prediction accuracy

	16-feature ELM (accuracy %)	24-feature ELM (accuracy %)
January	65.11	62.47
April	70.79	71
August	68.58	68.47
November	73.16	73.16

Table 3 Relative performance of ELM and NN (taingdm, traingd, traingdx) (training time in seconds) for the case of 10-min samples

	Training time (ELM)	Training time (NN-traingdm)	Training time (NN-traingd)	Training time (NN-traingdx)
January	0.0387	10.3978	10.6595	3.4758
April	0.0483	10.44	10.4609	3.9135
August	0.0394	10.5588	11.0594	3.5890
November	0.0460	10.6650	10.4136	3.4917

Table 4 Relative performance of ELM and NN (taingdm, traingd, traingdx) (testing time in seconds) for the case of 10-min samples

	Testing time (ELM)	Testing time (NN-traingdm)	Testing time (NN-traingd)	Testing time (NN-traingdx)
January	0.0138	0.0428	0.0389	0.0398
April	0.0127	0.0429	0.0406	0.0391
August	0.0136	0.0373	0.0422	0.04
November	0.0151	0.0376	0.0408	0.0423

6 Conclusions

Wind Power is undoubtedly the energy resource of the future. With evolving smart grids and preference for green energy, its prevalence will increase. The problem of wind ramp event classification is extremely vital to the entire project as discussed chiefly because of the requirement that consumption must match production.

Our work was aimed at studying various facets of ELM to evaluate its applicability to the problem of wind power ramp event classification in particular. Overall the study reveals a lot about competence of ELM as a tool in Wind Power Ramp forecasting. None of the applied algorithms showed outstanding performance. Further ELM was found to be mostly at par with the rest of neural network algorithms. However ELM provides distinct advantage over all other neural network algorithms in the field of computation time. There is scope of improvement of prediction by studying feature selection methodology and coming up with definitive inferences regarding the right approach to select historical data.

Acknowledgements This work was supported by the Scientists' Pool Scheme of the Council of Scientific and Industrial Research (CSIR), Government of India (No. 8741-A).

References

1. Global Wind Energy Council. Global Wind Statistics 2014. Available at www.gwec.net
2. Soman, S., Zareipour, H., Malik, O., Mandal, P.: A review of wind power and wind speed forecasting methods with different time horizons. In: 42nd North American Power Symposium (NAPS), Arlington, Texas, USA, 26–28 Sept 2010
3. Wang, X., Guo, P., Huang, X.: A review of wind power forecasting models. Energy Procedia **12**, 770–778 (2011)
4. Zareipour, H., Janjani, A., Leung, H., Motamedi, A., Schellenberg, A.: Classification of future222 electricity market prices. IEEE Trans. Power Syst. **26**(1), 165–173 (2011)
5. Shrivastava, N.A., Panigrahi, B.K., Lim, M.-H.: Electricity price classification using extreme learning machines, Neural Comput. Appl. **24**(1) (2014). doi:10.1007/s00521-013-1537-1
6. Rong, H.-J., Huang, G.-B., Ong, Y.-S.: Extreme learning machine for multi-categories classification applications. In: IEEE International Joint Conference on Neural Networks 2008 (IJCNN 2008), IEEE World Congress on Computational Intelligence, pp. 1709–1713 (2008)
7. Rong, H.-J., Ong, Y.-S., Tan, A.-H., Zhu, Z.: A fast prunedextreme learning machine for classification problem. Neurocomputing **72**(1–3), 359–366 (2008)
8. Wang, G., Zhao, Y., Wang, D.: A protein secondary structure prediction framework based on the extreme learning machine. Neurocomputing **72**(1–3), 262–268 (2008)
9. Zhao, X., Wang, G., Bi, X., Gong, P., Zhao, Y.: XML document classification based on ELM. Neurocomputing **74**(16), 2444–2451 (2011)
10. Chen, X., Dong, Z.Y., Meng, K., Xu, Y., Wong, K.P., Ngan, H.W.: Electricity price forecasting with extreme learning machine and bootstrapping. IEEE Trans. Power Syst. **27**(4), 2055–2062 (2012)
11. Huang, G.-B., Zhu, Q.-Y., Siew, C.-K.: Extreme learning machine: theory and applications. Neurocomputing **70**, 489–501 (2006)
12. Serre, D.: Matrices: theory and applications. Springer, New York (2002)

13. Rao, C., Mitra, S.: Generalized inverse of matrices and its applications. Wiley, New York (1971)
14. Zhao, L., Qi, J., Wang, J., Yao, P.: IOP Publishing Ltd 2012. The study of using an extreme learning machine for rapid concentration estimation in multi-component gas mixtures (2012)

Facial Expression Recognition Based on Ensemble Extreme Learning Machine with Eye Movements Information

Bo Lu, Xiaodong Duan and Ye Yuan

Abstract Facial expression recognition has become a very active research in computer vision, behavior interpretation of emotions, human computer interaction, cognitive science and intelligent control. Traditional facial expression analysis methods mainly focuses on the facial muscle movement and basic expression features of face image. In this paper, we propose a novel method for facial expression recognition based on ensemble extreme learning machine with eye movements information. Here, the eye movements information is regarded as explicit clue to improve the performance of facial expression recognition. Firstly, we extract eye movements features from eye movements information which recorded by Tobii eye tracker. The histogram of orientation gradient (HOG) features are simultaneously obtained from the face images by dividing it into a number of small cells. Secondly, we combine the eye movements features together with the HOG features of face images by using a tensor kernel. Finally, the fusion features are trained by ensemble extreme learning machine and a bagging algorithm is explored for producing the results. Extensive experiment on the two widely available datasets of facial expressions demonstrate that our proposal effectively improves the accuracy and efficiency of face expression recognition and achieve performance at extremely high speed.

Keywords Facial expression recognition · Ensemble extreme learning machine · Eye movements information

B. Lu · X. Duan(✉)
Dalian Key Lab of Digital Technology for National Culture,
Dalian Nationalities University, Dalian 116600, China
e-mail: duanxd@dlnu.edu.cn

B. Lu
e-mail: lubo@dlnu.edu.cn

Y. Yuan
School of Information Science and Engineering, Northeastern University,
Shenyang 110004, China

© Springer International Publishing Switzerland 2016
J. Cao et al. (eds.), *Proceedings of ELM-2015 Volume 2*,
Proceedings in Adaptation, Learning and Optimization 7,
DOI 10.1007/978-3-319-28373-9_25

1 Introduction

Facial expression recognition has attracted research community during the last few decades as it is the most common visual pattern in our environment. Facial expression is one of the most powerful, nature, and immediate means for human beings to communicate their emotions and intentions [1]. Traditional facial expression analysis methods are greatly depends on the facial muscle movement and accurate facial features [2]. However, there are some useful implicit information can be used to assist the task of facial expression recognition. Eye movements information provides an index of over attention and reveal the information strategically selected to categorize expressions. Eye is a vital organ of the human emotion, to some extent, it will help people to better understanding of nonverbal communication. We can understand a person's emotion and even the inner world through eyes. Eye movements and facial expressions have a close relationship. Therefore, the eye movements information plays an important role in facial expression recognition. Earlier studies of Hardoon and Kaski [3] explored the problem of where an implicit information retrieval query is inferred from eye movements measured during a reading task. However, there is few literature use eye movements information as a feature to classify the facial expression. In this study, we use the eye movements information as a particular source to improve the performance of facial expression recognition.

Moreover, we expect there to be small training errors during the training stage in the neural network, but neural network training usually suffers from overtraining, which might degrade the generalization performance of the network. In such cases, the neural network can classify the training data without any errors but it can not guarantee that there will be good classification performance on the validation datasets. Huang et al. [4, 5] proposed a new learning algorithm called extreme learning machine (ELM) for single-hidden-layer feedforward neural networks (SLFNs) which can be used in regression and classification applications with a fast running speed. In this paper, we have selected bagging to generate the ensemble of the ELM for recognizing facial expressions. There are several reasons behind choosing ELM as a base classifier. First, the ELM takes random weights between the input and hidden layer. We train the same dataset several times, which gives different classification accuracy with different output space. Second, the ELM is a much simpler learning algorithm for a feedforward neural network. Unlike the traditional neural networks (e.g. support vector machine), it does not need to calibrate the parameters, such as learning rate, learning epochs, etc. Another reason is that the learning speed of ELM is extremely fast.

As mentioned above, in this paper, we propose a novel method for facial expression recognition based on ensemble ELM with eye movements information. We firstly extract eye movements features from eye movements information which recorded by Tobii eye tracker. Then, The histogram of orientation gradient (HOG) features are simultaneously obtained from the face images by dividing it into a number of small cells. We further combine the eye movements features together with the HOG features of face images by using a tensor kernel. The fusion features are trained using ELM base classifiers in a bagging algorithm produce better results as compared with other methods in the literature of facial expression recognition. Here we applied the bagging algorithm to generate the number of training bags from the original training datasets. The fusion features are calculated for each newly generated datasets and facial expression are trained by using individual ELMs. Finally, the result from individual ELMs is combined by using a majority voting scheme. Instead of using a single ELM for expression classification, we used the ELM ensemble with bagging algorithm. This is because it improves the generalization capability of the whole system, even if the classification accuracy of the individual ELMs is smaller.

The rest of this paper is organized as follows. Section 2 describes the procedure of extracting eye movements features, we indicate that the eye movements features and HOG features are combined by using tensor kernel. In Sect. 3, we introduce the ensemble ELM classifier for the task of facial expression recognition. Section 4 demonstrate the performance of our proposal and give experiment results. We finally conclude the paper in Sect. 5.

2 Preliminaries

2.1 Eye Movements Feature Extraction

The eye movement features are computed using only on the eye trajectory and locations of the facial expression images. This type of features are general-purpose and are easily applicable to all application scenarios. The features are divided into two categories; the first category uses the raw measurements obtained from the eye tracker, whereas the second category is based on fixations estimated from the raw data. The following definitions are shown in order to clearly illustrate the whole procedure of extraction of eye movements features.

(i) Area of interest (AOI): area of visual environment which is focus of your attention, select the area for further processing.

(ii) Fixation duration: statistic the duration of each fixation point of the AOI. It is an important indicator of eye movements data, the longer the duration, is likely to mean that participants' access to information in the AOI is more important.

(iii) Fixation count: the number of fixations in an AOI. It is a sign to distinguish the interest area important degree, the more the count, indicates this area is more important.

(iv) Fixation order: order of the participants' fixations in AOI. To some extent, this indicator can reflect the attention degree of AOI.

(v) Regression times: participants observe AOI again formed the order of sequence.

(vi) Regression duration: the duration of AOI's fixation once again.

(vii) Scan path: a sequence of spatial arrangement about fixations.

Fig. 1 AOI's division
diagram

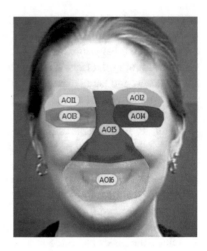

2.2 *Preprocessing of Eye Movements Features*

There is an important step that we have to reduce the size of the recorded eye movements features by eye movements tracking system. Therefore, we firstly preprocess the eye movements data by calculating the weights of AOI. When people observing the human face images, their fixations mainly focuses on eyes, eyebrows, nose and mouth [6]. In order to better analyze the subjects' AOI of fixation situation, the facial images are divided into six AOIs. Left eyebrow is AOI_1, right eyebrow is AOI_2, left eye is AOI_3, right eye is AOI_4, nose is AOI_5, mouth is AOI_6. Figure 1 shows the AOIs' division diagram.

We defined f_i as the duration of AOI_j on i, i is the sequence number of the participants' fixation, AOI_j represent participants in the ith order's fixation of AOI, j is interest area number (range is 1–6), w_1 is the difference between the order of importance of the order factor, w_1 value is in the range of [1, 2], the first fixation factor value is 2, after fixation factor in turn reduce the i/p, p is the number of all the fixation point of participant. According to the order of fixation we can get each AOI fixation time for T_1.

$$T_1 = f_i (2 - i/p) \tag{1}$$

When an AOI have a regression, this experiment is defined f_i as the ith regression duration of AOI_j, c_i is the ith position' in this AOI is the Nth regression, w_2 is regression factor, w_2 value is in the range of [1, 2], the first regression factor value is 1, after regression factor in turn increase the $c_i/20$. Since each AOI there may be have multiple regressions, the total regression time should be the sum of all of the regression values for this AOI. According to the above method get each AOI regressive value of time T_2.

$$T_2 = f_i \left(1 + c_i/20\right) \tag{2}$$

Finally, the AOI's weight is $(T_1 + T_2)$/total, total is the sum of all the fixation and regression time of AOI. By Eqs. (1) and (2) using the literature [7] obtain each AOI's $Weight_j$.

$$Weight_j = \frac{\sum_{i=0}^{i=p} f_i (2 - i/p) (1 + c_i/20)}{total} \qquad (3)$$

2.3 Features Fusion Based on Tensor Kernel

As mentioned above, we can obtain the eye movements features and HOG features of facial image from different data source. Then, we propose to construct a tensor kernel to combine image and eye movements features. Let $X \in R^{n \times m}$ and $Y \in R^{l \times m}$ be the matrix of sample vectors, x and y, for the HOG features and eye movements features respectively, where n is the number of image features and m are the total number of the samples. We continue to define K^x, K^y as the kernel matrices for the facial expression image and eye movements respectively. In our experiments we use linear kernels, i.e. $K^x = X'X$ and $K^x = Y'Y$. The resulting kernel matrix of the tensor $T = X \circ Y$ can be expressed as pair-wise product [8].

$$\bar{K}_{ij} = (T'T)_{ij} = K^x_{ij} K^y_{ij} \qquad (4)$$

3 Ensemble ELM for Facial Expression Recognition

In this section we will describe the classifier that we have employed in our method for facial expression recognition. In the following subsection we will give brief introduction of ELM and ensemble ELM for facial expression recognition.

3.1 Brief of ELM

ELM is a new algorithm based on Single-Hidden Layer Feedforward Networks (SLFNs) [9, 10]. Compared with traditional SVM, ELM not only tends to reach the smallest training error but also the smallest norm of the output weights. ELM is not very sensitive to user specified parameters and has fewer optimization constraints [11]. In addition, ELM tends to provide good generalization performance at extremely high learning speeds.

For the multi-categories classification problem, ELM classifier uses a network of multiple output nodes equal to the number of pattern classes m, as shown in

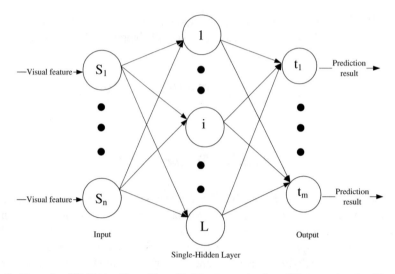

Fig. 2 Example of ELM classifier with multiple output nodes for multi-categories classification

Fig. 2. For each training sample s_i, the target output t_i is an m-dimensional vector $(t_1, t_2, \ldots, t_m)^T$.

The learning procedure of the ELM classifier is given below. For N arbitrary distinct shot samples $(s_i, t_i) \in R^n \times R^m$, if an SLFN with L hidden nodes can approximate these N samples with zero error, then we have

$$\sum_{i=1}^{L} \beta_i G(a_i, s_j, b_i) = t_j, \quad j = 1, \ldots, N \tag{5}$$

where a_i is the weight vector connecting the ith hidden node and the input nodes, β_i is the output weight linking the ith hidden node to the output node, and b_i is the threshold of the ith hidden node. $G(a_i, s_j, b_i)$ is the output of the ith hidden node with respect to the input s_j. In our simulations, sigmoid activation function of hidden nodes are used, we formulate it as $G(a_i, s_j, b_i) = 1/(1 + \exp(-a_i \cdot s_j + b_i))$. Under these conditions, Eq. (3) can be written compactly as

$$H\beta = T \tag{6}$$

where

$$H = \begin{bmatrix} G(a_1, s_1, b_1) & \cdots & G(a_L, s_1, b_L) \\ \vdots & \cdots & \vdots \\ G(a_1, s_N, b_1) & \cdots & G(a_L, s_N, b_L) \end{bmatrix}_{N \times L} \tag{7}$$

$$\beta = \begin{bmatrix} \beta_1^T \\ \vdots \\ \beta_L^T \end{bmatrix}_{L \times m}, \quad T = \begin{bmatrix} t_1^T \\ \vdots \\ t_N^T \end{bmatrix}_{N \times m} \tag{8}$$

While computing, the determination of the output weights β is estimated by the smallest norm least-squares solution and defined as

$$\hat{\beta} = H^\dagger T \tag{9}$$

where $H^\dagger$ is the Moore-Penrose generalized inverse [12] of the hidden layer output matrix H. The original algorithm of ELM proposed by Huang et al. [4] contains three steps as follows:

Algorithm ELM: Given a training set $\aleph = \{(s_i, t_i) | s_i \in R^n, t_i \in R^m, i = 1, \dots, N\}$, activation function $G(x)$, and hidden node number L, then training the ELM classifier takes several steps:

Step 1: Randomly assign hidden node parameters $(a_i, b_i), i = 1, \dots, L$.
Step 2: Calculate the hidden layer output matrix H.
Step 3: Calculate the output weight $\hat{\beta} = H^\dagger T$.

3.2 Ensemble ELM for Facial Expression Recognition

Let X denote the training samples for a K-class classification problem:

$$X = \bigcup_{i=1}^{K} X_i = \bigcup_{i=1}^{K} \{X_l^i\}_{l=1}^{L_i} \tag{10}$$

where X_i is the training sample set of the class C_i, X_l^i is the lth sample, and L_i is the number of samples in the class C_i.

This K-class classification problem can be divided into $K(K-1)/2$ smaller binary subproblems through the one-versus-one (OAO) strategy as follows:

$$T_{ij} = \{X_l^i, +1\}_{l=1}^{L_i} \cup \{(X_l^j, -1)\}_{l=1}^{L_i} \tag{11}$$

where the classes C_i and C_j are taken as positive and negative classes, respectively. Alternatively, a K-class classification problem can also be divided into K binary subproblems through the one-versus-all (OVA) strategy as follows:

$$T_i = \{X_l^i, +1\}_{l=1}^{L_i} \cup \left(\bigcup_{j=1, j \neq i}^{K} \{(X_l^j, -1)\}_{l=1}^{L_i} \right) \tag{12}$$

These binary subproblems defined by Eqs. (11) and (12) can be further divided. Assume that the sample set X_i is partitioned into N_i subsets in the form

$$X_i^\mu = \{X_l^{(i,\mu)}\}_{l=1}^{L_i^\mu} \tag{13}$$

where X_i^μ is the ath subset of X_i, $X_l^{(i,\mu)}$ is the lth sample, L_i^μ is the number of the samples, and $\bigcup_{\mu=1}^{N_i} X_i^\mu = X_i$.

After the partition of the sample sets defined by Eq. (13), each binary subproblem T_{ij} is divided as follows:

$$T_{ij}^{(\mu,v)} = \{(X_l^{(i,\mu)}, +1)\}_{l=1}^{L_i^\mu} \cup \{(X_l^{(j,v)}, -1)\}_{l=1}^{L_j^v} \tag{14}$$

where the sample sets X_i^μ and X_j^v are taken as positive and negative sets, respectively.

After all the binary subproblems defined by Eq. (14) have been learned, the trained classifiers are integrated through the M^3-network. The module combination for the task decomposition defined by Eq. (14) can be formulated as

$$g_{ij}(x) = \max_{\mu=1}^{N_i} \min_{v=1}^{N_j} h_{ij}^{(\mu,v)}(x) \tag{15}$$

where x is a sample, $g_{ij}(x)$ is the discriminant function of the binary problem T_{ij}, and $h_{ij}^{(\mu,v)}$ is the output of the module $M_{ij}^{(\mu,v)}$.

4 Experimental Evaluation

In this section, we evaluate the performance of our proposed approach on facial expression datasets through extensive experiments.

4.1 Experiments Setup

The performance of the proposed system is evaluated on two well-known facial expression datasets, which are the Japanese Female Facial Expression (JAFFE) dataset [13] and the Extended Cohn-Kanade (CK+) facial expression dataset [14]. All of the results presented in this paper for both the JAFFE and CK+ dataset belong to the validation set. Moreover, we define some symbol which include AN, DI, FE, HA, NE, SA, SU to represent 7 facial expressions include neural, angry, disgust, fear, happy, sadness and surprise respectively.

We chose 7 facial expression from the above two databases, each expression has 20 images, all the CK+ images were cropped into 300 * 340 pixels, the JAFFE images were cropped into 210 * 230. Participant sat in front of the computer approximately 60cm, each image was displayed 10s. Then using Tobii X2-30 eye tracker to record the participants' eye movements. Forty participants took part in this experiment, half males and half females. All participants were university students. They had normal or corrected-to-normal vision and provided no evidence of color blindness. There is a initial five points of calibration process in every experiment, everyone carried out the experiment three times, then we took the average as the final data.

4.2 Experiments Results

In this section, we demonstrate that several of the experimental results from both of the datasets what we used. First of all, we compared single HOG features with fusion feature (combination of HOG features and eye movements features). Figure 3 shows the performance of the ensemble classification system on the CK+ dataset with a different number of base ELM classifiers. We determined the HOG features by dividing the image into $8 \times 8, 6 \times 6$, and 4×4 cells, with 9 bins per cell, which resulted in 512, 328, and 128 dimensional HOG features. The image size taken for this experiment was 100×100 pixels. From Fig. 3 we can observe that classification performance using 328-D fusion features is better on others.

Moreover, we performed was on expression classification with different resolutions of input face images. The resolution of the input image is very important. If we want to develop a real time facial expression classification system, then in general, the resolution of the detected input face image can vary from low resolution to high resolution. In our experiment, we found that by reducing the image resolution, the performance of the individual base ELMs also decreased. However, the bagging

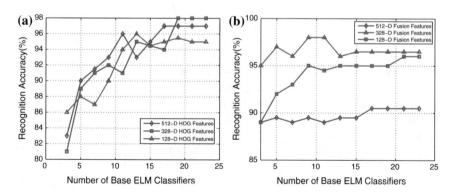

Fig. 3 Recognition performance of an ensemble ELM using bagging with HOG features and fusion features on CK+ dataset. **a** Performance of ensemble ELM use HOG features. **b** Performance of ensemble ELM use fusion features

Table 1 Average recognition rate (%) of the 23 base ELM classifiers with different image resolutions

Database	Resolution of image					
	96×96	84×84	72×72	60×60	48×48	32×32
CK+	87.90	87.72	82.67	81.22	79.50	75.54
JAFFE	74.51	70.46	70.29	65.45	63.14	58.02

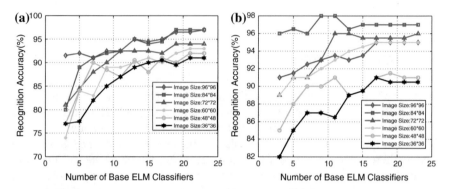

Fig. 4 Recognition performance of an ensemble ELM using bagging for different image resolutions with a different number of base ELM classifiers on dataset. **a** Performance of ensemble ELM use JAFFE dataset. **b** Performance of ensemble ELM on CK+ dataset

result was not significantly decreased, even if the classification performance of the individual base ELMs decreased. Table 1 shows the average classification accuracy of 23 base ELMs with a different resolution in both datasets. Regardless of the resolution of the image, we divided the face image into 6×6 cells and the 328-D fusion features were obtained.

From Table 1 we observed that the average classification accuracy was decreased from 74.51 to 58.02 % by reducing the image resolution from 96×96 to 36×36 in the case of the JAFFE dataset and that the classification accuracy was reduced from 87.90 to 75.54 % in the case of the CK+ dataset. Classifying facial expressions in the CK+ dataset is relatively easier than classifying expressions in the JAFFE dataset. There is big difference in the classification accuracy of a single ELM, as we reduced the image resolution. Figure 4 shows the result of an ELM ensemble using bagging with different input face image resolutions. For low-resolution images, even though the base classifier accuracy was smaller, the bagging performance was better. With 23 base ELMs in bagging algorithms, the lowest classification accuracy was around 90 % and the highest classification accuracy was around 97 %, in the case of both the JAFFE and CK+ facial expression datasets. The classification results are relatively more stable in the case of the CK+ dataset.

5 Conclusion

In this paper, we propose a novel method for facial expression recognition based on ensemble extreme learning machine with eye movements information. Here, the eye movements information is regarded as explicit clue to improve the performance of facial expression recognition. Firstly, we extract eye movements features from eye movements information which recorded by Tobii eye tracker. The histogram of orientation gradient (HOG) features are simultaneously obtained from the face images by dividing it into a number of small cells. Secondly, we combine the eye movements features together with the HOG features of face images by using a tensor kernel. Finally, the features of fusion are trained by ensemble extreme learning machine and a bagging algorithm is explored for producing the results. Extensive experiment on the two widely available datasets of facial expressions demonstrate that our proposal effectively improves the accuracy and efficiency of face expression recognition. Our future work will expand this experiment to six basic expressions plus neutral expression and invite more people to carry out this experiment, this will develop such dataset which could be more useful in developing computational model for visual attention, by analysis this dataset, the results would be more accurate and robust.

Acknowledgments This work was supported by the National Natural Science Foundation of China under Grant No. 61370146; National Natural Science Foundation of China under Grant No. 2013405003; Fundamental Research Funds for the Central Universities No. DC201501030401, DC201502030203.

References

1. Tian, Y., Brown, L., Hampapur, A., Pankanti, S., Senior, A., Bolle, R.: Real world real-time automatic recognition of facial expression. IEEE PETS, Australia (2003)
2. Whitehill, J., Bartlett, M.S., Movellan, J.: Automatic facial expression recognition. In: Social Emotions in Nature and Artifact. London: Oxford University Press (2013)
3. Hardoon, D., Kaski, K.: Information retrieval by inferring implicit queries from eye movements. In: Proceedings of the International Conference on Artificial Intelligence and Statistics, pp. 29–46 (2007)
4. Huang, G.B., Zhu, Q.Y., Siew, C.K.: Extreme learning machine: theory and applications. Nuerocomputing **70**(1–3): 489–501 (2006)
5. Huang, G.B., Zhu, Q.Y., Siew, C.K.: Etreme learning machine: a new learning scheme of feedforward neural networks. In: Proceedings of International Joint Conference on Neural Networks, vol. 2, pp. 985–990 (2004)
6. Schurgin, M.W., Nelson, J., Lida, S., Franconeri, S.L.: Eye movements during emotion recognition in faces. J. Vis. **14**(13), 1–16 (2014)
7. Chen, J.J., Yan, H.X., Xiang, J.: Study of decoding mental state based on eye tracks using SVM. Comput. Eng. Appl. **47**(11), 39–42 (2011)
8. Pulmannova, S.: Tensor products of hilbert space effect algebras. Rep. Math. Phys. **53**(2), 301–316 (2004)
9. Huang, G.B., Chen, L.: Convex incremental extreme learning machine. Neurocomputing **70**(1), 3056–3062 (2007)

10. Huang, G.B., Zhu, Q.Y., Siew, K.Z.: Can threshold networks be trained directly? IEEE Trans. Circuits Syst. II **53**(3), 187–191 (2006)
11. Huang, G.B., Ding, X.J., Zhou, H.: Optimization method based extreme learning machine for classification. Neurocomputing (2010)
12. Serre, D.: Matrieces: Theory and Applications. Springer, New York (2002)
13. Lyons, M., Akamatsu, S., Kamachi, M., Gyoba, J.: Coding facial expressions with Gabor wavelets. In: Proceedings of the IEEE International Conference on Face and Gesture Recognition, pp. 200–205 (1998)
14. Lucey, P., Cohn, J.F., Kanade, T., Saragih, J., Matthews, I.: The extended cohn-kanade dataset (CK+): a complete dataset for action unit and emotion-specific expressions. IEEE Computer Society Conference on Computer Vision and Pattern Recognition Workshops, pp. 94–101 (2010)

Correlation Between Extreme Learning Machine and Entorhinal Hippocampal System

Lijuan Su, Min Yao, Nenggan Zheng and Zhaohui Wu

Abstract In recent years there has been a considerable interest in exploring the nature of learning and memory system among artificial intelligence researchers and neuroscientists about the neural mechanisms, simulation and enhancement. While a number of studies have investigated the artificial neural networks inspired by biological learning and memory systems, for example the extreme learning machine and support vector machine, seldom research exists examining and comparing the recording neural data and these neural networks. Therefore, the purpose of this exploratory qualitative study is to investigate the extreme learning machine proposed by Huang as a novel method to analyze and explain the biological learning process in the entorhinal hippocampal system, which is thought to play an important role in animal learning, memory and spatial navigation. Data collected from multiunit recordings of different rat hippocampal regions in multiple behavioral tasks was used to analyze the relationship between the extreme learning machine and the biological learning. The results demonstrated that there was a correlation between the biological learning and the extreme learning machine which can contribute to a better understanding of biological learning mechanism.

Keywords Extreme learning machine · Biological learning · Entorhinal hippocampal system · Multiunit extracellular recordings · Local field potential · Spike

1 Introduction

In recent years, researchers in artificial intelligence as well as neuroscience have paid considerable attention to explore the learning and memory mechanisms to explain how the biological brain itself to process information [1–3]. The study

L. Su · M. Yao (✉) · N. Zheng · Z. Wu
College of Computer Science and Technology, Zhejiang University, Hangzhou, China
e-mail: myao@zju.edu.cn

L. Su · M. Yao · N. Zheng · Z. Wu
Qiushi Academy for Advanced Studies, Zhejiang University, Hangzhou, China

© Springer International Publishing Switzerland 2016
J. Cao et al. (eds.), *Proceedings of ELM-2015 Volume 2*,
Proceedings in Adaptation, Learning and Optimization 7,
DOI 10.1007/978-3-319-28373-9_26

of the learning and memory mechanisms is a prominent topic within the field of neuroscience. Considerations is most commonly given to the anatomical structure of hippocampal formation, which includes the dentate gyrus, the hippocampus proper, the subiculum and the entorhinal cortex. Interest has also been taken in the inter-action between hippocampal formation, in terms of the internal circuitry [4]. In the biological brain, the dendrite of a typical neuron is responsible to collect signals from other dentrites, and the axon of the neuron is a long and thin stand to deliver spikes of electrical activity. At the end of each axon, there are thousands of branches which were called synapses. A synapse is a structure of one neuron to pass an electrical or chemical signal to another neuron. When a neuron receives excitatory input that is enough large compared with the inhibitory input, it will send a spike of electrical activity along its axon. In this way, the learning and memory occur by changing the connection between the two neurons which is strengthened when both the neurons are active at the same time.

On the other hand, the artificial neural networks inspired by biological neural networks have drawn much attention and have been successfully used to solve a wide variety of tasks, especially in classification, regression and clustering [5]. Artificial neural networks are typically consisted of many interconnected nodes, which are the abstract models of neurons and can be divided into three types of input layer, hidden layer, and output layer [6]. And the synapse between two neurons is modelled as the connected weight between two nodes which can be modified. Each node in the artificial neural network can convert all its received signals into one single activity signal which is broadcasted to all other connected nodes. In this process, there are two important stages. First, one node multiples each input signal by the corresponding connected weight and adds all these weighted inputs together to produce the total input. Then, the node transforms the total input into the output activity signal with a transfer function, such as threshold transfer function, log-sigmoid transfer function, piecewise linear transfer function, and Gaussian transfer function. There are many different types of artificial neural networks, but they are generally classified into feedforward and feedback neural networks. A feedforward neural network is a non-recurrent neural network in which all the signals from the input layer, the hidden layer and the output layer can only deliver in one direction. Typical feedforward neural networks include perceptron, radial basic function, extreme learning machine [6–8] and so on. On the other hand, a recurrent artificial neural network can travel in two directions, for example Hopfield network, Elman network and Jordan network.

Although considerable research has been devoted to the learning and memory mechanisms as well as the artificial neural networks, rather less attention has been paid to the relationship between the artificial neural networks and the neural mech-anisms, especially from the neural data perspective. This lack of empirical stud-ies is somewhat surprising as in recent years there has been growing interest in the multiunit extracellular recordings in local field potential and action potential. Under-standing the correlation between the artificial neural network and the extracellular recording data would seem to be a promising attempt to fill the gap between biolog-ical learning and the artificial learning.

Our primary goal in this paper is to explore the correlation between the extreme learning machine and the entorhinal hippocampal learning system. In short, we will attempt to demonstrate that extreme learning machine can make a useful contribution to our understanding of the biological learning and the entorhinal hippocampal system. To achieve this aim the structure of this paper is as follows. We first review empirical work on the entorhinal hippocampal system in Sect. 2. Subsequently, we report the theoretical research of the study on extreme learning machine in Sect. 3. And then the evaluation and the correlation between extreme learning machine and the entorhinal hippocampal system are analysed and evaluated with experiments in Sect. 4. Finally the conclusion as well as the discussion of our current and future work are present in Sect. 5.

2 Extracellular Recordings in the Entorhinal Hippocampal System

Recent researches on lesion experiments with the involvement of the hippocampus in the animal learning and memory have revealed that the hippocampus plays an especially important role in processing and remembering the spatial and contextual information, however the other related hippocampal formation structures also make important contributions to learning and memory [4, 9].

The hippocampal formation in animal brain includes the hippocampus proper and some other adjacent closely associated cortical regions in the animal brain. The hippocampus proper consists of the cornu ammonis fields: the much-studied cornu ammonis field 1 and cornu ammonis field 3, and the smaller little-studied cornu ammonis field 2 [10]. Thus the main components of the hippocampal formation include: the entorhinal cortex which is divided into lateral and medial entorhinal cortex to constitute the major gateway between the hippocampal formation and the neocortex, dentate gyrus, cornu ammonis field 1, cornu ammonis field 3, subiculum, presubiculum and parasubiculum [11].

As shown in Fig. 1, classic pathway consists of projection from Entorhinal cortex (LEC: lateral entorhinal cortex; MEC: medial entorhinal cortex) to Dentate gyrus (DG), from DG to cornu ammonis field 3 (CA3), and from CA3 to cornu ammonis field 1 (CA1). The entorhinal input also consists of direct monosynaptic LEC and MEC projections to CA3, to CA1, and to subiculum (Sb). CA1 projection to Sb and to LEC/MEC, and Sb projections to LEC/MEC, complete the circuit. Other circuits involve projections from subiculum (Sb) to presubiculum (PreSb) and to parasubiculum (ParaSb), and projections from PreSb to MEC and ParaSb to LEC and MEC [12]. Among all the layers of entorhinal cortex, the superficial layers (ECs) are typically thought as the interconnected set of cortical cortex to deliver information to the hippocampus, on the other hand the deep layers (ECd) and the subiculum are regarded to provide the output from the hippocampal formation to a variety of other multimodal association areas of the cortex such as parietal, temporal, and prefrontal

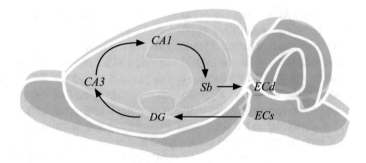

Fig. 1 Major regions and pathways in the entorhinal hippocampal system

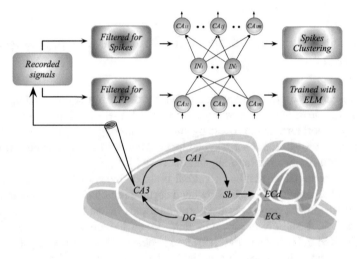

Fig. 2 Flowchart of processing of the extracellular recording data in the entorhinal hippocampal system

cortex in the brain [12]. It has been reported that some CA3 pyramidal cells also project to some other CA3 pyramidal cells. To distinct from the CA3 to CA1 projections proposed by Marr to underlie pattern completion, these projections was called the Schaffer collaterals [13].

In this paper, our data contains multiunit recordings from different rat hippocampal regions which includes EC2, EC3, EC4, EC5, CA1, CA3 and DG [11, 14]. In the experiments, the data was recorded using either 4 or 8 extracellular electrodes from 11 rats performing 14 different behavioral tasks, such as midSquare, bigSquare and bigSquarePlus. Each extracellular electrode has 8 recording sites, in this case in our database there are 32 or 64 recording sites. As shown in Fig. 2, the raw data recorded from multiple channels at 20,000 Hz sample rate is consisted of the spikes generated by one or more neurons and the local field potentials (LFPs) which represent the total synaptic current in the neuronal circuit. Typically, the spikes can be identified by high-pass filtering, thresholding, and subsequent sorting, whereas the LFPs can be extracted by low-pass filtering the wideband signal.

3 Spike Clustering with Extreme Learning Machine

To explore and explain the mechanisms of animal learning and memory, recording brain activity is an important method [15]. With the development of neuroscience, there are many technologies available to record brain activity, such as electroencephalography (EEG), magnetoencephalography (MEG), functional magnetic resonance imaging (fMRI), local field potential (LFP) and spikes [16]. Single or multiple unit recordings can be done intracellularly or extracellularly, and much of our mechanistic understanding of brain function comes from extracellular recordings, which provide a direct and effective measure of electrical activity near the tip of the recording electrode [17].

In this paper, our data contains multiple unit recordings from different rat hippocampal regions which includes EC2, EC3, EC4, EC5, CA1, CA3 and DG. In the experiments, the data was recorded using either 4 or 8 extracellular electrodes from 11 rats performing 14 different behavioral tasks, such as midSquare, bigSquare, and bigSquarePlus. Each extracellular electrode has 8 recording sites, so there are 32 or 64 recording sites in the data. The raw data recorded from channel at 20 KHz sample rate is composed of the spikes emitted by one or more neurons and the local field potentials which represent the total synaptic current in the neuronal circuit. Typically, the spikes can be identified by high-pass filtering, thresholding, and subsequent sorting, whereas the LFP can be extracted by low-pass filtering the wideband signal [18, 19].

The raw data was firstly processed by a high-pass filter at 800 Hz. And then the filtered data was used to detect possible spikes using the simple threshold trigger method, for the amplitude is the most prominent feature of the spike shape. When spikes are detected, extracting features with principal component analysis from multiple spike shapes will save much computational time as well as be helpful for the spikes clustering. Once the features are selected, the data can be used to group spikes with similar features into clusters which represent the different neurons. In this paper, we use extreme learning machine to do the spike sorting and compare the results with the other methods shown in Fig. 3 [20, 21].

In the experiments, the voltages from all the recording sites are recorded. When any one of the voltages is bigger than a threshold, a possible spike is detected and stored with the time of the possible spike and a window of data surrounding the spike. Then with the extreme learning machine we do the spike sorting on this possible spikes. The results computed by extreme learning machine mean that the spikes in the same cluster are most likely produced by the same neuron. We train the data with 50 times, and the average result is regarded as the total number of recorded neurons.

Fig. 3 Schematic diagram
of spikes clustering with
extreme learning machine

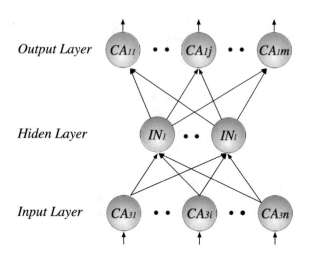

The total number computed by extreme learning machine on one session data is 21, which is similar as the total number computed by the KlustaKwick and Kclusters of 18.

4   Local Field Potentials Trained with Extreme Learning Machine

Understanding the relationships between single-neuron spiking and network activity is therefore of great importance and the latter can be readily estimated from low-frequency brain signals known as local field potentials(LFPs) [22]. The local field potential (LFP) is the electrical potential recorded with intracranial electrodes in the extracellular space around neurons which can be obtained by low pass-filtering usually with a cutoff low-pass frequency in the range of 100–300 Hz and can capture multitude of neural processes, such as synchronized synaptic potentials, afterpotentials of somatodendritic spikes, and voltage-gated membrane oscillations.

For this reason, the LFP can be widely used to investigate the dynamics and the functions of neural circuits in different conditions, which can offer unique windows onto integrative excitatory and inhibitory synaptic process at level of neural population activity [23]. And on other hand, the LFP is sensitive to subthreshold integrative processes and carries information about the state of the cortical network and local intracortical processing, including the activity of excitatory and inhibitory interneurons and the effect of neuromodulatory pathways. These contributions are almost impossible to capture using spiking activity from only a few neurons. Therefore, the combined recording and analysis of LFPs and spikes offers more insights into the circuit mechanism that cannot be obtained at present by examining spikes alone [24].

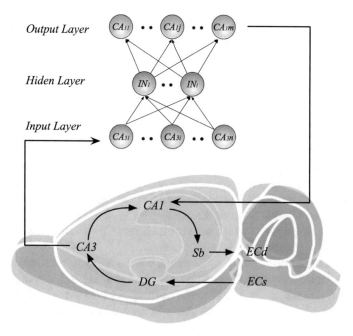

Fig. 4 Schematic diagram of LFPs trained with extreme learning machine

Here in this paper we train the electrical neural data on the projection from CA3 to CA1 with extreme learning machine. As shown in Fig. 4, the neural data is recorded from the CA3 and CA1 region in the entorhinal hippocampal system of the rat brain. By low pass-filtering in the range 100–300, we obtain the LFPs in these regions. The neural data from CA3 is regarded as inputs of extreme learning machine, which the neural data from CA1 is regarded as the output of the extreme learning machine. The hidden layer in this network is to model the interneurons (IN) between the projection from CA3 neurons to CA1 neurons. After the network is trained, we compare the output of the network with the neural data recorded from CA1 in the test dataset. The average accuracy of 50 trainings on one session data with 11,710 samples is 75 %, and the accuracy can be improved with the data preprocessing to 82 %.

5   Conclusion and Discussion

The paper has successfully demonstrated that the theory of the extreme learning machine has some similarities with the biological learning. In particular, it has been shown that the network has the similar feature as the neural data in entorhinal hippocampal system. Furthermore, this claim has been given experimental evidence by the local field potentials and the action potentials. We hope the theoretical framework and the correlation analysis will assist biologists to explore or verify the underlying mechanisms. Beyond in the theoretical science, the application or the improvements of the extreme learning machine can lead to industrial innovations, robotic control and physical rehabilitation.

Our studies add to the understanding of the correlation between the extreme learning machine and the biological learning [7]. We conducted two studies of the extreme learning machine methods in the neural data perspective from the entorhinal hippocampal system. Results across the studies indicate that the spike clustering with the extreme learning machine can obtain good results. Furthermore, the results across the studies imply that using the extreme learning machine to train the local field potentials data can have a stable neural network with a good generalization [25]. This finding, consistent with the previous studies [20, 26], indicated that there may be significant correlation between the extreme learning machine and the biological learning. This conclusion also fits with the findings of the similar random property and the stable performance about different tasks.

Another contribution of the present research is the use of feedforward neural network methods to investigate the complex nature of learning and memory system in entorhinal hippocampal system. Extreme learning machine is particularly valuable for such research because it is more stable in a wide range and is much easier in online sequential learning with fast speed and high accuracy.

However, a comparison between the related methods of spike clustering and network training is lacking, thus should also be one topic of our future research. Still, we did not measure the instructional practices in our studies, and doing so would add to the present research. Some related studies have shown the importance of the implementation of the neural network about the projection from CA3 to CA1. Another direction for our future work would be to further consider the consistency between the extreme learning machine and the biological machine. Clearly, the relations between artificial neural networks and the biological learning raised further question for future investigation on this topic. It is to be hoped that researchers will continue to pursue this research agenda.

Acknowledgments This work was supported by the National Key Basic Research Program of China (973 program, No. 2013CB329504) and partially supported by Zhejiang Provincial Natural Science Foundation of China (No. LZ14F020002) and the Natural Science Foundation of China (No. 61103185, No. 61572433 and No. 61472283).

References

1. Hartley, T., Lever, C., Burgess, N., O'Keefe, J.: Space in the brain: how the hippocampal formation supports spatial cognition. Philos. Trans. Roy. Soc. Lond. B: Biol. Sci. **369**(1635), 20120510 (2014)
2. Su, L., Zhang, N., Yao, M., Wu, Z.: A computational model of the hybrid bio-machine mpms for ratbots navigation. IEEE Intell. Syst. **29**(6), 5–13 (2014)
3. Zheng, N., Su, L., Zhang, D., Gao, L., Yao, M., Wu, Z.: A computational model for ratbot locomotion based on cyborg intelligence. Neurocomputing **170**, 92–97 (2015)

4. Yamaguchi, Y., Sato, N., Wagatsuma, H., Wu, Z., Molter, C., Aota, Y.: A unified view of theta-phase coding in the entorhinal-hippocampal system. Curr. Opin. Neurobiol. **17**(2), 197–204 (2007)

5. Huang, G.B., Zhu, Q.Y., Siew, C.K.: Extreme learning machine: a new learning scheme of feed-forward neural networks. In: 2004 IEEE International Joint Conference on Neural Networks. Proceedings, vol. 2, pp. 985–990. IEEE (2004)

6. Huang, G.B., Wang, D.H., Lan, Y.: Extreme learning machines: a survey. Int. J. Mach. Learn. Cybern. **2**(2), 107–122 (2011)

7. Huang, G.B.: What are extreme learning machines? Filling the gap between frank rosenblatts dream and john von neumanns puzzle. Cogn. Comput. **7**(3), 263–278 (2015)

8. Huang, G.B., Zhu, Q.Y., Siew, C.K.: Extreme learning machine: theory and applications. Neurocomputing **70**(1), 489–501 (2006)

9. Gluck, M.A., Meeter, M., Myers, C.E.: Computational models of the hippocampal region: linking incremental learning and episodic memory. Trends Cogn. Sci. **7**(6), 269–276 (2003)

10. Deshmukh, S.S., Knierim, J.J.: Representation of non-spatial and spatial information in the lateral entorhinal cortex. Front. Behav. Neurosci. **5** (2011)

11. Mizuseki, K., Sirota, A., Pastalkova, E., Buzsáki, G.: Theta oscillations provide temporal windows for local circuit computation in the entorhinal-hippocampal loop. Neuron **64**(2), 267–280 (2009)

12. Diba, K., Buzsáki, G.: Hippocampal network dynamics constrain the time lag between pyramidal cells across modified environments. J. Neurosci. **28**(50), 13448–13456 (2008)

13. Deuker, L., Doeller, C.F., Fell, J., Axmacher, N.: Human neuroimaging studies on the hippocampal ca3 region—integrating evidence for pattern separation and completion. Front. Cellular Neurosci. **8** (2014)

14. Mizuseki, K., Diba, K., Pastalkova, E., Teeters, J., Sirota, A., Buzsáki, G.: Neurosharing: large-scale data sets (spike, lfp) recorded from the hippocampal-entorhinal system in behaving rats. F1000Research **3** (2014)

15. Klauke, N., Smith, G.L., Cooper, J.: Extracellular recordings of field potentials from single cardiomyocytes. Biophys. J. **91**(7), 2543–2551 (2006)

16. Buzsáki, G., Anastassiou, C.A., Koch, C.: The origin of extracellular fields and currentseeg, ecog, lfp and spikes. Nat. Rev. Neurosci. **13**(6), 407–420 (2012)

17. Gold, C., Henze, D.A., Koch, C., Buzsáki, G.: On the origin of the extracellular action potential waveform: a modeling study. J. Neurophysiol. **95**(5), 3113–3128 (2006)

18. Belitski, A., Gretton, A., Magri, C., Murayama, Y., Montemurro, M.A., Logothetis, N.K., Panzeri, S.: Low-frequency local field potentials and spikes in primary visual cortex convey independent visual information. J. Neurosci. **28**(22), 5696–5709 (2008)

19. Quilichini, P., Sirota, A., Buzsáki, G.: Intrinsic circuit organization and theta-gamma oscillation dynamics in the entorhinal cortex of the rat. J. Neurosci. **30**(33), 11128–11142 (2010)

20. Huang, G., Huang, G.B., Song, S., You, K.: Trends in extreme learning machines: a review. Neural Netw. **61**, 32–48 (2015)

21. Lekamalage, C.K.L., Liu, T., Yang, Y., Lin, Z., Huang, G.B.: Extreme learning machine for clustering. In: Proceedings of ELM-2014, vol. 1, pp. 435–444. Springer, Berlin (2015)

22. Einevoll, G.T., Kayser, C., Logothetis, N.K., Panzeri, S.: Modelling and analysis of local field potentials for studying the function of cortical circuits. Nat. Rev. Neurosci. **14**(11), 770–785 (2013)

23. Zanos, T.P., Mineault, P.J., Pack, C.C.: Removal of spurious correlations between spikes and local field potentials. J. Neurophysiol. **105**(1), 474–486 (2011)

24. Mazzoni, A., Logothetis, N.K., Panzeri, S.: The information content of local field potentials: experiments and models. arXiv preprint arXiv:1206.0560 (2012)

25. Huang, G.B.: An insight into extreme learning machines: random neurons, random features and kernels. Cogn. Comput. **6**(3), 376–390 (2014)

26. Huang, G.B., Bai, Z., Kasun, L.L.C., Vong, C.M.: Local receptive fields based extreme learning machine. IEEE Comput. Intell. Mag. **10**(2), 18–29 (2015)

RNA Secondary Structure Prediction Using Extreme Learning Machine with Clustering Under-Sampling Technique

Tianhang Liu, Jiarun Lin, Chengkun Wu and Jianping Yin

Abstract This paper gives a machine learning method for the subject of RNA secondary structure prediction. The method is based on extreme learning machine for its outstanding performance in classification problem, and use under-sampling technique to solve the problem of data imbalance. Feature vector in the classifier includes covariation score and inconsistent sequence penalty. The proposed method is compared with SVM and ELM without under-sampling, as well as classical method RNAalifold in terms of sensitivity, specificity, Matthews correlation coefficient and G-mean. The training and testing data are 68 RNA aligned families from Rfam, version 11.0. The results show that the proposed method can achieve highest scores in sensitivity, MCC and G-mean, which means that it is an effective method for RNA secondary structure prediction.

1 Introduction

Determining RNA secondary structures is significant for creating drugs and understanding genetic diseases [1–4]. The structures can be achieved by experimental methods, such as X-ray crystallography [5, 6] and nuclear magnetic resonance (NMR) spectroscopy [7]. However, these methods are extremely expensive and time consuming [8]. Therefore, many computing methods have been proposed to predict RNA secondary structures, such as hierarchical networks [9], k-nearest neighbor classifier [9], genetic algorithm [10, 11], supporter vector machine (SVM) [12] and extreme learning machine (ELM) [13–15]. Using machine learning

The authors are with the College of Computer Science, National University of Defense Technology, 410073, Changsha, P.R. China.

This work was supported by the National Natural Science Foundation of China (Project No.61379145, 61170287, 61232016, 61070198,61402508,61303189).

T. Liu (✉) · J. Lin · C. Wu · J. Yin
College of Computer, National University of Defense Technology, Changsha, People's Republic of China
e-mail: liuth007@163.com

© Springer International Publishing Switzerland 2016
J. Cao et al. (eds.), *Proceedings of ELM-2015 Volume 2*,
Proceedings in Adaptation, Learning and Optimization 7,
DOI 10.1007/978-3-319-28373-9_27

techniques in comparative sequence analysis method has showed good performance in experiments, while in practice, the data imbalance in training set is a main problem that can effect the performance [13].

In this paper, we propose a machine learning method for RNA secondary structure prediction based on ELM with clustering under-sampling technique. Extreme learning machine is a rapidly developing machine learning technique in recent years, and clustering under-sampling technique reduce the number of negative samples.

The feature vector we use in this paper contains two parts: (1) covariation score and (2) fraction of complementary nucleotides. The training and testing sets are 68 RNA alignment families from Rfam (version 11.0) [16, 17].

The paper's structure is as follows. Section 2 states the preliminaries, including biological basics in RNA secondary structure prediction, covariation score, inconsistent sequence penalty and evaluation metrics. Section 3 gives the proposed method. Performance evaluation including data specification, parameters setting and experimental results are provided in Sect. 4. Finally discussions are given in Sect. 5.

2 Preliminary

2.1 Biological Basics in RNA Secondary Structure Prediction

Generally, the secondary structure is composed by A-U and C-G base pairs which are known as Waston-Crick base pairs [18] as well as G-U base pairs known as wobble base pairs [4, 19], and some other types of base pairs [20].

2.2 Feature Vector

Covariation Score Covariation score is a metric that quantifies the covariation of two columns in an alignment [21], which is defined as:

$$C(i,j) = \sum_{XY, X'Y'} f_{i,j}(XY) D(XY, X'Y') f_{i,j}(X'Y'), \tag{1}$$

where i and j represent the ith column and the jth column of the alignment respectively. For the ij column pair, X, Y belong to $D = \{A, U, G, C\}$. $f_{i,j}(XY)$ is the frequency of XY in the ith and the jth columns. $D(XY, X'Y')$ is the Hamming distance between XY and $X'Y'$. If both XY and $X'Y'$ belong to $B = \{GC, CG, AU, UA, GU, UG\}$, $D(XY, X'Y')$ equals to the distance, otherwise $D(XY, X'Y')$ equals to zero. The sum is taken over all possible combinations of X and Y.

Inconsistent Sequence Penalty The inconsistent sequence penalty was proposed to measure the degree of inconsistency between two columns [12], which is defined as

$$\hat{P}(i,j) = N_b(i,j)/N_{b-all},$$ (2)

where $N_b(i,j)$ is the number of sequences that can match in ith and jth nucleotides, and N_{b-all} is the number of all sequences.

2.3 Evaluation Metrics

We use sensitivity, specificity, the Matthews correlation coefficient (MCC) [9, 22] and *G-mean* as evaluation metrics.

$$sensitivity = \frac{tp}{tp + fn},$$ (3)

$$specificity = \frac{tp}{tp + fp},$$ (4)

$$MCC = \frac{tp \cdot tn - fp \cdot fn}{\sqrt{(tp + fp)(tp + fn)(tn + fp)(tn + fn)}}.$$ (5)

where tp, tn, fp, fn are true-positive, true-negative, false-positive and false-negative respectively,

$$G - mean = (\prod_{i=1}^{m} R_i)^{1/m},$$ (6)

where R_i is the recall of ith class.

3 Method

3.1 Extreme Learning Machine

Extreme learning machine (ELM) for single-hidden layer feed-forward neural networks (SLFNs) has been proved effective and efficient for both classification and regression problems [14, 23–26]. The essential superiority of ELM is that the hidden layer of SLFNs doesn't need to be tuned. The parameters of hidden neurons are randomly assigned, and the output weights can be calculated by the Moore-Penrose generalized inverse, which gives ELM better generalization performance at a faster speed with least human interventions. Given N training samples (x_i, t_i), the mathematical model of ELM is defined as

$$H\beta = T,$$ (7)

where β is the output weight, T is the target vector and H is the hidden layer output matrix:

$$H = [h(x_1); h(x_2); \dots ; h(x_N)],\tag{8}$$

where

$$h(x_i) = [h_1(x_i), \dots , h_L(x_i)]\tag{9}$$

is the hidden layer output row vector with L hidden nodes. $h_j(x) = h(\alpha_j, b_j, x)$ is the feature mapping function, where the parameters α_j, b_j are randomly generated according to any continuous probability distribution. Then the output weight β can be analytically determined as the least square solution by

$$\beta = H^\dagger T = \begin{cases} H^T \left(\dfrac{I}{C} + HH^T\right)^{-1} T & when\ N < L \\ \left(\dfrac{I}{C} + H^T H\right)^{-1} H^T T & when\ N \ge L \end{cases}\tag{10}$$

So given a new sample x, the output function of ELM is

$$f(x) = \begin{cases} h(x)H^T \left(\dfrac{I}{C} + HH^T\right)^{-1} T & when\ N < L \\ h(x)\left(\dfrac{I}{C} + H^T H\right)^{-1} H^T T & when\ N \ge L \end{cases}\tag{11}$$

3.2 Clustering Under-Sampling Technique

The problem of RNA secondary structure prediction can be considered as a binary classification problem: to judge whether two nucleotides can form a base pair or not. The clustering under-sampling technique clusters the negative samples using k-means, and generates N central points, which are regarded as new negative samples. This process relieves the problem of data imbalance by reducing the number of negative samples.

3.3 ELM with Clustering Under-Sampling

The algorithm ELM with clustering under-sampling is shown as below.

Algorithm 1 ELM with clustering under-sampling.

Input:

Ψ: Training set $\Psi = (x_i, t_i), x_i \in R^n, t_i \in R^m, i = 1, 2, \dots, N$.

$\tilde{N}$: The number of clusters for k-means.

C: The trade off constant C.

L: The number of hidden node L.

Output:

M: Trained ELM model;

1: Generate $\tilde{N}$ clusters from negative samples using k-means, and set the central points of the clusters as new negative samples.

2: Randomly generate α_j, b_j, $j = 1, 2, \dots, L$ for L nodes and trained ELM model with L, C.

3: **return** the parameters of the classifiers M.

Test:

Given an unlabeled sample x, output the predicted label.

---

4 Performance Evaluation

4.1 Data Specification

We use the "seed" alignments and consensus secondary structures provided by Rfam (version 11.0) [17]. The Rfam (version 11.0) contains 10 years RNA families, in the form of alignments and structures, which is a widely used benchmark of RNA secondary structure prediction. For insuring the quality of the features, the alignments with ≥ 30 sequences are adopted, resulting in 68 RNA sequence alignment families in our data set, which is shown in Table 1.

In the training set, a positive sample is a column pair of the alignment that can form a base pair and a negative sample is one that cannot. In the 68 families, there are total 2,057 positive samples and 720,013 negative samples. Then 5-fold crossover validation is performed on the data set.

Table 1 The 68 RNA sequence alignment families from Rfam used as training and testing sets

RF00001	RF00002	RF00003	RF00004	RF00005	RF00008	RF00011
RF00012	RF00015	RF00017	RF00019	RF00020	RF00026	RF00031
RF00032	RF00037	RF00041	RF00045	RF00048	RF00049	RF00050
RF00061	RF00062	RF00066	RF00094	RF00095	RF00097	RF00100
RF00102	RF00105	RF00106	RF00162	RF00163	RF00164	RF00167
RF00168	RF00169	RF00171	RF00175	RF00176	RF00181	RF00198
RF00199	RF00209	RF00210	RF00214	RF00215	RF00216	RF00233
RF00238	RF00250	RF00260	RF00309	RF00374	RF00375	RF00376
RF00386	RF00389	RF00436	RF00451	RF00465	RF00467	RF00468
RF00469	RF00480	RF00481	RF00486	RF00506		

Table 2 The compared results of ELM with under-sampling, ELM, SVM and RNAalifold

Metrics	Sensitivity	Specificity	MCC	G-mean
ELM with under-sampling	0.8447	0.7291	0.7565	0.9167
ELM	0.5603	0.9681	0.7073	0.8476
SVM	0.5587	0.9548	0.7011	0.8427
RNAalifold	0.5525	0.7062	0.6237	0.8101

4.2 Parameter Setting

Three parameters need to be tuned: the trade off constant C, the number of hidden node L, as well as the number of clusters $\widetilde{N}$. C and L effect the performance of ELM. And $\widetilde{N}$ is related with the number of positive samples, it should be close to the number of positive samples, and properly chosen to avoid over-fitting. C and L are determined through grid search, and $\widetilde{N}$ changes from 3000 to 10000, with step 500, according to the number of positive samples. Finally we got $C = 100$, $L = 16$, $\widetilde{N} = 4500$.

4.3 Experimental Results

The prediction performance was evaluated by five indexes: sensitivity, specificity, Matthews correlation coefficient (MCC) and G-mean. Besides our method, we run SVM and ELM without under-sampling, as well as RNAalifold as comparisons. The parameters of SVM and ELM are defined through grid search, SVM uses RBF kernel. For SVM, $\log_2 C = 13$, $\log_2 \gamma = -5$. For ELM, $C = 100$, $L = 16$. The results are shown in Table 2. We can see from Table 2 that, the proposed method has highest values in sensitivity, MCC and G-mean, which means that the weighted ELM has a outstanding performance in imbalanced data set.

5 Discussion

In this paper, we proposed a new machine learning method for RNA secondary structure prediction using ELM and under-sampling technique. For rfam 11.0 data set, there are few positive samples but much more negative samples, and ELM with under-sampling technique can solve the problem.

Experimental results on 68 families from Rfam, vesion 11.0, show that the proposed method can get highest scores in terms of sensitivity, MCC and G-mean, which means that the method has outstanding performance when the data is significantly imbalanced. Future works include considering the phylogenetic relationship of the aligned sequences, and further analyzing the effects of different parameters.

References

1. Lodish, H.: Molecular Cell Biology. Macmillan (2008)
2. Schultes, E.A., Bartel, D.P.: One sequence, two ribozymes: implications for the emergence of new ribozyme folds. Science **289**(5478), 448–452 (2000)
3. Tinoco, I., Bustamante, C.: How RNA folds. J. Mol. Biol. **293**(2), 271–281 (1999)
4. Varani, G., McClain, W.: The g x u wobble base pair. a fundamental building block of RNA structure crucial to RNA function in diverse biological systems. EMBO Rep. **1**(1), 18–23 (2000)
5. Batey, R.T., Gilbert, S.D., Montange, R.K.: Structure of a natural guanine-responsive riboswitch complexed with the metabolite hypoxanthine. Nature **432**(7015), 411–415 (2004)
6. Kim, S.H., Quigley, G., Suddath, F., Rich, A.: High-resolution x-ray diffraction patterns of crystalline transfer RNA that show helical regions. Proc. Natl. Acad. Sci. **68**(4), 841–845 (1971)
7. Ferentz, A.E., Wagner, G.: NMR spectroscopy: a multifaceted approach to macromolecular structure. Q. Rev. Biophys. **33**(1), 29–65 (2000)
8. Ray, S.S., Pal, S.K.: RNA secondary structure prediction using soft computing. IEEE/ACM Trans. Comput. Biol. Bioinform. **10**(1), 0002–17 (2013)
9. Bindewald, E., Shapiro, B.A.: RNA secondary structure prediction from sequence alignments using a network of k-nearest neighbor classifiers. RNA **12**(3), 342–352 (2006)
10. Benedetti, G., Morosetti, S.: A genetic algorithm to search for optimal and suboptimal RNA secondary structures. Biophys. Chem. **55**(3), 253–259 (1995)
11. Chen, J.-H., Le, S.-Y., Maizel, J.V.: Prediction of common secondary structures of RNAs: a genetic algorithm approach. Nucleic Acids Res. **28**(4), 991–999 (2000)
12. Yingjie, Z., Zhengzhi, W.: Consensus RNA secondary structure prediction based on SVMS. In: The 2nd International Conference on Bioinformatics and Biomedical Engineering, 2008. ICBBE 2008, pp. 101–104, IEEE (2008)
13. Li, K., Kong, X., Lu, Z., Wenyin, L., Yin, J.: Boosting weighted elm for imbalanced learning. Neurocomputing **128**, 15–21 (2014)
14. Liu, Q., Ile, Q., Shi, Z.: Extreme support vector machine classifier. In: Advances in Knowledge Discovery and Data Mining, pp. 222–233. Springer (2008)
15. Tianhang, L.I.U., Jianping, Y.I.N.: RNA secondary structure prediction using self-adaptive evolutionary extreme learning machine. J. Comput. Inform. Syst. **11**(3), 995–1004 (2015)
16. Burge, S.W., Daub, J., Eberhardt, R., Tate, J., Barquist, L., Nawrocki, E.P., Eddy, S.R., Gardner, P.P., Bateman, A.: Rfam 11.0: 10 years of RNA families. Nucleic Acids Res. **1005** (2012)
17. Griffiths-Jones, S., Moxon, S., Marshall, M., Khanna, A., Eddy, S.R., Bateman, A.: Rfam: annotating non-coding rnas in complete genomes. Nucleic Acids Res. **33**(suppl 1), 121–124 (2005)
18. Watson, J.D., Crick, F.H., et al.: Molecular structure of nucleic acids. Nature **171**(4356), 737–738 (1953)
19. Crick, F.: Codon-anticodon pairing: the wobble hypothesis (1965)
20. Leontis, N.B., Westhof, E.: Geometric nomenclature and classiffication of RNA base pairs. RNA **7**(04), 499–512 (2001)
21. Hofacker, I.L., Fekete, M., Stadler, P.F.: Secondary structure prediction for aligned RNA sequences. J. Mol. Biol. **319**(5), 1059–1066 (2002)
22. Gardner, P.P., Giegerich, R.: A comprehensive comparison of comparative RNA structure prediction approaches. BMC Bioinform. **5**(1), 140 (2004)

23. Cambria, E., Huang, G.-B., Kasun, L.L.C., Zhou, H., Vong, C.M., Lin, J., Yin, J., Cai, Z., Liu, Q., Li, K., et al.: Extreme learning machines trends and controversies. IEEE Intell. Syst. **28**(6), 30–59 (2013)
24. Huang, G.-B., Zhu, Q.-Y., Siew, C.-K.: Extreme learning machine: theory and applications. Neurocomputing **70**(1), 489–501 (2006)
25. Huang, G.-B., Ding, X., Zhou, H.: Optimization method based extreme learning machine for classification. Neurocomputing **74**(1), 155–163 (2010)
26. Huang, G.-B., Zhou, H., Ding, X., Zhang, R.: Extreme learning machine for regression and multiclass classification. IEEE Trans. Syst. Man Cybern. Part B: Cybern. **42**(2), 513–529 (2012)

Multi-instance Multi-label Learning by Extreme Learning Machine

Chenguang Li, Ying Yin, Yuhai Zhao, Guang Chen and Libo Qin

Abstract Multi-instance Multi-label learning is a learning framework, where every object is represented by a bag of instances and associated with multiple labels simultaneously. The existing degeneration strategy based methods often suffer from some common drawbacks: (1) the user-specific parameter for the number of clusters may incur the effective problem; (2) utilizing SVM as the classifiers builder may bring the high computational cost. In this paper, we propose an algorithm, namely MIML-ELM, to address the problems. To our best knowledge, we are the first utilizing ELM in MIML problem and conducting the comparison of ELM and SVM on MIML. Extensive experiments are conducted on the real datasets and the synthetic datasets. The results show that MIML-ELM tends to achieve better generalization performance at a higher learning speed.

Keywords Multi-instance multi-label · Extreme learning machine

1 Introduction

When utilizing machine learning to solve the practical problems, we often consider an object as a feature vector. Then, we get an instance of the object. Further, associating the instance with a specific class label of the object, we obtain an example. Given a large collection of examples, the task is to get a function mapping from the instance space to the label space. We expect that the learned function can predict the labels of unseen instances correctly. However, in some applications, a real-world object is often of ambiguity, which consists of multiple instances and corresponds to multiple different labels simultaneously.

For example, an image usually contains multiple patches each represented by an instance, while in image classification such an image can belong to several classes

C. Li · Y. Yin (✉) · Y. Zhao · G. Chen · L. Qin
College of Information Science and Engineering, Northeastern University,
Shenyang 110819, China
e-mail: yinying@ise.neu.edu.cn

© Springer International Publishing Switzerland 2016
J. Cao et al. (eds.), *Proceedings of ELM-2015 Volume 2*,
Proceedings in Adaptation, Learning and Optimization 7,
DOI 10.1007/978-3-319-28373-9_28

simultaneously, e.g. an image can belong to mountains as well as Africa [1]; Another example is text categorization [1], where a document usually contains multiple sections each of which can be represented as an instance, and the document can be regarded as belonging to different categories if it was viewed from different aspects, e.g. a document can be categorized as scientific novel, Jules Verne's writing or even books on travelling; The MIML problem also arises in the protein function prediction task [2]. A domain is a distinct functional and structural unit of a protein. A multi-functional protein often consists of several domains, each fulfilling its own function independently. Taking a protein as an object, a domain as an instance and each biological function as a label, the protein function prediction problem exactly matches the MIML learning task.

In this context, Multi-instance Multi-label learning was proposed [1]. Similar to two another multi-learning frameworks, i.e. Multi-instance learning (MIL) [3] and Multi-label learning (MLL) [4], the MIML learning framework also results from the ambiguity in representing the real-world objects. Differently, more difficult than two another multi-learning frameworks, MIML studies the ambiguity in terms of both the input space (i.e. instance space) and the output space (i.e. label space) while MIL just studies the ambiguity in the input space and MLL just studies the ambiguity in the output space, respectively. In [1], Zhou et al. proposed a degeneration strategy based framework for MIML, which consists of two phases. First, the MIML problem is degenerated into the single-instance multi-label (SIML) problem through a specific clustering process; Second, the SIML problem is decomposed into multiple independent binary classification (i.e. single-instance single-label) problem using SVM as the classifiers builder. This two-phase framework has been successfully applied to many real-world applications and has been shown effective [5]. However, it could be further improved if the following drawbacks are tackled. On one hand, the clustering process in the first phase requires a user-specific parameter for the number of clusters. Unfortunately, it is often trouble to determine the correct number of clusters in advance. The incorrect number of clusters may affect the accuracy of the learning algorithm; On the other hand, SIML is degenerated into SISL (i.e. single instance single label) in the second phase, as will increase the volume of data to be handled and thus burden the classifier building. Utilizing SVM as the classifiers builder in this phase may suffer from the high computational cost and require much number of parameters to be optimized.

In this paper, we propose to enhance the two-phase framework by tackling the two above issues and make the following contributions: (1) we utilize Extreme Learning Machine (ELM) [6] instead of SVM to improve the efficiency of the two-phase framework. To our best knowledge, we are the first utilizing ELM in MIML problem and conducting the comparison of ELM and SVM on MIML; (2) we design a method of theoretical guarantee to determine the number of clusters automatically while incorporating it into the improved two-phase framework for effectiveness.

The remainder of this paper is organized as follows. In Sect. 2, we give a brief introduction to MIML and ELM; Sect. 3 details the improvements of the two-phase framework; Finally, Sect. 5 concludes this paper.

2 The Preliminaries

This research is related to some previous work on multi-instance multi-label (MIML) learning and extreme learning machine (ELM). In what follows, we briefly review some preliminaries of the two related work in Sects. 2.1 and 2.2, respectively.

2.1 Multi-instance Multi-label Learning

In the traditional supervised learning, the relationships between an object and its description and its label are always one-to-one correspondence. That is, an object is represented by a single instance and associated with a single class label. In this sense, we refer to it as single-instance single-label learning (SISL for short). Formally, let X be the instance space (or say feature space) and Y the set of class labels. The goal of SISL is to learn a function $f_{SISL}:X \rightarrow Y$ from a given data set $\{(x_1, Y_1), (x_2, Y_2), \ldots, (x_m, Y_m)\}$, where $x_i \in X$ is a instance and $y_i \in Y$ is the label of x_i. This formalization is prevailing and successful. However, as mentioned in Sect. 1, a lot of real-world objects are complicated and ambiguous in their semantics. Representing these ambiguous objects with SISL may lose some important information and make the learning task problematic [1]. Thus, many real-world complicated objects do not fit in this framework well.

In order to deal with this problem, several multi-learning frameworks have been proposed, e.g. Multi-Instance Learning (MIL), Multi-Label Learning (MLL) and Multi-Instance Multi-Label Learning (MIML). MIL studies the problem where a real-world object described by a number of instances is associated with a single class label. Multi-instance learning techniques have been successfully applied to diverse applications including image categorization [7], image retrieval [8], text categorization [9], web mining [10], computer-aided medical diagnosis [11], etc. Differently, MLL studies the problem where a real-world object is described by one instance but associated with a number of class labels. The existing work of MLL falls into two major categories. The one attempts to divide multi-label learning to a number of two class classification problems [12] or transform it into a label ranking problem [13]; the other tries to exploit the correlation between the labels [14]. MLL has been found useful in many tasks, such as text categorization [15], scene classification [16], image and video annotation [17, 18], bioinformatics [19, 20] and even association rule mining [21, 22]. MIML is a generalization of traditional supervised learning, multi-instance learning and multi-label learning, where a real-world object may be associated with a number of instances and a number of labels simultaneously. MIML is more reasonable than (single-instance) multi-label learning in many cases. In some cases, understanding why a particular object has a certain class label is even more important than simply making an accurate prediction while MIML offers a possibility for this purpose.

2.2   A Brief Introduction to ELM

Extreme Learning Machine (ELM for short) is a generalized Single Hidden-layer Feedforward Network. In ELM, the hidden-layer node parameter is mathematically calculated instead of being iteratively tuned, thus it provides good generalization performance at thousands of times faster speed than traditional popular learning algorithms for feedforward neural networks [23].

As a powerful classification model, ELM has been widely applied in many fields, such as protein sequences classification in bioinformatics [24, 25], online social network prediction [26], XML document classification [23], Graph classification [27] and so on. How to classify the data quickly and correctly is an important thing. For example, in [28], ELM was applied for plain text classification by using the one-against-one (OAO) and one-against-all (OAA) decomposition scheme. In [23], an ELM based XML document classification framework was proposed to improve classification accuracy by exploiting two different voting strategies. A protein secondary prediction framework based on ELM was proposed in [29] to provide good performance at extremely high speed. Wang et al. [30] implemented the protein-protein interaction prediction on multi-chain sets and on single-chain sets using ELM and SVM for a comparable study. In both cases, ELM tends to obtain higher Recall values than SVM and shows a remarkable advantage in the computational speed. Zhang et al. [31] evaluated the multicategory classification performance of ELM on three microarray data sets. The results indicate that ELM produces comparable or better classification accuracies with reduced training time and implementation complexity compared to artificial neural networks methods and Support Vector Machine methods. In [32], the use of ELM for multiresolution access of terrain height information was proposed. Optimization method based ELM for classification was studied in [33].

Given N arbitrary distinct samples (x_i, t_i), where $x_i = \left[x_{i1}, x_{i2}, \ldots, x_{in}\right]^T \in \mathbf{R}^n$ and $t_i = \left[t_{i1}, t_{i2}, \ldots, t_{im}\right]^T \in \mathbf{R}^m$, standard SLFNs with L hidden nodes and activation function $g(x)$ are mathematically modeled as

$$f(x) = \sum_{i=1}^{L} \beta_i g\left(\mathbf{a}_i, b_i, \mathbf{x}\right) \tag{1}$$

where a_i and b_i are the learning parameters of hidden nodes and β_i is the weight connecting the ith hidden node to the output node. $g\left(\mathbf{a}_i, b_i, \mathbf{x}\right)$ is the output of the ith hidden node with respect to the input x. In our case, sigmoid type of additive hidden nodes are used. Thus, Eq. (1) is given by

$$f(x) = \sum_{i=1}^{L} \beta_i g\left(\mathbf{a}_i, b_i, \mathbf{x}\right) = \sum_{i=1}^{L} \beta_i g\left(\mathbf{w}_i \cdot \mathbf{x}_j + b_i\right) = \mathbf{o}_j, (j = 1, \ldots, N) \tag{2}$$

where $\mathbf{w}_i = \left[w_{i1}, w_{i2}, ..., w_{in}\right]^T$ is the weight vector connecting the i-th hidden node and the input nodes, $\beta_i = \left[\beta_{i1}, \beta_{i2}, ..., \beta_{im}\right]^T$ is the weight vector connecting the ith hidden node and the output nodes, b_i is the bias of the ith hidden node, and o_j is the output of the jth node [34].

If an SLFN with activation function $g(x)$ can approximate the N given samples with zero error that $\Sigma_{j=1}^{L} \left\| o_j - t_j \right\| = 0$, there exist β_i, a_i and b_i such that

$$\sum_{i=1}^{L} \beta_i g \left(\mathbf{w}_i \cdot \mathbf{x}_j + b_i\right) = \mathbf{t}_j, j = 1, \ldots, N \tag{3}$$

Equation (3) can be expressed compactly as follows

$$\mathbf{H}\beta = \mathbf{T} \tag{4}$$

where

$$\mathbf{H}\left(\mathbf{w}_1, \ldots, \mathbf{w}_L, b_1, \ldots, b_L, \mathbf{x}_1, \ldots, \mathbf{x}_N\right) = \begin{bmatrix} g\left(\mathbf{w}_1 \cdot \mathbf{x}_1 + b_1\right) & \cdots & g\left(\mathbf{w}_L \cdot \mathbf{x}_1 + b_L\right) \\ \vdots & \cdots & \vdots \\ g\left(\mathbf{w}_1 \cdot \mathbf{x}_N + b_1\right) & \cdots & g\left(\mathbf{w}_L \cdot \mathbf{x}_N + b_L\right) \end{bmatrix}_{N \times L},$$

$$\beta = \left[\beta_1^T, \ldots, \beta_L^T\right]_{m \times L}^T, \text{and } \mathbf{T} = \left[\mathbf{t}_1^T, \ldots, \mathbf{t}_L^T\right]_{m \times N}^T$$

H is called the hidden layer output matrix of the network. The ith column of **H** is the ith hidden nodes output vector with respect to inputs $x_1, x_2, \ldots, x_N$ and the jth row of **H** is the output vector of the hidden layer with respect to input x_j.

For the binary classification applications, the decision function of ELM [33] is

$$f(x) = sign\left(\sum_{i-1}^{L} \beta_i g(\mathbf{a}_i, b_i, \mathbf{x})\right) = sign(\beta \cdot h(x)) \tag{5}$$

$h(\mathbf{x}) = \left[g\left(\mathbf{a}_1, b_1, \mathbf{x}\right), \ldots, g\left(\mathbf{a}_L, b_L, \mathbf{x}\right)\right]^T$ is the output vector of the hidden layer with respect to the input $\mathbf{x}$. $h(\mathbf{x})$ actually maps the data from the d-dimensional input space to the L-dimensional hidden layer feature space **H**.

Algorithm 1: ELM

Input: DB:dataset, HN: Number of Hidden Layer nodes, AF: ActivationFunction
Output: Results
1 **for** $i=1$ *to* L **do**
2   randomly assign input weight w_i;
3   randomly assign bias b_i;
4 **end**
5 calculate **H**;
6 calculate $\beta = \mathbf{H}^{\dagger}\mathbf{T}$

In ELM, the parameters of hidden layer nodes, i.e. w_i and b_i, can be chosen randomly without knowing the training data sets. The output weight $\mathbf{L}$ is then calculated with matrix computation formula $\mathbf{L} = \mathbf{H}^\dagger \mathbf{T}$, where $\mathbf{H}^\dagger$ is the Moore-Penrose Inverse of $\mathbf{H}$. ELM not only tends to reach the smallest training error but also the smallest norm of weights [6]. Given a training set $\aleph = \{(\mathbf{x}_i, \mathbf{t}_i) | \mathbf{x}_i \in \mathbf{R}^n, \mathbf{t}_i \in \mathbf{R}^m, i = 1, \ldots, N\}$, activation function $g(x)$ and hidden node number L, the pseudo code of ELM [34] is given in Algorithm 1.

3 The Proposed Two-Phase MIMLELM Framework

MIMLSVM is a representative two-phase MIML algorithm successfully applied in many real-world tasks [2]. It was first proposed by Zhou et al. in [1], and recently improved by Li et al. in [5]. MIMLSVM solves the MIML problem by first degenerating it into single-instance multi-label problems through a specific clustering process and then decomposing the learning of multiple labels into a series of binary classification tasks using SVM. However, as ever mentioned, MIMLSVM may suffer from some drawbacks in either of the two phases. For example, in the first phase, the user-specific parameter for the number of clusters may incur the effective problem; in the second phase, utilizing SVM as the classifiers builder may bring the high computational cost and require much number of parameters to be optimized.

Algorithm 2: The MIMLELM Algorithm

Input: DB:dataset, HN: Number of Hidden Layer nodes, AF: ActivationFunction
Output: Results
1 $DB = \{(X_1, Y_1), (X_2, Y_2), \ldots, (X_m, Y_m)\}, \Gamma = X_1, X_2, \ldots, X_m$;
2 Determine the number of clusters, k, using AIC;
3 randomly select k elements from Γ to initialize the k medoids $\{M_1, M_2, \ldots, M_k\}$;
4 **repeat**
5 $\quad$ $\Gamma_t = \{M_t\}(t = 1, 2, \ldots, k)$;
6 $\quad$ **foreach** $X_u \in (\Gamma - \{M_t\})$ **do**
7 $\quad\quad$ $index = \arg \min_{t \in \{1, 2, \ldots, k\}} d_H(X_u, M_t)$;
8 $\quad\quad$ $\Gamma_{index} = \Gamma_{index} \cup \{X_u\}$
9 $\quad$ **end**
10 $\quad$ $M_t = \arg \min_{A \in \Gamma_t} \sum_{B \in \Gamma_t} d_H(A, B)(t = 1, 2, \ldots, k)$;
11 $\quad$ Transform (X_u, Y_u) into into an SIML example (z_u, Y_u), where $z_u = (d_H(X_u, M_1), d_H(X_u, M_2), \ldots, d_H(X_u, M_k))$;
12 **until** M_t $(t = 1, 2, \ldots, k)$ *don't change*;
13 **foreach** z_u $(u \in \{1, 2, \ldots, m\})$ **do**
14 $\quad$ **foreach** $y \in Y_u$ **do**
15 $\quad\quad$ decompose (z_u, Y_u) into $|Y_u|$ SISL examples
16 $\quad$ **end**
17 **end**
18 Train ELM_y for every class y

In this paper, we present another algorithm, namely MIMLELM, to make MIMLSVM more efficient and effective. The MIMLELM algorithm is outlined in Algorithm 2. It consists of four major elements: (1) determination of the number of clusters (line 2); (2) transformation from MIML to SIML (line 3–12); (3) transformation from SIML to SISL (line 13–17); (4) multi-label learning based on ELM (line 18).

4 Performance Evaluation

In this section, we study the performance of the proposed MIMLELM algorithm in terms of both efficiency and effectiveness. The experiments are conducted on a HP PC with 2.33 GHz Intel Core 2 CPU, 2 GB main memory running Windows 7 and all algorithms are implemented in MATLAB 2013.

Fig. 1 The efficiency comparison on Image data set. **a** The comparison of training time, **b** the comparison of testing time

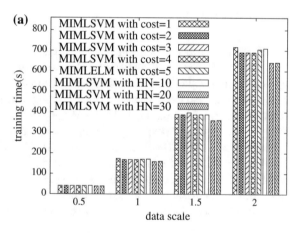

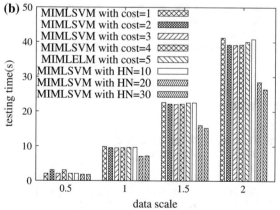

Four real datasets are utilized in our experiments. The data set is *Image* [1], which comprises 2000 natural scene images and 5 classes. The percent of images of more than one class is over 22%. On average, each image is of 1.24 ± 0.46 class labels and 1.36 ± 0.54 instances.

In the next experiments, we study the efficiency of MIMLELM by testing its scalability. That is, the data set is replicated different number of times, and then we observe how the training time and the testing time vary with the data size increasing. Again, MIMLSVM+ is utilized as the competitor. Similarly, the MIMLSVM+ algorithm is implemented with a Gaussian kernel while the penalty factor Cost is set to be 1, 2, 3, 4 and 5, respectively. The experimental results are given in Figs. 1. The Image data set is replicated 0.5–2 times with step size set to be 0.5. When the number of copies is 2, the efficiency improvement could be up to one 92.5% (from about 41.2 s down to about 21.4 s). As we observed, as the data size increasing, the superiority of MIMLELM becomes more and more significant.

5 Conclusion

MIML is a framework for learning with complicated objects, and has been proved to be effective in many applications. However, the existing two-phase MIML approaches may suffer from the effectiveness problem arising from the user-specific clusters number and the efficiency problem arising from the high computational cost. In this paper, we propose the MIMLELM approach to learn with MIML examples fast. On one hand, the efficiency is highly improved by integrating Extreme Learning Machine into the MIML learning framework. To our best knowledge, we are the first utilizing ELM in MIML problem and conducting the comparison of ELM and SVM on MIML. On the other hand, we develop a method of theoretical guarantee to determine the number of clusters automatically and exploit a genetic algorithm based ELM ensemble to further improve the effectiveness.

Acknowledgments National Natural Science Foundation of China (61272182, 61100028, 61572117, 61173030, 61173029), State Key Program of National Natural Science of China (61332014,U1401256), New Century Excellent Talents (NCET-11-0085) and the Fundamental Research Funds for the Central Universities under grants (No.130504001).

References

1. Zhou, Z.H., Zhang, M.L.: Multi-instance multi-label learning with application to scene classification. In: Schölkopf, B., Platt, J., Hoffman, T. (eds.) Advances in Neural Information Processing Systems 19, pp. 1609–1616. MIT Press, Cambridge (2007)
2. Wu, J., Huang, S., Zhou, Z.: Genome-wide protein function prediction through multi-instance multi-label learning. IEEE/ACM Trans. Comput. Biol. Bioinform. **11**, 891–902 (2014)
3. Dietterich, T.G., Lathrop, R.H., Lozano-Pérez, T.: Solving the multiple instance problem with axis-parallel rectangles. Artif. Intell. **89**, 31–71 (1997)

4. Schapire, R.E., Singer, Y.: Boostexter: a boosting-based system for text categorization. Mach. Learn. **39**, 135–168 (2000)
5. Li, Y., Ji, S., Kumar, S., Ye, J., Zhou, Z.: Drosophila gene expression pattern annotation through multi-instance multi-label learning. IEEE/ACM Trans. Comput. Biol. Bioinform. **9**, 98–112 (2012)
6. Huang, G.B., Zhu, Q.Y., Siew, C.K.: Extreme learning machine: a new learning scheme of feedforward neural networks. In: Proceedings of International Joint Conference on Neural Networks (IJCNN2004), vol. 2, pp. 985–990, Budapest, Hungary, 25–29 July 2004
7. Chen, Y., Bi, J., Wang, J.Z.: MILES: multiple-instance learning via embedded instance selection. IEEE Trans. Pattern Anal. Mach. Intell. **28**, 1931–1947 (2006)
8. Yang, C., Lozano-Pérez, T.: Image database retrieval with multiple-instance learning techniques. In: ICDE, pp. 233–243 (2000)
9. Andrews, S., Tsochantaridis, I., Hofmann, T.: Support vector machines for multiple-instance learning. In: Advances in Neural Information Processing Systems 15 [Neural Information Processing Systems, NIPS 2002, December 9–14, 2002, Vancouver, British Columbia, Canada], pp. 561–568 (2002)
10. Zhou, Z., Jiang, K., Li, M.: Multi-instance learning based web mining. Appl. Intell. **22**, 135–147 (2005)
11. Fung, G., Dundar, M., Krishnapuram, B., Rao, R.B.: Multiple instance learning for computer aided diagnosis. In: Advances in Neural Information Processing Systems 19, Proceedings of the Twentieth Annual Conference on Neural Information Processing Systems, pp. 425–432, Vancouver, British Columbia, Canada, 4–7 Dec 2006
12. Joachims, T.: Text categorization with suport vector machines: learning with many relevant features. In: Proceedings Machine Learning: ECML-98, 10th European Conference on Machine Learning, pp. 137–142, Chemnitz, Germany, 21–23 April 1998
13. Elisseeff, A., Weston, J.: A kernel method for multi-labelled classification. In: Advances in Neural Information Processing Systems 14 [Neural Information Processing Systems: Natural and Synthetic, NIPS 2001, December 3–8, 2001, Vancouver, British Columbia, Canada], pp. 681–687 (2001)
14. Liu, Y., Jin, R., Yang, L.: Semi-supervised multi-label learning by constrained non-negative matrix factorization. In: Proceedings, The Twenty-First National Conference on Artificial Intelligence and the Eighteenth Innovative Applications of Artificial Intelligence Conference, pp. 421–426, Boston, Massachusetts, USA, 16–20 July 2006
15. Godbole, S., Sarawagi, S.: Discriminative methods for multi-labeled classification. In: Proceedings Advances in Knowledge Discovery and Data Mining, 8th Pacific-Asia Conference, PAKDD 2004, 22–30, Sydney, Australia, 26–28 May 2004
16. Boutell, M.R., Luo, J., Shen, X., Brown, C.M.: Learning multi-label scene classification. Pattern Recogn. **37**, 1757–1771 (2004)
17. Kang, F., Jin, R., Sukthankar, R.: Correlated label propagation with application to multi-label learning. In: 2006 IEEE Computer Society Conference on Computer Vision and Pattern Recognition (CVPR 2006), pp. 1719–1726, New York, NY, USA, 17–22 June 2006
18. Qi, G., Hua, X., Rui, Y., Tang, J., Mei, T., Zhang, H.: Correlative multi-label video annotation. In: Proceedings of the 15th International Conference on Multimedia 2007, 17–26, Augsburg, Germany, 24–29 Sept 2007
19. Barutçuoglu, Z., Schapire, R.E., Troyanskaya, O.G.: Hierarchical multi-label prediction of gene function. Bioinformatics **22**, 830–836 (2006)
20. Brinker, K., Fürnkranz, J., Hüllermeier, E.: A unified model for multilabel classification and ranking. In: Proceedings ECAI 2006, 17th European Conference on Artificial Intelligence, August 29–September 1, 2006, Riva del Garda, Italy, Including Prestigious Applications of Intelligent Systems (PAIS 2006), pp. 489–493 (2006)
21. Rak, R., Kurgan, L.A., Reformat, M.: Multi-label associative classification of medical documents from MEDLINE. In: Fourth International Conference on Machine Learning and Applications, ICMLA 2005, Los Angeles, California, USA, 15–17 Dec 2005

22. Thabtah, F.A., Cowling, P.I., Peng, Y.: MMAC: a new multi-class, multi-label associative classification approach. In: Proceedings of the 4th IEEE International Conference on Data Mining (ICDM 2004), pp. 217–224, Brighton, UK, 1–4 Nov 2004

23. Zhao, X., Wang, G., Bi, X., Gong, P., Zhao, Y.: Xml document classification based on elm. Neurocomputing **74**, 2444–2451 (2011)

24. Zhao, Y., Wang, G., Yin, Y., Li, Y., Wang, Z.: Improving elm-based microarray data classification by diversified sequence features selection. Neural Comput. Appl. **2014**, 1–12 (2014)

25. Zhao, Y., Wang, G., Zhang, X., Yu, J.X., Wang, Z.: Learning phenotype structure using sequence model. IEEE Trans. Knowl. Data Eng. **26**, 667–681 (2014)

26. Sun, Y., Yuan, Y., Wang, G.: An on-line sequential learning method in social networks for node classification. Neurocomputing **149**, 207–214 (2015)

27. Wang, Z., Zhao, Y., Wang, G., Li, Y., Wang, X.: On extending extreme learning machine to non-redundant synergy pattern based graph classification. Neurocomputing **149**, 330–339 (2015)

28. Zhang, R., Huang, G.B., Sundararajan, N., Saratchandran, P.: Multi-category classification using an extreme learning machine for microarray gene expression cancer diagnosis. IEEE/ACM Trans. Comput. Biol. Bioinform. **4**, 485–495 (2007)

29. Wang, G., Zhao, Y., Wang, D.: A protein secondary structure prediction framework based on the extreme learning machine. Neurocomputing **72**, 262–268 (2008)

30. Wang, D.D., Wang, R., Yan, H.: Fast prediction of protein-protein interaction sites based on extreme learning machines. Neurocomputing **128**, 258–266 (2014)

31. Zhang, R., Huang, G.B., Sundararajan, N., Saratchandran, P.: Multicategory classification using an extreme learning machine for microarray gene expression cancer diagnosis. IEEE/ACM Trans. Comput. Biol. Bioinform. **4**, 485–495 (2007)

32. Yeu, C.W.T., Lim, M.L., Huang, G.B., Agarwal, A., Ong, Y.S.: A new machine learning paradigm for terrain reconstruction. IEEE Geosci. Remote Sens. Lett. **3**, 382–386 (2006)

33. Huang, G.B., Ding, X., Zhou, H.: Optimization method based extreme learning machine for classification. Neurocomputing **74**, 155–163 (2010)

34. Huang, G.B., Zhu, Q.Y., Siew, C.K.: Extreme learning machine: theory and applications. Neurocomputing **70**, 489–501 (2006)

A Randomly Weighted Gabor Network for Visual-Thermal Infrared Face Recognition

Beom-Seok Oh, Kangrok Oh, Andrew Beng Jin Teoh,
Zhiping Lin and Kar-Ann Toh

Abstract In this paper, a novel three-layer Gabor-based network is proposed for heterogeneous face recognition. The input layer of our proposed network consists of pixel-wise image patches. At the hidden layer, a set of Gabor features are extracted by a projection operation and a magnitude function. Subsequently, a non-linear activation function is utilized after weighting the extracted Gabor features with random weight vectors. Finally, the output weights are deterministically learned similarly to that in extreme learning machine. Some experimental results on private BERC visual-thermal infrared database are observed and discussed. The proposed method shows promising results based on the average test recognition accuracy.

Keywords Heterogeneous face recognition · Gabor features · Extreme learning machine · Random weighting

B.-S. Oh · K. Oh · A.B.J. Teoh · K.-A. Toh (✉)
School of Electrical and Electronic Engineering, Yonsei University,
50 Yonsei-ro, Seodaemun-gu, Republic of Korea
e-mail: katoh@yonsei.ac.kr,katoh@ieee.org

B.-S. Oh
e-mail: a-bullet@yonsei.ac.kr,bsoh@ntu.edu.sg

K. Oh
e-mail: kangrok.oh@yonsei.ac.kr

A.B.J. Teoh
e-mail: bjteoh@yonsei.ac.kr

B.-S. Oh · Z. Lin
School of Electrical and Electronic Engineering, Nanyang Technological University,
50 Nanyang Avenue, Singapore, Singapore
e-mail: ezplin@ntu.edu.sg

© Springer International Publishing Switzerland 2016
J. Cao et al. (eds.), *Proceedings of ELM-2015 Volume 2*,
Proceedings in Adaptation, Learning and Optimization 7,
DOI 10.1007/978-3-319-28373-9_29

1 Introduction

According to [1], heterogeneous face recognition refers to face matching across different visual domains such as visual, near-infrared, thermal infrared, sketch, 3D, and etc. To the best of our knowledge, [2] was among the pioneering work to have introduced the term based on earlier works in applications such as face sketch recognition [3] and illumination invariant face recognition [4]. Due to the demand of matching face images captured under different conditions for practical reasons, heterogeneous face recognition became an interesting research topic recently. Particularly, heterogeneous face recognition is important when the modality gap between a registered image and a query image is significantly large. For example, consider a forensic application which requires matching between visual face images and non-visual face images. Under this situation, it is hard to exploit established techniques in homogeneous face recognition without resolving problems caused by the modality gap between different visual domains.

Based on the ways to deal with the modality gap, existing works in heterogeneous face recognition can be grouped into four approaches: (i) common subspace learning approach, (ii) invariant feature extraction approach, (iii) synthesis-based approach, and (iv) classification-based approach. Related works with common subspace learning approach aim to learn a common subspace from heterogeneous face images [5–8] to narrow the modality gap. The key idea of invariant feature extraction approach is to represent heterogeneous face images by designing a method to extract domain invariant features [9–11]. Methods with synthesis-based approach attempted to synthesize face images of a modality with those of other modalities [12, 13] to narrow the modality gap. For these three approaches, majority of existing works adopted a nearest neighbor (NN) classifier based on a similarity measure at recognition stage.

Different from the above approaches, classification-based approach utilizes a classifier learned from the training data [14] to narrow down the modality gap. Comparing with the first three approaches, it is possibly beneficial to utilize a learning based classifier (for example, partial least square (PLS) [14]) with an appropriate parameter setting instead of using a prototype based NN. Our work belongs to classification-based approach.

In this paper, we propose a Gabor-based extreme learning machine for heterogeneous face recognition. The main contributions of this paper can be enumerated as follows: (i) proposal of a novel scheme based on a pixel-wise image patch and random weight vectors for extracting randomly weighted Gabor features, (ii) proposal of a novel network structure which propagate randomly weighted Gabor features to a hidden layer. This paper is organized as follows: some pre-requisite background knowledge is provided in Sect. 2. Section 3 presents the proposed network. The experimental results and discussions are presented in Sect. 4. Some concluding remarks are provided in Sect. 5.

2 Preliminaries

In this section, we provide brief descriptions on Gabor filtering [15, 16] and extreme learning machine (ELM) [17] as background knowledge of the proposed network.

2.1 Gabor Filtering

According to [15, 16], a Gabor kernel can be defined as follows

$$\mathbf{K}_{\mu,v}(\mathbf{c}) = \frac{\left\|\mathbf{v}_{\mu,v}\right\|^2}{\sigma^2} \cdot \exp\left(-\frac{\left\|\mathbf{v}_{\mu,v}\right\|^2 \|\mathbf{c}\|^2}{2\sigma^2}\right) \cdot \left[\exp\left(i\mathbf{v}_{\mu,v}^T\mathbf{c}\right) - \exp\left(-\frac{\sigma^2}{2}\right)\right], \quad (1)$$

where $\mathbf{c} = (x, y)^T$ denotes an image coordinate consisting of row(x) and column(y), $\mathbf{K}_{\mu,v} \in \mathbb{C}^{H \times W}$ is a complex Gabor kernel. Here, μ and v are the orientation and scale of the Gabor kernel, and σ is the standard deviation. We utilize $\|\cdot\|$ and i to denote the L_2-norm operator and the imaginary axis. The wave vector $\mathbf{v}_{\mu,v}$ is defined as $\mathbf{v}_{\mu,v} = \frac{k_{max}}{f^v} \exp\left(i\frac{\mu\pi}{8}\right)$ where k_{max} is the maximum frequency, f is a factor regarding the space between kernels in frequency domain. Given an image matrix $\mathbf{X} \in \mathbb{R}^{P \times Q}$, the Gabor kernel $\mathbf{K}_{\mu,v}$ defined in (1) is convolved with $\mathbf{X}$. Subsequently, magnitude or phase information are extracted from the convolution results [16, 18].

2.2 Extreme Learning Machine (ELM) for Binary Classification

Given M pairs of training sample and corresponding class label, $(\mathbf{x}_i, y_i), i = 1, \ldots, M$, where $\mathbf{x}_i \in \mathbb{R}^{D \times 1}$ is an input vector and $y_i \in \{0, 1\}$ is a corresponding class label, a single hidden layer feedforward neural network (SLFN) with N hidden nodes and an activation function $h(\cdot)$ can be written as

$$\sum_{j=1}^{N} \beta_j h\left(\mathbf{w}_j^T\mathbf{x}_i + b_j\right) = o_i, \quad i = 1, \ldots, M, \quad (2)$$

where $\mathbf{w}_j \in \mathbb{R}^{D \times 1}$ is a randomly assigned weight vector connecting the input nodes to the jth hidden node, $\beta_j \in \mathbb{R}$ is a weight value connecting the jth hidden node to the output node, $b_j \in \mathbb{R}$ is a randomly assigned threshold for the jth hidden node, and $o_i \in \mathbb{R}$ is a network output value.

For M training samples, (2) can be written compactly as

$$\mathbf{H}\boldsymbol{\beta} = \mathbf{y}, \tag{3}$$

where

$$\mathbf{H} = \begin{bmatrix} h\left(\mathbf{w}_1^T\mathbf{x}_1 + b_1\right) & \cdots & h\left(\mathbf{w}_N^T\mathbf{x}_1 + b_N\right) \\ \vdots & \ddots & \vdots \\ h\left(\mathbf{w}_1^T\mathbf{x}_M + b_1\right) & \cdots & h\left(\mathbf{w}_N^T\mathbf{x}_M + b_N\right) \end{bmatrix}_{M \times N}, \tag{4}$$

and $\boldsymbol{\beta} = \left[\beta_1, \dots, \beta_N\right]^T$ and $\mathbf{y} = \left[y_1, \dots, y_M\right]^T$.

According to [17], a closed-form solution for $\boldsymbol{\beta}$ to (3) can be written as

$$\hat{\boldsymbol{\beta}} = \mathbf{H}^\dagger \mathbf{y} = \left(\mathbf{H}^T\mathbf{H}\right)^{-1}\mathbf{H}^T\mathbf{y}, \tag{5}$$

where $\mathbf{H}^\dagger$ is the Moore-Penrose generalized inverse of $\mathbf{H}$. Given a test sample $\mathbf{x}_t$, the corresponding network output value o_t can be estimated as $o_t = \mathbf{h}_t^T \hat{\boldsymbol{\beta}}$ where $\mathbf{h}_t = \left[h\left(\mathbf{w}_1^T\mathbf{x}_t + b_1\right), \dots, h\left(\mathbf{w}_N^T\mathbf{x}_t + b_N\right)\right]^T$. This algorithm for SLFN is called an extreme learning machine (ELM) in [17]. A similar network structure can be found in [19–21].

3 Gabor-Based Extreme Learning Machine

Suppose that we have M pairs of training sample and class label $\left(\mathbf{X}_i, y_i\right)$, $i = 1, \dots, M$, where $\mathbf{X}_i \in \mathbb{R}^{P \times Q}$ and $y_i \in \{1, \dots, C\}$ denote a face image and a corresponding class label respectively. Here, C is the number of subjects existing in the training data. Let $\mathbf{Z}_{x,y}^i \in \mathbb{R}^{H \times W}$ denote an image patch centered at an image coordinate (x, y) of the ith face image and $\mathbf{K}_{\mu,v} \in \mathbb{C}^{H \times W}$ denote the Gabor kernel with an orientation μ and a scale v defined in (1). By defining an index j as $j = Q \times (x - 1) + y$ for $x = 1, \dots, P$ and $y = 1, \dots, Q$ and changing matrix representations into vector representations, we obtain new representations $\mathbf{z}_j^i \in \mathbb{R}^{S \times 1}$ and $\mathbf{k}_{\mu,v} \in \mathbb{C}^{S \times 1}$ instead of $\mathbf{Z}_{x,y}^i \in \mathbb{R}^{H \times W}$ and $\mathbf{K}_{\mu,v} \in \mathbb{C}^{H \times W}$, respectively, where $S = H \times W$. Then the proposed GaborELM can be written as

$$\sum_{j=1}^{L} \beta_j h\left(\mathbf{w}_j^T \left|\mathbf{G}\mathbf{z}_j^i\right| + b_j\right) = o_i, \quad i = 1, \dots, M, \tag{6}$$

where $\mathbf{G} = \left[\mathbf{k}_{0,0}, \dots, \mathbf{k}_{0,v_{max}}, \dots, \mathbf{k}_{\mu_{max},0}, \dots, \mathbf{k}_{\mu_{max},v_{max}}\right]^T \in \mathbb{C}^{T \times S}$ is a matrix stacking Gabor kernels, $T = \left(\mu_{max} + 1\right) \times \left(v_{max} + 1\right)$ stands for the number of Gabor kernels, $\mathbf{w}_j \in \mathbb{R}^{T \times 1}$ denotes a randomly assigned weight vector connecting image patches to jth hidden node, $|\cdot|$ is an element-wise magnitude function, and $L = P \times Q$.

In order to learn a class specific output weights vector $\boldsymbol{\beta}_c \in \mathbb{R}^{L \times 1}$, $c = 1, \ldots, C$, we further adopt a total error rate (TER) minimization [22, 23] to exploit its efficient learning capability on a discriminative model. Since the resolution of training images is high and the number of training samples is low, the TER solution under a dual solution space [23] is adopted. According to [23], the output weight vector for cth class (subject) can be learned as

$$\hat{\boldsymbol{\beta}}_c = \mathbf{H}_c^T \left(\lambda \mathbf{I} + \mathbf{W}_c \mathbf{H}_c \mathbf{H}_c^T \right)^{-1} \mathbf{W}_c \mathbf{y}_c, \quad c = 1, \ldots, C, \tag{7}$$

where

$$\mathbf{H}_c = \begin{bmatrix} \mathbf{H}_c^- \\ \mathbf{H}_c^+ \end{bmatrix}_{M \times L}, \tag{8}$$

$$\mathbf{H}_c^- = \begin{bmatrix} h\left(\mathbf{w}_1^T \left| \mathbf{Gz}_1^1 \right| + b_1 \right) & \cdots & h\left(\mathbf{w}_L^T \left| \mathbf{Gz}_L^1 \right| + b_L \right) \\ \vdots & \ddots & \vdots \\ h\left(\mathbf{w}_1^T \left| \mathbf{Gz}_1^{M_c^-} \right| + b_1 \right) & \cdots & h\left(\mathbf{w}_L^T \left| \mathbf{Gz}_L^{M_c^-} \right| + b_L \right) \end{bmatrix}_{M_c^- \times L}, \tag{9}$$

and

$$\mathbf{H}_c^+ = \begin{bmatrix} h\left(\mathbf{w}_1^T \left| \mathbf{Gz}_1^1 \right| + b_1 \right) & \cdots & h\left(\mathbf{w}_L^T \left| \mathbf{Gz}_L^1 \right| + b_L \right) \\ \vdots & \ddots & \vdots \\ h\left(\mathbf{w}_1^T \left| \mathbf{Gz}_1^{M_c^+} \right| + b_1 \right) & \cdots & h\left(\mathbf{w}_L^T \left| \mathbf{Gz}_L^{M_c^+} \right| + b_L \right) \end{bmatrix}_{M_c^+ \times L}. \tag{10}$$

Here, M_c^- and M_c^+ denote the number of samples in negative and positive class respectively ($M = M_c^+ + M_c^-$), $\mathbf{W}_c = \mathbf{W}_c^- + \mathbf{W}_c^+ \in \mathbb{R}^{M \times M}$ indicates a class-specific diagonal weighting matrix, $\mathbf{W}_c^- = \mathrm{diag}(1, \ldots, 1, 0, \ldots, 0)/M_c^-$, and $\mathbf{W}_c^+ = \mathrm{diag}(0, \ldots, 0, 1, \ldots, 1)/M_c^+$. The target vector $\mathbf{y}_c$ is defined as $\mathbf{y}_c = \begin{bmatrix} \mathbf{y}_c^- \\ \mathbf{y}_c^+ \end{bmatrix} = \begin{bmatrix} (\tau - \eta)\mathbf{1}_- \\ (\tau + \eta)\mathbf{1}_+ \end{bmatrix} \in \mathbb{R}^{M \times 1}$, where $\mathbf{1}_- = [1, \cdots, 1]^T \in \mathbb{N}^{M_c^- \times 1}$, $\mathbf{1}_+ = [1, \cdots, 1]^T \in \mathbb{N}^{M_c^+ \times 1}$, τ is a threshold value and η is an offset value [22].

Given an unseen test face image $\mathbf{X}_t$, the class label can be predicted based on one-versus-all technique as

$$y_t = \arg\max_c \left(\mathbf{h}_t^T \left[\hat{\boldsymbol{\beta}}_1, \ldots, \hat{\boldsymbol{\beta}}_C \right] \right), \tag{11}$$

where $\mathbf{h}_t = \left[h\left(\mathbf{w}_1^T \left| \mathbf{Gz}_1^t \right| + b_1 \right), \ldots, h\left(\mathbf{w}_L^T \left| \mathbf{Gz}_L^t \right| + b_L \right) \right]^T$ is a feature vector extracted by GaborELM from $\mathbf{X}_t$, and $y_t \in \{1, \ldots, C\}$ is a predicted class label.

4 Experiment

4.1 Database and Preprocessing

In our experiments, private BERC VIS-TIR database [24] has been utilized to show the usefulness of our proposed GaborELM for heterogeneous face recognition task. The BERC VIS-TIR database consists of 10,368 images captured from 96 subjects. The number of images captured under the visual (VIS) and the thermal infrared (TIR) spectrums are exactly the same (5,184 images). Images in the database contains variation in pose, expression, illumination, and appearance (with and without glasses) [24]. Among the 10,368 images, only 288 VIS images (size = 640×480) and 288 TIR images (size = 320×240) which were acquired under frontal pose and neutral expression conditions were utilized in the experiments to exclude pose and expression variation. Hence, each subject has three VIS images and three TIR images.

The VIS and TIR images were geometrically normalized (size = 150×130) based on the center of eyes and lips. Subsequently, the geometrically normalized images were photometrically normalized using preprocessing sequences (PS) technique [25]. It includes a variety of well-known image preprocessing techniques such as gamma correction, difference of Gaussian (DoG), masking and equalization of variation. The adopted techniques for PS are performed in the listed order. Figure 1 shows the original and preprocessed images (VIS and TIR) of one subject.

4.2 Experimental Setup

In order to configure the training and test data, we follow a learning protocol for heterogeneous face recognition which was recently proposed by [14]. In the protocol, a database is firstly divided into two sets, namely Set1 and Set2. At training phase, training positive class samples for cth subject consist of only VIS images of this cth subject in Set1 while training negative class samples for cth subject consist of VIS

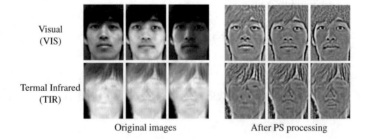

Original images   After PS processing

Fig. 1 Samples images of one subject in BERC VIS-TIR database and preprocessed images by the geometrical and photometrical normalization

Table 1 Two conducted experiments for performance evaluation of the proposed GaborELM

Name	Purpose
Experiment I	Evaluation of accuracy and CPU time performance of GaborELM under a heterogeneous face recognition scenario (three VIS images/subject for training & three TIR images/subject for test)
Experiment II	Investigation into the effect of the number of random weight vectors for hidden nodes on recognition accuracy

images of other subjects in Set1 and some non-visual (NVIS) images (TIR images in our experiments) of Set2. The NVIS images is called training cross examples in [14]. Note that the size of training negative class samples is now $M_c^- + U$, where M_c^- is the number of negative class samples of the cth subject in Set1, and U is the number of training cross examples in Set2. At testing phase, the protocol utilizes only NVIS images of each subject in Set1 as test samples. The size of Set1 and Set2 can be controlled by a parameter (image set partitioning ratio) s (which is the number of subjects of Set1 / the number of entire subjects).

The accuracy and computational (CPU) time performance of the proposed GaborELM were evaluated on BERC VIS-TIR database. As the accuracy performance measure, we adopted test recognition accuracy (which is the number of correctly classified samples / the number of entire test samples). Table 1 shows a brief description on the two experiments. In Experiment I, the accuracy and CPU time performance of GaborELM is evaluated under a heterogeneous face recognition scenario using three VIS and TIR images per subject. Here, we compared the proposed method with a method from [14] which was based on histogram of oriented gradients (HOG) features and a partial least square (PLS) regression. We will denote HOG-PLS for this compared method hereafter. In Experiment II, we investigate into the effect of the number of random weight vectors for hidden nodes on accuracy performance.

Under Experiment I, we set $k_{max} = \pi/2$, $\sigma = 2\pi$, $\mu_{max} = 7$, $v_{max} = 4$, $H = 21$ and $W = 21$ for values of parameters regarding Gabor kernels according to [16, 26]. This results in 40 Gabor kernels each with the size of 21×21 pixels. For the image patches near the four sides of an image, zero values were padded to prevent information loss. For the HOG-PLS, we followed settings in [14] (cell size = 8×8, block size = 2×2, window overlapping size = 8). A set of discrete values with the range of $\{0.1, 0.2, \ldots, 0.9\}$ was utilized for the image set partitioning ratio s. For GaborELM, the accuracy and CPU time performance were averaged from 100 iterations resulted from 10 iterations for random image set partitioning and 10 iterations for random settings regarding ELM. The accuracy and CPU time performance of HOG-PLS was averaged from 10 random settings regarding the image set partitioning ratio. Under Experiment II, we followed the settings regarding Gabor kernels as in Experiment I and we set the image set partitioning ratio $s = 0.5$. For the number of random weight vectors for hidden nodes, we utilized a set of values with the range of $\{1, 10, 100, 500, 1000, 3000, 5000, 8000, 10000, 13500\}$.

4.3   Results

Experiment I Figure 2 shows the average test recognition accuracy (%) of GaborELM and HOG-PLS algorithms on BERC VIS-TIR database. As illustrated in the figure, GaborELM showed slightly better average test accuracy performance than HOG-PLS. Both GaborELM and HOG-PLS showed relatively good average test accuracy performance as the image set partitioning ratio decreases. The CPU time performance of GaborELM and HOG-PLS algorithms on BERC VIS-TIR database are illustrated in Fig. 3. As shown in Fig. 3a, HOG-PLS showed better training CPU time performance than GaborELM, When three images per subject were utilized for training and test, GaborELM consumed around 10 seconds more than HOG-PLS.

Fig. 2 Average test recognition accuracy of GaborELM and HOG-PLS algorithms on BERC VIS-TIR database

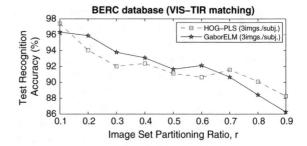

Fig. 3 Training CPU time, and test CPU time of GaborELM and HOG-PLS algorithms on BERC VIS-TIR database.
a Training CPU time (sec.)
b Test CPU time (sec.)

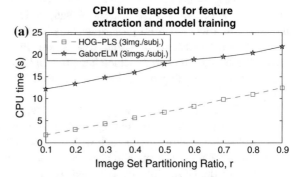

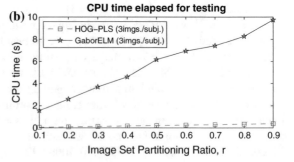

Fig. 4 Average test recognition accuracy of GaborELM on BERC VIS-TIR database by changing the number of random weight vectors for hidden nodes

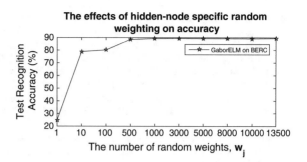

As shown in Fig. 3b, test CPU time performance showed a similar trend with training CPU time performance. Both GaborELM and HOG-PLS showed relatively good training and test CPU time performance as the image partitioning ratio decreases.

Experiment II Figure 4 shows the average test recognition accuracy of GaborELM by changing the number of random weight vectors for hidden nodes. As shown in the figure, the accuracy performance increases as the number of random weight vectors increases until reaching to 1000. The average test recognition accuracy performance was saturated when 1000 random weight vectors were utilized.

5   Conclusion

In this paper, we proposed a novel Gabor-based extreme learning machine for heterogeneous face recognition. The proposed input layer consists of pixel-wise image patches. The hidden layer outputs randomly weighted Gabor magnitude features through a non-linear activation function. The output weights were deterministically learned by a total error rate minimizer. The proposed Gabor-based extreme learning machine showed promising results on private BERC visual-thermal infrared database based on average test recognition accuracy.

Acknowledgments This research was supported by Basic Science Research Program through the National Research Foundation of Korea (NRF) funded by the Ministry of Education, Science and Technology (Grant number: NRF-2012R1A1A2042428).

References

1. Ouyang, S., Hospedales, T., Song, Y.Z., Li, X.: A Survey on Heterogeneous Face Recognition: Sketch, Infra-red, 3D and Low-resolution (2014). arXiv:1409.5114
2. Lin, D., Tang, X.: Inter-modality face recognition. In: European Conference on Computer Vision, pp. 13–26. Springer (2006)
3. Tang, X., Wang, X.: Face sketch recognition. IEEE Trans. Circuits Syst. Video Technol. **14**(1), 50–57 (2004)

4. Gross, R., Brajovic, V.: An image preprocessing algorithm for illumination invariant face recognition. In: Audio-and Video-Based Biometric Person Authentication, pp. 10–18 (2003)
5. Klare, B.F., Jain, A.K.: Heterogeneous face recognition using kernel prototype similarities. IEEE Trans. Pattern Anal. Mach. Intell. **35**(6), 1410–1422 (2013)
6. Yi, D., Liu, R., Chu, R., Lei, Z., Li, S.Z.: Face matching between near infrared and visible light images. In: IAPR/IEEE International Conference on Biometrics, pp. 523–530 (2007)
7. Liao, S., Yi, D., Lei, Z., Qin, R., Li, S.Z.: Heterogeneous face recognition from local structures of normalized appearance. In: IAPR/IEEE International Conference on Biometrics, pp. 209–218 (2009)
8. Lei, Z., Zhou, C., Yi, D., Jain, A.K., Li, S.Z.: An improved coupled spectral regression for heterogeneous face recognition. In: IAPR International Conference on Biometrics, pp. 7–12, IEEE (2012)
9. Liu, S., Yi, D., Lei, Z., Li, S.Z.: Heterogeneous face image matching using multi-scale features. In: IAPR/IEEE International Conference on Biometrics, pp. 79–84 (2012)
10. Huang, L., Lu, J., Tan, Y.P.: Learning modality-invariant features for heterogeneous face recognition. In: International Conference on Pattern Recognition, pp. 1683–1686 (2012)
11. Chen, C., Ross, A.: Local Gradient Gabor Pattern (LGGP) with applications in face recognition, cross-spectral matching, and soft biometrics. In: SPIE Defense, Security, and Sensing, International Society for Optics and Photonics, pp. 87120R–87120R (2013)
12. Chen, J., Yi, D., Yang, J., Zhao, G., Li, S.Z., Pietikainen, M.: Learning mappings for face synthesis from near infrared to visual light images. In: IEEE Conference on Computer Vision and Pattern Recognition, pp. 156–163 (2009)
13. Zhang, Z., Wang, Y., Zhang, Z.: Face synthesis from near-infrared to visual light via sparse representation. In: International Joint Conference on Biometrics, pp. 1–6 (2011)
14. Hu, S., Choi, J., Chan, A.L., Schwartz, W.R.: Thermal-to-visible face recognition using partial least squares. J. Opt. Soc. Am. A **32**(3), 431–442 (2015)
15. Daugman, J.G.: Two-dimensional spectral analysis of cortical receptive field profiles. Vis. Res. **20**(10), 847–856 (1980)
16. Liu, C., Wechsler, H.: Gabor feature based classification using the enhanced fisher linear discriminant model for face recognition. IEEE Trans. Image Process. **11**(4), 467–476 (2002)
17. Huang, G.B., Zhu, Q.Y., Siew, C.K.: Extreme learning machine: theory and applications. Neurocomputing **70**(1–3), 489–501 (2006)
18. Su, Y., Shan, S., Chen, X., Gao, W.: Hierarchical ensemble of global and local classifiers for face recognition. IEEE Trans. Image Process. **18**(8), 1885–1896 (2009)
19. Broomhead, D.S., Lowe, D.: Multivariable functional interpolation and adaptive networks. complex Syst. **2**, 321–355 (1988)
20. Schmidt, W.F., Kraaijveld, M., Duin, R.P.: Feed forward neural networks with random weights. In: IAPR International Conference on Pattern Recognition, Conference B: Pattern Recognition Methodology and Systems, pp. 1–4, IEEE (1992)
21. Pao, Y.H., Takefji, Y.: IEEE functional-link net computing. Comput. J. **25**(5), 76–79 (1992)
22. Toh, K.A., Eng, H.L.: Between classification-error approximation and weighted least-squares learning. IEEE Trans. Pattern Anal. Mach. Intell. **30**(4), 658–669 (2008)
23. Toh, K.A.: Deterministic neural classification. Neural Comput. **20**(6), 1565–1595 (2008)
24. Kim, S.K., Lee, H., Yu, S., Lee, S.: Robust face recognition by fusion of visual and infrared cues. In: IEEE Conference on Industrial Electronics and Applications, pp. 1–5, IEEE (2006)
25. Tan, X., Triggs, B.: Enhanced local texture feature sets for face recognition under difficult lighting conditions. IEEE Trans. Image Process. **19**(6), 1635–1650 (2010)
26. Lei, Z., Liao, S., Pietikainen, M., Li, S.Z.: Face recognition by exploring information jointly in space, scale and orientation. IEEE Trans. Image Process. **20**(1), 247–256 (2011)

Dynamic Adjustment of Hidden Layer Structure for Convex Incremental Extreme Learning Machine

Yongjiao Sun, Yuangen Chen, Ye Yuan and Guoren Wang

Abstract Extreme Learning Machine (ELM) is a learning algorithm based on generalized single-hidden-layer feed-forward neural network. Since ELM has an excellent performance on regression and classification problems, it has been paid more and more attention recently. The determination of structure of ELM plays a vital role in ELM applications. Essentially, determination of the structure of ELM is equivalent to the determination of the hidden layer structure. Utilizing a smaller scale of the hidden layer structure can promote faster running speed. In this paper, we propose algorithm PCI-ELM (Pruned-Convex Incremental Extreme Learning Machine) based on CI-ELM (Convex Incremental Extreme Learning Machine). Furthermore, we also present an improved PCI-ELM algorithm, EPCI-ELM (Enhanced Pruned-Convex Incremental Extreme Learning Machine), which introduces a filtering strategy for PCI-ELM during the neurons adding process. In order to adjust the single-hidden-layer feed-forward neural network more flexibly and achieve the most compact form of the hidden layer structure, in this paper, we propose a algorithm which can dynamically determine hidden layer structure, DCI-ELM (Dynamic Convex Incremental Extreme Learning Machine). At the end of this paper, we verify the performance of PCI-ELM, EPCI-ELM and DCI-ELM. The results show that PCI-ELM, EPCI-ELM and DCI-ELM control hidden layer structure very well and construct the more compact single-hidden-layer feed-forward neural network

Keywords Extreme learning machine · Dynamic adjustment · Feed-forward neural network · Convex optimal increment

1 Introduction

ELM [1], which is now an important branch of neural networks, gains high testing accuracy, extremely fast learning speed and good generalization performance. Increasing attention has been drawn to ELM in both industrial and academic fields.

Y. Sun (✉) · Y. Chen · Y. Yuan · G. Wang
Northeastern University, Shenyang 110819, Liaoning, China
e-mail: sunyongjiao@ise.neu.edu.cn

© Springer International Publishing Switzerland 2016
J. Cao et al. (eds.), *Proceedings of ELM-2015 Volume 2*,
Proceedings in Adaptation, Learning and Optimization 7,
DOI 10.1007/978-3-319-28373-9_30

In the computing theory of ELM, the input weights and bias are generated randomly. The output weights are calculated using input data matrix, input weights and bias. Then the whole structure of neural networks are generated. From this point of view, we draw a conclusion that the calculation cost are directly related to the structure of the neural network. Researchers have done some work trying to simplify the structure of ELM on the premise of keeping the advantages of extremely fast learning speed and generalization performance, so that the overfitting problem can be avoided.

Many researchers focused on improvement of the testing accuracy, generalization performance and training speed. Hai-Jun Rong et al. proposed P-ELM (Pruned Extreme Learning Machine) [2, 7–9], which analyzes the relativity of neurons with mathematical statistics methods and eliminates the neurons insensitive of class labels. P-ELM reduces the calculation cost and is capable of real-time structure adjustment. Yoan Miche et al. focused on the influence of the incorrect training data and proposed OP-ELM (Optimally Pruned Extreme Learning Machine) [3]. Different from P-ELM, OP-ELM takes the influence of irrelative training data into consideration and increases the generalization performance and robustness using pruning strategies. Guorui Feng et al. proposed E-ELM (Error Minimized Extreme Learning Machine) [4] to realize increment of single or multiple neurons by minimizing error. Rui Zhang et al. analyzed the weights of the hidden layer nodes and proposed D-ELM (Dynamic Extreme Learning Machine) [5] to further evaluate the changes after the increment of neurons. Guorui Feng et al. proposed DA-ELM [6] (Dynamic Adjustment Extreme Learning Machine) based on theory of application circle expectation minimization to reduce errors.

Given a specific application, how to determine the structure of the neural network remains an open question. In this paper, we try to solve the problem in the following aspects.

- Generate appropriate neural network structures automatically for various applications. Increasing strategies of hidden layer neurons have to be proposed to minimize training errors.
- Evaluate neurons to organize hidden layer with higher performance. Neurons with less contribution have to be eliminated to generate a reduced but efficient neural network.
- Alter the complexity of the hidden layer structure. Adjustment strategies of hidden layer make the alteration more reasonable.

2 Improved Convex Incremental ELM

During the adjustment of the ELM structure, the increment of hidden layer nodes without consideration of the effectiveness will cause the redundant of the ELM structure. Furthermore, the additive node will also affect the effectiveness of the neurons already added into the hidden layer. Therefore, when eliminating the neurons with low effectiveness, the measure of neurons faces great challenges.

2.1 Pruned Convex Incremental Extreme Learning Machine

The pruning ELMs usually measure and sort the neurons by some criteria to eliminate the neurons with low effectiveness. Based on the same idea, we propose Pruned Convex Incremental Extreme Learning Machine (PCI-ELM), which measures the neurons by the output weights. In the case of multiple output nodes, PCI-ELM measures the neurons by the norm of the output weights:

$$\|\beta_i\| = \sqrt{\beta_{i1}^2 + \cdots + \beta_{im}^2} \tag{1}$$

where $\beta_i = [\beta_{i1}, \ldots, \beta_{im}]^T$ is the output weights of the ith neurons; m is the number of output nodes. However, even if the output weight is not small, but the output of the activation function, taking sigmoid as example, is small, thus, in this case, the output weight cannot represent the importance of the neuron. In CI-ELM [10], the weights are updated as

$$\beta_i = (1 - \beta_L)\beta_i \tag{2}$$

If we view β_i as the function of additive weights β_L, β_i is a monotonical decreasing function. $o = [o_1, \ldots, o_m]^T$ is the output of ELM, $h_i(x_j)$ is the output of the jth sample on the ith neuron:

$$o_i = h_1(\mathbf{x}_i)\beta_1 + \cdots + h_L(\mathbf{x}_i)\beta_L \tag{3}$$

Since the parameters of the existing neurons do not change, the outputs of hidden layer neurons $h(\mathbf{x})$ do not change either. Given any training sample $\mathbf{x}$, the influence of the existing neurons $h_i(\mathbf{x})\beta_i$, $i \in (1, L - 1)$ will change due to the Lth additive node. Therefore, in CI-ELM, the output weights indicate the effectiveness of hidden layer nodes very well. According to the output weights, we sort the hidden layer nodes and eliminate the ones with little effectiveness. A threshold of the weights γ is set, which is initiated as

$$\gamma' = \gamma = \frac{1}{L} \sum_{i=1}^{L} |\beta_i| \tag{4}$$

The elimination is determined by the training accuracy: if the training accuracy $\varepsilon' < \varepsilon$ after the elimination, the neurons have to be kept in the network; if $\varepsilon' > \varepsilon$ after the elimination, the network has to be pruned.

In order to provide better adjustment, PCI-ELM eliminates the neurons with low effectiveness to construct a neural network with simplest structure.

2.2 Enhanced Pruned Convex Incremental Extreme Learning Machine

The pruning after the construction of the ELM simplifies the network structure. However, the calculation of the effectiveness of the neurons and the performance variation lead to much more calculation cost. Therefore, a better way is to verify the effectiveness of the neurons before they are added into the network.

Since the smaller the norm of the output weights is, the better generalization performance the network gains [11], we also use the norm of the output weights $\|\beta\|$ as a criterion to choose additive neurons. After the output weights are updated as $\beta_i = (1 - \beta_L)\beta_i$, the neuron with smaller $\|\beta\|$ will be added into the network. The selection is summarized as $\Psi_{L+1} = \begin{cases} \Psi_{L+1}^{(1)} \\ \Psi_{L+1}^{(2)} \; if \; \| E\left(\Psi_{L+1}^{(2)}\right) \| < \| E\left(\Psi_{L+1}^{(1)}\right) \| \; and \; \| \beta^{\Psi_{L+1}^{(2)}} \| < \| \beta^{\Psi_{L+1}^{(1)}} \| \end{cases}$

where $\Psi_{(L+1)}$ is the network with $L+1$ hidden layer neurons. The output weight of the Lth neuron is calculated as:

$$\left\|\beta^{L-1}\right\| = \sqrt{\beta_1^2 + \cdots + \beta_{L-1}^2} \tag{5}$$

When the Lth neuron is added, the output weights is calculated as:

$$\begin{aligned} \|\beta^L\| &= \sqrt{\left(1 - \beta_L\right)^2 \cdot \left(\beta_1^2 + \cdots + \beta_{L-1}^2\right) + \beta_L^2} \\ &= \sqrt{\left(1 - \beta_L\right)^2 \|\beta^{L-1}\|^2 + \beta_L^2} \\ &\geq \sqrt{\frac{\|\beta^{L-1}\|^2}{\|\beta^{L-1}\|^2 + 1}} = \frac{\|\beta^{L-1}\|}{\sqrt{\|\beta^{L-1}\|^2 + 1}} \end{aligned} \tag{6}$$

As to $\|\beta^L\|$, $\|\beta^{L-1}\|$ is a constant, thus, $\|\beta^L\|$ can be viewed as a dependent variable, β_L as an independent variable. Equation 6 can be viewed as function $f(\mathbf{x}) = \sqrt{(1 - \mathbf{x})^2 a^2 + \mathbf{x}^2}$, where a is a constant, $\mathbf{x} \in R$, thus, $f(\mathbf{x})$ is monotonic decreasing in the interval $\left[-\infty, \frac{a^2}{1+a^2}\right]$, monotonic increasing in the interval $\left[\frac{a^2}{1+a^2}, +\infty\right]$.

Note that the method mentioned above only tries to guarantee that the existing neurons have certain effectiveness. Although in each step we try to make $\|\beta\|$ as small as possible, the output weight $\|\beta\|$ is not smaller than $\|\beta\|$ in CI-ELM for sure.

The redundancy of the network cannot be avoided, which is because the remaining training error is calculated as:

$$
\begin{aligned}
\Delta &= \left\| e_{L-1} \right\|^2 - \left\| e_L \right\|^2 \\
&= \left\| e_{L-1} \right\|^2 - \left\| e_{L-1} - \beta_L \left(\mathbf{H} - \mathbf{F}_{L-1} \right) \right\|^2 \\
&= 2\beta_L \left\langle e_{L-1}, \mathbf{H}_L - \mathbf{F}_{L-1} \right\rangle - \beta_L^2 \left\| \mathbf{H}_L - \mathbf{F}_{L-1} \right\|^2 \\
&= \left\| \mathbf{H}_L - \mathbf{F}_{L-1} \right\|^2 \left(\frac{\left\langle e_{L-1}, \mathbf{H}_L - \mathbf{F}_{L-1} \right\rangle^2}{\left\| \mathbf{H}_L - \mathbf{F}_{L-1} \right\|^4} - \left(\beta_L - \frac{\left\langle e_{L-1}, \mathbf{H}_L - \mathbf{F}_{L-1} \right\rangle}{\left\| \mathbf{H}_L - \mathbf{F}_{L-1} \right\|^2} \right)^2 \right)
\end{aligned}
\tag{7}
$$

When $\beta_L = \frac{\langle e_{L-1}, \mathbf{H}_L - \mathbf{F}_{L-1} \rangle}{\| \mathbf{H}_L - \mathbf{F}_{L-1} \|^2}$, Δ is maximum, $\Delta_{max} = \frac{\langle e_{L-1}, \mathbf{H}_L - \mathbf{F}_{L-1} \rangle^2}{\| \mathbf{H}_L - \mathbf{F}_{L-1} \|^2}$. The greater β_L is, the greater Δ_{max} is.

From Eq. 6 we can see that, if the output weight of the additive node is $\beta^L = \frac{\|\beta^{L-1}\|^2}{\|\beta^{L-1}\|^2 + 1}$, the norm $\|\beta\|$ is minimum. Therefore, we try to choose larger $\beta_L = \frac{\|\beta^{L-1}\|^2}{\|\beta^{L-1}\|^2 + 1}$, which means larger $\|\beta^{L-1}\|$.

If a large enough $\|\beta^1\|$ is set in the neural network, the output of the additive node can also be large. In the initiate phase, K neurons are generated randomly as $h_1, \ldots, h_K$, $\beta_i = \frac{\mathbf{E} \cdot [\mathbf{E} - (\mathbf{F} - \mathbf{H}_i)]^\mathrm{T}}{[\mathbf{E} - (\mathbf{F} - \mathbf{H}_i)] \cdot [\mathbf{E} - (\mathbf{F} - \mathbf{H})_i]^\mathrm{T}}$, $i = 1, \ldots, K$, larger β will be chosen as the initiate neuron.

Based on the random search strategy, the probability of finding a neuron with output weight $\frac{\|\beta^{L-1}\|^2}{\|\beta^{L-1}\|^2 + 1}$ is nearly zero. Thus, it is not necessary to pursue the minimum $\|\beta^L\|$. Like the random search in EI-ELM, a maximum search time k is set, the neurons with more training error decrement and the least norm is added into the network. In other words, these neurons must exists in the circle with the center located as $\beta^L = \frac{\|\beta^{L-1}\|^2}{\|\beta^{L-1}\|^2 + 1}$, γ as the radius. In most cases, the output weights near center $\frac{\|\beta^{L-1}\|^2}{\|\beta^{L-1}\|^2 + 1}$ are more likely to be chosen.

3   Dynamic Convex Incremental Extreme Learning Machine

In this section, we introduce the Convex Incremental Extreme Learning Machine (DCI-ELM). We merge the pruning of useless neurons into the process of adding new neurons. So we can delete the useless or inefficient neurons from the ELM network earlier, simplify the neuron network structure as early as possible, and make the most effort to get the most compact and efficient ELM hidden structure.

The less error the new added neurons using CI-ELM bring to the ELM network training process, the more efficient the network will be. But this may lead to the decrease of the effectiveness of other neurons in the hidden layer. Meanwhile, the method that CI-ELM construct the front-feedback single hidden layer network can be treated as ordering a sequence of neurons in the hidden layer. So trying some kinds of ordering method for each new added neuron can maximally mining its effectiveness. Thus, for the ELM network with i neurons in the hidden layer, there would exist a set Φ_i with size V_{max} recording the network hidden layers with currently smallest training error for the i neurons in the hidden layer.

Before the processing of DCI-ELM, all the sets can be treated as not existing. At this time, no ELM network with any size of hidden layer is constructed. The first step of DCI-ELM is to construct an ELM network $\Psi_1^{(1)}$ with only one neuron in its hidden layer. Using the same training method of the first neuron with CI-ELM, we get its corresponding weight in the output layer. Then, $\Psi_1^{(1)}$ is treated as an element and add to the set Φ_1. We update the set Φ_1 with the maximal training error $\left\| E(\Phi_1) \right\|_{max}$ and size $V_1 = 1$.

If the constructed single-hidden-layer feed-forward neural network $\Psi_1^{(1)}$ cannot satisfy the given training target, then DCI-ELM randomly generate a neuron $\mathbf{H}_2$, $\mathbf{H}_2 = [G(\mathbf{a}_2, b_2, \mathbf{x}_1), G(\mathbf{a}_2, b_2, \mathbf{x}_2), \ldots, G(\mathbf{a}_2, b_2, \mathbf{x}_N)]^T$. At this time, for the neuron $\mathbf{H}_2$, there exist two choices: (1) construct a new ELM network $\Psi_1^{(2)}$ containing only $\mathbf{H}_2$; (2) add $\mathbf{H}_2$ into ELM $\Psi_1^{(1)}$ and construct a new network $\Psi_2^{(1)}$ containing two neurons in the hidden layer. To mostly utilize the generated neurons to construct the optimal ELM network, and mine the effectiveness of each neuron in the hidden layer, $\mathbf{H}_2$ will conduct these two processes. At the same time, to make each set full faster and drive the compete among them to select preferable middle networks, the update strategy of the sets is bottom up, i.e., the adding order of the new generated neurons is $\Phi_1, \Phi_2, \ldots$ Then, when there is an ELM network reaches the training accuracy, the algorithm halts. This ensures that the simpler and efficient ELM networks are chosen in prior. Thus, h_2 firstly constructs the ELM network, and then is added to the set Φ_1:

$$\Psi_1^{(2)} \in \Phi_1, V_1 = V_1 + 1 \tag{8}$$

where, $\Phi_1 = \{\Psi_1^{(1)}, \Psi_1^{(2)}\}$. If $\exists \Psi_1 \in \Phi_1$, makes $\| E(\Psi_1) \| < \varepsilon$, then it means we have constructed an appropriate ELM network, and the algorithm halts. Otherwise, based on the elements in Φ_1, we add neuron $\mathbf{H}_2$ to these ELM networks. We use the following equations to calculate the corresponding output weight β_2 of $\mathbf{H}_2$, and the training error E of each new generated network:

$$\beta_2 = \frac{\mathbf{E} \cdot [\mathbf{E} - (\mathbf{F} - \mathbf{H}_2)]^T}{[\mathbf{E} - (\mathbf{F} - \mathbf{H}_2)] \cdot [\mathbf{E} - (\mathbf{F} - \mathbf{H})_2]^T} \tag{9}$$

$$\mathbf{E} = (1 - \beta_2)\mathbf{E} + \beta_2(\mathbf{F} - \mathbf{H}_2) \tag{10}$$

Thus, we can obtain two more complex networks $\Psi_2^{(1)} = \Psi_1^{(1)} + \beta_2^{(1)}\mathbf{H}_2$ and $\Psi_2^{(2)} = \Psi_1^{(2)} + \beta_2^{(2)}\mathbf{H}_2$, and add them to the set Φ_2, $\Phi_2 = \{\Psi_2^{(1)}, \Psi_2^{(2)}\}$.

When the Lth neurons are generated if the volume of each set $V_{max} < L$, all the sets are full, and the set containing the most complex ELM network is Φ_{L-1}, in which the hidden layer of each network has L-1 neurons. For the sets $\Phi_1, \Phi_2, ..., \Phi_{L-1}$, the neuron $\mathbf{H}_L$ are added to the middle network of these sets in order. Thus, compared to the original network, in the new generated middle network, the more accurate network remains in the set and continue to be increased, and the ones with less accuracy are pruned. Thus, for each time we add neurons and generate a new single-hidden-layer feed-forward neural network, it has the following comparison process with the existing sets that contain the same size of neurons in the hidden layer:

If $V_i = V_{max}, i < L$, and let $k=1$, when $k \le V_{max}$,

$$\Psi_i^{max} = \begin{cases} \tilde{\Psi}_i^{(k)}, if \left\| \mathbf{E}\left(\Psi_i^{max}\right) \right\| > \left\| \mathbf{E}\left(\tilde{\Psi}_i^{(k)}\right) \right\| \\ \Psi_i^{max}, if \left\| \mathbf{E}\left(\Psi_i^{max}\right) \right\| \le \left\| \mathbf{E}\left(\tilde{\Psi}_i^{(k)}\right) \right\| \end{cases} \tag{11}$$

Here, Ψ_i^{max} presents the element whose training error is the largest in the set Φ_i; $\tilde{\Psi}_i^{(k)}$ presents the kth result in the single-hidden-layer feed-forward neural network. Thus, this kind of comparison or competition makes the accuracy of any neuron in each set keeps increasing, so that we can obtain the network meeting the training target as soon as possible. When the Lth neuron is added to the set Φ_{L-1}, the set Φ_L is empty. So after adding a neuron to the set Φ_{L-1}, all the generated V_{L-1} single-hidden-layer feed-forward neural networks will be added to the set Φ_L.

Therefore, when any single-hidden-layer feed-forward neural network added into any set will active the comparison with the training target ε. If the ones whose training error is smaller than the training target ε have existed in the set, the algorithm halts. To consolidate the generalization ability of ELM, for the situation in which more than one single-hidden-layer feed-forward neural networks reach the target, we will choose the one in which $\|\beta\|$ is the smallest. In order words, the final chosen structure of ELM is as follows:

$$\Psi = \Psi_i, if \left\| \mathbf{E}(\Psi_i) \right\| \le \varepsilon \ and \ \forall\Psi \in \Phi_i, min\|\beta\| \tag{12}$$

4   Performance Evaluation

In this Section, we test the algorithms proposed in the former using extensive experiments. The tests focus on three targets: proceeding time, generalization ability, and the number of neurons in the hidden layer. Meanwhile, we compare the algorithms, PCI-ELMEPCI-ELM, and DCI-ELM, introduced in the former, with the existing typical ELM algorithms, such as I-ELMEI-ELM and CI-ELM. All the evaluations were carried out in MATLAB R2009a in a Intel Core $i3$ processor with 3.3 GHz and 4GB RAM. A group of real datasets [12] about the regression problem are used to test the performance the PCI-ELM, EPCI-ELM, and DCI-ELM algorithms.

To ensure the effectiveness of the training results, each dataset is separated in to training data and testing data using a proportion of 2:1, so that the training data and testing data are independent and not repeated. At the same time, before using these data to train the networks, the input data in all dimensions are normalized into a range of $[-1,1]$ according to the following equation.

$$Input\,(:,i) = \frac{Input\,(:,i) - \min\,(Input\,(:,i))}{\max\,(Input\,(:,i)) - \min\,(Input\,(:,i))} \times 2 - 1 \qquad (13)$$

Here, Input means the matrix of the input data, and Input$(:,i)$ means a line of the input matrix, i.e., all the input data in the same attribute or dimension.

We will analyze the experiment results of I-ELM, CI-ELM, PCI-ELM, and DCI-ELM algorithms. We compare the training time of different algorithms over the same dataset, predicting time, and the number of neurons in the hidden layer w.r.t the learning problems.

To avoid the occasionality of a single experiment result, in this paper, all our experiments over each dataset are conducted 30 times and calculate the average value as the final result. Moreover, for each algorithm, the number of neurons is counted from 0 to 1000. Meanwhile, when EI-ELM and EPCI-ELM searching for prior neurons, we set the maximal searching times and the size of sets both as 5. Thus, for the given training target, if the training result cannot reach , the result is also recorded when the number of neurons in the hidden layer reaches 1000.

Figures 1 and 2 show the results of using sigmoid function and RBF function in different algorithms respectively. Even though the effect would be somehow different when using different excitation function, in general, the training time has obvious similarity between the two experiments. As we delete useless or inefficient neurons during the process of training, the training time of PCI-ELM is longer than that of CI-ELM conducted over the same dataset. As EPCI-ELM as to PCI-ELM is just like EI-ELM as to I-ELM, there exists a more strategy to select the neurons, and thus the training time of EPCI-ELM is longer than that of PCI-ELM. However, this is not absolutely. Because the searching of neurons can speed up the convergence of algorithm, there exists such situation that EPCI-ELM is faster than PCI-ELM or even CI-ELM. For the same reason, similar situation also happen in the process of EI-ELM and I-ELM.

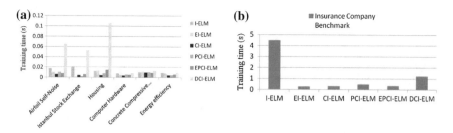

Fig. 1 Comparison chart of training time by using sigmoid function. **a** Without insurance company bench-mark. **b** Insurance company benchmark

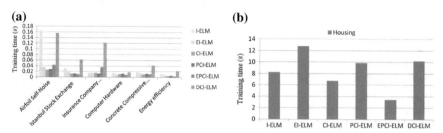

Fig. 2 Comparison chart of training time by using RBF function. **a** Without housing. **b** Housing

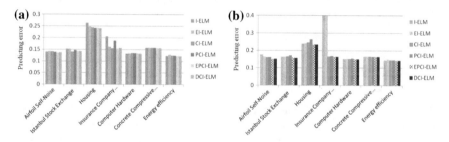

Fig. 3 Comparison chart of predicted results. **a** Sigmoid function. **b** RBF function

From the experiment results, we can see that the training time has similarity when using sigmoid or RBF function. Usually, EI-ELM and EPCI-ELM enhance the selection of neurons in the hidden layer, so it will spend more training time. But this is not absolute. As EI-ELM and EPCI-ELM speed up the convergence of algorithms, they may be faster sometimes. So as to DCI-ELM, even though training more than one networks spends more training time, it also speed up the convergence when constructing the hidden layer structures.

Figure 3 shows the predicting results of the algorithms conducted over different datasets. Firstly, we can see that the pruning of hidden layer structures in CI-ELM influences little in the approximate ability. Secondly, as optimization methods for generalization ability exist in EPCI-ELM and DCI-ELM, the predicting results of these two algorithms is not worse or even better than those of CI-ELM.

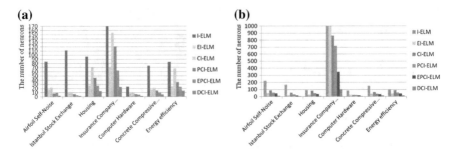

Fig. 4 Comparison chart of hidden layer scale. **a** Sigmoid function. **b** Sigmoid function

From Fig. 4, we can see that the pruning that PCI-ELM providing to CI-ELM largely decrease the redundancy when constructing the networks suing CI-ELM, so that the hidden layer is more compact. EPCI-ELM improves the searching strategy of the neurons in the hidden layer, so that it further avoiding the attendance of redundant neurons. In other words, it optimize the constructed single-hidden-layer feed-forward neural networks from the very beginning. The DCI-ELM makes an adjustment to the network construction in a largest degree. In other words, it mines the most compact structure constructed by the generated neurons in a largest degree, so that the needed neurons is fewer and make the ELM predicting efficient.

5   Conclusion

In this paper, we firstly introduce the ELM theory and its typical dynamic construction algorithms, and analyze the corresponding idea of structure selection. Then, we propose two hidden layer structure optimization algorithms: PCI-ELM and EPCI-ELM. We further adjust the hidden layer structure constructed by CI-ELM to make it more compact and efficient. After this, we propose a DCI-ELM algorithm, and provide a more dynamic hidden layer structure adjustment algorithm, so that it can allocate the pruning to the process of adding neurons. Finally, we take extensive experiments over real datasets and analyze the effectiveness of our proposed algorithms.

References

1. Huang, G.B., Zhu, Q.Y., Siew, C.K.: Extreme learning machine: theory and applications. Neurocomputing **70**(1), 489–501 (2006)
2. Rong, H.J., Ong, Y.S., et al.: A fast pruned-extreme learning machine for classification problem. Neurocomputing **72**(1), 359–366 (2008)

3. Miche, Y., Sorjamaa, A., et al.: OP-ELM: optimally pruned extreme learning machine. Neural Netw. **21**(1), 158–162 (2010)
4. Feng, G., Huang, G.B., et al.: Error minimized extreme learning machine with growth of hidden nodes and incremental learning. Neural Netw. **20**(8), 1352–1357 (2009)
5. Zhang, R., Lan, Y., Huang, G.B., et al.: Dynamic extreme learning machine and its approximation capability. Cybernetics **43**(6), 2054–2065 (2013)
6. Feng, G., Lan, Y., et al.: Dynamic adjustment of hidden node parameters for extreme learning machine. Cybernetics **45**(2), 279–288 (2015)
7. Huang, G.B., Zhou, H., et al.: Extreme learning machine for regression and multiclass classification. Cybernetics **42**(2), 513–529 (2012)
8. Liang, N.Y., Huang, G.B., et al.: A fast and accurate online sequential learning algorithm for feedforward networks. Neural Netw. **17**(6), 1411–1423 (2006)
9. Huang, G.B., Chen, L.: Enhanced random search based incremental extreme learning machine. Neurocomputing **71**(16), 3460–3468 (2008)
10. Huang, G.B., Chen, L.: Convex incremental extreme learning machine. Neurocomputing **70**(16), 3056–3062 (2007)
11. Bartlett, P.L.: The sample complexity of pattern classification with neural networks: the size of the weights is more important than the size of the network. Inform. Theory **44**(2), 525–536 (1998)
12. Blake, C., Merz, C.: UCI Repository of Machine Learning Databases, Department of Information and Computer Sciences, University of California, Irvine, USA, 1998. http://archive.ics.uci.edu/ml/datasets.html

ELMVIS+: Improved Nonlinear Visualization Technique Using Cosine Distance and Extreme Learning Machines

Anton Akusok, Yoan Miche, Kaj-Mikael Björk, Rui Nian,
Paula Lauren and Amaury Lendasse

Abstract This paper presents ELMVIS+, a significant improvement in ELMVIS methodology that enables faster computation, more stable results and a wider application range. The novel cost function and a fast way of estimating it speeds up the method compared to ELMVIS, especially in large-dimensional datasets. The included Genetic Algorithms add global optimization that helps ELMVIS+ to find a better optimum. The improved methodology shows state-of-the-art performance in three different benchmark datasets.

Keywords Visualization · Nonlinear dimensionality reduction · Machine learning · Neural network · Genetic algorithms · Cosine distance · Extreme learning machines · Big data · Big dimensionality · Projection

1 Introduction

High-dimensional data is omnipresent in the modern world, but it stays virtually impenetrable for human analysis, except for images or audio. Thus data visualization [1] stays a demanded area of research. For analysing or exploration of an

A. Akusok (✉) · A. Lendasse
The Iowa Informatics Initiative, The University of Iowa, Iowa City, USA
e-mail: anton-akusok@uiowa.edu

A. Lendasse
e-mail: amaury-lendasse@uiowa.edu

Y. Miche
Nokia Solutions and Networks Group, Espoo, Finland

K.-M. Björk
Arcada University of Applied Sciences, Helsinki, Finland

R. Nian
Ocean University of China, Qingdao, China

P. Lauren
Oakland University, Rochester, USA

© Springer International Publishing Switzerland 2016
J. Cao et al. (eds.), *Proceedings of ELM-2015 Volume 2*,
Proceedings in Adaptation, Learning and Optimization 7,
DOI 10.1007/978-3-319-28373-9_31

arbitrary high dimensional data, a suitable visualization should be created. It is commonly restricted to 2 or 3 dimensions, which are easier to show, but for the visualization to be useful it must be representative of the original data.

The naive dimensionality reduction method is variable (feature) selection, but a few selected variables could present only a part of the data structure, if any. Other dimensionality reduction methods optimize a selected criterion, with different criteria resulting in two different algorithms.

Linear dimensionality reduction methods are PCA [2] and linear MDS [3], which however yield the same results, as proven in [1]. Their criterion is variance maximization which works for datasets with linear dependencies, but the general performance may be poor.

If the variables are relevant but correlated (which is often the case), the dimensionality of data is higher than necessary. Then the same data could be explained by a smaller set of transformed variables, and is said to lie on a manifold [1]. As an example, one can imagine a camera rotating around an object at a fixed distance, then the pictures of that camera would lie on a 2-dimensional manifold (sphere), while their actual dimension would be much higher. Many nonlinear dimensionality reduction methods, including those listed in the next section, aim to find and unfold such a manifold using various cost functions and training algorithms. Even PCA would find a manifold in the data, if that is linear. Manifolds are commonly found by preserving the neighbourhood in original and reduced spaces. Topology-preserving methods that use graph distances, like CDA [4], normally provide excellent results for un-foldable manifolds.

In a very high dimensional space, neighbourhood rank is a weak metric [5]. This is caused by an empty space phenomenon [6] and the curse of dimensionality, studied thoroughly in [5]. The problem comes from the change of distances distribution between points in space as the dimensionality goes up. Distances between points in a dataset are typically normally distributed. With the increase of a space dimensionality, the mean of that normal distribution increases whereas the variance stays the same. It causes the distribution to concentrate around some value, and reduces distance differences between various ranked neighbours, making the nearest neighbour unstable already at 10–20 dimensions [5]. These cases require a nonlinear dimensionality reduction method with general cost function without other assumptions. The Extreme Learning Machine (ELM)-based visualization methods ELMVIS [7] and its improvement ELMVIS+ use Mean Squared Error (MSE) or cosine distance or ELM-reconstructed data accordingly, while the non-linearity of ELM provides the desired nonlinear projection.

The ELMVIS method links data points with the given visualization points. An ELM model learns the de-projection of visualization points back into the original space, where a cost function is calculated. The optimization task is to find the best order of data points for a given fixed order of visualization points.

This task is similar to an open loop travelling salesman problem [8] and its optimization is challenging. In the ELMVIS+ methodology, two improvements are proposed: a new cost function with a very fast way of updating it when exchanging positions of two data samples, and a global optimization step with Generic

Algorithms [9] (GA). In total, they provide a fast and useful method of data visualization onto arbitrary fixed set of points in the visualization space. The method has only one hyper-parameter that is the number of neurons in ELM model, and with the proposed cost function it works for very high-dimensional data.

The rest of the paper is organized as following. Section 2 describes the ELM algorithm and its adaptation for computation and fast update of a cost function for visualization, and a use of GA for a first stage. Section 3 presents experiments on various datasets in comparison with other methods, performance and convergence analysis, and evaluates an effect of Genetic Algorithms stage. Section 4 concludes on the work done, discusses about improvements compared to the original ELMVIS, benefits and drawbacks.

2 Methodology

This paper presents an improved way of creating a visualization using Extreme Learning Machines (ELM). The data is projected onto fixed visualization points. ELM learns a reverse projection of visualization points to the original data space, and an error is computed in the original space. Thus ELM is utilized as a nonlinear metric for the reconstruction error which stands for recall, or continuity [10] in data visualization field. A relation between original and visualization data spaces is shown on Fig. 1.

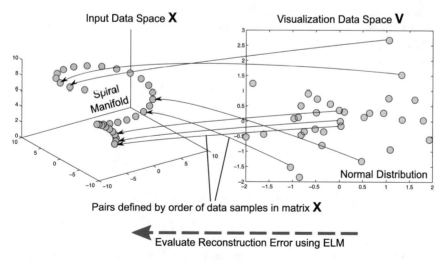

Fig. 1 ELMVIS+ finds an optimal order, or permutation, of data samples $\mathbf{x}_i$ in matrix $\mathbf{X}$ for the given visualization space points $\mathbf{v}_i$ in matrix $\mathbf{V}$. Points $\mathbf{V}$ are fixed, and are chosen in any suitable way—for instance from a normal distribution, or on a regular grid. An ELM learns a projection $\mathbf{V} \rightarrow \mathbf{X}$ and estimates $\hat{\mathbf{X}}$ from $\mathbf{V}$. The visualization cost function is an error between $\mathbf{X}$ and $\hat{\mathbf{X}}$, which is a negative cosine similarity for ELMVIS+

While the visualization points and their order is fixed, the order of data samples is not defined. ELMVIS+ method computes an error (the visualization cost function) with some order of data samples. The cost function is optimized by changing that order and updating the error; an order of samples which results in a lower error is kept.

The ELM method and all steps of ELMVIS+ are explained in detail in the rest of the Methodology section.

2.1 Extreme Learning Machine

The Extreme Learning Machine algorithm was originally proposed by Guang-Bin Huang et al. in [11–14], and is originally a regression method [15]. The method is proven to be a universal approximator given enough hidden neurons [16]. It works as following:

Consider a set of N distinct samples $(\mathbf{x}_i, \mathbf{t}_i)$ with $\mathbf{x}_i \in \mathbb{R}^d$ and $\mathbf{t}_i \in \mathbb{R}^c$. Then a SLFN with L hidden neurons is modelled as:

$$\sum_{i=1}^{L} \beta_i \phi(\mathbf{w}_i \mathbf{x}_j + b_i), \ j \in [\![1, N]\!] \tag{1}$$

with ϕ being an activation function, $\mathbf{w}_i$ the input weights, b_i the biases and β_i the output weights.

In case the SLFN would perfectly approximate the data, the errors between the estimated outputs y_i and the targets t_i are zero, and the relation between inputs, weights and targets is then:

$$\sum_{i=1}^{L} \beta_i \phi(\mathbf{w}_i \mathbf{x}_j + b_i) = \mathbf{t}_j, \ j \in [\![1, N]\!] \tag{2}$$

which can be written compactly as $\mathbf{H}\beta = \mathbf{T}$, with

$$\mathbf{H} = \begin{pmatrix} \phi(\mathbf{w}_1 \mathbf{x}_1 + b_1) & \cdots & \phi(\mathbf{w}_L \mathbf{x}_1 + b_L) \\ \vdots & \ddots & \vdots \\ \phi(\mathbf{w}_1 \mathbf{x}_N + b_1) & \cdots & \phi(\mathbf{w}_L \mathbf{x}_N + b_L) \end{pmatrix} \tag{3}$$

$$\beta = (\beta_1^T \ldots \beta_L^T)^T, \ \mathbf{T} = (\mathbf{t}_1^T \ldots \mathbf{t}_N^T)^T. \tag{4}$$

Solving the output weights β from the hidden layer representation of inputs $\mathbf{H}$ and targets $\mathbf{T}$ is done using the Moore-Penrose generalized inverse of the matrix $\mathbf{H}$, denoted as $\mathbf{H}^\dagger$ [17]. The training of ELM requires no iterations, and the most computationally cost part is the calculation of a pseudo-inverse of the matrix $\mathbf{H}_{(d \times L)}$, which makes ELM an extremely fast artificial neural network method.

2.2 Visualization with ELM

In visualization, the data points $\mathbf{x}_i \in \mathbb{R}^d$, $i \in [\![1, N]\!]$ are projected to the corresponding visualization points $\mathbf{v}_i \in \mathbb{R}^{\tilde{d}}$, $i \in [\![1, N]\!]$ in a smaller dimensional space $\tilde{d}$, usually $\tilde{d} = 2$ or $\tilde{d} = 3$.

An ELM learns the reverse projection model, that is ELM projects visualization space points $\mathbf{V}$ into the original data space $\mathbb{R}^d$, see Fig. 2.

An ELM provides an approximated data samples $\hat{\mathbf{X}} = f(\mathbf{V})$ projected from the input points $\mathbf{V}$. If there is a smooth relation between $\mathbf{V}$ and $\mathbf{X}$, the approximation $\hat{\mathbf{X}}$ learned by ELM is close to the true data $\mathbf{X}$. If visualization points $\mathbf{V}$ are located arbitrary and do not relate to $\mathbf{X}$, an ELM with a limited number of neurons fails to learn an accurate projection model, and samples $\hat{\mathbf{X}}$ are far from the original samples in $\mathbf{X}$. Note that sample similarity for the visualization is computed in the original data space $\mathbb{R}^d$, which provides more accurate results that finding it in a reduced visualization space $\mathbb{R}^2$ or $\mathbb{R}^3$.

An ELM in ELMVIS+ is trained with input data $\mathbf{V}$ and target data $\mathbf{X}$. The data is arranged in pairs $(\mathbf{v}_i, \mathbf{x}_j) \mid i = j$, $i \in [\![1, N]\!]$, $j \in [\![1, N]\!]$. The indexes i and j are given implicitly by the position of a particular sample $\mathbf{v}_i$ or $\mathbf{x}_j$ in the corresponding data matrix $\mathbf{V}$, $\mathbf{X}$. The optimization is done by creating a random permutation $\mathbf{p}$ of index j, exchanging pairs of values in $\mathbf{p}$, and applying the permutation $\mathbf{p}$ to samples (rows) in $\mathbf{X}$ before computing the cost function.

After the training completes, the final permutation $\mathbf{p}$ stores the best order of data samples for the given order of visualization points in $\mathbf{V}$.

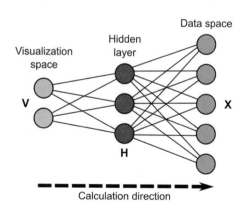

Fig. 2 ELM learns a reverse projection of visualization points $\mathbf{V}$ in to the data space $\mathbb{R}^d$ with the original samples $\mathbf{X}$

2.3 Fast Cost Function from ELM

In the original ELMVIS method, the cost function was the MSE (Mean Squared Error) between the original data samples $\mathbf{x}$ and the projected visualization points $f(\mathbf{v}) = \hat{\mathbf{x}}$. The cost function of ELMVIS+ method is the negative cosine similarity between $\mathbf{x}$ and $\hat{\mathbf{x}}$. The cosine similarity can be used because the absolute value of a cost function is irrelevant for the optimization, while the cosine similarity provides a convenient formula and good speedup over MSE. Because the optimization problem is formulated as minimization of an error, a mean negative cosine similarity is used as an error.

The negative cosine similarity has a compact and fast formula to use in the reverse-projecting ELM framework. A dot product between two vectors is defined as:

$$\mathbf{a} \cdot \mathbf{b} = \|a\| \|\mathbf{b}\| \cos \theta \tag{5}$$

Assume the input data is normalized to $\|\mathbf{x} = 1$, then $\|\hat{\mathbf{x}} \approx 1$ and

$$\text{similarity} = \cos \theta = \frac{\mathbf{x} \cdot \hat{\mathbf{x}}}{\|\mathbf{x}\|\hat{\mathbf{x}}} = \mathbf{x} \cdot \hat{\mathbf{x}} = \mathbf{x}^T \hat{\mathbf{x}} \tag{6}$$

For the whole data matrices $\mathbf{X}$, $\hat{\mathbf{X}}$ the cost function s, which is a mean negative cosine similarity, is

$$s = -\frac{1}{d}\text{trace}(\mathbf{X}^T \hat{\mathbf{X}}) \tag{7}$$

Here an $\hat{\mathbf{X}}$ has to be computed using ELM. But because the visualization points $\mathbf{V}$ are fixed, the output $\mathbf{H}$ of an ELM hidden layer never changes and needs to be computed only once. A formula based on $\mathbf{H}$ is derived from the ELM solution:

$$\hat{\mathbf{X}} = \mathbf{H}\beta \tag{8}$$

$$\beta = \mathbf{H}^\dagger \mathbf{X} = (\mathbf{H}^T \mathbf{H})^{-1} \mathbf{H}^T \mathbf{X} \tag{9}$$

$$\hat{\mathbf{X}} = \mathbf{H}(\mathbf{H}^T \mathbf{H})^{-1} \mathbf{H}^T \mathbf{X} \tag{10}$$

$$\mathbf{X}^T \hat{\mathbf{X}} = \mathbf{X}^T \mathbf{H}(\mathbf{H}^T \mathbf{H})^{-1} \mathbf{H}^T \mathbf{X} \tag{11}$$

$$\mathbf{H}(\mathbf{H}^T \mathbf{H})^{-1} \mathbf{H}^T = \mathbf{A} = \text{const} \tag{12}$$

$$s = -\frac{1}{d}\text{trace}(\mathbf{X}^T \mathbf{A} \mathbf{X}) \tag{13}$$

The matrix $\mathbf{A}$ in Eq. (13) depends only on $\mathbf{H}$ and needs to be computed once after an ELM model is built. It is re-used to update the cost function s for every change in permutation $\mathbf{p}$ of rows in $\mathbf{X}$.

In fact, a cost function update formula exists, which allows a fast update of s if only a few rows in $\mathbf{X}$ change. If a row $\mathbf{x}_k$ in $\mathbf{X}$ changes by $\delta = \mathbf{x}_k^{new} - \mathbf{x}_k \in \mathbb{R}^{1,d}$, the update formula is:

$$(\mathbf{X}_{new}^T \mathbf{A} \mathbf{X}_{new})_{k,m} = (\mathbf{X}^T \mathbf{A} \mathbf{X})_{k,m} + 2[\mathbf{A}_k^T \mathbf{X}]_m \delta_m + \mathbf{A}_{k,k} \delta_m^2, \; m \in [\![1, d]\!] \qquad (14)$$

$$s_{new} = -\frac{1}{d} \mathrm{trace}(\mathbf{X}_{new}^T \mathbf{A} \mathbf{X}_{new}) \qquad (15)$$

The second term in Eq. (14) is derived from the fact that diagonal elements in $\mathbf{X}^T \mathbf{A} \mathbf{X}$ are computed from $(\mathbf{X}^T \mathbf{A})_{k,:} \cdot \mathbf{X}_{k,:}$ and $\mathbf{X}_{k,:}^T \cdot (\mathbf{A} \mathbf{X})_{,k}$, which are the same due to symmetric matrix $\mathbf{A}$. The third term is a correction for the diagonal element of $\mathbf{A}$ which gets computed twice. In fact, Eq. (14) correctly updates only the diagonal of matrix $(\mathbf{X}_{new}^T \mathbf{A} \mathbf{X}_{new})$, but because an error function is a trace of that matrix, only diagonal elements matter.

For a change of two values i, j in p, the update formula is applied twice: for $\delta^1 = \mathbf{x}_j - \mathbf{x}_i$, $k = i$ and $\delta^2 = \mathbf{x}_i - \mathbf{x}_j$, $k = j$.

2.4 ELMVIS+ Algorithm

The updated ELMVIS+ method starts by building an ELM and computing matrix $\mathbf{A}$ from Eq. (13). It needs to be done only once when ELMVIS+ method starts. A random permutation p of data samples in $\mathbf{X}$ is taken. The ELMVIS+ method is optimized by randomly selecting and changing two indexes i, j from p. If the change results in a lower cost function it is kept, otherwise it is reverted. With a fast cost function update equation from Eq. (14), even a laptop with 1,4 GHz dual-core CPU runs 5 millions updates per minute for a small dataset with 100 samples. The only parameter of the ELMVIS+ method is a number of neurons L in an ELM. The optimal L depends on a task, and controls variation across the visualization space. Too few neurons fail a visualization, while too many neurons create an overly variative visualization pattern. An good number is found by trial. The ELMVIS+ algorithm for data visualization is presented on Algorithm 1.

Algorithm 1 ELMVIS+ Algorithm

---

given input data $\mathbf{X} \in \mathbb{R}^d$ with N samples
given visualization points $\mathbf{V} \in \mathbb{R}^c$, typically $c = 2$ or $c = 3$
create a random permutation p and apply it to the samples of $\mathbf{X}$
train an ELM $\mathbf{V} \rightarrow \mathbf{X}$ and obtain $\mathbf{H}$
compute $\mathbf{A} = \mathbf{H}(\mathbf{H}^T\mathbf{H})^{-1}\mathbf{H}^T$

compute s^{best} with initial p as in Eq. (13)
while no improvement in s^{best} during many iterations **do**
    randomly choose $i, j \mid i \neq j$
    compute s_{new} as in Eq. (14)
    if $s_{\text{new}} < s^{\text{best}}$ **then**
        $s^{\text{best}} = s_{\text{new}}$
        swap values p_i, p_j in p
        reset *no improvements in s^{best}* counter
    end if
end while

initialize GA with p
run GA to globally optimize $p \rightarrow p^*$ using the cost function from Eq. (13)
optimize $p*$ with another iterative part

order samples in $\mathbf{X}$ according to $p*$

---

3 Experimental Results

The ELMVIS+ visualization method is tested on the same three datasets as the original ELMVIS for comparable results. Reference methods are PCA as the baseline, Self-Organizing Maps (SOM) [18, 19] as another method which uses fixed visualization points, and NeRV [20] as a state-of-the-art nonlinear visualization method. A fast ELM model is provided by a toolbox from [21].

The overview of performance of all methods is given by the MSE of a data reconstruction from visualization. The reconstruction (a reverse projection) already exists in ELMVIS+ and ELMVIS; for other methods it is learned by a separate model as in [1]. This separate model is another ELM; the lowest error over 100 retrains is presented. The errors for all methods are gathered in Table 1.

ELMVIS+ method always performs better and runs faster than the original ELMVIS. It is much better than any other method in *Sculpture faces* because it is able to process the huge original dimensionality of the data, and in *Real faces* because it achieves a better optimization with millions of iterations run in an hour with the new cost function. On *Spiral* dataset it is second only to NeRV, but the errors are similarly small.

Table 1 Reconstruction MSE for all methods; the lowest error of 100 initializations is shown

Dataset	PCA	SOM	NeRV	ELMVIS	**ELMVIS+**
Spiral	0.482	0.054	0.011	0.049	0.017
Sculpture faces	0.980	0.916	0.769	0.718	0.712 (compressed) 0.292 (original)
Real faces	0.724	0.511	0.501	0.449	0.156

ELMVIS+ on *Sculpture faces* dataset is run twice: with the compressed ($d = 240$) and an original ($d = 4096$) image representations

3.1 Artificial Faces Dataset

A set of 698 face images is proposed in [22], and then widely used for benchmark purposes, for instance in [1, 20]. These images are computer renderings of a 3D sculpture head under different poses and lighting directions. Examples of faces are shown on Fig. 3.

Each image consists of an array of 64 by 64 brightness values of pixels, giving the input data dimensionality of 4096. A preprocessing step is applied to reduce dimensionality with PCA: the first 240 principal components keep over 99 % of the global variance [1]. ELMVIS uses 100 randomly selected samples at a time and is repeated 100 times to get a true estimate. ELM and SOM use 20 neurons for visualization. ELMVIS+ method uses the original 4096-dimensional data, as the speed of the new cost function update rule allows processing such high-dimensional datasets; it also uses the compressed data for comparison purposes.

The results are presented in Table 1. The PCA does not change the data which is already an output of another PCA. Two first principal components keep only 2 % of variance. SOM and NeRV both perform poorly, although it may be explained by a bad estimation of a reverse projection MSE obtained with ELM. ELMVIS and ELMVIS+ with the compressed data give better results, but still the error is too high for the results to be good in absolute numbers. Only ELMVIS+ with the original data results in a significantly lower error, even though the reconstruction ELM has the same 20 hidden neurons for prediction of a much higher dimensional data.

An example visualization is shown on Figs. 4 and 5. Both NeRV and ELMVIS+ show a clear organisation of sculpture faces. ELMVIS+ maps those faces on a

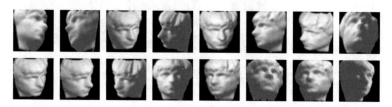

Fig. 3 Some examples from the 698 sculpture face pictures from [22]

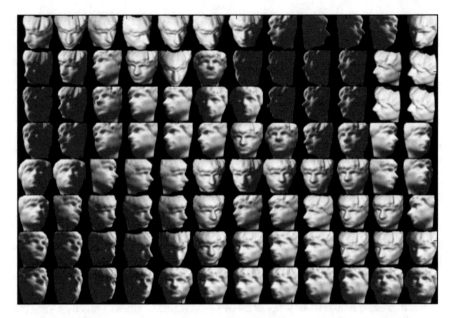

Fig. 4 ELMVIS+ visualization using the original 4096-dimensional data and 20 neurons. A clear manifold structure is visible, similar to NeRV mapping

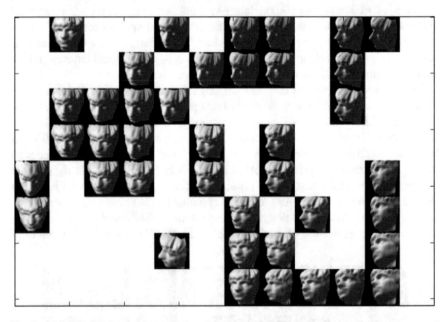

Fig. 5 Sculpture face images mapped to a grid using the NeRV results. If several faces correspond to the same grid cell, a random one is displayed

regular grid; NeRV is unable to achieve that so there are holes in NeRV projection. A poor MSE error of NeRV probably refers to an inability of ELM to learn an inverse projection with PCA-compressed data.

3.2 Computational Time

On a laptop with 1.4 GHz dual-core CPU, a visualization of the Spiral dataset runs at 850,000 iterations per minute for ELMVIS and 5,000,000 for ELMVIS+. The speedup is less than 6 times, but the processed variables are so small that execution overhead takes the most time.

On a desktop workstation with 4-core 3.6GHz CPU, the visualization of 400 real faces runs at 20,000 iterations per minute for ELMVIS and 1,000,000 iterations per minute for ELMVIS+. The speedup is thus roughly 50 times for large tasks. Even the slowest processing of 4096-dimensional Sculpture faces dataset with 100 samples and 20 neurons runs at 500,000 iterations per minute on the same desktop. In general, the cosine similarity criterion in ELMVIS+ provides a notable speedup over the original ELMVIS, especially in cases of high dimensional data. Thus ELMVIS+ has a wider application area.

3.3 Convergence Speed

Theoretically, the ELMVIS+ method has a complexity of $\mathcal{O}(N!)$ for a dataset with N samples because it considers all random permutations of data samples. In practice, only two samples are changed at a time before the cost function is updated. This leads to a local optimum, but decreases the complexity to $\mathcal{O}(N^2)$ as only two samples are changed (swapped) at a time—the same complexity as NeRV. This is observed in practice—ELMVIS+ typically takes $10N^2$ iterations to converge, and a better solution is found in no more than N^2 iterations. The convergence time for a dataset with $N = 100$ samples is less than a minute, and for the largest tested dataset with all faces it is about 1 h.

4 Conclusion

This paper describes the ELMVIS+, a significant improvement of ELMVIS method of data visualization. The method is based on a random search of an optimal order of data samples for the pre-defined visualization points. It uses ELM as a nonlinear metric, with a cosine similarity between the true and estimated (by ELM) data samples computed in the original data space. This metric is accurate and allows for millions of iterations per minute. Combined with GA, ELMVIS+ optimizes a data visualization both globally and locally. The ELMVIS+ method is tested on three datasets. It showed a decrease of error and a significant improvement in runtime ranging from 6 to 50 times. It even allowed for processing of an extremely high dimensional data, which provided greatly improved visualization results for image data samples. The performance of an improved ELMVIS+ visualization method is similar or better than other visualization methods, including the state-of-the-art NeRV [22]. In practice, it also has the same $\mathcal{O}(N^2)$ complexity as NeRV, with N being the number of data samples. Thus an ELMVIS+ is a new way of creating useful data visualizations in a wide application range. In order to target visualization of Big Data, in the future work we will investigate the GPU acceleration of ELMVIS+ iterative optimization. The new fast update formula of a cosine similarity cost function from Eq. (14) uses vector operations, which are conveniently parallelizable on a suitable hardware. This will allow ELMVIS+ method process Big Data, specifically the emerging Big Dimensionality [23] data.

References

1. Lee, J.A., Verleysen, M.: Nonlinear Dimensionality Reduction. Springer, New York (2007)
2. Pearson, K.: LIII. On lines and planes of closest fit to systems of points in space. London Edinburgh Dublin Philos. Mag. J. Sci. **2**(11), 559–572 (1901)
3. Kruskal, J.B.: Multidimensional scaling by optimizing goodness of fit to a nonmetric hypothesis. Psychometrika **29**(1), 1–27 (1964)
4. Lee, J.A., Lendasse, A., Donckers, N., Verleysen, M.: A robust nonlinear projection method. In: Proceedings of ESANN. pp. 13–20 (2000)
5. Beyer, K.S., Goldstein, J., Ramakrishnan, R., Shaft, U.: In: ICDT
6. Scott, D.W., Thompson, J.R.: Probability density estimation in higher dimensions. In: Computer Science and Statistics: Proceedings of the Fifteenth Symposium on the Interface. vol. 528, pp. 173–179. North-Holland, Amsterdam (1983)
7. Cambria, E., et al.: Extreme learning machines. IEEE Intell. Syst. (6), 30–59
8. Gutin, G., Punnen, A.P. (eds.): The Traveling Salesman Problem and its Variations. Combinatorial Optimization. Kluwer Academic, Dordrecht, London
9. Mitchell, M.: An Introduction to Genetic Algorithms. The MIT Press (1996)
10. Kaski, S., Peltonen, J.: Informative discriminant analysis. In: Proceedings of the Twentieth International Conference on Machine Learning (ICML-2003). vol. 20, p. 329 (2003)
11. Huang, G.B., Zhou, H., Ding, X., Zhang, R.: Extreme learning machine for regression and multiclass classification. IEEE Trans. Syst. Man Cybern. Part B: Cybern. **42**(2), 513–529 (2012)
12. Huang, G.B., Zhu, Q.Y., Mao, K.Z., Siew, C.K., Saratchandran, P., Sundararajan, N.: Can threshold networks be trained directly? IEEE Trans. Circuits Syst. II Express Briefs **53**(3), 187–191 (2006)
13. Huang, G.B., Zhu, Q.Y., Siew, C.K.: Extreme learning machine: theory and applications. Neurocomputing **70**(1), 489–501 (2006)
14. Miche, Y., Sorjamaa, A., Bas, P., Simula, O., Jutten, C., Lendasse, A.: Op-elm: Optimally pruned extreme learning machine. IEEE Trans. Neural Netw. **21**(1), 158–162 (2010)
15. Yu, Q., Miche, Y., Eirola, E., van Heeswijk, M., Sverin, E., Lendasse, A.: Regularized extreme learning machine for regression with missing data. Neurocomputing **102**, 45–51 (2013)
16. Huang, G.B., Chen, L., Siew, C.K.: Universal approximation using incremental constructive feedforward networks with random hidden nodes. IEEE Trans. Neural Netw. **17**(4), 879–892 (2006)

17. Rao, C.R., Mitra, S.K.: Generalized Inverse of a Matrix and its Applications. Wiley (1971)
18. Lendasse, A., Cottrell, M., Wertz, V., Verleysen, M.: Prediction of electric load using kohonen maps—application to the polish electricity consumption. In: Proceedings of the 2002 American Control Conference. vol. 5, pp. 3684–3689 (2002)
19. Merlin, P., Sorjamaa, A., Maillet, B., Lendasse, A.: X-SOM and L-SOM: a double classification approach for missing value imputation. Neurocomputing **73**(7–9), 1103–1108 (2010)
20. Venna, J., Peltonen, J., Nybo, K., Aidos, H., Kaski, S.: Information retrieval perspective to nonlinear dimensionality reduction for data visualization. J. Mach. Learn. Res. **11**, 451–490 (2010)
21. Akusok, A., Björk, K.M., Miché, Y., Lendasse, A.: High-performance extreme learning machines: a complete toolbox for big data applications. IEEE Access, pp. 1011–1025
22. Tenenbaum, J.B., De Silva, V., Langford, J.C.: A global geometric framework for nonlinear dimensionality reduction. Science **290**(5500), 2319–2323 (2000)
23. Zhai, Y., Ong, Y.S., Tsang, I.W.: The emerging "Big Dimensionality". IEEE Comput. Intell. Mag. **9**(3), 14–26 (2014)

On Mutual Information over Non-Euclidean Spaces, Data Mining and Data Privacy Levels

Yoan Miche, Ian Oliver, Silke Holtmanns, Anton Akusok, Amaury Lendasse and Kaj-Mikael Björk

Abstract In this paper, we propose a framework for measuring the impact of data privacy techniques, in information theoretic and in data mining terms. The need for data privacy and anonymization is often hampered by the fact that the privacy functions alter the data in non-measurable amounts and details. We propose here to use Mutual Information over non-Euclidean spaces as a means of measuring this distortion. In addition, and following the same principle, we also propose to use Machine Learning techniques in order to quantify the impact of the data obfuscation in terms of further data mining goals.

1 Introduction

There is nowadays a strong need for data anonymization, be it for legal purposes or commercial ones: data sharing for advertisement purposes, e.g., could probably do without the full extent of the data, which reveals private and personal information that is not only irrelevant to the task being carried out, but also sometimes too sensitive to be handed out.

Companies or legal bodies who have gathered such data should be allowed to share the data once a certain set of criteria have been fulfilled regarding the anonymity of the data. The problem typically lies in the choice of the techniques used on the data at hand, as well as the extent to which they are used.

The need for data privacy is well known [1], however the techniques for ensuring privacy are less well established [2, 3]. While techniques such as encryption

Y. Miche (✉) · I. Oliver · S. Holtmanns
Nokia Solutions and Networks, Espoo, Finland
e-mail: yoan.miche@nokia.com

A. Akusok · A. Lendasse
Department of Mechanical and Industrial Engineering
and The Iowa Informatics Initiative, The University of Iowa, Iowa City, USA

K.-M. Björk
Arcada University of Applied Sciences, Helsinki, Finland

© Springer International Publishing Switzerland 2016
J. Cao et al. (eds.), *Proceedings of ELM-2015 Volume 2*,
Proceedings in Adaptation, Learning and Optimization 7,
DOI 10.1007/978-3-319-28373-9_32

and hashing can be considered to some extent as privacy preserving techniques, we decide in this work to focus more on techniques that do not merely alter the syntax and the format of the data, but attempt to modify the semantic of the data. Note that the framework proposed in this paper is not excluding such techniques as hashing and encryption, but it is devised in order to look at techniques that affect more the semantics than only the format. The "popular" techniques such as data suppression, tokenisation, κ-anonymity [4] including its relatives ℓ-diversity and t-closeness [5], and the latest ideas of differential privacy [6] all alter data such that its analytics usage is compromised as little as possible without revealing enough information for sufficient cross-referencing to reconstruct the original data.

The application of such techniques is further compounded by the fact that they are generally insufficiently used or even mis-used [7]. For example, insufficient diversity removal, insufficient randomness of continuous values and failure to apply the techniques together across multiple fields in a data set.

One difficulty regarding the practical use of such techniques is in the setting of the various parameters involved in the anonymization: too much added noise or diversity removal will irremediably render such fields unusable, and in this sense, the anonymization has failed to prove useful for the further steps that will use this dataset.

One of the problems this paper studies is in the influence of these parameters, and to define a framework to control and quantify the amount of distortion generated to anonymize the data to "acceptable" levels within a machine learning context. Thus, we propose here two points of view, each with a slightly different goal, but hopefully reconcilable:

- An information theoretic point of view, in which the approach aims at establishing a set of standards in quantifiable and mathematically sound ways using mutual information estimation;
- A data mining point of view, which focuses on providing measurable guarantees on the data mining task to be performed after the anonymization, by using various performance metrics.

The following Sect. 2 first proposes a visual and high level motivation and explanation of the problem. We then, in Sect. 3 introduce the general notations for the framework and the subsequent discussion. Section 4 then proposes the first approach, Mutual Information based; and Sect. 5, the second one, Machine Learning based.

2 High-Level Motivation for Quantifying Data Privacy

In this section, we propose a high level description of the problem tackled in this paper. The next sections then describe the two proposed means of doing so.

In an ideal situation, data mining in general, and classification or partition of data in particular, can be made in an unambiguous manner; meaning that, for example, a classification of the data can be made and the number of border cases is minimal as depicted in Fig. 1.

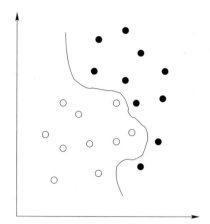

Fig. 1 Depiction of an example case in which the data is easily separable, with little border cases

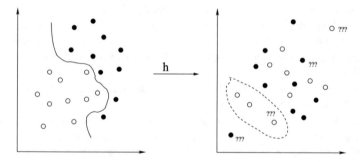

Fig. 2 Depiction of the data after the use of a privacy function h: the data has become much more difficult to separate. In this case, the data space remains the same

Application of algorithms that increase the privacy (or the entropy) of a system distort this in some known manner, in terms of the direct effects on the data fields

For example κ-anonymity [4] and ℓ-diversity [5] reduce the distribution and amounts of unique values in the discrete valued cases; differential privacy [6] adds noise in the continuous valued cases, for example, speeds, distances etc. A privacy function h (defined more precisely in the next section) distorts a system such that classification either can not be made or becomes difficult to make in a reasonable manner, as illustrated on Fig. 2. In this figure, the application of the privacy function over the data has modified it, but the data has remained in the same space.

In this work, we consider as well the case of such privacy functions that modify the data in such a way as to change the "format" of the data, and thus the underlying space in which the data lies. Indeed, another way to consider this is that the privacy functions alter the underlying space or topology rather than moving the elements themselves. This altering of the topology in the best case involves continuous (metric preserving or homotopy preserving) stretching and shrinking, but may also include tearing and creasing of the space such that the resolution of the original metric function is no longer possible. For example in Fig. 3 (depicting this in a two dimensional

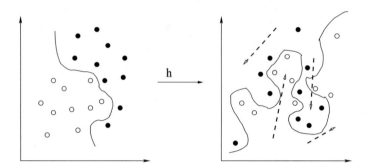

Fig. 3 Depiction of the topological stretching case: the *arrows* depict the areas in which the space is not longer "continuous" due to the application of the privacy functions

plot), one can imagine that the underlying space, in terms of the distance defined on it, is no longer continuous, but presents something like singularities.

The challenge here is then to "accommodate" for the deformation of the space to a degree such that classification of information in the deformed space is as good as classification made with the original information and topology. Or at least as close as possible to it.

However as we do not work in an ideal world, establishing commutativity between the original, deformed and reconstructed spaces becomes a probabilistic exercise rather than a definitive one. Unless we have access to the original data or have *some* knowledge—which is more likely—of it then we have no metric for measuring how close our reconstruction is. Indeed in most cases a high percentile level of confidence is normally sufficient.

The work in this paper is aimed at this problem: proposing two points of view for quantifying the effects of the privacy/obfuscation functions, and thus have a means of measuring in "practically useful" terms, how much the data has been altered.

3   General Notations and a Note About Non-Euclidean Spaces

Let us first define the notations for a metric space and sets.

3.1   *Notations*

Let $\mathcal{X}_i = (\mathbb{X}_i, d_i)$ be a metric space on the set $\mathbb{X}_i$ with the distance function d_i : $\mathbb{X}_i \times \mathbb{X}_i \longrightarrow \mathbb{R}_+$. The $\mathcal{X}_i$ need not be Euclidean spaces, and in the cases discussed in the later sections, are not.

Parting slightly from the notations of the data privacy literature, define by $\mathbf{X} = \left[\mathbf{x}^1, \ldots, \mathbf{x}^n\right]$ a $N \times n$ matrix, with each column vector $\mathbf{x}^i \in \mathbb{X}_i^{N \times 1}$. The $\mathbf{x}_i$ are thus discrete random variables representing a set of samples over the set of all the possible samples from the attribute represented here by $\mathbb{X}_i$. And $\mathbf{X}$ is a table over these attributes.

The fact that the $\mathcal{X}_i$ are not necessarily Euclidean spaces in this work poses the problem of the definition of the distance function associated, d_i. Indeed, most data mining and machine learning tools rely on the Euclidean distance and its properties; and even if the learning of the model does not require the use of Euclidean distances directly, the evaluation criterion typically relies on it, for example as a Mean Square Error for regression problems.

Similarly, as described in Sect. 4, information theory metrics estimators such as mutual information estimators typically rely on the construction of the set of nearest neighbours, and therefore also on the Euclidean distance.

3.2 Distances over Non-Euclidean Spaces

The argument for considering the use of distances over non-Euclidean spaces in this work, is that it is possible to tweak and modify such non-Euclidean distances so that their distribution and properties will be "close enough" to that of the original Euclidean distance.

More formally, let us assume that we have two metric spaces $\mathcal{X}_i = (\mathbb{X}_i, d_i)$ and $\mathcal{X}_j = (\mathbb{X}_j, d_j)$, with $\mathcal{X}_i$ the canonical Euclidean space (i.e. $\mathbb{X}_i = \mathbb{R}^d$ and d_i the Euclidean norm) and $\mathcal{X}_j$ a non-Euclidean metric space endowed with a non-Euclidean metric. Drawing uniformly samples from the set $\mathbb{X}_j$, we form $\mathbf{X}_j = \left[\mathbf{x}_j^1, \ldots, \mathbf{x}_j^n\right]$, a set of random variables, with $\mathbf{x}_j^l$ having values over $\mathbb{X}_j$. Denoting then by f_{d_j} the distribution of pairwise distances over all the samples in $\mathbf{X}_j$, we assume that it is possible to modify the non-Euclidean metric d_j such that

$$\lim_{n \to \infty} f_{d_j} = f_{d_i}, \tag{1}$$

where f_{d_i} is the distribution of the Euclidean distances d_i over the Euclidean space $\mathcal{X}_i$. The limit here is over n as the distribution f_{d_j} is considered to be estimated using a limited number n of random variables, and we are interested in the limit case where we can "afford" to draw as many random variables as possible to be as close to the Euclidean metric as possible. That is, that we can make sure that the non-Euclidean metric behaves over its non-Euclidean space, as would a Euclidean metric over a Euclidean space.

This assumption is "theoretically reasonable", as it comes down to being able to transform a distribution into another, given both. And while this may not be simple nor possible using linear transformation tools, most Machine Learning techniques are able to fit a continuous input to another different continuous output.

3.3 Learning the Mapping Between Distances Using ELM

In this work, we propose to use Extreme Learning Machines (ELM) [8, 9] as the mapping tool between distance functions. The reasons for using this specific Machine Learning technique are threefold: first, it lies among the techniques with the best performance/computational time ratio, as the model is simple and involves a minimal amount of computations. Second, given that we are here working in terms of the limit from Eq. 1, we need a model that can learn the mapping in reasonable time from large amounts of data, if such a need arises. Third, the theory behind ELM (and behind single layer feed-forward neural networks in general) states that it is a universal function approximator (per the universal approximation theorem [10]), and can therefore fit any continuous function, to a $\varepsilon > 0$.

The Extreme Learning Machine algorithm was originally proposed by Huang et al. in [9] (and further developed, e.g. in [11–13] and analysed in [14]). It uses the structure of a Single Layer Feed-forward Neural Network (SLFN). The main concept behind ELM is the replacement of a computationally costly procedure of training the hidden layer, by its random initialisation. Then an output weights matrix between the hidden representation of inputs and the outputs remains to be found. The ELM is proven to be a universal approximator given enough hidden neurons [9]. It works as following:

Consider a set of N distinct samples $(\mathbf{x}_i, \mathbf{y}_i)$ with $\mathbf{x}_i \in \mathbb{R}^d$ and $\mathbf{y}_i \in \mathbb{R}^c$. Then a SLFN with n hidden neurons is modelled as $\sum_{j=1}^{n} \beta_j \phi(\mathbf{w}_j \mathbf{x}_i + b_j)$, $i \in [1, N]$, with $\phi : \mathbb{R} \to \mathbb{R}$ being the activation function, $\mathbf{w}_j$ the input weights, b_j the biases and β_j the output weights.

In case the SLFN would perfectly approximate the data, the errors between the estimated outputs $\hat{\mathbf{y}}_i$ and the actual outputs $\mathbf{y}_i$ are zero, and the relation between inputs, weights and outputs is then $\sum_{j=1}^{n} \beta_j \phi(\mathbf{w}_j \mathbf{x}_i + b_j) = \mathbf{y}_i$, $i \in [1, N]$ which can be written compactly as $\mathbf{H}\beta = \mathbf{Y}$, with $\beta = (\beta_1^T \ldots \beta_n^T)^T$, $\mathbf{Y} = (\mathbf{y}_1^T \ldots \mathbf{y}_N^T)^T$.

Solving the output weights β from the hidden layer representation of inputs $\mathbf{H}$ and true outputs $\mathbf{Y}$ is achieved using the Moore-Penrose generalised inverse of the matrix $\mathbf{H}$, denoted as $\mathbf{H}^\dagger$ [15]. The training of ELM requires no iterations, and the most computationally costly part is the calculation of a pseudo-inverse of the matrix $\mathbf{H}$, which makes ELM an extremely fast Machine Learning method.

Therefore, using ELM, we propose to learn the transformation between the previously defined f_{d_j}, the distribution of pairwise distances over a non-Euclidean space, and f_{d_i}, the distribution of pairwise Euclidean distances over the canonical Euclidean space.

With this assumption made, we propose to first take a look at the Information Theoretic approach, that is, to use the Mutual Information to quantify the data privacy level.

4 Mutual Information for Data Privacy Quantification

4.1 Mutual Information Estimation

With the previous notations from Sect. 3, we take here the definition of mutual information $I(\mathbf{x}_i, \mathbf{x}_j)$ between two discrete random variables $\mathbf{x}_i, \mathbf{x}_j$ as

$$I(\mathbf{x}_i, \mathbf{x}_j) = \sum_{x_i \in \mathbf{x}_i} \sum_{x_j \in \mathbf{x}_j} p(x_i, x_j) \log \left(\frac{p(x_i, x_j)}{p(x_i)p(x_j)} \right). \tag{2}$$

Obviously, the marginals $p(x_i)$ and $p(x_j)$ as well as the joint $p(x_i, x_j)$ are unknown, and we resort to estimators of the mutual information.

Most of the mutual information estimators (Kraskov's [16], Pal's [17, 18], e.g.) make use of the canonical distance defined in the metric space in which lies the data. Typically, this is defined and computable for a Euclidean space, with the traditional Euclidean distance used as the distance function.

We give some details in the following about the two most famous (arguably) mutual information estimators, merely to illustrate their strong dependencies on distances. This is mainly to make the point that mutual information can thus be estimated using non-Euclidean distances over non-Euclidean spaces, given the precautions mentioned in the previous Sect. 3.2.

4.1.1 Kraskov's Estimator

In [16], Kraskov et al. propose a mutual information estimator relying on counts of nearest neighbours, as follows. To be precise, two estimators are proposed in the original paper, but we only present the second one here, as we have found it to be more reliable in practice.

The mutual information estimator $I^{(2)}$ between two random variables $\mathbf{x}_j^l$ and $\mathbf{x}_j^m$ is defined as

$$I^{(2)}\left(\mathbf{x}_j^l, \mathbf{x}_j^m\right) = \Psi(k) - 1/k - \langle \Psi(\mathbf{n}_{\mathbf{x}_j^l}) + \Psi(\mathbf{n}_{\mathbf{x}_j^m}) \rangle + \Psi(N), \tag{3}$$

with Ψ the digamma function, k the number of neighbours to use (to be decided by the user), and $\mathbf{n}_{\mathbf{x}_j^l} = \left[n_{\mathbf{x}_j^l}(1), \dots, n_{\mathbf{x}_j^l}(N-1) \right]$ the vector holding counts of neighbours $n_{\mathbf{x}_j^l}(i)$ defined as

$$n_{\mathbf{x}_j^l}(i) = \text{Card}\left(\left\{x_i \in \mathbf{x}_j^l : d_j(x_j - x_i) \leq \varepsilon_{\mathbf{x}_j^l}(i)/2\right\}\right) \tag{4}$$

where $\varepsilon_{\mathbf{x}_j^l}(i)/2 = ||z_i - z_{k\text{NN}(i)}||_{\max}$ is the distance between sample z_i and its kth nearest neighbour in the joint space $\mathbf{z} = (\mathbf{x}_j^l, \mathbf{x}_j^m)$, and the distance $|| \cdot ||_{\max}$ defined as $||z_q - z_r||_{\max} = \max\left\{||x_j^l(q) - x_j^l(r)||, ||x_j^m(q) - x_j^m(r)||\right\}$, where $x_j^l(q)$ clunkily denotes the qth sample from the random variable $\mathbf{x}_j^l$.

Basically, the calculation requires calculating the nearest neighbours of points in a joint space, and counting how many lie in a certain ball.

Note that while we have adapted the notations to our needs, here, the original article relies on the Euclidean distance, and not on arbitrary distances on non-Euclidean distances. But as discussed earlier, this is not necessarily a problem.

4.1.2 Pal's Estimator

More recently, in [17, 18], David Pal et al. have proposed another estimator of the more general Rényi Entropy, as well as of the associated Rényi mutual information. The estimator relies on the use of nearest neighbours graphs, as follows.

The authors define what they name the *nearest-neighbour graph* $NN_S(\mathbf{x}_i)$ as a directed graph on the values that the discrete random variable $\mathbf{x}_i$ takes, and S is a finite set of integers of which the maximum value is denoted by k. For each $j \in S$, there is an edge in $NN_S(\mathbf{x}_i)$ from each vertex $x \in \mathbf{x}_i$ to its jth nearest neighbour.

We must then define $L_p(\mathbf{x}_i)$, the sum of the pth powers of the Euclidean lengths of the edges of the nearest neighbour graph:

$$L_p(\mathbf{x}_i) = \sum_{(x,y) \in E(NN_S(\mathbf{x}_i))} d_i(x, y), \tag{5}$$

with $E(NN_S(\mathbf{x}_i))$ the edge set of the nearest neighbour graph $NN_S(\mathbf{x}_i)$. It so happens that the following constant value γ, depending on $L_p(\mathbf{x}_i)$, is required in the following calculations of entropy and mutual information. γ is defined as

$$\gamma = \lim_{n \to \infty} \frac{L_p(\mathbf{X_j})}{n^{1-p/d}}, \tag{6}$$

where in this equation, $\mathbf{X}_j = (\mathbf{x}_j^1, \dots, \mathbf{x}_j^n)$ is the formerly defined set of n random variables, where each $\mathbf{x}_j = (x_j^1, \dots, x_j^d)$. γ has to be estimated beforehand, empirically.

The Rényi entropy $H_\alpha(\mathbf{X}_j)$ is then estimated as

$$H_\alpha(\mathbf{X}_j) = \frac{1}{1 - \alpha} \log \frac{L_p(\mathbf{X}_j)}{\gamma n^{1-p/d}}, \tag{7}$$

where $p = d(1 - \alpha)$.

The authors then show that the mutual information can be calculated by using directly the entropy estimator from Eq. (7), but on the marginal distributions of the $(\mathbf{x}_j^1, \ldots, \mathbf{x}_j^n)$. These marginals being unknown, they are estimated as well.

Thus, Kraskov's estimators from [16] make use of nearest neighbours counts within certain vicinities, while Pal's [17, 18] estimator (of the Rényi entropy and mutual information) use also the nearest neighbour graphs, but this time with the normalised ratio between the sum of the pth powers of the nearest neighbour graph distances and the limit of this same sum.

Therefore, both estimators require the use of the distance functions $d = \{d_i\}_{1 \leq i \leq n}$, where the distance function d_i is defined as before on the set $\mathbb{X}_i$.

4.2 Using Mutual Information Estimators

Introducing the notation $x_i \underset{NN}{\leq} x_j$ for convenience: $x_i \underset{NN}{\leq} x_j$ states that x_j is the nearest neighbour for x_i in terms of the distance defined on that space. We can then write $x_i \underset{NN}{\leq} x_j \underset{NN}{\leq} x_k \underset{NN}{\leq} x_l$, which signifies that the 3 nearest neighbours of x_i are, in order of increasing distance to x_i, respectively x_j, x_k and x_l. We decide to break the possible ties arbitrarily.

We then want to create a family $h = \{h_i\}_{1 \leq i \leq n}$ of obfuscating functions which intuitively "preserves" the distances, and in a sense also the mutual information between the random variables.

More precisely, defining $d^{(h)} = \left\{ d_i^{(h)} \right\}_{1 \leq i \leq n}$ a family of distance functions such that $d_i^{(h)} : \mathbb{X}_i^{(h)} \times \mathbb{X}_i^{(h)} \longrightarrow \mathbb{R}_+$, with $\mathbb{X}_i^{(h)}$ denoting the set of possible values obtained by obfuscating the elements in $\mathbb{X}_i$ using the function h_i, i.e. $\mathbb{X}_i^{(h)} = \{h_i(x), x \in \mathbb{X}_i\}$, the family $d^{(h)}$ and the obfuscating functions family h should be so that neighbourhoods are preserved, i.e., with $x, y, z \in \mathbb{X}_i$ three elements,

$$\text{if } x \underset{NN}{\leq} y \underset{NN}{\leq} z, \text{ then } h_i(x) \underset{NN}{\leq} h_i(y) \underset{NN}{\leq} h_i(z), \tag{8}$$

in which the distance functions d_i and $d_i^{(h)}$ are hidden in the notation $\underset{NN}{\leq}$.

If this is the case, and even if the distances in $\mathbb{X}_i$ and $\mathbb{X}_i^{(h)}$ are not in the same "range" of values (remember we map the distances from non-Euclidean spaces to fit the distribution of the Euclidean one), the nearest neighbour graphs will be preserved. We can then look at the matrix holding the pairwise mutual information values, $\mathbf{I}(\mathbf{X})$, defined as

$$\mathbf{I}(\mathbf{X}) = \left\{ I(\mathbf{x}^i, \mathbf{x}^j) \right\}, 1 \leq i, j \leq n, \tag{9}$$

where $\mathbf{I}(\mathbf{X}) \in \mathbb{R}_+^{n \times n}$, $\mathbf{X} = [\mathbf{x}^1, \ldots, \mathbf{x}^n]$ and $I(\mathbf{x}^i, \mathbf{x}^j)$ is the mutual information between the two random variables $\mathbf{x}^i$ and $\mathbf{x}^j$, using one of the estimators presented.

We want to get, in terms of mutual information, similar matrices between the obfuscated data and the original data.

More precisely, if we have a measure $\mu(p_j, p_k)$ between two probability densities p_j, p_k (e.g. the Kullback-Leibler divergence [19] or Wasserstein metric [20]), we want to make sure that the distribution of the pairwise mutual information values between the original discrete random variables $\mathbf{x}_i$ and the obfuscated ones $\mathbf{x}_i^{(h)}$ is such that

$$\mu\left(\text{hist}(\mathbf{I}(\mathbf{X})), \text{hist}(\mathbf{I}(\mathbf{X}^{(h)}))\right) \leq \varepsilon, \tag{10}$$

with $\varepsilon \geq 0$ as small as possible, and hist($\mathbf{I}(\mathbf{X})$) denotes the distribution of the pairwise mutual information values.

Thus, to conclude on this part using the Mutual Information as a way to quantify the Data Privacy, we propose to define (μ, h, d) as the tuple holding the objects defining respectively the measure μ, the obfuscation function family h and the distances (both over the original space and the obfuscated space) families d. The spaces are themselves defined by the data that is being obfuscated, and some of the required distance functions might be obvious, over such spaces.

This tuple defines the *obfuscation functions* and *measures* needed to quantify their impact.

Given (μ, h, d), we can then propose privacy levels in terms of thresholds ε:

- Small thresholds imply a low distortion due to the obfuscation, in terms of the pairwise mutual information, and thus force the obfuscation functions (or their parameters) to be limited in their effect;
- While a high threshold allows for high distortions, and thus allows for high levels of obfuscation, at the possible cost of unusable data.

It is worth noting here that while this criterion of Mutual Information preservation is important, it might not convey sufficient restrictions over the obfuscation functions to make sure the data is still "usable" for further data mining, e.g.

For example, if a certain low threshold ε is respected in terms of this proposed mutual information criterion, it is possible that the data has become much more difficult to process, in terms of Machine Learning techniques: the information that has been kept between the several random variables considered is still present, but extracting and using it for meaningful purposes is not guaranteed by this mutual information criterion.

5   Machine Learning for Data Privacy Quantification

This former section presented a criterion on the level of the mutual information. We ultimately want to see how much the obfuscation affects a clustering/classification/regression algorithm, as well.

Using the same notations as for the case of the Mutual Information, in the previous section, we assume this time, that we use a machine learning technique class $\text{ML}(\cdot, \cdot)$

with a certain instance $ML_j(\mathbf{X}, d)$ to perform a data mining task requiring the distance function family d on the data $\mathbf{X}$.

We want to devise/have a set of L performance criteria or metrics $m = \{m_i\}_{1 \leq i \leq L}$ and have

$$||m_i \left(ML_j(\mathbf{X}, d)\right) - m_i \left(ML_j \left(\mathbf{X}^{(h)}, d^{(h)}\right)\right) ||_2 \leq \varepsilon, 1 \leq i \leq L, \tag{11}$$

with $|| \cdot ||_2$ the Euclidean norm.

Note that in this definition, we have two varying quantities to use to measure the distortion introduced by the obfuscating function family h: the Machine Learning techniques ML_j and the performance metrics m.

As such, and for the same reasons as in Sect. 3.3, we propose to use the ELM as the class of Machine Learning techniques. Another reason to do so, here, is that the ELMs, by their non-deterministic nature, ensure that the same model class with the same hyper-parameters (i.e. in this case, the number of neurons and the activation functions used in each) can be used, while still having different models in terms of the inner coefficients.

Thus, we propose for this case to use the Mean Square Error (or the classification accuracy, depending on the problem at hand) as the performance measure m (meaning $L = 1$ for this proposition), and the ELM with various random initialisation as the ML_j.

In practice, if the data mining techniques to be used are fixed, then the metrics and the means to generate diversity in the Machine Learning models need to be arranged.

In the same way as for the previous case with the Mutual Information, we can then proceed, for a fixed set $(m, h, \{ML_j(\cdot)\})$, to define appropriate thresholds ε that quantify the amount of data distortion introduced by the obfuscation functions, but this time in terms of the "final goal" of data mining.

6 Conclusions and Future Work

This work proposes means of quantifying and measuring the impact of data privacy functions on the data itself, as well as on the further processing on such data, in data mining terms. We propose a framework that allows to use Mutual Information estimators on non-Euclidean spaces, thus allowing to use the mutual information as a measure of the distortion introduced by the obfuscation functions. We discuss the assumptions made and the necessary transformations required on the distance functions over non-Euclidean spaces to allow for such computations. While measuring this distortion using the mutual information is valuable, it might not reflect fully the difficulty introduced in terms of further processing of the data, and, specifically, in terms of data mining on such data.

We thus propose the second approach in this framework, by measuring the performance of several instances of Machine Learning techniques and comparing the performance on the original data and on the obfuscated data. This approach, while less "precise", maybe, allows for direct insight into the effects of the data obfuscation; and thus is useful in terms of tuning the privacy functions parameters.

Future work on this topic is clearly to provide large experimental results on the application of this framework, for various data formats, privacy functions and Machine Learning techniques. In addition, we hope to be able to quantify, given a fixed data format and privacy functions, a set of "reasonable" thresholds ε for both the mutual information and machine learning criterion. Reasonable meaning in this instance that such thresholds could set up standards for data sharing and exchange: if we consider network traces, e.g. as the data format, and a fixed set of privacy functions over the various fields included in these traces, we want to be able to give precise recommendations as to how much the data needs to be obfuscated to ensure user privacy, while allowing further data mining based processing, for instance.

References

1. Nissenbaum, H.: A contextual approach to privacy online. Daedalus, **140**(4):32–48 (Fall 2011)
2. Gürses, S., Troncoso, C.G., Diaz, C.: Engineering privacy by design. In: Computers, Privacy and Data Protection (2011)
3. Oliver, I.: Privacy Engineering: A Data Flow and Ontological Approach. CreateSpace (2014)
4. Ciriani, V., Capitani di Vimercati, S., Foresti, S., Samarati, P.: κ-anonymity. In: Ting, Y., Jajodia, S. (eds.) Secure Data Management in Decentralized Systems. Advances in Information Security, vol. 33, pp. 323–353. Springer, New York (2007)
5. Machanavajjhala, A., Gehrke, J., Kifer, D., Venkitasubramaniam, M.: ℓ-diversity: Privacy beyond κ-anonymity. In: 2013 IEEE 29th International Conference on Data Engineering (ICDE), 0:24 (2006)
6. Dwork, C.: Differential privacy: a survey of results. In: Theory and Applications of Models of Computation. Lecture Notes in Computer Science, vol. 4978, pp. 1–19. Springer, Berlin (2008)
7. The UK Cabinet Office. Security policy framework (April 2013)
8. Huang, G., Chen, L., Siew, C.-K., Huang, G.-B., Chen, L., Siew, C.-K.: Universal approximation using incremental constructive feedforward neural networks with random hidden nodes. IEEE Trans. Neural Networks **17**(4), 879–892 (2006)
9. Huang, G.-B., Zhu, Q.-Y., Siew, C.-K.: Extreme learning machine: theory and applications. Neurocomputing **70**(1), 489–501 (2006)
10. Cybenko, G.: Approximations by superpositions of sigmoidal functions. Math. Control Signals Syst. **2**(4), 303–314 (1989)
11. Miche, Y., Sorjamaa, A., Bas, P., Simula, O., Jutten, C., Lendasse, A.: OP-ELM: optimally-pruned extreme learning machine. IEEE Trans. Neural Networks **21**(1), 158–162 (2010)
12. Miche, Y., van Heeswijk, M., Bas, P., Simula, O., Lendasse, A.: TROP-ELM: a double-regularized ELM using LARS and Tikhonov regularization. Neurocomputing **74**(16), 2413–2421 (2011)
13. Van Heeswijk, M., Miche, Y., Oja, E., Lendasse, A.: GPU-accelerated and parallelized ELM ensembles for large-scale regression. Neurocomputing **74**(16), 2430–2437 (2011)
14. Cambria, E., Huang, G.-B., Kasun, L.L.C., Zhou, H., Vong, C.M., Lin, J., Yin, J., Cai, Z., Liu, Q., Li, K., Leung, V.C.M., Feng, L., Ong, Y.-S., Lim, M.-H., Akusok, A., Lendasse, A., Corona, F., Nian, R., Miche, Y., Gastaldo, P., Zunino, R., Decherchi, S., Yang, X., Mao, K., Oh, B.-S., Jeon, J., Toh, K.-A., Teoh, A.B.J., Kim, J., Yu, H., Chen, Y., Liu, J.: Extreme learning machines (trends and controversies). IEEE Intell. Syst. **28**(6), 30–59 (2013)
15. Radhakrishna Rao, C., Mitra, S.K.: Generalized Inverse of Matrices and Its Applications. Wiley, New York (1972)

16. Kraskov, A., Stögbauer, H., Grassberger, P.: Estimating mutual information. Phys. Rev. E **69**, 066138 (2004)
17. Pál, D., Póczos, B., Szepesvári, C.: Estimation of Rényi entropy and mutual information based on generalized nearest-neighbor graphs. ArXiv e-prints (2010)
18. Pál, D., Póczos, B., Szepesvári, C.: Estimation of rényi entropy and mutual information based on generalized nearest-neighbor graphs. In: Lafferty, J.D., Williams, C.K.I., Shawe-Taylor, J., Zemel, R.S., Culotta, A. (eds.) Advances in Neural Information Processing Systems 23, pp. 1849–1857. Curran Associates, Inc. (2010)
19. Kullback, S., Leibler, R.A.: On information and sufficiency. Ann. Math. Stat. **22**(1), 79–86 (1951)
20. Bogachev, V.I., Kolesnikov, A.V.: The Monge-Kantorovich problem: achievements, connections, and perspectives. Russ. Math. Surv. **67**, 785–890 (2012)

Probabilistic Methods for Multiclass Classification Problems

Andrey Gritsenko, Emil Eirola, Daniel Schupp,
Edward Ratner and Amaury Lendasse

Abstract In this paper, two approaches for probability-based class prediction are presented. In the first approach, the output of Extreme Learning Machines algorithm is used as an input for Gaussian Mixture models. In this case, ELM performs as dimensionality reduction technique. The second approach is based on ELM and a newly proposed Histogram Probability method. Detailed description and analysis of these methods are presented. To evaluate these methods five datasets from UCI Machine Learning Repository are used.

Keywords Classification · Machine learning · Neural networks · Extreme learning machines · Gaussian mixture model · Naive Bayesian classifier · Multiclass classification · Probabilistic classification · Histogram distribution · Leave-one-out cross-validation · PRESS statistics

1 Introduction

There are a lot of algorithms that can be successfully used for multiclass classification problems. For many real world problems, it would be preferable to be able to build probabilistic models for confidence determination, yet many methods yield binary classifications without a probability.

The Extreme Learning Machines (ELM) [1–6] and Neural Networks in general, as well as other classification methods, have a successful history of being used to solve multi-class classification problems. The standard procedure is to convert the class

A. Gritsenko · A. Lendasse (✉)
Department of Mechanical and Industrial Engineering and the Iowa
Informatics Initiative, The University of Iowa, Iowa City, USA
e-mail: amaury-lendasse@uiowa.edu

E. Eirola · A. Lendasse
Arcada University of Applied Sciences, Helsinki, Finland

A. Gritsenko · D. Schupp · E. Ratner
Lyrical Labs LLC, Iowa City, USA

© Springer International Publishing Switzerland 2016
J. Cao et al. (eds.), *Proceedings of ELM-2015 Volume 2*,
Proceedings in Adaptation, Learning and Optimization 7,
DOI 10.1007/978-3-319-28373-9_33

labels into numerical 0/1 binary variables, effectively transforming the situation into a regression task. When a new sample is fed through the network to produce a result, the class is assigned based on which numerical value it is closest to. While this leads to good performance in terms of classification accuracy and precision, the network outputs as such are not always meaningful. This paper presents two methods, which convert the outputs into more interpretable probabilities by using Gaussian Mixture Models (GMM) [7] and building histograms of hits and misses.

Most classifiers provide results which can not directly be interpreted as probabilities. Probabilities are useful for understanding the confidence in classification, and evaluating the possibility of misclassification. In a multiclass problem, for instance, certain misclassification results may be considerably more harmful or expensive than others. One example is in website filtering based on user-defined categories, where neural networks are used to classify previously uncategorized sites [8, 9].

It is true that the optimal least-squares estimator for ELM output values is equivalent to the conditional probability:

$$\hat{y}(x) = E[Y \mid x] = p(Y = 1 \mid x). \tag{1}$$

In practice, however, the results can be outside the range 0–1, and this interpretation is not very reliable or directly useful.

The proposed probabilistic methods can be used to transform the output values in the output layer to more interpretable probabilities. The detailed explanation of these methods is given in Sects. 4.2 and 4.3.

The remainder of this paper is structured as follows: Sect. 2 describes previously done research in the related area, Sect. 3 describes the problem of calculating and comprehending probabilistic outputs, Sect. 4 reviews the overall process of obtaining probabilistic outputs using ELM as baseline algorithms. All methods used in this process are described in the corresponding subsections. An experimental comparison on a variety of datasets is provided in Sect. 5 that also has a note on specific implementation of methods. Section 6 presents conclusions and further works.

2   Previous Works

Regardless of great benefit that could be obtained with probabilistic classification, only early work has been done. The most popular method is called Naive Bayes Classifier (NBC) [10, 11] and is based on Bayes' Theorem to compute conditional probabilities

$$p\left(C_k|x\right) = \frac{p\left(C_k\right) p\left(x|C_k\right)}{p\left(x\right)}, \tag{2}$$

where x is a sample vector of features, $p\left(C_k\right)$ is the *prior* probability of class C_k, $p\left(x|C_k\right)$ is the likelihood of x for a given class C_k, and $p\left(x\right) = \sum_k p\left(C_k\right) p\left(x|C_k\right)$ is the marginal probability of x. Though this theorem works well in probability theory, NBC cannot be considered as a reliable probability classifier as it assumes input variables are conditionally independent given the class label.

There are a lot of measures to estimate the accuracy of non-probabilistic learning algorithms [12, 13], while the lack of interest in probabilistic classification methods results in poor developed methodology to evaluate probabilistic outputs. The most well known Mean Square Error estimation [13] is usually used to evaluate accuracy of probabilistic methods, though this approach may discard the meaning of probability output.

Here for each example x, the square error (SE) is defined as

$$\mathbf{SE} = \sum_i \left(T\left(c_i \mid x\right) - P\left(c_i \mid x\right)\right)^2 , \tag{3}$$

where $P(c_i|x)$ is the probability estimated for example x and class c_i and $T(c_i|x)$ is defined to be 1 if the actual label of x is c and 0 otherwise. The drawback of this method, which is true for all MSE-based estimators, is that they have heavily weighted outliers. First of all, that means that large errors have a bigger impact on the estimation than small errors. Secondly, but what is more important in case of probabilities, is that a lot of small errors would influence the evaluation and not let us to estimate properly how far our prediction from the correct class.

3 Description of the Problem

Consider the following problem of multiclass classification of an object depicted on an image, where each object could belong to any of three classes: class 1—'Dog', class 2—'Cat', Class 3—'Bird'. Assume that for three different samples, the following output values have been received (see Table 1). Suppose that for each sample, output value $\hat{t}_i$ for the corresponding class i belongs to the interval $[0; 1]$ and $\sum_{i=1}^{3} \hat{t}_i = 1$. The discussion of what classification algorithm has been used to obtain these output values is out of the scope of this example.

When using standard classification estimators (e.g., mean square error estimator [14]) the class with the highest output value should be picked as a prediction. Though, in terms of the given example, the mentioned approach is meaningful only for the first data sample. For samples 2 and 3 it is more intuitively to use the terms of

Table 1 Possible outputs for multiclass classification problem

Sample	Class 1 (Dog)	Class 2 (Cat)	Class 3 (Bird)
Sample 1	99	1	0
Sample 2	49	51	0
Sample 3	33.3	33.3	33.4

probability when discussing and estimating the corresponding outputs. With this it can be stated, that sample 2 most probably belongs to either class 1 or 2 with slightly better hand of class 2 and definitely does not belong to class 3; and sample 3 has a more or less equal probability of belonging to any of three classes, that can indicate, for example, that the sample belongs to a class of unseen during the training step data, or the sample possess features of all three classes.

Though it is more intuitively to comprehend and interpret probabilistic outputs, additional methods are required to convert raw outputs into probabilistic ones (see Sects. 4.2 and 4.3 for more information).

4 Global Methodology

4.1 Extreme Learning Machines

Extreme Learning Machines (ELMs) [1] are single hidden-layer feed-forward neural networks where *only* the output weights are optimised, and all the weights between the input and hidden layer are assigned randomly (Fig. 1). Training this model is simple, as the optimal output weights β can be calculated by ordinary least squares or various regularised alternatives.

In the following, a multi-class classification task is assumed. The data is a set of N distinct samples $\{x_i, y_i\}$ with $x_i \in \mathbb{R}^d$ and $y_i \in \{1, \ldots, C\}$ where C is the number of distinct classes. Encode classification targets as one binary variable for each class (one-hot encoding). $\mathbf{T}$ is the matrix of targets such that $\mathbf{T}_{ij} = 1$ if and only if $y_i = j$, i.e., sample i belongs to class j. Otherwise, $\mathbf{T}_{ij} = 0$. $\hat{\mathbf{T}}$ is the output matrix of the method. In the case of two classes, a single output variable is sufficient. Ideally,

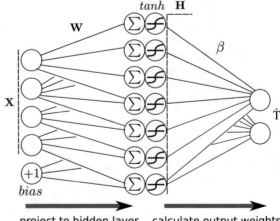

Fig. 1 Extreme learning machine with multiple outputs. Bias is conveniently included as an additional constant +1 input. Hidden layer weights **W** are fixed, only output layer weights β are calculated

values of output matrix should be $\hat{\mathbf{T}}_{ij} \in \{0, \ldots, 1\}$, but in practice the bounds are usually extended to some extent depending on the model random initialization.

A single (hidden) layer feedforward neural network (SLFN) with d input nodes, C output nodes, and M neurons in the hidden layer can be written as

$$f(x) = \sum_{k=1}^{M} \beta_k h \left(w_k \cdot x \right) , \tag{4}$$

where w_k are randomly assigned d-dimensional weight vectors, the output layer weights β_k are C-dimensional vectors, and $h(\cdot)$ an appropriate nonlinear activation function, e.g., the sigmoid function. The output of f is a C-dimensional vector, and class assignment is determined by which component is the largest.

In terms of matrices, the training of the network can be re-written as finding the least-squares solution to the matrix equation

$$\mathbf{H}\beta = \mathbf{T}, \quad \text{where} \quad H_{ik} = h\left(w_k \cdot x_i \right) . \tag{5}$$

Constant bias terms are commonly included by appending a 1 to each x_i and concatenating a column of 1 s to $\mathbf{H}$.

4.1.1 PRESS Optimization for Number of Neurons

The number of hidden neurons is the only tunable hyperparameter in an ELM model. It is selected using a Leave-One-Out (LOO) Cross-Validation error. The LOO method is usually a costly approach to optimize a parameter since it requires to train the model on the whole dataset but one sample, and evaluate on this sample repeatedly for all the samples of the dataset. However, the output layer is linear for the ELM model, and the LOO error has a closed form given by Allen's Prediction Sum of Squares (PRESS) [15]. This closed form allows for fast computation of the LOO Mean Square Error, which gives an estimate of the generalization error of ELM. The optimal number of hidden neurons is found as the minimum of that Meas Squared Error.

The Allen's PRESS formula written with the multi-output notations of the paper is

$$\text{MSE}_{\text{LOO}}^{\text{PRESS}} = \frac{1}{Nc} \sum_{n=1}^{N} \sum_{k=1}^{c} \left(\frac{\mathbf{T} - \mathbf{H}\mathbf{H}^{\dagger}\mathbf{T}}{\left[\mathbf{1}_N - \text{diag}\left(\mathbf{H}\mathbf{H}^{\dagger} \right) \right] \mathbf{1}_c^T} \right)_{ik}^{2} , \tag{6}$$

where $\mathbf{H}^{\dagger}$ denotes the Moore-Penrose pseudo-inverse [16] of $\mathbf{H}$, and the division and square operations are applied element-wise. Additionally, for the particular implementation of PRESS optimization Tikhonov regularization [17] was also used in the paper when computing the pseudo-inverse matrix of $\mathbf{H}$

$$\mathbf{H}^{\dagger} = \left(\mathbf{H}^T\mathbf{H} + \alpha\mathbf{I}\right)^{-1}\mathbf{H}^T , \tag{7}$$

where $\alpha = 10^{-5}$ is a chosen regularization parameter and $\mathbf{I}$ is an identity matrix. Tikhonov regularization is a well-known regularization technique for ill-posed problems [18], which allows to avoid possible numerical and computational issues when finding the solution of the system described by Eq. (5).

4.2 Gaussian Mixture Models

Mixtures of Gaussians can be used for a variety of applications by estimating the density of data samples [19, 20]. A *Gaussian Mixture Model* can approximate any distribution by fitting a number of components, each representing a multivariate normal distribution.

The model is defined by its parameters, which consist of the mixing coefficients π_k, the means $\boldsymbol{\mu}_k$, and covariance matrices $\boldsymbol{\Sigma}_k$ for each component k ($1 \leq k \leq K$) in a mixture of K components. The combination of parameters is represented as $\boldsymbol{\theta} = \{\pi_k, \boldsymbol{\mu}_k, \boldsymbol{\Sigma}_k\}_{k=1}^K$.

The model specifies a distribution in $\mathbb{R}^d$, given by the probability density function

$$p(\boldsymbol{x} \mid \boldsymbol{\theta}) = \sum_{k=1}^{K} \pi_k \mathcal{N}(\boldsymbol{x} \mid \boldsymbol{\mu}_k, \boldsymbol{\Sigma}_k), \tag{8}$$

where $\mathcal{N}(\boldsymbol{x} \mid \boldsymbol{\mu}, \boldsymbol{\Sigma})$ is the probability density function of the multivariate normal distribution

$$\mathcal{N}(\boldsymbol{x} \mid \boldsymbol{\mu}, \boldsymbol{\Sigma}) = \frac{1}{\sqrt{(2\pi)^d \det(\boldsymbol{\Sigma})}} \exp\left(-\frac{1}{2}(\boldsymbol{x} - \boldsymbol{\mu})^T \boldsymbol{\Sigma}^{-1}(\boldsymbol{x} - \boldsymbol{\mu})\right) . \tag{9}$$

The standard procedure for fitting a Gaussian Mixture Model to a dataset is maximum likelihood estimation by the Expectation-Maximisation (EM) algorithm [20–22]. The E-step and M-step are alternated until convergence is observed in the log-likelihood. Initialisation before the first E-step is arbitrary, but a common choice is to use the clustering algorithm K-means to find a reasonable initialisation [19].

The only parameter to tune select is the number of components K. This can be done by separately fitting several models with different values for K, and using the BIC criterion [23] to select the best model. In the proposed methodology, we are using the BIC criterion to select the value of K. Several further criteria are discussed in [24, Chap. 6].

4.2.1 ELM as Dimensionality Reduction Technique for GMM

In [25] a method was proposed that combines ELM and GMM algorithms. The core of this method is to train a standard ELM for classification at first and then apply GMM to the output of ELM to obtain more interpretable probabilistic results. As usually the number of classes in data is less then number of features/variables, ELM could be intuitively considered as a dimensionality reduction technique that decreases number of training parameters N_{param} of GMM:

$$N_{param} = N_{comp}N_{dim} + \frac{N_{comp}N_{dim}\left(N_{dim} + 1\right)}{2} + N_{comp} - 1 \,, \tag{10}$$

where N_{dim} is the dimension of input data for GMM and N_{comp} is number of mixture components.

The main requirement to fit a representative mixture model is that the number of observations should be greater than the number of parameters. That means, that aside from the fact that using ELM as data preprocessing reduces computational complexity of GMM algorithm, it also increases the possibility to basically building a GMM.

In the current situation, GMM is applied in conjunction with Bayes' theorem to find estimates for the posterior probabilities of each class for a sample. Specifically, it is used to estimate the term $p\left(x|C_k\right)$ in Eq. (2). This requires that a separate GMM is built for each class.

Algorithm 1 Training the model and finding the conditional class probabilities for unseen data.

▷ **Training step**
Require: Input data $\mathbf{X}$, targets $\mathbf{T}$
1: Randomly assign input vectors w_k and form $\mathbf{H}$
2: Calculate β as the least squares solution to Eq. (5)
3: Calculate outputs on training data: $\mathbf{Y} = \mathbf{H}\beta$
4: **For each** class C **do**
5:     Fit a GMM_C to the rows of $\mathbf{Y}$ corresponding to the class C
6: **End for**
7: Calculate $p(C)$ based on proportions of each class
8: **Return** w_k, β, GMM_C, $p(C)$

▷ **Testing step**
Require: Test data $\mathbf{X}_t$, weights w_k, β, GMM_C and $p(C)$ for each class C
1: Form $\mathbf{H}_t$ by using the weights w_k
2: Calculate outputs: $\mathbf{Y}_t = \mathbf{H}_t\beta$
3: **For each** class C **do**
4:     Use GMM_C to calculate $p(Y_t \mid C)$ for each sample
5: **End for**
6: Calculate $p(C \mid Y_t) \propto p(Y_t \mid C)p(C)$ for each sample
7: **Return** Conditional probabilities $p(C \mid Y_t)$ for each class for each sample

---

Considering all of the above and according to [25], the procedure of combining ELM and GMM in classification method with probabilistic outputs was summarized in Algorithm 1.

4.3   New Proposed Histogram Probability Method

4.3.1   Overview

In broad terms, the histogram method is a simple characterization of the continuous result from a classification algorithm. In short, the proposed approach is to take advantage of a continuous decision function (as opposed to a binary decision maker). The proposed approach then bins the resulting decision space, and uses training data to build a model of the class distributions [26]. Bayesian theory can then be applied for robust decision making.

4.3.2   Methodology

Hereafter the steps of performing the Histogram Probability method of obtaining probabilistic output from raw output of any multi-class classification algorithm are described.

At first, some classification algorithm should be trained in order to obtain outputs (for the sake of unambiguousness, called raw outputs) that would serve as a basis for probabilistic outputs for a certain multi-class problem. In order to build histograms, the range of raw training sct outputs is computed as $\left[\min \hat{\mathbf{T}}_{train}; \max \hat{\mathbf{T}}_{train}\right]$. The range is then divided into bins, and a certain number of bins can differ for different classification problems with respect to their complexity. It should be noted that too small number of bins would result in loosing features of distribution, while too large number could be a reason for sparse histograms.

Then, for each class two types of histograms are built. At the first stage, samples in training set are split up into different sets according to the corresponding correct class. Then for each sample i in separated sets the raw output value of the correct class is added to the set of IN-class values, output values for other classes are added to the set of OUT-class values.

$$\hat{T}_{ij} \mapsto \begin{cases} IN_{C_k} & \text{if } j = k \\ OUT_{C_k} & \text{if } j \neq k \end{cases}, \quad k = correct\,class\,(i)\,, \tag{11}$$

where $\hat{T}_{ij}$—is the raw output value of sample i for the corresponding class C_j, k identifies correct class for sample i, IN_{C_k} and OUT_{C_k} are respectively sets of IN-class and OUT-class values for class C_k.

After all, samples are processed, histograms of *IN*-class and *OUT*-class values are constructed on the selected bins. Obviously, number of values used to build OUT-class histogram is bigger than number of raw output values used to build IN-class histogram for the same class. Moreover, for real-world multi-class classification problems, number of samples in training set for different classes could be different. In order to compensate these differences, histograms are normalized in the following way

$$\overline{IN}_{C_i} = \frac{IN_{C_i}}{size\left(C_i\right)} \quad \text{and} \tag{12}$$

$$\overline{OUT}_{C_i} = \frac{OUT_{C_i}}{\sum_{j=1}^{N_{classes}} size\left(C_j\right)} , j \neq i , \tag{13}$$

where IN_{C_i} and OUT_{C_i} are respectively IN-class and OUT-class histograms built for class C_i, $size\left(C_j\right)$ represents the number of samples in the training set, whose correct class is class C_j, $N_{classes}$ is number of total classes in a certain multi-class classification problem. For each class, resulting histograms can be considered as histograms of probability density functions that for a certain output value define the probability of either correct (in case of IN-class histogram) or wrong (in case of OUT-class histogram) classification, if that output value would be picked as the resulting value (the highest output value) for a sample.

With the respect to the used classification algorithm, probabilistic output for a given sample i, which has a raw output value $\hat{T}_{ij}$ for j class respectively, is computed as follows

$$p\left(C_j \mid \hat{T}_{ij}\right) = \frac{\overline{IN}_{C_j}\left(\hat{T}_{ij}\right)}{\overline{IN}_{C_j}\left(\hat{T}_{ij}\right) + \overline{OUT}_{C_j}\left(\hat{T}_{ij}\right)} . \tag{14}$$

Equation (14) does not guarantee that for each sample probabilities would sum up to 1. The main reason is that for each class these probabilities are calculated regardless of information about other classes. In order to treat results of Eq. (14) as probabilities for each sample they should be normalized.

5 Experiments

In the following subsections, presented earlier probabilistic classification methodologies are compared, using a number of multiclass datasets. These compared methods include the original ELM algorithm, the two variants of the proposed combination of ELM and GMM: ELM-GMM and ELM-GMMr, and also ELM+HP.

Performance of the mentioned algorithms was compared in terms of both training time and accuracy—percentage of correct predictions, where class with the highest output value or probability (for probability-output methods) is kept as prediction.

5.1 Implementation

All methods described above were implemented using OpenCV v.2.4.9—an open source computer vision and machine learning software library [27]. At first, ELM was trained on the original input data. The number of neurons for each dataset has also been chosen before running experiments. For each dataset PRESS Leave-One-Out cross-validation technique [15, 28] has been performed 1000 times and number of neurons that gives the highest average accuracy in predictions has been chosen.

Gaussian Mixture Model was implemented using built-in OpenCV functions realizing Expected Maximization (EM) algorithm [29]. For each class, Gaussian Mixture Model was built with the number of mixture components equal to 2 for each model. Amount of mixture components was chosen empirically as the number, for which the accuracy of ELM-GMM method has the smallest error.

5.2 Datasets

Five different datasets have been used for the experiments. Four datasets have already been used to compare results of ELM-GMM probabilistic classification methodology in [25], and for this paper have mainly been chosen to provide an overall comparison of two different probabilistic classification techniques. In addition, *Iris* dataset has also been included in the comparison, because it is a well-known dataset usually used to compare performance of multi-class classification algorithms. All datasets have been collected from the University of California at Irvine Machine Learning Repository [30].

Table 2 summarizes the different attributes for all datasets. The datasets have been preprocessed in the same way: two thirds of the points are used to create the training set and the remaining third is used as the test set. For all datasets, proportions of classes have been kept balanced in both training and test sets. For all datasets, the training set is standardized to zero mean and unit variance, and the test set is also standardized using the same mean and variance calculated and used for the training set.

Table 2 Information about used datasets

Dataset	Variables	Classes	Samples		Neurons	
			Train	Test	Max	Mean
Iris	4	3	100	50	150	30
Wine	13	3	118	60	150	40
Image segmentation	18	7	1540	770	1000	410
First-order theorem proving	51	6	4078	2040	2000	960
Cardiotocography	21	10	1417	709	1000	400

Table 3 Accuracy and training time comparison for presented methods

Dataset	ELM		ELM-GMM		ELM-GMMr		ELM-HP	
	Time	Acc.(%)	Time	Acc.(%)	Time	Acc.(%)	Time	Acc.(%)
Iris	0.003	96.36	0.069	94.59	0.071	95.70	0.007	96.42
Wine	0.007	93.89	0.124	93.54	0.124	93.56	0.013	93.81
Image Seg.	3.240	94.54	4.491	93.37	4.557	94.08	3.351	94.33
F.-O.T.P.	49.03	53.66	52.39	51.59	51.40	52.67	49.28	53.23
Card.	2.743	74.18	3.865	74.95	3.839	74.49	3.004	73.67

5.3 Results

All experiments have been run on the same computational server, single threaded execution, for the sake of comparison. Technical characteristics of the Windows server are the following: 32 Intel Xeon v2 processors with 2.8 GHz each, 90 GB of RAM and 100 GB of hard drive memory, no swapping has been used for any of the experiments.

Because ELM is a single hidden-layer feed-forward neural network with randomly assigned weights w_k (see Sect. 4.1) that influence the accuracy of the method, each ELM-based method was run 1000 times for each dataset and average performance was calculated.

In [25] a method to refine the training of GMM was presented. The suggestion was to use as inputs for GMM only correct predictions outputs, because GMM trained on wrong predictions inherit the error. For datasets, for which ELM made predictions with high accuracy, the proposed method (dispayed as ELM-GMMr in Table 3) shows better results in comparison to GMM trained on ELM outputs for both correct and wrong predictions.

Table 3 contain results of performing probabilistic classification for five different datasets and four methods as mentioned in Sect. 5.

6 Conclusions and Further Works

In this paper, two approaches for probability-based class prediction are presented: both are using Extreme Learning Machines algorithm as a first stage, then the output is transformed into a probability in the second stage. The second stage was performed using Gaussian Mixture models or a new proposed Histogram Probability method.

In the future, the proposed methodologies have to be improved in terms of computational time in order to target Big Data problems. For example, comparison with k-nearest neighbors will be investigated since they can also be used to provide classification probabilities. Furthermore, to fairly evaluate performances when probability outputs are provided by the classifiers and since no traditional criteria can easily be used in that case, a new criteria has be proposed instead of traditional 'training time-accuracy' comparison approach.

References

1. Huang, G.B., Zhu, Q.Y., Siew, C.K.: Extreme learning machine: theory and applications. Neurocomputing **70**(1–3), 489–501 (2006)
2. Huang, G.B., Chen, L., Siew, C.K.: Universal approximation using incremental constructive feedforward networks with random hidden nodes. IEEE Trans. Neural Netw. **17**(4), 879–892 (2006)
3. Akusok, A., Björk, K.M., Miché, Y., Lendasse, A.: High-performance extreme learning machines: a complete toolbox for big data applications. IEEE Access **3**, 1011–1025 (2015)
4. Cambria, E., et al.: Extreme learning machines. IEEE Intell. Syst. **28**(6), 30–59 (2013)
5. Miche, Y., Sorjamaa, A., Bas, P., Simula, O., Jutten, C., Lendasse, A.: OP-ELM: optimally pruned extreme learning machine. IEEE Trans. Neural Netw. **21**(1), 158–162 (2010)
6. Yu, Q., Miche, Y., Eirola, E., van Heeswijk, M., Séverin, E., Lendasse, A.: Regularized extreme learning machine for regression with missing data. Neurocomputing **102** (2013) 45–51 cited By 9
7. Dinov, I.D.: Expectation maximization and mixture modeling tutorial. Statistics Online Computational Resource (2008)
8. Qi, X., Davison, B.D.: Web page classification: features and algorithms. ACM Comput. Surv. **41**(2), 12:1–12:31 (2009)
9. Patil, A.S., Pawar, B.: Automated classification of web sites using naive bayesian algorithm. In: Proceedings of the International MultiConference of Engineers and Computer Scientists, vol. 1 (2012)
10. Rennie, J.D.M., Shih, L., Teevan, J., Karger, D.R.: Tackling the poor assumptions of naive bayes text classifiers. In: In Proceedings of the Twentieth International Conference on Machine Learning, pp. 616–623 (2003)
11. In: Emerging Intelligent Computing Technology and Applications. Communications in Computer and Information Science, vol. 304 (2012)
12. Westgard, J.O., Carey, R.N., Wold, S.: Criteria for judging precision and accuracy in method development and evaluation. Clin. Chem. **20**(7), 825–833 (1974)
13. Bermejo, S., Cabestany, J.: Oriented principal component analysis for large margin classifiers. Neural Netw. **14**(10), 1447–1461 (2001)
14. Mood, A., Graybill, F.: Introduction to the Theory of Statistics. International Student Edition: McGraw-Hill Series in Probability and Statistics. McGraw-Hill Book Company, Incorporated (1963)
15. Allen, D.M.: The relationship between variable selection and data agumentation and a method for prediction. Technometrics **16**(1), 125–127 (1974)
16. Rao, C.R., Mitra, S.K.: Generalized Inverse of Matrices and Its Applications. Wiley (1971)
17. Tikhonov, A.: Numerical Methods for the Solution of Ill-Posed Problems. Current Plant Science and Biotechnology in Agriculture. Springer (1995)
18. Bell, J.B.: Math. Comput. **32**(144), 1320–1322 (1978)
19. Bishop, C.M.: Pattern Recognition and Machine Learning. Springer (2006)
20. Eirola, E., Lendasse, A., Vandewalle, V., Biernacki, C.: Mixture of gaussians for distance estimation with missing data. Neurocomputing **131**, 32–42 (2014)
21. Dempster, A.P., Laird, N.M., Rubin, D.B.: Maximum likelihood from incomplete data via the EM algorithm. J. R. Stat. Soc. Ser. B (Methodol.) **39**(1), 1–38 (1977)
22. McLachlan, G., Krishnan, T.: The EM Algorithm and Extensions. Wiley Series in Probability and Statistics. Wiley, New York (1997)

23. Schwarz, G.: Estimating the dimension of a model. Ann. Stat. **6**(2), 461–464 (1978)
24. McLachlan, G.J., Peel, D.: Finite Mixture Models. Wiley Series in Probability and Statistics. Wiley, New York (2000)
25. In: Advances in Computational Intelligence. Lecture Notes in Computer Science, vol. 9095 (2015)
26. Schupp, D., Ratner, E., Gritsenko, A.: U.S. Provisional Patent Application No. 7062320: object categorization using statistically-modeled classifier outputs (08 2015)
27. Bradski, G.: The opencv library. Dr. Dobb's J. Softw. Tools (2000)
28. Myers, R.: Classical and Modern Regression with Applications. Bookware Companion Series. PWS-KENT (1990)
29. Bilmes, J.: A gentle tutorial of the em algorithm and its application to parameter estimation for gaussian mixture and hidden markov models. Technical report, International Computer Science Institute and Computer Science Division, University of California at Berkeley (1998)
30. Lichman, M.: UCI machine learning repository. http://archive.ics.uci.edu/ml (2013)

A Pruning Ensemble Model of Extreme Learning Machine with $L_{1/2}$ Regularizer

Bo He, Tingting Sun, Tianhong Yan, Yue Shen and Rui Nian

Abstract Extreme learning machine (ELM) as an emerging branch of machine learning has shownits good generalization performance at a fast learning speed. Nevertheless, the preliminary ELM and other evolutional versions based on ELM cannot provide the optimal solution of parameters between the hidden and output layer and cannot determine the suitable number of hidden nodes automatically. In this paper, a pruning ensemble model of ELM with $L_{1/2}$ regularizer (PE-ELMR) is proposed to solve above problems. It involves two stages. First, we replace the original solving method of the output parameter in ELM to a minimum squared-error problem with sparse solution by combining ELM with $L_{1/2}$ regularizer. In addition, $L_{1/2}$ regularizerguarantees the sparse solution with less computational cost. Second, in order to get the required minimum number for good performance, we prune the nodes in hidden layer with the ensemble model, which reflects the superiority in searching the reasonable hidden nodes. Experimental results present the performance of L_1 and $L_{1/2}$ regularizers used in our model PE-ELMR, compared with ELM and OP-ELM, for regression and classification problems under a variety of benchmark datasets.

Keywords Neural networks · Extreme learning machine · $L_{1/2}$ regularizer · Ensemble models · Pruning methods

B. He (✉) · T. Sun · Y. Shen · R. Nian
School of Information and Engineering, Ocean University of China, Qingdao, Shandong, China
e-mail: bhe@ouc.edu.cn

T. Yan
School of Mechatronic Engineering, China Jiliang University, Hangzhou, Zhejiang, China
e-mail: yanth@163.com

© Springer International Publishing Switzerland 2016
J. Cao et al. (eds.), *Proceedings of ELM-2015 Volume 2*,
Proceedings in Adaptation, Learning and Optimization 7,
DOI 10.1007/978-3-319-28373-9_34

1   Introduction

Extreme learning machine (ELM) [1, 2] proposed by Huang et al. has proved to be of good generalization performance at a fast learning speed. ELM is designed based on the generalized single-hidden layer feedforward networks (SLFNs) [3, 4] with a variety of hidden nodes. SLFNs have a strong ability of nonlinear approximation for regression problem, and forming disjoint decision regions with arbitrary dimension data for classification problem. In addition, except manually setting the number of nodes in hidden layer and the activation function, ELM needs no human intervention. The above characteristics of ELM give contribution to a wide range of applications, such as feature learning, clustering, regression and classification problems.

However, the preliminary ELM and other evolutional versions based on ELM have two main drawbacks. On one hand, it does not provide an optimal solution of parameters in the output layer. On the other hand, the suitable number of hidden nodes must be determined by a trial and error method. In addition, it can achieve the better performance to select the required minimum number of hidden nodes than other numbers. To overcome the mentioned problems, researchers have proposed several effective schemes.

Hansen and Salamon presented that generalization performance of using an ensemble of neural networks with a plurality consensus scheme is better than that of a single network [5]. Then, Zhou et al. suggested that ensemble several models were better than ensemble all of them, and a selective neural network ensemble based on genetic algorithm (GASEN) was presented later [6]. However, the proposed GASEN method has lower speed than other ensemble algorithms because genetic algorithm is employed to select an optimum set of individual networks. Moreover, Sun et al. [7] and Liu et al. [8] combined the ensemble method with ELM and a new model of the ELM ensemble was proposed to get stable good performance. In addition, Xu et al. [9] proposed a genetic ensemble of ELM on the basis of Liu's method with genetic algorithm. In general, ensemble methods can achieve better performance than the single ELM network.

Some pruning methods have been proposed to handle with the number of the hidden nodes. Rong et al. presented a pruned ELM (P-ELM) for classification problems [10]. The method initializes a large network and then eliminates the hidden nodes with low relevance to the class labels by using statistical criteria. Miche et al. proposed another pruning algorithm named optimally-pruned ELM (OP-ELM) for both regression and binary classification problems [11]. Since both P-ELM and OP-ELM are evolutional versions of the preliminary ELM, structural drawbacks also exist in these algorithms and they do not have satisfactory performance. Moreover, it is hard to determine the time when to start the pruning process in these two methods.

In this paper, inspired by the existing methods, a pruning ensemble model of ELM with $L_{1/2}$ regularizer (PE-ELMR) is proposed to overcome the drawbacks of the preliminary ELM. And it can be divided into two stages. First, we replace the original solving method of the output parameters in the preliminary ELM to a minimum squared-error problem with sparse solution by combining ELM with $L_{1/2}$ regularizer [12, 13] (ELMR). Second, in order to get the required minimum number for good performance, the hidden nodes are pruned by the ensemble model. Therefore, our approach has a great potential for achieving better performance with less human intervention.

The rest of paper is organized as follows. Section 2 briefly discusses previous work on the preliminary ELM and minimum squared-error problem with regularization methods. Section 3 describes the proposed method named a pruning ensemble model of ELM with $L_{1/2}$ regularizer. Section 4 reports experimental results on our method and the others for regression and classification problems under a variety of benchmark datasets. In Sect. 5, conclusions are summarized on our recent research and future work.

2 ELM and Regularization Methods

2.1 The Preliminary ELM

ELM [1, 2] is an algorithm designed based on the generalized SLFNs. The weights and bias between the input and hidden nodes are generated randomly, and then the output weights can be calculated by solving a linear system analytically without iteration.

Given N pairs of training samples $\{x_i, y_i\}_{i=1}^{N}$, where $x_i = [x_{i1}, \cdots, x_{in}]^T \in \mathbb{R}^n$ is the input data and $y_i = [y_{i1}, \cdots, y_{im}]^T \in \mathbb{R}^m$ is the output target. The standard model of ELM with L hidden neurons and activation function $g(w_j, b_j, x_i)$ can be mathematically expressed as

$$\sum_{j=1}^{L} \beta_j g(w_j, b_j, x_i) = o_i, \quad i = 1, 2, \cdots, N \tag{1}$$

where the weight vector $w_j = [w_{j1}, \cdots, w_{jn}]^T \in \mathbb{R}^n$ connects the j_th hidden nodes and the input nodes, andbias b_j is of the j_th hidden nodes. The weight vector $\beta_j = [\beta_{j1}, \cdots, \beta_{jm}]^T \in \mathbb{R}^m$ connects the j_th hidden nodes and the output nodes.

ELM can approximate these N samples with zero error means that $\sum_{i=1}^{N} \|o_i - y_i\| = 0$, and it can be modified as

$$\mathbf{H}\beta = \mathbf{Y} \tag{2}$$

where **H** is called the hidden layer output matrix of the neural network

$$\mathbf{H} = \begin{bmatrix} g(w_1, b_1, x_1) & \cdots & g(w_L, b_L, x_1) \\ & \vdots & \\ g(w_1, b_1, x_N) & \cdots & g(w_L, b_L, x_N) \end{bmatrix}_{N \times L} \tag{3}$$

$$\beta = [\beta_1^T, \cdots, \beta_L^T]_{L \times m}^T \tag{4}$$

$$\mathbf{Y} = [y_1^T, \cdots, y_N^T]_{N \times m}^T \tag{5}$$

H can be obtained. And then the output weights β can be generated as

$$\beta = \mathbf{H}^\dagger \mathbf{Y} \tag{6}$$

where $\mathbf{H}^\dagger$ denotes the Moore–Penrose generalized inverse of matrix **H**.

2.2 Regularization Methods

In statistics and machine learning, regularization methods [14, 15] are used for model selection. The general form of the regularization methods can be modeled as

$$\min \left\{ \frac{1}{n} \sum_{i=1}^n l(y_i, f(x_i)) + \lambda \|f\|_k \right\} \tag{7}$$

where $l(\cdot)$ represents a loss function, $\{x_i, y_i\}_{i=1}^N$ is a dataset and $f(x_i)$ is the function with regard to x_i. λ is the regularization parameter, and $\|\cdot\|_k$ denotes $k-norm$. The equation is the general form of L_k regularizer. Given a $M \times N$ matrix **A** and a target y, then an observation can be obtained by

$$y = \mathbf{A}x + \varepsilon \tag{8}$$

where x is the weight and ε is the observation noise. In order to solve the coefficients x of the equation, L_k regularization methods are used for minimizing the observation noise term. And then $l(\cdot)$ and $f(\cdot)$ in (7) can be replaced by (8), and the solution of the regression problem is in the following form.

$$x = \arg \min_x \left\{ \|y - \mathbf{A}x\|_2^2 + \lambda \|x\|_k \right\} \tag{9}$$

Four forms of regularization methods are presented as follows. When $k = 0$, it is referred to as the L_0 regularizer [16]. The L_0 regularizer can generate the sparsest solutions, however, solving L_0 regularizer in the minimization problem is a NP-hard problem. When $k = 1$ (L_1 regularizer), it transforms to the Lasso problem [17]. The solution of Lasso is derived from solving a quadratic programming problem [18, 19]. But it is less sparse than the L_0 regularizer. When $k = 2$ (L_2 regularizer), it transforms to the ridge regression [20]. The solutions of L_2 regularizer have the properties of being smooth, but have no properties of sparse.

3 Proposed PE-ELMR

3.1 ELM with $L_{1/2}$ Regularizer

In order to obtain the optimal solution of parameters between the hidden and output layer and prevent over-fitting in the learning procedure. Also, $L_{1/2}$ regularizer proposed by Xu et al. [12] can produce sparser solutions than L_1 regularizer and easier computation than L_0 regularizer. Therefore, a structure of combining ELM with $L_{1/2}$ regularizer (ELMR) is proposed. The important prerequisite of ELMR is that assuming the errors between the output of the network and the target are not zero. A problem of a minimum squared-error with sparse solution is addressed and then the error $\delta(\delta \geq 0)$ is defined as

$$\delta = \|\mathbf{Y} - \mathbf{H}\beta\|. \tag{10}$$

Our aim is to calculate the output weight β of (10) with the constraint that the error δ is equal to or approximate to zero. The output weight β can be calculated by $L_k(k = 1/2)$ regularization methods.

$$\beta = \arg\min_{\beta}\left\{\|\mathbf{Y} - \mathbf{H}\beta\|_2^2 + \lambda\|\beta\|_{1/2}\right\} \tag{11}$$

Although the solutions of $L_{1/2}$ regularizer prove to be much sparser than those of L_1 regularizer, the solving process of $L_{1/2}$ regularizer seems to be more difficult. Then an iterative method is proposed to solve of $L_{1/2}$ regularizer problem, whose main principle is to transform $L_{1/2}$ regularizer into a series of weighted L_1 regularizer [12]. In order to solve the L_1 regularizer problem, we adopt the *l1_lstoolbox* proposed by Koh [21].

3.2 The Pruning Methods

The preliminary ELM cannot select the number of the nodes in hidden layer without human intervention, so a pruning method is need to decrease the hidden nodes number to a suitable value for optimization problem. First, sort the output weight β from large to small and the sorted β can be represented by $\bar{\beta}(\bar{\beta}_1 \geq \bar{\beta}_2 \geq \cdots \geq \bar{\beta}_L)$. Second, select the hidden nodes by a threshold γ. The ratio of the first l accumulation coefficients to the sum coefficients can be represented by

$$\gamma_l = \frac{\sum_{i=1}^{l} \bar{\beta}_i}{\sum_{i=1}^{L} \bar{\beta}_i}, \quad l = 1, 2, \cdots, L \tag{12}$$

According to the set threshold γ, the necessary number S of hidden nodes can be defined as

$$S = \min\{l | \gamma_l \geq \gamma\}, \quad l = 1, 2, \cdots, L \tag{13}$$

The output weight refers to coefficients of the regularization problems, which can reveal the relativity between the variables of hidden nodes and the output targets. The rest of the work is to set a threshold for selecting the hidden nodes. We resolve it without human participation by employing the ensemble model of ELM.

3.3   A Pruning Ensemble Model of ELMs with $L_{1/2}$ Regularizer

Since the process of pruning is needed to repeat to determine the optimal number, and the multiple processes of pruning increase the computational cost, ensemble model is then proposed to handle the multiple processes at the same time. With the ensemble model of ELMs, we can randomly set different thresholds on pruning process at the same time and then select the individual ELM under the best performance.

The structure of the ensemble ELMs model is shown in Fig. 1. We adopt the ensemble model of ELMs for training process, and employ the single ELM for testing process with the suitable number of hidden nodes selected by the training process.

After pruning the hidden nodes in each of ELMs, we need to update the weights and bias between the input and hidden nodes. Define the subset of the deleting hidden nodes as $D = \{d_1, d_2, \cdots, d_{L-S}\}$. The weight vector w_j is assigned to $w_j^{old} = \left[w_{j1}^{old}, \cdots, w_{jn}^{old} \right]^T \in \mathbb{R}^n, j = 1, 2, \cdots, L$. And the old bias is assigned to $b_j^{old} = b_j$, $j = 1, 2, \cdots, L$. Then the new weights are updated by the following formula

$$w_{s,i}^{new} = w_{s,i}^{old} + \frac{1}{L-S} \sum_{t=1}^{L-S} w_{d_t, i} \quad i = 1, 2, \cdots n, \quad s = 1, 2, \cdots, S \tag{14}$$

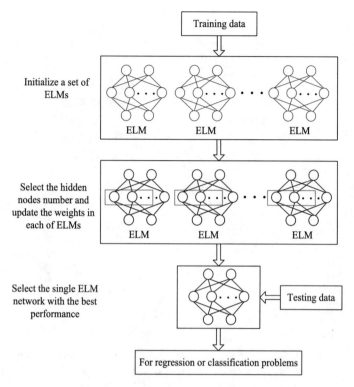

Fig. 1 The structure of the pruning ensemble ELMs model

The new bias are updated as

$$b_s^{new} = b_s^{old} + \frac{1}{L-S} \sum_{t=1}^{L-S} b_{d_t}, \quad s = 1, 2, \cdots, S \qquad (15)$$

Therefore, the proposed method can be described as Algorithm 1.

Algorithm 1 The PE-ELMR algorithm

Input Trainingdataset $\{x_i, y_i\}_{i=1}^N$, initial hidden nodes number L, the activation function $g(\cdot)$, the parameter λ, the ELMs number K of the ensemble model.

Output The $\bar{k} - th$ single ELM network, its suitable hidden nodes number $S^{\bar{k}}$.

Step 1 Construct a set of K ELM networks and randomly assign the weight w_j^k and bias b_j^k between the input and hidden layer in each of the ELMs.

$$f_L(x_i) = \sum_{j=1}^{L} \beta_j^k g\left(w_j^k, b_j^k, x_i\right), \quad i = 1, 2, \cdots, N \quad k = 1, 2, \cdots K$$

Step 2   Calculate the coefficients β^k of the L_1 regularizer or $L_{1/2}$ regularizer, and then rank them in each of the ELMRs, respectively.
For L_1 regularizer,

$$\beta^k = \arg\min_{\beta^k}\left\{\left\|\mathbf{Y} - \mathbf{H}\beta^k\right\|_2^2 + \lambda\left\|\beta^k\right\|_1\right\}, \quad k = 1, 2, \cdots, K$$

For $L_{1/2}$ regularizer,

$$\beta^k = \arg\min_{\beta^k}\left\{\left\|\mathbf{Y} - \mathbf{H}\beta^k\right\|_2^2 + \lambda\left\|\beta^k\right\|_{1/2}\right\}, \quad k = 1, 2, \cdots, K$$

Randomly generate K thresholds $\gamma^k \in [0, 1]$, and prune the hidden nodes.

$$S^k = \arg\min_{l}\left\{\gamma_l^k \geq \gamma^k\right\}, \quad l = 1, 2, \cdots, l, \quad k = 1, 2, \cdots, K$$

Step 3   Update the weights and bias between the input and hidden nodes, and the output matrix $\mathbf{H}$ can be recalculated by

$$(\mathbf{H}^{new})^k = \begin{bmatrix} g\left(\left(w_1^{new}\right)^k, \left(b_1^{new}\right)^k, x_1\right) & \cdots & g\left(\left(w_{S^k}^{new}\right)^k, \left(b_{S^k}^{new}\right)^k, x_1\right) \\ \vdots & & \\ g\left(\left(w_1^{new}\right)^k, \left(b_1^{new}\right)^k, x_N\right) & \cdots & g\left(\left(w_{S^k}^{new}\right)^k, \left(b_{S^k}^{new}\right)^k, x_N\right) \end{bmatrix}_{N \times S^k}$$

Step 4   Calculate the actual output $\mathbf{O}^k$ using (1) and the residuals r^k compared with the target $\mathbf{Y}$.

$$\mathbf{O}^k = (\mathbf{H}^{new})^k \beta^k, \quad k = 1, 2, \cdots, K$$

$$r^k = \left\|\mathbf{Y} - \mathbf{O}^k\right\|_2^2, \quad k = 1, 2, \cdots, K$$

Select the $\bar{k} - th$ individual ELM network with the best performance, which has the suitable number of hidden nodes. Then save the corresponding pruned hidden nodes number $S^{\bar{k}}$.

$$\bar{k} = \arg\min_{k}\left\{r^k\right\}$$

4 Experimental Results

4.1 Experimental Setup

All the simulations for the algorithms are carried out in Matlab R2010b environment running in an Intel (R) Core (TM) i5-3470 3.20 GHz CPU. Before the experiments get started, each of the dataset is split into 70 % of the total for training samples and 30 % of the total for the testing samples. The attributes of the input samples have been normalized into the range $[-1, 1]$ for implementing the ELMs. In the experiments of ELM part, the activation function is set to 'sig' representing the sigmoidal function, which tends to perform better than the other activation functions. For the preliminary ELM, the hidden nodes number L_{ELM} is set using trial and error method. For our method PE-ELMR, OP-ELM and ELM-L1 (our structure with L_1 regularizer), the hidden nodes number L is set to 400. And the number of ELMs is set to 20 in the ensemble model.

The datasets used in our experiments are collected from the University of California at Irvine (UCI) Machine Learning Repository. We choose the datasets based on the overall heterogeneity in terms of variable numbers and the nonlinearity. Tables 1 and 2 show the information and characteristics of the datasets for regression and classification problems, respectively.

Table 1 Specification of regression datasets

Dataset	Number of samples		#Attributes
	Training	Testing	
Sinc	5000	5000	1
Abalone	772	331	8
Housing	354	152	13
Bodyfat	70	30	9
Redwine	1119	480	11

Table 2 Specification of classification datasets

Dataset	Number of samples		#Attributes	#Classes
	Training	Testing		
Diabetes	576	192	8	2
Ionosphere	246	105	34	2
Iris	100	50	4	3
Glass	150	64	10	6
Landsat	4435	2000	36	6
Wine	125	53	13	3
Segment	147	63	19	7

4.2 Parameters Selection

As previously described, the nodes number in the hidden layer has effect on the performance as Fig. 2 shows. The experiments were carried out under the Boston Housing for regression problem and the Diabetes for classification problem. Moreover, the Root Mean Square Error (RMSE) was regarded as the measure of performance for the regression problem. Similarly, the accuracy was regarded as the measure of performance for the classification problem. Figure 2 also shows that the proposed method gives a better performance at the same node number.

4.3 Performance Comparison for Regression Problems

We evaluated the performance of different datasets for regression problem with four algorithms, including ELM, OP-ELM, ELM-L1 and PE-ELMR. In this section, every experiment was repeated 20 times, and then the average training RMSE and testing RMSE under different datasets were recorded in Table 3. 'Tr' means training, 'Te' means testing and STD is the abbreviation of standard deviation.

As Table 3 shows, both ELM-L1 and PE-ELMR methods perform well in the training RMSE and are superior to the existing ELM and OP-ELM methods, but for the testing RMSE, our method achieves the best performance over the others in the testing RMSE. In addition, the small STD value means the algorithm has the ability of stability. It is obvious that our method has the least STD among the algorithms under all the regression datasets except the 'Sinc' dataset.

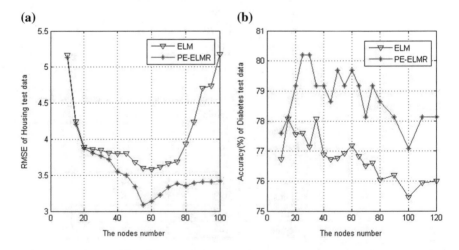

Fig. 2 The performance of regression and classification testing data with different nodes number in hidden layer. **a** RMSE of housing testing data versus the nodes number. **b** Accuracy of diabetes testing data versus the nodes number

Table 3 Performance comparison for regression problems

Dataset	Algorithm	Tr_accuracy	Te_accuracy	STD	#Nodes
Sinc	ELM	0.1158	0.0081	6.0989×10^{-5}	30
	OP-ELM	0.1163	0.0126	5.3541×10^{-4}	**15**
	ELM-L1	0.0187	0.0079	$\mathbf{7.1362 \times 10^{-6}}$	38
	PE-ELMR	**0.0094**	**0.0075**	8.4324×10^{-5}	38
Abalone	ELM	2.4533	2.5546	0.0282	25
	OP-ELM	2.5882	2.5761	0.0523	**20**
	ELM-L1	**2.4153**	2.5009	0.0258	23
	PE-ELMR	2.4351	**2.4924**	**0.0249**	23
Housing	ELM	3.8315	3.6471	0.1866	**50**
	OP-ELM	5.2309	4.0759	0.5895	**50**
	ELM-L1	**3.7325**	3.1339	0.1386	55
	PE-ELMR	3.7839	**3.0809**	**0.1147**	55
Bodyfat	ELM	3.6873	3.7252	0.1203	15
	OP-ELM	4.8098	3.8515	0.0454	15
	ELM-L1	3.8234	3.6621	0.0245	**14**
	PE-ELMR	**3.8232**	**3.6272**	**0.0198**	**14**
Redwine	ELM	0.6028	0.6851	0.0073	50
	OP-ELM	0.6429	0.6953	0.0398	**30**
	ELM-L1	**0.6063**	0.6680	0.0033	48
	PE-ELMR	0.6093	**0.6635**	**0.0032**	48

4.4 Performance Comparison for Classification Problems

The proposed learning structure is appropriate for regression problem, but it must be modified whilst confronting with classification problem, especially multi-classification problem. As the definition $y_i = [y_{i1}, \cdots, y_{im}]^T \in \mathbb{R}^m$, when $m = 1$, it can solve the regression and binary classification problems and when $m > 1$, it can solve the multi-classification problem. Since the multi-classification problem has multiple outputs, we deal with the number m of the outputs is equal to the classes C. Then the coefficients $\beta_c \in \mathbb{R}^L$, $c = 1, 2, \cdots, C$ of each class can be calculated and all the coefficients form a new matrix $B = [\beta_1, \beta_2, \cdots, \beta_C]_{L \times C}$.

It shows that ELM-L1 performs a high testing accuracy nearly as our approach PE-ELMR in Table 4. However, our method can get the least STD value compared with other methods. On some datasets, the hidden number of OP-ELM is less than that of PE-ELMR, but OP-ELM cannot achieve the best performance as our approach does.

Table 4 Performance comparison for classification problems

Dataset	Algorithm	Tr_accuracy	Te_accuracy	STD	#Nodes
Diabetes	ELM	**0.8056**	0.7646	0.0077	50
	OP-ELM	0.7778	**0.8021**	0.0196	**25**
	ELM-L1	0.8038	**0.8021**	0.0057	30
	PE-ELMR	0.7969	**0.8021**	**0.0046**	30
Ionosphere	ELM	0.9557	0.8848	0.0337	50
	OP-ELM	0.9593	0.9143	0.0181	**45**
	ELM-L1	0.9593	0.9333	0.0064	51
	PE-ELMR	**0.9715**	**0.9429**	**0.0113**	51
Iris	ELM	**0.9900**	0.9620	0.0474	**50**
	OP-ELM	0.9800	0.9600	0.0341	30
	ELM-L1	0.9700	**0.9800**	**0.0032**	**21**
	PE-ELMR	0.9700	**0.9800**	**0.0032**	**21**
Glass	ELM	0.7188	0.7438	0.0233	100
	OP-ELM	0.9200	0.7656	0.0204	**20**
	ELM-L1	0.9333	**0.7813**	0.0084	24
	PE-ELMR	**0.9400**	**0.7813**	**0.0083**	24
Landsat	ELM	0.8186	0.7883	0.0028	100
	OP-ELM	0.6489	0.6355	0.0224	85
	ELM-L1	**0.8169**	0.7925	**0.0046**	**80**
	PE-ELMR	0.8147	**0.7945**	0.0054	**80**
Wine	ELM	0.9912	0.8981	0.0239	50
	OP-ELM	0.8880	0.9245	0.0430	15
	ELM-L1	**0.9440**	**0.9811**	0.0348	**8**
	PE-ELMR	**0.9440**	**0.9811**	**0.0215**	**8**
Segment	ELM	0.9918	0.7556	0.0438	100
	OP-ELM	0.8571	0.7143	0.0205	**30**
	ELM-L1	**0.9600**	**0.8413**	0.0056	80
	PE-ELMR	**0.9660**	**0.8413**	**0.0045**	80

5 Conclusions

In this paper, a pruning ensemble model of ELM with $L_{1/2}$ regularizer is proposed, which can handle with over-fitting of the preliminary ELM. Moreover, it can prune the nodes in the hidden layer and select the suitable number of hidden nodes automatically. For the section of ELMR, we replace the original solving method of the output parameter in the preliminary ELM to a minimum squared-error problem with sparse solution by combining ELM with $L_{1/2}$ regularizer. For the section of ensemble model building, we prune the hidden nodes with the ensemble model of ELM, in order to get the required minimum number for good performance. With these two stages of our approach, we manage to solve the regression and classification problems at a better performance. We have also shown through numbers of experiments on the benchmark datasets, that the proposed method is efficient and performs very well in comparison with other methods.

References

1. Huang, G.B., Zhu, Q.Y., Siew, C.K.: Extreme learning machine: theory and applications. Neurocomputing **70**, 489–501 (2006)
2. Huang, G.B., Zhou, H., Ding, X., Zhang, R.: Extreme learning machine for regression and multiclass classification. IEEE Trans. Syst. Man Cybern. Part B Cybern. **42**, 513–528 (2012)
3. Hornik, K., Stinchcombe, M., White, H.: Multilayer feedforward networks are universal approximators. Neural Netw **2**(5), 359–366 (1989)
4. Huang, G.B., Zhu, Q.Y., Siew, C.K.: Extreme learning machine: a new learning scheme of feedforward neural networks. Proc. Int. Joint Conf. Neural Netw. **70**, 25–29 (2004)
5. Zhou, Z.H., Chen, S.F.: Neural network ensemble. Chin. J. Comput. **25**, 1–8 (2002)
6. Zhou, Z.H., Wu, J.X., Jiang, Y.: Genetic algorithm based selective neural network ensemble. In: Proceedings of the 17th International Joint Conference on Artificial Intelligence, vol. 2 (2001)
7. Sun, Z.L., Choi, T.M.: Sales forecasting using extreme learning machine with applications in fashion retailing. Decis. Support Syst. **46**, 411–419 (2008)
8. Liu, N., Wang, H.: Ensemble based extreme learning machine. IEEE Trans. Signal Proc. Lett. **17**(8), 754–757 (2010)
9. Xue, X., Yao, M., Wu, Z., Yang, J.: Genetic ensemble of extreme learning machine. Neurocomputing **129**, 175–184 (2014)
10. Rong, H.J., Ong, Y.S., Tan, A.H., Zhu, Z.: A fast pruned-extreme learning machine for classification problem. Neurocomputing **72**, 359–366 (2008)
11. Miche, Y., Sorjamaa, A., Bas, P., Simula, O.: OP-ELM: optimally pruned extreme learning machine. IEEE Trans. Netw. **21**, 158–162 (2010)
12. Xu, Z.B., Guo, H.L., Wang, Y., Zhang, H.: Representative of L 1/2 Regularization among L q $(0 < q \leq 1)$ regularizations: an experimental study based on phase diagram. Acta Automatica Sin. **38**, 1225–1228 (2012)
13. Zeng, J., Lin, S., Wang, Y., Xu, Z.: L1/2 regularization: convergence of iterative half thresholding algorithm. IEEE Trans. Signal Proc. **62**(9) (2014)
14. Hankeand, M, Hansen, P.C.: Regularization methods for large-scale problem. Surv. Math. Ind. **3** (1993)
15. Argyriou, A., Baldassarre, L., Miccheli, C.A., Pontil, M.: On Sparsity Inducing Regularization Methods for Machine Learning. Empirical Inference, pp. 205–216. Springer, Berlin (2013)
16. Akaike, H.: Information theory and an extension of the maximum likelihood principle. In: Selected Papers of Hirotugu Akaike, pp. 199–213. Springer, New York (1998)
17. Tibshirani, R.: Regression shrinkage and selection via the lasso. J. R. Stat. Soc. Ser. B (Methodol) **58**(1), 267–288 (1996)
18. Kukreja, S.L., Lofberg, J., Brenner, M.J.: A least absolute shrinkage and selection operator (LASSO) for nonlinear system identification (2006)
19. Berkin, B., Fan, A.P., Polimeni, J.R.: Fast quantitative susceptibility mapping with L1 regularization and automatic parameter selection. Magn. Reson. Med. **72**, 1444–1459 (2014)
20. Saunders, C., Gammerman, A., Vovk, V.: Ridge regression learning algorithm in dual variables. In: Proceedings of the 15th International Conference on Machine Learning (ICML-1998). Morgan Kaufmann, Burlington (1998)
21. Koh, K., Kim, S.J., Boyd, S.: l1_ls: Simple Matlab Solver for L1-regularized Least Squares Problems. Available: http://stanford.edu/~boyd/l1_ls/

Evaluating Confidence Intervals for ELM Predictions

Anton Akusok, Yoan Miche, Kaj-Mikael Björk, Rui Nian, Paula Lauren and Amaury Lendasse

Abstract This paper proposes a way of providing more useful and interpretable results for ELM models by adding confidence intervals to predictions. Unlike a usual statistical approach with Mean Squared Error (MSE) that evaluates an average performance of an ELM model over the whole dataset, the proposed method computed particular confidence intervals for each data sample. A confidence for each particular sample makes ELM predictions more intuitive to interpret, and an ELM model more applicable in practice under task-specific requirements. The method shows good results on both toy and a real skin segmentation datasets. On a toy dataset, the predicted confidence intervals accurately represent a variable magnitude noise. On a real dataset, classification with a confidence interval improves the precision at the cost of recall.

Keywords Extreme learning machines · Confidence · Confidence interval · Regression · Image segmentation · Skin segmentation · Classification · Interpretability · Big data

A. Akusok (✉) · A. Lendasse
Department of Mechanical and Industrial Engineering and the Iowa Informatics Initiative,
The University of Iowa, Iowa City, USA
e-mail: anton-akusok@uiowa.edu

A. Lendasse
e-mail: amaury-lendasse@uiowa.edu

Y. Miche
Nokia Solutions and Networks Group, Espoo, Finland

Y. Miche
Aalto University School of Science, 00076 Helsinki, Finland

K.-M. Björk
Arcada University of Applied Sciences, Helsinki, Finland

R. Nian
Ocean University of China, Qingdao, China

P. Lauren
Oakland University, Rochester, USA

© Springer International Publishing Switzerland 2016
J. Cao et al. (eds.), *Proceedings of ELM-2015 Volume 2*,
Proceedings in Adaptation, Learning and Optimization 7,
DOI 10.1007/978-3-319-28373-9_35

413

1 Introduction

Extreme Learning Machines [1–3] (ELM) are fast [4] and robust [5, 6] methods of training feed-forward networks, which have the universal approximation property [7] and have numerous applications in regression [8–10] and classification [11] problems. They are an active research topic with numerous extensions and improvements proposed over the last decade.

ELMs are powerful non-linear methods, but they share one common drawback of non-linear methods in practical applications, which is a non transparency of results (predictions). A prediction made by a linear model from input data is easily explained and interpreted by observing the coefficients at input data features. Results which have an explanation are easier to trust and apply for people outside a Machine Learning field. Non-linear models lack such transparency, so their results are hard to be trusted, and thus non-linear methods (including ELM) are sometimes denied despite a supreme performance compared to linear methods.

This paper proposes a way of providing more useful and interpretable results for ELM models by adding confidence intervals [12–15] to predictions. Unlike a usual statistical approach with Mean Squared Error (MSE) [16] that evaluates an average performance of an ELM model over the whole dataset, the proposed method computed particular confidence intervals for each data sample. These intervals are small for samples on which a model is accurate, and large for samples where a model is unstable and inaccurate. A confidence for each particular sample makes ELM predictions more intuitive to interpret, and an ELM model more applicable in practice under task-specific requirements to precision and recall of predictions.

The next Sect. 2 introduces the method of input-specific confidence intervals. The experimental Sect. 3 presents the examples of confidence intervals on artificially made toy dataset and on a real image segmentation task. In the conclusion, Sect. 4 the method is summarised, and further research directions are discussed.

2 Methodology

Confidence intervals are estimated boundaries of a stochastic output sample for a given input sample and confidence level, in a regression or classification task. They provide a measure of confidence for a prediction result of an ELM. This information is practically important in many ELM applications, and is useful in complex systems which utilize ELM as their part.

A simple way of estimating confidence interval of ELM predictions is to use Mean Squared Error (MSE), which is a variance of error between model predictions and true output values. But this method provides constant confidence intervals for the whole dataset, while predictive performance of ELM may vary depending on the input. More useful confidence intervals are defined in the input space, as described hereafter.

Confidence intervals are estimated from the variance (or standard deviation), however predictions of a single ELM are deterministic. To obtain stochastic predictions, this work considers a family of ELM models. Input weights of these ELMs are randomly sampled, but each model has the same parameters including random weights distribution, projection function and network structure. As ELM is a very fast training method, hundreds to millions of ELMs can be trained in a few minutes (depending on training data and model size), providing adequately precise estimation of outputs distribution of an ELM model family.

2.1 Confidence Interval for Regression

A data set is a limited set of N samples $\{\mathbf{x}_i, \mathbf{t}_i\}$, $i \in [\![1, N]\!]$ which represents an unknown projection function $F : \mathbf{X} \to \mathbf{T}$. An ELM approximates that function F by a smooth function f such as $f(\mathbf{x}_i) = \mathbf{y}_i = \mathbf{t}_i + \epsilon$. The noise ϵ comes from an imperfect approximation of the true projection function, noise in the dataset, and uncertainty of the dataset itself.

An assumption is made that a model prediction $\mathbf{y}_i$ is normally distributed $\mathcal{N}(\mu_i, \sigma_i^2)$ where $\sigma_i = \sigma(\mathbf{x}_i)$ is defined in the input space. The confidence intervals for an input sample $\mathbf{x}_i$ are computed from $\sigma(\mathbf{x}_i)$ at the desired confidence level. However, evaluating $\sigma(\mathbf{x})$ for arbitrary $\mathbf{x}$ is complicated because the dataset input samples do not cover all input space.

In fact, the $\sigma(\mathbf{x})$ needs to be evaluated only for the given input points, not the whole input space. These evaluations are obtained using ELM models, which cover the whole input space (an ELM produces an output for any input sample) and can evaluate $\sigma(\mathbf{x})$ for any given input sample directly.

The standard deviation $\sigma(\mathbf{x}_i)$ is evaluated by training multiple ELMs with the same network structure but different randomly sampled hidden layer weights. The obtained $\sigma(\mathbf{x}_i)$ is influenced by a local data outputs distribution and a model structure $\sigma_1(\mathbf{x}_i) = \sigma^{\text{data}}(\mathbf{x}_i) + \sigma^{\text{model}}(\mathbf{x}_i)$. Unfortunately, the model structure influence is dominant and cannot be removed by training a large number of ELM models, see Fig. 1.

The following method is proposed to remove model influence $\sigma^{\text{model}}(\mathbf{x}_i)$. Each model in the ELM family is trained again, but with a smaller random subset of data samples $\{\mathbf{x}_j, \mathbf{t}_j\}$, $j \in [\![1, M < N]\!]$. The model component of $\sigma(\mathbf{x})$ will be the same because the ELM models are the same, and the data component will increase because with less training samples ELMs have worse fit and larger variance of predictions. The result will be $\sigma_2(\mathbf{x}_i) = (1 + \beta)\sigma^{\text{data}}(\mathbf{x}_i) + \sigma^{\text{model}}(\mathbf{x}_i)$ with some positive $\beta > 0$ (see Fig. 2). The model-independent estimation is obtained as $\sigma(\mathbf{x}_i) = \sigma_2(\mathbf{x}_i) - \sigma_1(\mathbf{x}_i) = \beta\sigma^{\text{data}}(\mathbf{x}_i) \propto \sigma^{\text{data}}(\mathbf{x}_i)$ (see Fig. 3).

The scale α of an estimate $\sigma(\mathbf{x}_i) \propto \sigma^{\text{data}}(\mathbf{x}_i) = \alpha\sigma(\mathbf{x}_i) = \alpha[\sigma_2(\mathbf{x}_i) - \sigma_1(\mathbf{x}_i)]$ is not defined. It is obtained using a validation dataset, and a desired confidence level. For c confidence level, α is adjusted such that $1 - c$ validation samples are outside

Fig. 1 Estimated standard deviation of data (*solid line*) with a family of 10,000 ELM models, and true standard deviation of data (*dotted line*)

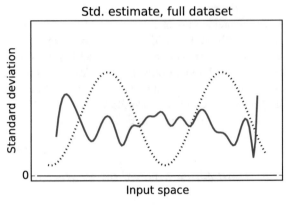

Fig. 2 Estimated standard deviation of data (*solid line*) with a family of 10,000 ELM models, using 70 % random training samples

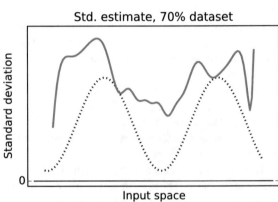

Fig. 3 Estimated standard deviation of data (*solid line*) with a family of 10,000 ELM models, as a difference between ELM model family trained on a full dataset and the same family trained on randomly selected 70 % training data (each)

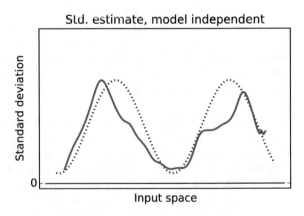

$\mathbf{y}_i \pm \alpha[\sigma_2(\mathbf{x}_i) - \sigma_1(\mathbf{x}_i)]$ interval. The computed α is then used for calculating the confidence intervals for test data samples.

2.2 ELM Confidence Intervals Algorithm

An algorithm of confidence intervals method for ELM is presented on Algorithm 1.

Algorithm 1 ELM Confidence Intervals Algorithm

---

given data sets $(\mathbf{X}, \mathbf{T})_{\text{train}}$, $(\mathbf{X}, \mathbf{T})_{\text{validation}}$ and $\mathbf{X}_{\text{test}}$
for all ELM model m in ELM family **do**
    initialize random input weights of m
    train m with full training set $(\mathbf{X}, \mathbf{T})_{\text{train}}$
    compute $\mathbf{Y}_{1,m}$ from m for validation and test sets
    train m again on a subset of $(\mathbf{X}, \mathbf{T})_{\text{train}}^{\text{subset}}$
    compute $\mathbf{Y}_{2,m}$ from re-trained m for validation and test sets
end for
compute σ_1 from $\mathbf{Y}_{1,m}^{\text{validation}}, m \in$ models
compute σ_2 from $\mathbf{Y}_{2,m}^{\text{validation}}, m \in$ models
compute $\sigma^{\text{validation}} = \sigma_2 - \sigma_1$ for validation set
compute α scaling coefficient for desired percentage on a validation set
compute test set confidence intervals $\sigma^{\text{test}} = \alpha(\sigma_2 - \sigma_1)$
report $\mathbf{Y}^{\text{test}} = \frac{1}{|\text{models}|} \sum_m \mathbf{Y}_{1,m}$ and σ^{test}

---

3 Experimental Results

3.1 Artificial Dataset

An artificial dataset (Fig. 4a, b) has one-dimensional input and target data, for the ease of visualization. The data is a sum of two sine functions. Noise has been added to data samples, with varying magnitude. Because the noise is added artificially, the exact 1σ and 2σ confidence boundaries are known, and can be compared with the estimated boundaries by the proposed method. The method uses 1000 training, 9000 validation and 1000 test samples.

The confidence boundaries are shown on Fig. 5a, b for variable magnitude noise and Fig. 5c, d for constant magnitude noise. All experiments train 1000 different OP-ELM models with 25 hidden neurons, which takes between 37 and 40 s on 1.4 GHz dual-core laptop using a toolbox from [17].

As a performance measure, an integral of absolute difference between two boundaries is evaluated, using the test samples. This integral is divided by an integral of the true boundary.

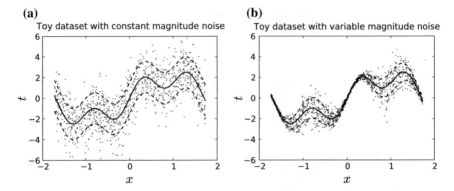

Fig. 4 Artificial datasets with added variable and constant magnitude noise. Dots are training samples, *solid line* is the true function, and *dash lines* show 1σ confidence boundaries. **a** Constant magnitude noise. **b** Variable magnitude noise

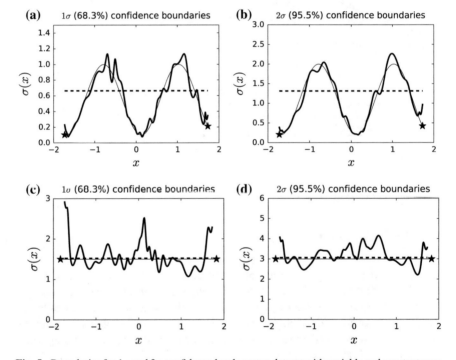

Fig. 5 Boundaries for 1σ and 2σ confidence levels on toy dataset with variable and constant magnitude noise. True shape of noise magnitude is shown by a *thin curve* ending with stars, ELM-estimated boundary by a *thick curve*, and MSE boundary by a *dash line*. **a** Variable noise, 1σ boundary. **b** Variable noise, 2σ boundary. **c** Constant noise, 1σ boundary. **d** Constant noise, 2σ boundary

(a)   **(b)**

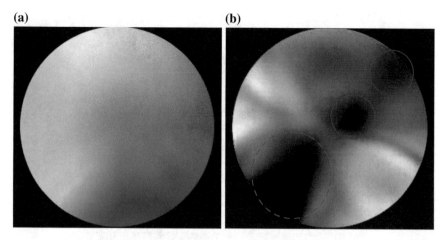

Fig. 6 Skin color confidence interval map, shown on a color wheel. Small confidence intervals (*black*) show high confidence regions—for skin (drawn by *red solid circles*) and non-skin (drawn by a *blue dashed circle*). Large confidence intervals (*white*) show low confidence for skin/non-skin classification of that color

$$\delta = \frac{\sum_{\mathbf{x}_{\text{test}}} |\sigma(\mathbf{x}) - \hat{\sigma}(\mathbf{x})|}{\sum_{\mathbf{x}_{\text{test}}} \sigma(\mathbf{x})} \times 100\% \tag{1}$$

With variable noise, delta of an ELM-estimated boundary is $\delta^{\text{ELM}} = 9\% \ldots 11\%$, while the delta of MSE boundary is $\delta^{\text{MSE}} = 50\%$. For constant noise, $\delta^{\text{ELM}} = 12\% \ldots 16\%$ and an MSE provides almost perfect boundary estimation with $\delta^{\text{MSE}} < 2\%$. ELM does not provide a smooth boundary in case of a constant noise, because while noise is being constant, the data itself is highly varying which is reflected by an ELM estimation. Also, ELM-estimation is not accurate at the edges of a dataset.

3.2 Skin Color Dataset

Confidence intervals for ELM predictions are tested on a Face/Skin Detection dataset [18], a useful benchmark [19] Big Data dataset. It includes 4000 photos of people under various real-world conditions, as well as manually created masks for faces and skin. The dataset is split into 2000 training and 2000 test images.

Confidence intervals are estimated for a simple task of classifying a pixel into skin/non-skin, based on its RGB color. 500 random skin and non-skin pixels are taken from each training image; 500 training set images are used for training and 1500 for validation. All pixels of a single test set image are added as test samples, for which the classification and confidence intervals are computed. A total of 1000 ELM models are built for confidence interval estimation.

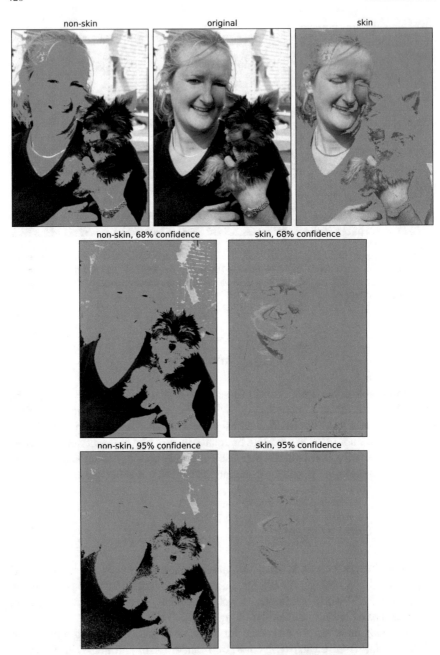

Fig. 7 Skin segmentation by pixel classification with ELM based on their RGB color values, using a threshold. *Top row* original image and thresholded skin/non-skin pixels. *Middle row* thresholded pixels with predicted values larger than their 68 % confidence intervals. *Bottom row* thresholded pixels with predicted values larger than their 95 % confidence intervals. Thresholding with confidence interval improves the precision at a cost of recall

The results are shown on Fig. 7. An original image is split into skin and non-skin with a good accuracy using a simple threshold, however some parts are misclassified. Thresholding pixels with more than 68 % confidence (roughly 1σ) significantly reduces the recall, but provides almost perfect precision. Increasing a threshold to 95 % confidence (or 2σ) provides perfect precision at a cost of even smaller recall.

Another experiment is performed in a similar setup, but a color wheel image is used instead of a test picture. The confidence intervals for different colors are shown on Fig. 6a, b. Confidence intervals are organized according to color values—they are small for skin colors and clearly non-skin colors, and are large in between these two. An MSE confidence interval would be constant over all colors in a color wheel, providing less accurate results.

4 Conclusion

A method for evaluating input-dependent confidence intervals for ELM model is proposed in the paper. It is based on estimation of a standard deviation of output (target) noise for a fixed ELM model trained on a different subsets of the training set. Then the confidence intervals are scaled using a validation set, and evaluated for the given test input samples. The method can compute confidence intervals even for a high dimensional inputs, because they are computed for the given test samples and not for the whole input space (Fig. 7).

The method shows good results on both toy and real datasets. On a toy dataset, the predicted confidence intervals accurately represent a variable magnitude noise. On a real dataset, classification with a confidence interval improves the precision at the cost of recall.

Future work on confidence intervals for ELM will be focused on computing confidence intervals for classification tasks, in a classification-specific way without value threshold. Other directions are creating more smooth confidence intervals for different models, and formulating rules for choosing parameters of the proposed method that ensure good and stable results in any application scenarios. Targeting Big Data is also an important future works topic, as the confidence interval estimations will add more value to ELM ability of handle Big Data.

References

1. Huang, G.B., Zhu, Q.Y., Siew, C.K.: Extreme learning machine: a new learning scheme of feedforward neural networks. In: Proceedings of the 2004 IEEE International Joint Conference on Neural Networks, vol. 2, pp. 985–990, IEEE (2004)
2. Huang, G.B., Zhou, H., Ding, X., Zhang, R.: Extreme learning machine for regression and multiclass classification. IEEE Trans. Syst. Man Cybern. Part B: Cybern. **42**(2), 513–529 (2012)
3. Huang, G.B.: What are Extreme learning machines? Filling the gap between Frank Rosenblatts Dream and John von Neumanns Puzzle. Cogn. Comput. **7**(3), 263–278 (2015)

4. van Heeswijk, M., Miche, Y., Oja, E., Lendasse, A.: Gpu-accelerated and parallelized ELM ensembles for large-scale regression. Neurocomputing **74**(16), 2430–2437 (2011)
5. Miche, Y., Sorjamaa, A., Bas, P., Simula, O., Jutten, C., Lendasse, A.: OP-ELM: optimally pruned extreme learning machine. IEEE Trans. Neural Netw. **21**(1), 158–162 (2010)
6. Miche, Y., van Heeswijk, M., Bas, P., Simula, O., Lendasse, A.: TROP-ELM: a double-regularized ELM using LARS and Tikhonov regularization. Neurocomputing **74**(16), 2413–2421 (2011)
7. Huang, G.B., Zhu, Q.Y., Siew, C.K.: Extreme learning machine: theory and applications. Neurocomputing **70**(1), 489–501 (2006)
8. Yu, Q., Miche, Y., Eirola, E., van Heeswijk, M., Séverin, E., Lendasse, A.: Regularized extreme learning machine for regression with missing data. Neurocomputing **102**, 45–51 (2013)
9. Miche, Y., Bas, P., Jutten, C., Simula, O., Lendasse, A.: A methodology for building regression models using extreme learning machine: OP-ELM. In: ESANN 2008 Proceedings
10. Cambria, E., et al.: Extreme learning machines. IEEE Intell. Syst. **28**(6), 30–59 (2013)
11. Akusok, A., Miche, Y., Karhunen, J., Björk, K.M., Nian, R., Lendasse, A.: Arbitrary category classification of websites based on image content. IEEE Comput. Intell. Mag. **10**(2), 30–41 (2015)
12. Shang, Z., He, J.: Confidence-weighted extreme learning machine for regression problems. Neurocomputing **148**, 544–550 (2015)
13. Lendasse, A., Ji, Y., Reyhani, N., Verleysen, M.: LS-SVM Hyperparameter Selection with a Nonparametric Noise Estimator. In: Artificial Neural Networks: Formal Models and Their Applications ICANN 2005. Volume 3697 of Lecture Notes in Computer Science. Springer, Berlin Heidelberg, pp. 625–630 (2005)
14. Pouzols, F.M., Lendasse, A., Barros, A.B.: Autoregressive time series prediction by means of fuzzy inference systems using nonparametric residual variance estimation. Fuzzy Sets Sys. **161**(4), 471–497 (2010). Theme: Forecasting, Classification, and Learning
15. Guillén, A., Sovilj, D., Lendasse, A., Mateo, F., Rojas, I.: Minimising the delta test for variable selection in regression problems. Int. J. High Perform. Syst. Archit. **1**(4), 269–281 (2008)
16. Bishop, C.M.: Pattern Recognition and Machine Learning. Volume 4 of Information science and statistics. Springer (2006)
17. Akusok, A., Björk, K.M., Miché, Y., Lendasse, A.: High-performance extreme learning machines: a complete toolbox for big data applications. IEEE Access **3**, 1011–1025 (2015)
18. Phung, S.L., Bouzerdoum, A., Chai D.: Skin segmentation using color pixel classification: analysis and comparison. IEEE Trans. Pattern Anal. Mach. Intell. **27**(1), 148–154 (2005)
19. Swaney, C., Akusok, A., Björk, K.M., Miche, Y., Lendasse, A.: Efficient skin segmentation via neural networks: HP-ELM and BD-SOM. **53**, 400–409 (2015)

Real-Time Driver Fatigue Detection Based on ELM

Hengyu Liu, Tiancheng Zhang, Haibin Xie, Hongbiao Chen and Fangfang Li

Abstract Driver fatigue is a serious road safety issue that results in thousands of road crashes every year. Image-based fatigue monitoring is one of the most important methods of avoiding fatigue-related accidents. In this paper, a vision-based real-time driver fatigue detection system based on ELM is proposed. The system has three main stages. The first stage performs facial features localization and tracking, by using the Viola–Jones face detector and the KLT algorithm. The second stage is the judgement of facial and fatigue status, applying twice ELM with an extremely fast learning speed. The last one is online learning, which can continuously improve ELM accuracy according to the user's feedback. Multiple facial features (including the movement of eyes, head and mouth) are used to comprehensively assess the driver vigilance state. In comparison to backpropagation (BP), the experimental results showed that applying ELM has a better performance with much faster training speed.

Keywords Extreme learning machine (ELM) · Fatigue detection · Driver monitoring system · Real-time learning

The work is partially supported by the National Narural Science Foundation of China (No. 61272180, 61202086, 61272179, 61472071) and the Fundamental Research Funds for the Central Universities (No. N140404013).

H. Liu · T. Zhang (✉) · H. Xie · H. Chen · F. Li
College of Information Science and Engineering, Northeastern University, Shenyang 110819, Liaoning, China
e-mail: tczhang@mail.neu.edu.cn

H. Xie
e-mail: l372511387@gmail.com

1 Introduction

Driver fatigue has become an important factor of resulting in thousands of traffic accidents. According to the statistical data from the NHTSA, about 100,000 road accidents involve driver fatigue and cause over 1500 fatalities and 71,000 injuries each year [1]. Driver fatigue detection system can detect a driver's level of vigilance and alert the driver promptly before drowsiness, which is beneficial to prevent fatigue-related crashes. Therefore, it's essential to develop a powerful monitoring system.

The information that is usually used as the basis for existing driver fatigue detection systems by researchers can be divided into the following types.

- Information on the driver's physiological status [2–6].
- Information on vehicle performance [7].
- Information on video images of the driver monitoring [8–11].

Driver drowsiness detection systems that identify the driver fatigue through driver's physiological signals such as brainwaves can achieve high detection accuracy. And systems through vehicle performance parameters such as the movements of the vehicle straying from the lane markers on the roadway can acquire steady detection accuracy. But they all produce high costs and need bulky equipment. In contrast with the previous two methods, systems through facial features including eye movement (e.g., eyelid closures) [8–11], head movement (e.g., nodding) [12] and mouth movement (e.g., yawning) [13], which can reflect driver fatigue directly, are high-efficiency and non-intrusive methods. But existing image-based operator fatigue monitoring systems mainly work by capturing single facial feature, which reduces system performance in a complex environment.

Extreme learning machine (ELM) has been widely applied in many fields, especially in image recognition, as ELM has extremely fast learning speed and excellent generalization performance. Using ELM on image-based fatigue-monitoring systems can correct the miscalculation of driver's vigilance state in real time, which could significantly increase the detection accuracy.

In this paper, ELM is used to infer the driver's state of vigilance comprehensively from multiple facial features. We propose an ELM-based online learning algorithm that is adaptable to various situations of different operators, which achieved a high degree of accuracy and dramatically enhance the training speed.

2 Extreme Learning Machine

ELM is a single-hidden layer feedforward neural networks (SLFNs) learning algorithm [14]. ELM only need to set the number of hidden nodes of the network, the algorithm execution process does not need to adjust the input weights and hidden element of bias, and only optimal solution, so it has the advantages of fast learning speed and good generalization performance.

Suppose we train an ELM with n input neurons, m output neurons, L hidden neurons and an activation function $g(x)$.

The train data is (x, y), where

$$x = [x_1, x_2, x_3, x_4, \ldots, x_N] \text{ and } y = [y_1, y_2, y_3, y_4, \ldots, y_N]$$

$$x_i = [x_{i1}, x_{i2}, x_{i3}, x_{i4}, \ldots, x_{iN}] \text{ and } y = [y_{i1}, y_{i2}, y_{i3}, y_{i4}, \ldots, y_{iN}] \tag{1}$$

The weight vector connecting the input nodes and the ith hidden node is

$$w_i = [w_{i1}, w_{i2}, w_{i3}, \ldots, w_{in}]. \tag{2}$$

And define $w = [w_1, w_2, \ldots, w_L]^T$.

The weight vector linking the ith hidden node to the output nodes is

$$\beta_i = [\beta_{i1}, \beta_{i2}, \beta_{i3}, \ldots, \beta_{in}]. \tag{3}$$

The offset of the ith hidden node which is randomly generated is

$$b = [b_1, b_2, \ldots, b_n]. \tag{4}$$

The ELM can be mathematically modeled as the matrix equation with forms, which is

$$H\beta = Y, \text{ where} \tag{5}$$

$$
H(w_1, w_2, w_3, \ldots, w_L; b_1, b_2, b_3, \ldots, b_L; x_1, x_2, x_3, \ldots, x_L)
$$
$$
= \begin{bmatrix}
g(w_1 * x_1 + b_1) & g(w_2 * x_1 + b_2) & \cdots & g(w_L * x_1 + b_L) \\
g(w_1 * x_2 + b_1) & g(w_2 * x_2 + b_2) & \cdots & g(w_L * x_2 + b_L) \\
\cdots & \cdots & \cdots & \cdots \\
g(w_1 * x_N + b_1) & g(w_2 * x_N + b_2) & \cdots & g(w_L * x_N + b_L)
\end{bmatrix}
$$

$$\beta = [\beta_1^T, \beta_2^T, \ldots, \beta_L^T]_{M*L}^T \text{ and } Y = [Y_1^T, Y_2^T, \ldots, Y_L^T]_{M*N}^T. \tag{6}$$

So we have the model as shown below in Fig. 1.

When we assign random w and b, the smallest norm least squares solution of the above linear system is

$$\beta = H^{-1}Y, \text{ where} \tag{7}$$

H^{-1} is the Moore–Penrose generalized inverse of matrix H.

Algorithm 1: ELM training

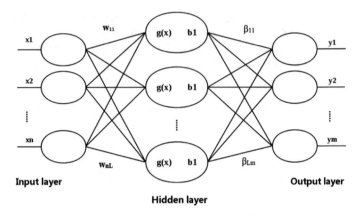

Fig. 1 The model of ELM

1 Determine the number of hidden neurons **L**, generate the
   hidden node parameters w and b randomly;
2 Choose a function $g(x)$ as the activation function which is
   infinitely differentiable in any interval;
3 Calculate the hidden layer output matrix **H**;
4 Calculate the output weight vector $\beta = H^{-1}Y$.

3   Fatigue Detection Based on ELM

3.1   An Overview of Framework

In order to get higher accuracy rate, we present a comprehensive fatigue detection
system that could determine the user's status by face, mouth and eyes feature, and
could optimize itself according to the user's feedback by online learning.

Figure 2 shows a general system framework, which can be divided into three
parts:

- Locate and track the face and facial feature.
- Judge the status of eyes, mouth and face by using ELM, and then use ELM
  again to determine the fatigue status on the basis of the previous ELM output.
- Adjust ELM itself in real time according to the user's feedback.

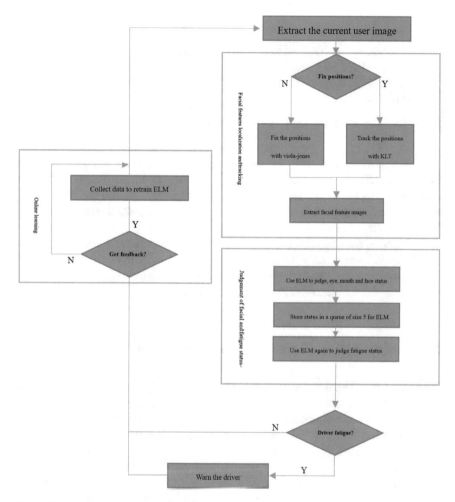

Fig. 2 The general framework of the system

3.2 Locate and Track the Face and Facial Feature

In order to judge the status of face and facial feature, we need to locate and track their position. And we adopt the Viola–Jones detector to detect face and facial features. Viola–Jones detector has three core parts: Haar-like features, Adaboost classifier and Cascade classifier. To calculate the Haar-like features quickly, it uses an intermediate representation for the image which is called integral image. What's more, it uses Cascade classifier to connect several Adaboost classifiers together where each of the Adaboost classifiers has outstanding ability in feature selection. A negative outcome from any classifier will result in an immediate rejection.The effect is shown in Fig. 3.

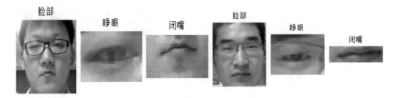

Fig. 3 The effect of locating and tracking the face and facial feature

As we need to deal with the video in real time, the time complexity of the Viola–Jones algorithm is unacceptable. So the Kanade–Lucas–Tomasi (KLT) algorithm is adopted to do most of the feature tracking. We use the Viola–Jones algorithm to locate the first set of features and then use the KLT algorithm for the feature tracking.

KLT makes use of spatial intensity information to search the best match positon. Briefly, good features are located by examining the minimum eigenvalue of each 2 by 2 gradient matrix, and features are tracked using a Newton–Raphson method of minimizing the difference between the two windows.

3.3 Judgement of Features and Fatigue Status

Compression of the feature vector In order to judge the user's fatigue status, we need to further determine the status of face, mouth, and eyes. We treat each pixel in the picture as an one-dimensional feature in order to determine the status. For these high dimension features, we use principal component analysis method to reduce variables to improve process speed.

Principal Component Analysis (PCA), is a statistical procedure that uses an orthogonal transformation to convert a set of observations of possibly correlated variables into a set of values of linearly uncorrelated variables called principal components.

The main target of PCA is to reduce feature dimension under the principle of minimum loss. In other words, we use lesser number of variables (600) than original data (3000) by using linear transformation while guarantee the minimum data information.

Judging characteristic position state by using ELM* Orthogonal projection method can be used efficiently in the ELM [15], If HH^T is a non-singular matrix, $H^T = H^T(HH^T)^{-1}$. According to the ridge regression theory, adding appropriate regularization term $\frac{1}{\lambda}$ to HH^T or H^TH oblique diagonal will make the final classifier more stable and have better generalization performance [15]. In large data applications, we can easily see that $L \ll N$. Therefore, the size of H^TH is much smaller than HH^T. We can get the following formula:

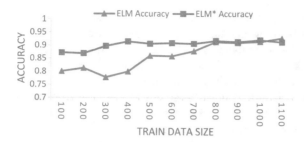

Fig. 4 The comparsion of accuracy between ELM* and ELM

$$\beta = \left(\frac{I}{\lambda} + H^T H\right)^{-1} H^T Y \tag{8}$$

The output function of the ELM will be

$$f(x) = h(x)\beta = h(x)\left(\frac{I}{\lambda} + H^T H\right)^{-1} H^T Y. \tag{9}$$

As shown in Fig. 4, ELM* performs better than ELM in generalization ability, accuracy and stability.

Fatigue judgment After the above steps, we are able to get the state of eyes, mouth and face at any time. However, the fatigue status cannot be judged only by a certain moment. We need to determine the status in continuous period of time. So we collect five consecutive frames (including the current frame) to judge the user's fatigue status.

The first ELM will generate 6 features (eyes, mouth and face each have 2 features) for each frame. So we have a total of 5 * 6 = 30 dimension features to judge fatigue status at a certain moment.

3.4 Online Learning

We can fulfill the online learning according to the feedback of users since ELM has the advantage of high learning speed. As the detection system adopts two layers of ELM (The first layer is used to determine the status of eyes, mouth and face, the second uses the output from the first layer to determine the fatigue status), misjudgment could be caused by the first layer or the second layer as follows.

1. The training set of first layer has no relevant data, which results in the identical output for both true and false data. In this case, we will collect new data for the user in order to retrain the first layer of ELM. When a new user uses the detection system, we will ask him/her to help the system collect new training data in order to achieve higher accuracy.

2. Due to the data in the critical range of true and false state, the detection system misjudges the fatigue status. For this case, we corrected it by training the second layer of ELM. When the system misjudges the fatigue state, the user feeds it back to the system. Then, the detection system records the wrong data and adds the wrong data into the specified queue. The second layer of ELM will retrain if misjudgments reach a certain amount or a certain time since the last training.

4 Performance Evaluation

4.1 Experimental Settings

We collected over 4000 volunteers' images, which were captured by the proposed detection system using a camera under good illumination conditions. The volunteers are Chinese between the ages of 20 and 40. For each image, we use the Viola/Jones face detector and the KLT Algorithm to extract the face, eyes and mouth images.

4.2 Evaluation Indexes

The proposed driver fatigue detection system uses ELM twice, which can be replaced by Back-Propagation Neural Networks. So it's necessary to use some objective evaluation indexes to evaluate systems' capabilities. We use four indexes—accuracy, precision, recall and training time.

The positive examples are fatigue state and the negative examples are non-fatigue state. Accuracy describes the percentage of true results (both true positives and true negatives) among the total number of examples examined.

Precision measures the percentage of positive results in diagnostic examples that are true positive. Recall represents the proportion of the positive examples that are correctly identified as fatigue state. Training time denotes the required time of the training classifier. Table 1 shows the relevant formulas to compute these indexes.

Table 1 Evaluation indexes

Indexes	Formulas
Accuracy	(TP + TN)/(TP + TN + FP + FN)
Recall	TP/(TP + FN)
Precision	TP/(TP + FP)
Training	The required time of training classifier

The meaning of parameters in the above table are:

(1) TP: The number of correctly identified images which have positive examples.
(2) TN: The number of correctly identified images which have negative examples.
(3) FP: The number of inaccurately classified images that have negative examples.
(4) FN: The number of inaccurately classified images which have positive examples.

When other parameters are at the same level, the less training time the more efficient it is.

4.3 Parameters Optimization

In the experiments, the authors apply ELM and BP separately as the detector of driver fatigue detection system so as to compare their capabilities. But first we need to determine and optimize the parameters that ELM and BP used, including the number of hidden neurons L and the regularization coefficient λ. We randomly selected 70 % of the data as the training set, and the remainder are regarded as the test set. The training set and test set were both used in the process of determining and optimizing the parameters.

The process of determining and optimizing the number of hidden nodes L and the regularization coefficient λ has the following two steps. First, we obtain a variety of L at a fixed interval values within a certain scope and do the same thing for the regularization coefficient λ, so we have multiple combinations (L, λ). Then we train and test by ELM using different combinations (L, λ), obtain the accuracies by every combination (L, λ) within the training set and the test set, and do the same thing to find the optimized parameters of BP. The above result is shown in Fig. 5.

As Fig. 5a shows, the optimal detection performance of ELM emerged when L = 290 and $\lambda = 1$. And as Fig. 5b shows, the optimal detection performance of BP emerged when L = 300 and $\lambda = 5$. Therefore, these optimized parameters will be used in the following steps of the experiment.

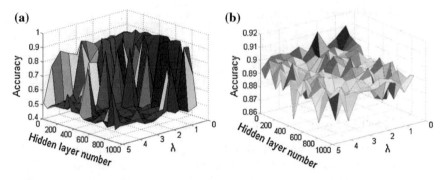

Fig. 5 Parameters determination (**a**) ELM and (**b**) BP

4.4 Experimental Results

The results of facial feature state judgment After we made a comparison of performance between ELM and BP using the same sets of different train data sizes, we got the results as shown below. The degree of accuracy, precision, recall and training time of ELM and BP were compared independently. For each evaluation index, the facial features (eyes, face and mouth) were compared independently.

As shown above, most evaluation indexes of BP go near to ELM's (including accuracy, precision and recall) as shown in Figs. 6, 7 and 8, which both perform well. But it is obvious that the training time of ELM is significantly less than that of BP as shown in Fig. 9, which means ELM has better abilities than BP.

Fig. 6 The accuracy of each facial feature

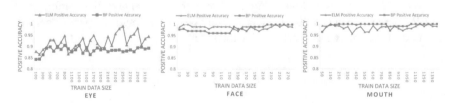

Fig. 7 The precision of each facial feature

Fig. 8 The recall of each facial feature

Fig. 9 The training time of each facial feature

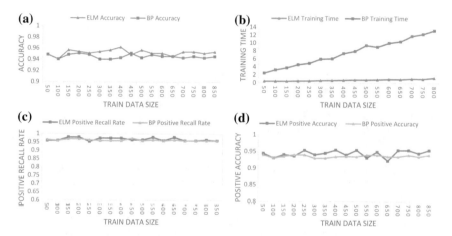

Fig. 10 Performance of ELM and BP

The results of comprehensive fatigue judgment As Fig. 10 shows, ELM and BP are both able to detect the fatigue state accurately, but the training speed of ELM is much better than BP, which enable ELM to study real-time driver's feedback online so as to optimize the system in time for higher adaptability.

5 Conclusions

This paper presents a driver fatigue detection system based on extreme learning machine. The system adopts the Viola/Jones face detector and the KLT Algorithm to detect and track the images of the driver's face, eyes and mouth. Then we use ELM to extract the fatigue features from those images. Finally, ELM is used again to judge the driver's level of vigilance based on the features.

The results of the experiments showed that this system is more accurate in fatigue detection and more adaptable in a complex environment compared with other existing systems as it synthesizes multiple facial features. Meanwhile, this system apply ELM twice with an extremely learning speed to studying in an online fashion through optimizing the parameters continuously from the driver's feedback, which enables the system to fulfill the job in real time.

In the future work we will synthesize more elements to improve fatigue detection accuracy, such as the driver physiological data and vehicle performance parameters.

References

1. Ji, Q., Zhu, Z., Lan, P.: Real-time nonintrusive monitoring and prediction of driver fatigue. Veh. Technol. IEEE Trans. **53**(4), 1052–1068 (2004)
2. Jap, B.T., Lal, S., Fischer, P., Bekiaris, E.: Using EEG spectral components to assess algorithms for detecting fatigue. Expert Syst. Appl. **36**(2), 2352–2359 (2009)
3. Lal, S.K., Craig, A., Boord, P., Kirkup, L., Nguyen, H.: Development of an algorithm for an EEG-based driver fatigue countermeasure. J. Saf. Res. **34**(3), 321–328 (2003)
4. Kar, S., Bhagat, M., Routray, A.: EEG signal analysis for the assessment and quantification of driver's fatigue. Transp. Res. Part F Traffic Psychol. Behav. **13**(5), 297–306 (2010)
5. Chang, B.C., Lim, J.E., Kim, H.J., Seo, B.H.: A study of classification of the level of sleepiness for the drowsy driving prevention. In: Annual Conference on SICE, 2007. IEEE, NJ, pp. 3084–3089
6. Redmond, S., Heneghan, C.: Electrocardiogram-based automatic sleep staging in sleep disordered breathing. In: Computers in Cardiology, 2003. IEEE, NJ, pp. 609–612
7. Dong, Y., Hu, Z., Uchimura, K., Murayama, N.: Driver inattention monitoring system for intelligent vehicles: a review. Intell. Transp. Syst. IEEE Trans. **12**(2), 596–614 (2011)
8. Horng, W.B., Chen, C.Y., Chang, Y., Fan, C.H.: Driver fatigue detection based on eye tracking and dynamk, template matching. In: IEEE International Conference on Networking, Sensing and Control 2004, vol 1. IEEE, NJ, pp. 7–12
9. Dong, W., Wu, X.: Fatigue detection based on the distance of eyelid. In: Proceedings of 2005 IEEE International Workshop on VLSI Design and Video Technology. IEEE, NJ, pp. 365–368
10. Devi, M.S., Bajaj, P.R.: Driver fatigue detection based on eye tracking. In: ICETET'08. First International Conference on Emerging Trends in Engineering and Technology, 2008. IEEE, NJ, pp. 649–652
11. Zhang, Z., Zhang, J.S.: Driver fatigue detection based intelligent vehicle control. In: ICPR 2006. 18th International Conference on Pattern Recognition, 2006, vol 2. IEEE, NJ, pp. 1262–1265
12. Popieul, J.C., Simon, P., Loslever, P.: Using driver's head movements evolution as a drowsiness indicator. In: Proceedings of IEEE Conference on Intelligent Vehicles Symposium, 2003. IEEE, NJ, pp. 616–621

13. Saradadevi, M., Bajaj, P.: Driver fatigue detection using mouth and yawning analysis. Int. J. Comput. Sci. Netw. Secur. **8**(6), 183–188 (2008)
14. Huang, G.B., Zhu, Q.Y., Siew, C.K.: Extreme learning machine: theory and applications. Neurocomputing **70**(1), 489–501 (2006)
15. Huang, G.B., Zhou, H., Ding, X., Zhang, R.: Extreme learning machine for regression and multiclass classification. Syst. Man Cybern. Part B Cybern. IEEE Trans. **42**(2), 513–529 (2012)

A High Speed Multi-label Classifier Based on Extreme Learning Machines

Meng Joo Er, Rajasekar Venkatesan and Ning Wang

Abstract In this paper a high speed neural network classifier based on extreme learning machines for multi-label classification problem is proposed and discussed. Multi-label classification is a superset of traditional binary and multi-class classification problems. The proposed work extends the extreme learning machine technique to adapt to the multi-label problems. As opposed to the single-label problem, both the number of labels the sample belongs to, and each of those target labels are to be identified for multi-label classification resulting in increased complexity. The proposed high speed multi-label classifier is applied to six benchmark datasets comprising of different application areas such as multimedia, text and biology. The training time and testing time of the classifier are compared with those of the state-of-the-arts methods. Experimental studies show that for all the six datasets, our proposed technique have faster execution speed and better performance, thereby outperforming all the existing multi-label classification methods.

Keywords Classification · Extreme learning machine · High-speed · Multi-label

1 Introduction

In recent years, the problem of multi-label classification is gaining much importance motivated by increasing application areas such as text categorization [1–5], marketing, music categorization, emotion, genomics, medical diagnosis [6], image

International Conference on Extreme Learning Machines, 2015.

M.J. Er · R. Venkatesan (✉)
Nanyang Technological University, Nanyang, Singapore
e-mail: raja0046@e.ntu.edu.sg

M.J. Er · N. Wang
Dalian Maritime University, Dalian, China

and video categorization, etc. Recent realization of the omnipresence of multi-label prediction tasks in real world problems has drawn increased research attention [7].

Classification in machine learning is defined as "Given a set of training examples composed of pairs $\{x_i, y_i\}$, find a function $f(x)$ that maps each attribute vector x_i to its associated class y_i, $i = 1, 2,\ldots, n$, where n is the total number of training samples" [8]. These classification problems are called single-label classification. Single-label classification problems involve mapping each of the input vectors to its unique target class from a pool of target classes. However, there are several classification problems in which the target classes are not mutually exclusive and the input samples belong to more than one target class. These problems cannot be classified using single-label classification thus resulting in the development of several multi-label classifiers to mitigate this limitation. By the recent advancements in technology, the application areas of multi-label classifiers spread across various domains such as text categorization, bioinformatics [9, 10], medical diagnosis, scene classification [11, 12], map labeling [13], multimedia, biology, music categorization, genomics, emotion, image and video categorization and so on. Several classifiers are developed to address the multi-label problem and are available in the literature. Multi-label problems are more difficult and more complex compared to single-label problems due to its generality [14]. In this paper, we propose a high-speed multi-label classifier based on extreme learning machines (ELM). The proposed ELM-based approach outperforms all existing multi-label classifiers with respect to training time and testing time and other performance metrics.

The rest of the paper is organized as follows. A brief overview of different types of multi-label classifiers available in the literature is discussed in Sect. 2. Section 3 describes the proposed approach for multi-label problems. Different benchmark metrics for multi-label datasets and experimentation specifications are discussed in Sect. 4. In Sect. 5, a comparative study of the proposed method with existing methods and related discussions are carried out. Finally, concluding remarks are given in Sect. 6.

2   Multi-label Classifier

The definition for multi-label learning as given by [15] is; "Given a training set, $S = (x_i, y_i)$, $1 \leq i \leq n$, consisting of n training instances, $(x_i \in X, y_i \in Y)$ drawn from an unknown distribution D, the goal of multi-label learning is to produce a multi-label classifier $h:X \to Y$ that optimizes some specific evaluation function or loss function".

Let p_i be the probability that the input sample is assigned to ith class from a pool of M target classes. For single-label classification such as binary and multi-class classification the following equality condition holds true.

$$\sum p_i = 1 \tag{1}$$

This equality does not hold for multi-label problems as each sample may have more than one target class. Also, it can be seen that the binary classification problems, the multi-class problems and ordinal regression problems are specific instances of the multi-label problems with the number of labels corresponding to each data sample restricted to 1 [16].

The multi-label learning problem can be summarized as follows:

- There exists an input space that contains tuples (features or attributes) of size D of different data types such as Boolean, discrete or continuous. $x_i \in X$, $x_i = (x_{i1}, x_{i2},...x_{iD})$.
- A label space of tuple size M exists which is given as, $L = \{\zeta_1, \zeta_2,..., \zeta_M\}$.
- Each data sample is given as a pair of tuples (input space and label space respectively). $\{(x_i, y_i) \mid x_i \in X, y_i \in Y, Y \subseteq L, 1 \le i \le N\}$ where N is the number of training samples.
- A training model that maps the input tuple to the output tuple with high speed, high accuracy and less complexity.

Several approaches for solving multi-label problem are available in the literature. Earlier categorization of the multi-label (ML) methods [17] classify the methods into two categories, namely, Problem Transformation (PT) and Algorithm Adaptation (AA) methods. This categorization is extended to include a third category of methods by Gjorgji Madjarov et al. [18] called Ensemble methods (EN). Several review articles are available in the literature that describe various methods available for multi-label classification [7, 8, 15, 17, 18]. As adapted from [18], an overview of multi-label methods available in the literature is given in Fig. 1.

Based on the machine learning algorithm used, the multi-label techniques can be categorized as shown in Fig. 2, adapted from [18]. This paper proposes a high speed multi-label learning technique based on ELM, which outperforms all the existing techniques based on speed and performance.

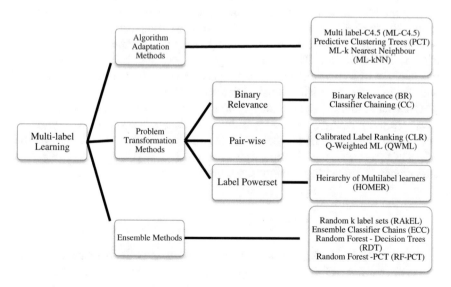

Fig. 1 Classification of multi-label methods

Fig. 2 Machine learning
algorithms for multi-label
problems

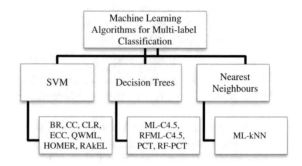

3   Proposed Approach

The extreme learning machine is a learning technique that operates on a single-layer feedforward neural network. The key advantage of the ELM over the traditional backpropagation (BP) neural network is that it has the smallest number of parameters to be adjusted and it can be trained with very high speed. The traditional BP network needs to be initialized and several parameters tuned and improper selection of which can result in local optima. On the other hand, in ELM, the initial weights and the hidden layer bias can be selected at random and the network can be trained for the output weights in order to perform the classification [19–22]. The key steps in extending the ELM to multi-label problems is in the pre-processing and post-processing of data. In multi-label problems, each input sample may belong to one or more samples. The number of labels an input sample belongs to is not previously known. Therefore, both the number of labels and the target labels are to be identified for the test input samples and also the degree of multi-labelness varies among different datasets. This results in increased complexity of the multi-label problem resulting in much longer training and testing time of the multi-label classification technique. The proposed algorithm exploits the inherent high speed nature of the ELM resulting in both high speed and superior performance compared with the existing multi-label classification techniques.

Consider N training samples of the form $\{(x_i, y_i)\}$ where x_i in the input denoted as $x_i = [x_{i1}, x_{i2},\ldots, x_{in}]^T \in R^n$ and y_i is the target label set, $y_i = [y_{i1}, y_{i2},\ldots y_{im}]^T$. As opposed to traditional single-label case, the target label is not a single label but is a subset of labels from the label space given as $Y \subseteq L$, $L = \{\zeta_1, \zeta_2,\ldots, \zeta_M\}$. Let $\overline{N}$ be the number of hidden layer neurons, the output 'o' of the SLFN is given by

$$\sum_{i=1}^{\overline{N}} \beta_i g_i(x_j) = \sum_{i=1}^{\overline{N}} \beta_i g(w_i.x_j + b_i) = o_j \qquad (2)$$

where, $\beta_i = [\beta_{i1}, \beta_{i2},\ldots \beta_{im}]^T$ is the output weight, $g(x)$ is the activation function, $w_i = [w_{i1}, w_{i2},\ldots w_{in}]T$ is the input weight and bi is the hidden layer bias.

For the ELM, the input weights w_i and the hidden layer bias bi are randomly assigned. Therefore, the network must be trained for βi such that the output of the network is equal to the target class so that the error difference between the actual output and the predicted output is 0.

$$\sum_{j=1}^{\overline{N}} \left\| o_j - y_j \right\| = 0 \tag{3}$$

Thus, the ELM classifier output can be as follows:

$$\sum_{i=1}^{\overline{N}} \beta_i g\left(w_i.x_j + b_i\right) = y_j \tag{4}$$

The above equation can be written in following matrix form:

$$H\beta = Y \tag{5}$$

The output weights of the ELM network can be estimated using the equation

$$\beta = H^+Y \tag{6a}$$

where H^+ is the Moore-Penrose inverse of the hidden layer output matrix H and it can be calculated as follows:

$$H^+ = \left(H^TH\right)^{-1}H^T \tag{6b}$$

The theory and mathematics behind the ELM have been extensively discussed in [23–25] and hence are not re-stated here. The steps involved in multi-label ELM classifier are given below.

Initialization of Parameters. Fundamental parameters such as the number of hidden layer neurons and the activation function are initialized.

Processing of Inputs. In the multi-label case, each input sample can be associated with more than one class labels. Hence, each of the input samples will have the associated output label as a m-tuple with 0 or 1 representing the belongingness to each of the labels in the label space L. The label set denoting the belongingness for each of the labels is converted from unipolar representation to bipolar representation.

ELM Training. The processed input is then supplied to the basic batch learning ELM. Let H be the hidden layer output matrix, β be the output weights and Y be the target label, the ELM can be represented in a compact form as $H\beta = Y$ where $Y \subseteq L, L = \{\zeta_1, \zeta_2, ..., \zeta_M\}$. In the training phase, the input weights and the hidden layer bias are randomly assigned and the output weights β are estimated as $\beta = H^+Y$, where $H^+ = (H^T H)^{-1}H^T$ gives the Moore-Penrose generalized inverse of the hidden layer output matrix.

ELM Testing. In the testing phase, the test data sample is evaluated using the values of β obtained during the training phase. The network then predicts the target output using the equation $Y = H\beta$. The predicted output Y obtained is a set of real numbers of dimension equal to the number of labels.

Post-processing and Multi-label Identification. The key challenge in multi-label classification is that the input sample may belong to one or more than one of the target labels. The number of labels that the sample corresponds to is completely unknown. Hence, a thresholding-based label association is proposed. The L dimensioned raw-predicted output is compared with a threshold value. The index values of the predicted output Y which are greater than the threshold fixed represents the belongingness of the input sample to the corresponding class.

Setting the threshold value is of critical importance. Threshold setting has to be made in such a way that it maximizes the difference between the values of the label the data belongs to and the labels the data does not. The distribution of the raw output values is categorized into a range of values that represent the belongingness of the label and the range of values that represent the non-belongingness of the label to a particular sample. From the distribution, a particular value is chosen that maximizes the separation between the two categories of the labels. It is to be highlighted that there are no ELM-based multi-label classifier in the literature thus far. The proposed method is the first to adapt the ELM for multi-label problems and make extensive experimentation and results comparison and analysis with the state-of-the-arts multi-label classification techniques.

4 Experimentation

This section describes the different multi-label dataset metrics and gives the experimental design used to evaluate the proposed method. Multi-label datasets have a unique property called the degree of multi-labelness. The number of labels, the number of samples having multiple labels, the average number of labels corresponding to a particular sample varies among different datasets. Two dataset metrics are available in the literature to quantitatively measure the multi-labelness of a dataset. They are Label Cardinality (LC) and Label Density (LD). Consider there are N training samples and the dataset is of the form $\{(x_i, y_i)\}$ where x_i in the input data and y_i is the target label set. The target label set is a subset of labels from the label space with M elements given as $Y \subseteq L$, $L = \{\zeta_1, \zeta_2... \zeta_M\}$.

Definition 4.1 Label Cardinality of the dataset is the average number of labels of the examples in the dataset [17].

$$Label - Cardinality = \frac{1}{N}\sum_{i=1}^{N}|Y_i| \tag{7}$$

Label Cardinality signifies the average number of labels present in the dataset.

Definition 4.2 Label Density of the dataset is the average number of labels of the examples in the dataset divided by |L| [17].

$$Label - Density = \frac{1}{N} \sum_{i=1}^{N} \frac{|Y_i|}{|L|} \tag{8}$$

Label density takes into consideration the number of labels present in the dataset. The properties of two datasets have same label cardinality, but different label density can vary significantly and may result in different behavior of the training algorithm [14]. The influence of label density and label cardinality on multi-label learning is analyzed by Bernardini et al. [26] in 2013. The proposed method is experimented with six benchmark datasets comprising of different application areas and its results are compared with 9 existing state-of-the-art methods. The datasets are chosen in such a way that they exhibit diverse nature of characteristics and the wide range of label density and label cardinality. The datasets are obtained from KEEL multi-label dataset repository and the specifications of the dataset are given in Table 1. The details of state-of-the-arts multi-label techniques used for result comparison are given in Table 2.

Table 1 Dataset specifications

Dataset	Domain	No. of features	No. of samples	No. of labels	LC	LD
Emotion	Multimedia	72	593	6	1.87	0.312
Yeast	Biology	103	2417	14	4.24	0.303
Scene	Multimedia	294	2407	6	1.07	0.178
Corel5 k	Multimedia	499	5000	374	3.52	0.009
Enron	Text	1001	1702	53	3.38	0.064
Medical	Text	1449	978	45	1.25	0.027

Table 2 Methods used for comparison

Method name	Method category	Machine learning category
Classifier chain (CC)	PT	SVM
QWeighted approach for multi-label learning (QWML)	PT	SVM
Hierarchy of multi-label classifiers (HOMER)	PT	SVM
Multi-label C4.5 (ML-C4.5)	AA	Decision trees
Predictive clustering trees (PCT)	AA	Decision trees
Multi-label k-nearest neighbors (ML-kNN)	AA	Nearest neighbors
Ensemble of classifier chains (ECC)	EN	SVM
Random forest predictive clustering trees (RF-PCT)	EN	Decision trees
Random Forest of ML-C4.5 (RFML-C4.5)	EN	Decision trees

5   Results and Discussions

This section discusses the results obtained by the proposed method and compares it with the existing methods. The results obtained from the proposed method are evaluated for consistency, performance and speed.

5.1   Consistency

Consistency is a key feature that is essential for any new technique proposed. The proposed algorithm should provide consistent results with minimal variance. Being an ELM based algorithm, since the initial weights are assigned in random, it is critical to evaluate the consistency of the proposed technique. The unique feature of multi-label classification is the possibility of partial correctness of the classifier, i.e. one or more of the multiple labels to which the sample instance belongs and/or the number of labels the sample instance belongs can be identified partially correctly. Therefore, calculating the error rate for multi-label problems is not same as that of traditional binary or multi-class problems. In order to quantitatively measure the correctness of the classifier, the hamming loss performance metric is used. To evaluate the consistency of the proposed method, a 5 fold and a 10 fold cross validation of hamming loss metric is evaluated for each of the six datasets and is tabulated.

From the Table 3, it can be seen that the proposed technique is consistent in its performance over repeated executions and cross validations thus demonstrating the consistency of the technique.

5.2   Performance Metrics

As foreshadowed, the unique feature of multi-label classification is the possibility of partial correctness of the classifier. Therefore, a set of quantitative performance evaluation metrics is used to validate the performance of the multi-label classifier. The performance metrics are hamming loss, accuracy, precision, recall and

Table 3 Consistency table—cross validation

Dataset	Hamming loss—5-fcv	Hamming loss—10-fcv
Emotion	0.2492($\pm$0.0058)	0.2509($\pm$0.0050)
Yeast	0.1906($\pm$0.0025)	0.1911($\pm$0.0031)
Scene	0.0854($\pm$0.0029)	0.0851($\pm$0.0033)
Corel5 k	0.0086($\pm$0.0005)	0.0090($\pm$0.0006)
Enron	0.0474($\pm$0.0022)	0.0472($\pm$0.0015)
Medical	0.0108($\pm$0.0008)	0.0109($\pm$0.0009)

Table 4 Hamming loss comparison

Dataset	CC	QWML	HOMER	ML-C4.5	PCT	ML-kNN	ECC	RFML-C4.5	RF-PCT	ELM
Emotion	0.256	0.254	0.361	0.247	0.267	0.294	0.281	0.198	0.189	**0.251**
Yeast	0.193	0.191	0.207	0.234	0.219	0.198	0.207	0.205	0.197	**0.191**
Scene	0.082	0.081	0.082	0.141	0.129	0.099	0.085	0.116	0.094	**0.085**
Corel5 k	0.017	0.012	0.012	0.01	0.009	0.009	0.009	0.009	0.009	**0.009**
Enron	0.064	0.048	0.051	0.053	0.058	0.051	0.049	0.047	0.046	**0.047**
Medical	0.077	0.012	0.012	0.013	0.023	0.017	0.014	0.022	0.014	**0.011**

Bold value denotes the performance of proposed method

Table 5 Accuracy comparison

Dataset	CC	QWML	HOMER	ML-C4.5	PCT	ML-kNN	ECC	RFML-C4.5	RF-PCT	ELM
Emotion	0.356	0.373	0.471	0.536	0.448	0.319	0.432	0.488	0.519	**0.412**
Yeast	0.527	0.523	0.559	0.48	0.44	0.492	0.546	0.453	0.478	**0.514**
Scene	0.723	0.683	0.717	0.569	0.538	0.629	0.735	0.388	0.541	**0.676**
Corel5 k	0.03	0.195	0.179	0.002	0	0.014	0.001	0.005	0.009	**0.044**
Enron	0.334	0.388	0.478	0.418	0.196	0.319	0.462	0.374	0.416	**0.418**
Medical	0.211	0.658	0.713	0.73	0.228	0.528	0.611	0.25	0.591	**0.715**

Bold value denotes the performance of proposed method

Table 6 Precision comparison

Dataset	CC	QWML	HOMER	ML-C4.5	PCT	ML-kNN	ECC	RFML-C4.5	RF-PCT	ELM
Emotion	0.551	0.548	0.509	0.606	0.577	0.502	0.58	0.625	0.644	**0.548**
Yeast	0.727	0.718	0.663	0.62	0.705	0.732	0.667	0.738	0.744	**0.718**
Scene	0.758	0.711	0.746	0.592	0.565	0.661	0.77	0.403	0.565	**0.685**
Corel5 k	0.042	0.326	0.317	0.005	0	0.035	0.002	0.018	0.03	**0.144**
Enron	0.464	0.624	0.616	0.623	0.415	0.587	0.652	0.69	0.709	**0.668**
Medical	0.217	0.697	0.762	0.797	0.285	0.575	0.662	0.284	0.635	**0.774**

Bold value denotes the performance of proposed method

Table 7 Recall comparison

Dataset	CC	QWML	HOMER	ML-C4.5	PCT	ML-kNN	ECC	RFML-C4.5	RF-PCT	ELM
Emotion	0.397	0.429	0.775	0.703	0.534	0.377	0.533	0.545	0.582	**0.491**
Yeast	0.6	0.6	0.714	0.608	0.49	0.549	0.673	0.491	0.523	**0.608**
Scene	0.726	0.709	0.744	0.582	0.539	0.655	0.771	0.388	0.541	**0.709**
Corel5 k	0.056	0.264	0.25	0.002	0	0.014	0.001	0.005	0.009	**0.043**
Enron	0.507	0.453	0.61	0.487	0.229	0.358	0.56	0.398	0.452	**0.508**
Medical	0.754	0.801	0.76	0.74	0.227	0.547	0.642	0.251	0.599	**0.744**

Bold value denotes the performance of proposed method

Table 8 F1 measure comparison

Dataset	CC	QWML	HOMER	ML-C4.5	PCT	ML-kNN	ECC	RFML-C4.5	RF-PCT	ELM
Emotion	0.461	0.481	0.614	0.651	0.554	0.431	0.556	0.583	0.611	**0.518**
Yeast	0.657	0.654	0.687	0.614	0.578	0.628	0.67	0.589	0.614	**0.658**
Scene	0.742	0.71	0.745	0.587	0.551	0.658	0.771	0.395	0.553	**0.697**
Corel5 k	0.048	0.292	0.28	0.003	0	0.021	0.001	0.008	0.014	**0.033**
Enron	0.484	0.525	0.613	0.546	0.295	0.445	0.602	0.505	0.552	**0.577**
Medical	0.337	0.745	0.761	0.768	0.253	0.56	0.652	0.267	0.616	**0.759**

Bold value denotes the performance of proposed method

Table 9 Comparison of training time (in seconds)

Dataset	CC	QWML	HOMER	ML-C4.5	PCT	ML-kNN	ECC	RFML-C4.5	RF-PCT	ELM
Emotion	6	10	4	0.3	0.1	0.4	4.9	1.2	2.9	**0.04**
Yeast	206	672	101	14	1.5	8.2	497	19	25	**0.2**
Scene	99	195	68	8	2	14	319	10	23	**0.12**
Corel5 k	1225	2388	771	369	30	389	10,073	385	902	**0.6**
Enron	440	971	158	15	1.1	6	1467	25	47	**0.26**
Medical	28	40	16	3	0.6	1	103	7	27	**0.11**

Bold value denotes the performance of proposed method

Table 10 Comparison of testing time (in seconds)

Dataset	CC	QWML	HOMER	ML-C4.5	PCT	ML-kNN	ECC	RFML-C4.5	RF-PCT	ELM
Emotion	1	2	1	0	0	0.4	6.6	0.1	0.3	**0**
Yeast	25	64	17	0.1	0	5	158	0.5	0.2	**0**
Scene	25	40	21	1	0	14	168	2	1	**0**
Corel5 k	31	119	14	1	1	45	2077	1.8	2.5	**0.06**
Enron	53	174	22	0.2	0	3	696	1	1	**0**
Medical	6	25	1.5	0.1	0	0.2	46	0.5	0.5	**0**

Bold value denotes the performance of proposed method

F1-measure. A comparison of performance metrics such as hamming loss, precision, recall, accuracy and F1 measure of the proposed technique is shown in Tables 4, 5, 6, 7 and 8. The performance of state-of-the-art techniques is adapted from [18]. From the tables, it is clear that the proposed method works uniformly well on all datasets. The proposed method outperforms all the existing methods in most cases and remains one of the top classification techniques in other cases.

5.3 Speed

The performance of the proposed method in terms of execution speed is evaluated by comparing the training time and the testing time of the algorithm used. The proposed method is applied to 6 datasets of different domains with a wide range of label density and label cardinality and the training time and the testing time are compared with other state-of-the-art techniques. The comparison table of training time and testing time is given in Tables 9 and 10 respectively.

In summary, the proposed method outperforms all existing multi-label learning techniques in terms of training and testing time by several orders of magnitude. From the results, it can be seen that the proposed method is the fastest multi-label classifier when compared to the current state-of-the-arts techniques. The speed of the proposed classifier is many-fold greater than existing methods. Also, from the comparison results of other performance metrics such as hamming loss, accuracy, precision, recall and F1-measure, it can be seen that the proposed method remains one of the top positions in each case. Also, the F1-measure of the proposed approach outperforms the most recent method which uses canonical correlation analysis (CCA) with ELM for multi-label problems [27] in most cases. The key advantage of the proposed method is that it surpasses all existing state-of-the-arts methods in terms of speed and simultaneously while remaining one of the top learning techniques in terms of other 5 performance metrics.

6 Conclusion

The proposed high speed multi-label classifier executes with both fast speed and high accuracy. It is to be highlighted that there are no extreme-learning-machine-based multi-label classifiers existing in the literature thus far. The proposed method is applied to 6 benchmark datasets of different domains and a wide range of label density and label cardinality. The results are compared with 9 state-of-the-arts multi-label classifiers. It can be seen from the results that the proposed method surpasses all state-of-the-arts methods in terms of speed and remain one of the top techniques in terms of other performance metrics. Thus, the proposed ELM-based multi-label classifier can be a better alternative for a wide range of multi-label classification techniques in order to achieve greater accuracy and very high speed.

Acknowledgements This work is supported by the National Natural Science Foundation of P. R. China (under Grants 51009017 and 51379002), Applied Basic Research Funds from Ministry of Transport of P. R. China (under Grant 2012-329-225-060), China Postdoctoral Science Foundation (under Grant 2012M520629), Program for Liaoning Excellent Talents in University (under Grant LJQ2013055). The second author would like to thank Nanyang Technological University for supporting this work by providing NTU RSS.

References

1. Gonclaves, T., Quaresma, P.: A preliminary approach to the multi-label classification problem of Portuguese juridical documents progress in artificial intelligence, pp. 435–444. Springer, Berlin (2003)
2. Joachims, T.: Text categorization with support vector machines: learning with many relevant features. In: Nedellec, C., Rouveirol, C. (eds.) ECML, LNCS, vol. 1938, pp. 137–142. Springer, Heidelberg (1998)
3. Luo, X., Zincir Heywood, A.N.: Evaluation of two systems on multi-class multi-label document classification. In: Hacid M.S., Murray N.V., Ras Z.W., Tsumoto S. (eds.) ISMIS 2005, LNCS (LNAI), vol. 3488, pp. 161–169, Springer, Heidelberg (2005)
4. Tikk, D., Biro, G.: Experiments with multi-label text classifier on the Reuters collection. Proceedings of the International Conference on Computational Cybernetics (ICCC 2003), Hungary, pp. 33–38 (2003)
5. Yu, K., Yu, S., Tresp, V.: Multi-label informed latent semantic indexing. Proceedings of the 28th Annual International ACM SIGIR Conference on Research and Development in information retrieval, pp. 258–265 (2005)
6. Karalic, A., Pirnat, V.: Significance level based multiple tree classification. Informatica **15**(5), 12 (1991)
7. Tsoumakas, G., Katakis, I, Vlahavas, I.: Mining multi-label data, data mining and knowledge discovery handbook. In: Maimon, O., Rokach, L. (ed.) Springer, 2nd edn. (2010)
8. de Carvalho, A.C., Freitas, A.A.: A tutorial on multi-label classification techniques. Found. Comput. Intell. **5**, 177–195 (2009)
9. Elisseeff, A., Weston, J.: A kernel method for multi-labelled classification. Neural Information Processing Systems, NIPS, vol. 14 (2001)
10. Zhang, M.L., Zhou, Z.H.: A k-nearest neighbour based algorithm for multi-label classification. Proceedings of the 1st IEEE International Conference on Granular Computing, pp. 718–721. Beijing, China (2005)
11. Boutell, M., Shen, X., Luo, J., Brouwn, C.: Multi-label semantic scene classification, Technical report. Department of Computer Science University of Rochester, USA (2003)
12. Shen X., Boutell, M., Luo, J., Brown, C.: Multi-label machine learning and its application to semantic scene classification. In: Yeung, M.M., Lienhart, R.W., Li, C.S. (eds.) Storage and retrieval methods and applications for multimedia. Proceedings of the SPIE, vol. 5307, pp. 188–199 (2003)
13. Zhu, B., Poon, C.K.: Efficient approximation algorithms for multi-label map labelling. In: Algorithms and computation, pp. 143–152. Springer, Heidelberg (1999)
14. Zhang, M.L., Zhou, Z.H.: ML-kNN: a lazy learning approach to multi-label learning. Pattern Recogn. **40**(7), 2038–2048 (2007)
15. Sorower, M.S.: A literature survey on algorithms for multi-label learning. Oregon State University, Corvallis (2010)
16. Elisseeff, A., Weston, J.: Kernel methods for multi-labelled classification and categorical regression problems, Technical report, BIOwulf Technologies (2001)
17. Tsoumakas, G., Katakis, I.: Multi-label Classification: an overview. Int. J. Data Warehouse. Min. **3**(3), 1–13 (2007)

18. Madjarov, G., Kocev, D., Gjorgjevikj, D., Dzeroski, S.: An extensive experimental comparison of methods for multi-label learning. Pattern Recogn. **45**, 3084–3104 (2012)
19. Wang, N., Sun, J.C., Er, M.J., Liu, Y.C.: A novel extreme learning control framework of unmanned surface vehicles. IEEE Transactions on Cybernetics. Accepted for Publication (2015).
20. Wang, N., Er, M.J., Han, M.: Generalized single-hidden layer feedforward networks for regression problems. IEEE Transac. Neural Networks Learn. Syst. **26**(6), 1161–1176 (2015)
21. Wang, N., Er, M.J., Han, M.: Parsimonious extreme learning machine using recursive orthogonal least squares. IEEE Transac. Neural Networks Learn. Syst. **25**(10), 1828–1841 (2014)
22. Wang, N., Han, M., Dong, N., Er, M.J.: Constructive multi-output extreme learning machine with application to large tanker motion dynamics identification. Neurocomputing **128**, 59–72 (2014)
23. Huang, G.B., Wang, D.H., Lan, Y.: Extreme learning machines: a survey. Int. J. Mach. Learn. Cybern. **2**, 107–122, 06/01 (2011)
24. Huang, G.B., Zhu, Q.Y., Siew, C.K.: Extreme learning machine: theory and applications. Neurocomputing **70**, 489–501, 12 (2006)
25. Ding, S., Zhao, H., Zhang, Y., Xu, X., Nie, R.: Extreme learning machine: algorithm, theory and applications. Artif. Intell. Rev. 1–13 (2013)
26. Bernardini, F.C., da Silva, R.B., Meza, E.M., das Ostras–RJ–Brazil, R.: Analyzing the influence of cardinality and density characteristics on multi-label learning (2009)
27. Kongsorot, Y., Horata, P.: Multi-label classification with extreme learning machine. International Conference on Knowledge and Smart Technology, pp. 81–86

Author Biography

Meng Joo Er is a Chair Professor with Marine Engineering College, Dalian Maritime University, Dalian 116026, China, and together with Rajasekar Venkatesan are with the School of Electrical and Electronics Engineering in NTU, Singapore; Ning Wang is with Marine Engineering College, Dalian Maritime University, Dalian 116026, China.

Image Super-Resolution by PSOSEN of Local Receptive Fields Based Extreme Learning Machine

Yan Song, Bo He, Yue Shen, Rui Nian and Tianhong Yan

Abstract Image super-resolution aims at generating high-resolution images from low-resolution inputs. In this paper, we propose a novel learning-based and efficient image super-resolution approach called particle swarm optimization based selective ensemble (PSOSEN) of local receptive fields based extreme learning machine (ELM-LRF). ELM-LRF is locally connected ELM, which can directly process information including strong correlations such as images. PSOSEN is a selective ensemble used to optimize the output of ELM-LRF. This method constructs an end-to-end mapping of which the input is a single low-resolution image and the output is a high resolution image. Experiments show that our method is better in terms of accuracy and speed with different magnification factors compared to the *state-of-the-art* methods.

Keywords Super-resolution · Particle swarm optimization based selective ensemble · Local receptive fields based extreme learning machine

1 Introduction

Due to the influence of various factors in environment, raw images obtained are always blurred and fuzzy. It is difficult to obtain details from these images. Accordingly, image super-resolution [1] processing is essential to enhance low-resolution images for extracting exact information. In recent years, important

Y. Song · B. He (✉) · Y. Shen · R. Nian
School of Information and Engineering, Ocean University of China, Qingdao,
Shandong, China
e-mail: bhe@ouc.edu.cn

T. Yan
School of Mechatronic Engineering, China Jiliang University, Hangzhou,
Zhejiang, China
e-mail: yanth@163.com

© Springer International Publishing Switzerland 2016
J. Cao et al. (eds.), *Proceedings of ELM-2015 Volume 2*,
Proceedings in Adaptation, Learning and Optimization 7,
DOI 10.1007/978-3-319-28373-9_38

455

progress has been made in single image super-resolution [2]. Many machine learning methods are used for single image super-resolution processing.

Nowadays, there are many ways to achieve image super-resolution. Methods for super-resolution (SR) can be classified into three types: (i) interpolation-based approach [3–5], (ii) reconstruction-based approach [6–8], (iii) learning-based approach [9–16]. Interpolation-based approach such as bicubic method and bilinear method often cause over-smooth. The first two methods get higher resolution images from a set of low-resolution input images. The third method is an end-to-end mapping which can directly output high-resolution images after learning the relationship between low-resolution images and high-resolution images. The interpolation-based approach includes registering low-resolution images first and then interpolating to get a high-resolution image and finally deblurring to enhance the high-resolution image. However, this method has high computational cost. The reconstruction-based SR approach constructs high-resolution images by using some heuristics or specific interpolation functions. The quality of the reconstructed image degrades when the magnification factor becomes large. Learning-based approach aims to enhance the high frequency information of the low-resolution image by learning to retrieve the most likely high-frequency information from the training image samples based on the local features of the input low-resolution image. A SR technique of learning-based approach is sparse representation [12, 15]. This method makes use of a sparse representation for each patch in a set of low-resolution images and then utilizes this representation to generate the high resolution output. By jointly training two dictionaries for the low- and high-resolution image patches, this method gets the similarity of sparse representations between the low-resolution and high-resolution image patch pair with respect to their own dictionaries. Recently, a deep learning method for single image super-resolution called Super-Resolution Convolutional Neural Network (SRCNN) [14] has been proposed. This method directly learns an end-to-end mapping between the low- and high-resolution images. The mapping is represented as a deep convolutional neural network (CNN) [17] that takes the low-resolution image as input and outputs high-resolution one. While results of it are appealing, SRCNN is time consuming in training step. Besides, super-resolution based on Extreme Learning Machine (SRELM) [13, 18] is another method which makes use of low-resolution images and their 1st and 2nd order derivatives. This method focuses on recovering the high-frequency (HF) components of the low-resolution image efficiently and accurately. During the training process, pixel intensity values in a local image patch and the 1st and 2nd order derivative are extracted from low-resolution images. Then ELM learns the relationship between pixel intensity values of high-resolution images and these features. However, features of super-resolution based on Extreme Learning Machine are selected manually. The prior knowledge we have about HF components of images has much effect on experimental results.

In this paper, we propose to use Particle Swarm Optimization based Selective Ensemble (PSOSEN) [19] of Local Receptive Fields Based Extreme Learning Machine (ELM-LRF) [20] to carry out image super resolution motivated by SRCNN and SRELM. Instead of making use of conventional Convolutional Neural Network (CNN) and Extreme Learning Machine (ELM) alone, we aim at using ELM-LRF that takes the advantages of these two algorithms both into account to achieve image super-resolution. The result is an end-to-end mapping between low- and high-resolution images. In order to improve the accuracy, PSOSEN is performed. PSOSEN is proposed in [19] which employ particle swarm optimization to choose good learners and combine these predictions for results better than genetic algorithms [21]. The method proposed in this paper has several advantages. First, the time it spends on training and testing is much shorter than the state-of-the-art methods. Second, the performance of this method is better than other learning-based methods.

The remainder of this paper is organized as follows. Section 2 presents some previous work about SRCNN, PSOSEN and ELM-LRF. Proposed method is described in Sect. 3. Section 4 is various results of our experiment. In Sect. 5, we conclude our method and efficiency and accuracy of the experiment. Besides, we discuss future work in the final section.

2 Review of Related Works

2.1 Super Resolution Convolutional Neural Network

Convolutional neural network is widely applied to image process for its local correlation. Recently, Dong et al. have proposed a method called Super Resolution Convolutional Neural Network (SRCNN). In this method, a deep convolutional neural network is used to realize super-resolution. And [14] makes a connection between SRCNN and sparse-coding-based SR method and concludes that the latter is one kind of SRCNN. The low-resolution training images of SRCNN are obtained from high-resolution images which are Gaussian-blurred and down-sampled and then bicubic interpolated. SRCNN can be divided into three steps:

(A) Patch extraction and representation. In this step, a set of feature images are extracted from low-resolution input images. Each feature map has the same weight and bias. Each patch of images has a high-dimensional vector at this step.
(B) Non-linear mapping. The high-dimensional vectors are mapped onto another high-dimensional vector in this step.
(C) Reconstruction. The high-dimensional vectors of step 2 are used to generate the final high-resolution images.

2.2 Local Receptive Fields Based Extreme Learning Machine

ELM has very fast learning speed and provides efficient results in the applications such as feature learning, clustering, regression and classification. The difference between conventional single-hidden layer feedforward neural networks (SLFNs) and ELM is that the latter confirms that the number of hidden neurons is important but the hidden neurons of ELM need not to be tuned iteratively.

Recently, Huang et al. have proposed a new theory called Local Receptive Fields Based Extreme Learning Machine. The connection of input layer and one node of hidden layer are generated according to continuous probability distribution. These random connections constitute local receptive fields. When it is applied to image processing and other similar tasks, ELM-LRF learns the local structure of images and generates more meaningful expression in the hidden layer.

In order to get the thorough representations of input, ELM-LRF uses K different random input weights to obtain K different feature maps. Hidden layers are composed by random convolution nodes. The same feature map shares the same input weight and different feature maps have different input weights. The input weights are random generated and orthogonal. ELM-LRF can extract more complete feature by using orthogonal input weights.

Assume that the initial input weights are $\hat{A}^{init}$, the size of each input weight is $r \times r$ and the size of each input image is $d \times d$. So the size of feature map is $(d - r + 1) \times (d - r + 1)$. Then

$$\hat{A}^{init} \in R^{r^2 \times K}, \quad \hat{A}^{init} = [\hat{a}_1^{init}, \hat{a}_2^{init}, \ldots, \hat{a}_K^{init}]$$
$$\hat{a}_K^{init} \in R^{r^2}, \quad k = 1, \ldots, K \tag{1}$$

$\hat{A}^{init}$ is orthogonalised by using singular value decomposition (SVD). The Orthogonalised input weights are $\hat{A}$, of which each column $\hat{a}_k$ is orthogonal basis of $\hat{A}^{init}$. If $r^2 < K$, $\hat{A}^{init}$ should be transposed at first and then orthogonalised and transposed again at last. The weight of the Kth feature map is $a_k \in R^{r \times r}$ and it is aligned of $\hat{a}_k$ by column. The convolutional node (i, j) of the Kth feature map is $c_{i,j,k}$:

$$c_{i,j,k}(x) = \sum_{m=1}^{r} \sum_{n=1}^{r} (x_{i+m-1,j+n-1} \cdot a_{m,n,k})$$
$$i, j = 1, \ldots, (d - r + 1) \tag{2}$$

Pooling size e is the distance between the center and the edge of pooling. And the size of pooling map is the same as feature map $(d - r + 1) \times (d - r + 1)$. $c_{i,j,k}$, $h_{p,q,k}$ stand for node (i, j) of kth feature map and node (p, q) of kth pooling map respectively:

$$h_{p,q,k} = \sqrt{\sum_{i=p-e}^{p+e} \sum_{j=q-e}^{q+e} c_{i,j,k}^2}$$
$$p, q = 1, \cdots, (d - r + 1) \tag{3}$$

If (i, j) is out of bound, the $c_{i,j,k} = 0$.

The connections of pooling layer and output layer are full. β is output weight and calculated in [23]. For every sample x, calculating its feature map using Eq. (2) and its pooling map using Eq. (3). Combinatorial layer matrix $H \in R^{N \times K \cdot (d-r+1)^2}$ is obtained by simply combining all combinatorial nodes into a row vector and putting the row vector of N samples into together. And

$$\beta = H^T(\frac{1}{C} + HH^T)^{-1}T$$
$$\text{if } N \leq K \cdot (d-r+1)^2$$

(4)

$$\beta = (\frac{1}{C} + HH^T)^{-1}H^T T$$
$$\text{if } N > K \cdot (d-r+1)^2$$

(5)

2.3 Particle Swarm Optimization Selective Ensemble

The parameters of ELM-LRF are generated randomly and the performance of it cannot be guaranteed. Therefore, selective ensemble [24] is utilized in our method. PSOSEN is proved to be more accurate and faster than the state-of-art selective ensemble method. PSOSEN assumes each learner can be assigned a weight and could characterize the fitness of including this learner in the ensemble. We will explain the principle of PSOSEN from the context of regression. Suppose the task is to employ learners to learn from the ensemble $f: R^m \rightarrow R^n$. Each learner $f_1, f_2, \ldots, f_N$ has its own weight $\omega_i (i = 1, 2, \ldots, N)$, which satisfying the following equations:

$$0 \leq \omega_i < 1$$

(6)

$$\sum_{i=1}^{N} \omega_i = 1$$

(7)

The output of lth output of the ensemble is calculated as:

$$\hat{f}_l = \sum_{i=1}^{N} \omega_i f_{i,l}$$

(8)

where $f_{i,l}$ is the lth output of the lth learner. For simplification, we suppose each learner has one output (but this method can be generalized to more outputs of any other learners). Suppose that $x \in R^m$ is random selected from its distribution $p(x)$ with expected output $d(x)$ and actual output $f_i(x)$. Then the output of the simple ensemble on x is

$$\hat{f}(x) = \sum_{i=1}^{N} \omega_i f_i(x) \tag{9}$$

The generalization error $E_i(x)$ and $\hat{E}(x)$ of the ith learner f_i and the simple ensemble respectively are

$$E_i(x) = (f_i(x) - d(x))^2 \tag{10}$$

$$\hat{E}(x) = (\hat{f}(x) - d(x))^2 \tag{11}$$

So the generalization error E_i and $\hat{E}$ of the ith learner f_i and the simple ensemble $p(x)$ respectively are

$$E_i = \int dx p(x) E_i(x) \tag{12}$$

$$\hat{E} = \int dx p(x) \hat{E}(x) \tag{13}$$

The correlation of the ith learner f_i and the jth learner f_j is defined as follows:

$$C_{ij} = \int dx p(x)(f_i(x) - d(x))(f_j(x) - d(x)) \tag{14}$$

Obviously, C_{ij} satisfies the following equations:

$$C_{ii} = E_i \tag{15}$$

$$C_{ij} = C_{ji} \tag{16}$$

Combine (9) and (11) and we can get

$$\hat{E}(x) = (\sum_{i=1}^{N} \omega_i f_i(x) - d(x))(\sum_{j=1}^{N} \omega_j f_j(x) - d(x)) \tag{17}$$

Combine (13), (14), (17) and we can get

$$\hat{E}(x) = \sum_{i=1}^{N} \sum_{j=1}^{N} \omega_i \omega_j C_{ij} \tag{18}$$

According to (18), to minimize the generalization error of the simple ensemble, the optimum weight vector ω_{opt} can be calculated as

$$w_{opt} = \arg_{\omega} \min \left(\sum_{i=1}^{N} \sum_{j=1}^{N} \omega_i \omega_j C_{ij} \right) \qquad (19)$$

The kth component $\omega_{opt.k}$ of ω_{opt} can be calculated by using Lagrange multiplier methods and the result is

$$\frac{\partial \left(\sum_{i=1}^{N} \sum_{j=1}^{N} \omega_i \omega_j C_{ij} - 2\lambda \left(\sum_{i=1}^{N} \omega_i - 1 \right) \right)}{\partial w_{opt.k}} = 0 \qquad (20)$$

Equation (20) can be simplified to

$$\sum_{\substack{j=1 \\ j \neq k}}^{N} w_{opt.k} C_{ij} = \lambda \qquad (21)$$

$\omega_{opt.k}$ satisfies (7) and

$$\omega_{opt.k} = \frac{\sum_{j=1}^{N} C_{kj}^{-1}}{\sum_{i=1}^{N} \sum_{j=1}^{N} C_{ij}^{-1}} \qquad (22)$$

Equation (22) can be viewed as an optimization problem. Particle swarm optimization is proved to be a powerful optimization tool [21] and PSOSEN is proposed for this. Weights are evolved by employing PSOSEN and can characterize the fitness of the learners in joining the ensemble.

3 Proposed Approach

This paper uses a new learning-based method to realize the super-resolution of image which utilizes PSOSEN of ELM-LRF. There are three procedures of our method which comprised of image preprocessing, the process of ELM-LRF and the process of PSOSEN. As is introduced in Sect. 2, there are three layers of ELM-LRF: hidden layer, combinatorial layer and output layer. An overview of our method is shown in Fig. 1.

3.1 Image Pre-processing

Images obtained from real environment always have low resolution because of noise or poor light. In order to get this kind of low resolution images in our method,

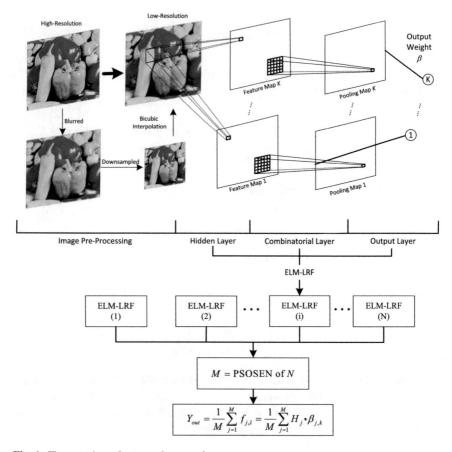

Fig. 1 The overview of proposed approach

the luminance channel of one high-resolution image is Gaussian blurred first and then down-sampled and then processed with bicubic interpolation. These low-resolution images and the original high-resolution images have the same size because scale of down-sampled is equivalent to scale of bicubic interpolation. We carry out this new method on the luminance channel for the reason that people are more sensitive to the change of this channel. These blurred images are as input of ELM-LRF.

3.2 Process of ELM-LRF

In hidden layer, we obtain feature maps by making convolution of the patches of low-resolution image. There are K random input weights with orthogonalization of hidden layer which are $W_1, W_2, \ldots, W_K \in \mathrm{R}^{n \times n}$. As is proved in [20], test error rates will decrease if the input weights are orthogonalized after random generating. And the size of low-resolution image I_0 is $N \times N$ and the size of each weight is $n \times n$. So node (i, j) in the kth feature map $C(i, j, k)$ is calculated as

$$C(i,j,k) = W_k * I_0'　　　　　　　　　(23)$$

The size of the kth feature map is $(N - n + 1) \times (N - n + 1)$. I_0' is one patch of I_0 with a size $n \times n$ and has a location from pixel (i, j) to pixel $(i + n - 1, j + n - 1)$.

The low-resolution image has K different feature maps, that is to say, each pixel of this image has K different features. These features are random selected by making convolution rather than prior knowledge. The weights need not to be tuned which is in contrast to [14]. The process of hidden layer is illustrated in Fig. 2.

In combinatorial layer, there are K pooling maps combined with K feature maps respectively and these two kinds of maps share the same size. The pooling size is n'. The node (p, q) of Kth pooling map $h_{p,q,k}$ is a square of features calculated in (3). As is shown in Fig. 3, if (i, j) of a feature map in hidden layer is out of bound, $C(i, j, k) = 0$. The structure that a convolutional layer followed by a square pooling layer is frequency selective and translational invariant as proved in [25].

In output layer, ELM-LRF transforms the feature map of combinatorial layer into a row vector and puts the rows of M different input images together to get combinatorial layer matrix $H \in R^{M \times K \cdot (N - n + 1)^2}$. $T \in R^{M \times N^2}$ is the ground truth matrix of training images. Finally, β is calculated according to (4) or (5).

While the weights of hidden layer need not to be tuned in ELM-LRF, there are four parameters that need to be tuned by using root mean square error (RMSE) between the luminance channel of input high-resolution images and the output of ELM-LRF. These four parameters include the number of feature maps K, the size of convolution n, the size of pooling n' and value of C. Then

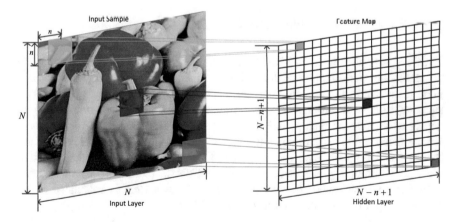

Fig. 2 The process of hidden layer. The *left image* is an input sample and the *right* is one of K feature maps. By making convolution of the patches of sample image, such as the colored patches of the sample image, we get corresponding features (with the same color) in feature map

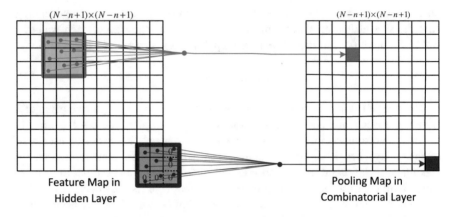

Fig. 3 The process between hidden layer and combinatorial layer

$$RMSE = \sqrt{\frac{1}{M}\sum_{i=1}^{M}\|T_i - f(y_i)\|^2} \qquad (24)$$

where y_i is ith low-resolution image and T_i is the ground truth high-resolution image of it. $f(y_i)$ is the output of ELM-LRF. M is the number of training inputs. We choose the parameters which can minimize RMSE.

After training, this method generates an end-to-end model and can output a high resolution image when the input is a low resolution image. ELM-LRF for super resolution is faster than other training-based methods for its no tuned weights in hidden layer.

3.3 Selective Ensemble Particle Swarm Optimization

PSOSEN is applied to ELM-LRF to improve the generalization performance. The generalization performance of the ELM-LRF mentioned above may not be quite good results from the random generation of weights in hidden layer. PSOSEN is superior to simple ensemble. The framework for the new algorithm can be explained as follows: first, N individual ELM-LRFs are initialized. The number of feature maps and the size of convolution map are the same for each ELM-LRF and the weights of hidden layer for each ELM-LRF are randomly generated. Second, the actual output $f_{i,L}$ is calculated and PSOSEN is performed and a selective ensemble M is obtained. At last, the output of the system is calculated as the average output of the individual in the ensemble set:

$$Y_{\text{out}} = \frac{1}{M} \sum_{j=1}^{M} f_{j,L} = \frac{1}{M} \sum_{j=1}^{M} H_j \cdot \beta_j \tag{25}$$

where H_j is the output matrix of the jth ELM-LRF, and β_j is the output weight calculated by the jth ELM-LRF.

4 Experiments

We use the same training images and test images as in [14] to compare our method with other methods fairly. 8 high-resolution images with size 512×512 are used in pre-processing before ELM-LRF. The methods for pre-processing are as follows. For training set, first select the luminance channel of these images (people's eyes are more sensitive to the change of luminance channel), and then these images are Gaussian-blurred with patch size 5×5 and standard deviation 1. Then these blurred images are down-sampled with scale 2, 4 respectively and finally done with bicubic interpolation with the same scale. These images are divided into sub-images with size 32×32 and we have a training set of 2048 sub-images. But for testing set, we didn't carry out the operation of Gaussian-blurred; the rest processing procedures are the same as training set. One image with size 512×512 can produce 256 sub-images. We also try to divide these images into more sub-images, but the performance didn't get more improvement.

We use sub-images with the size 32×32 processed by the above method to train ELM-LRF, and then save the value of β obtained by the train process. The pre-processed test sub-images are put into ELM-LRF, and β obtained in training step is utilized to calculate the output. The parameters of ELM-LRF including the number of feature maps, the size of convolution, the size of pooling and the value of C are 74, 9×9, 3×3 and 100 respectively. The output of ELM-LRF is sub-images at size 32×32, and then we use these output as the input of PSOSEN. Finally, the sub-images will be integrated into a whole image for visualization.

Our method is compared with other SR methods in this paper including bicubic, SRCNN (Super-resolution Convolutional Neural Network) and ANR (Anchored Neighbourhood Regression) [26] method. These codes are available on the authors' website.

Experimental results are shown in Fig. 4. As can be seen in Fig. 4, compared with the other method such as bicubic and super-resolution convolutional neural network (SRCNN), our methods perform better, and have a better visual effect. Our proposed methods and SRCNN both can produce the details of images. But our methods still work better and have more details when magnification factor is 4. The proposed methods are very close to SRCNN when the magnification factor is at 2, but it is easy to find that the proposed methods show more details. Furthermore, when the magnification factor is 4 and the images have low-resolution, our method is still able to achieve good performance.

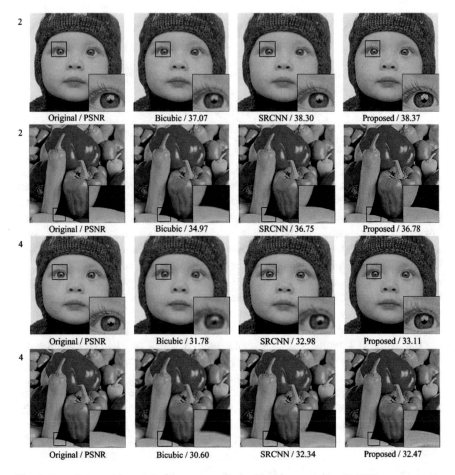

Fig. 4 From *left* to *right*: original images, results by bicubic, results by SRCNN, results by the proposed method. The *first two rows* are the results at down-sampled scale 2 and the *last two rows* are the results at down-sampled scale 4

To evaluate the performance quantitatively, Peak Signal to Noise Ratio (PSNR) [27] is calculated as shown in Table 1. PSNR has a relationship with RMSE, that

$$PSNR = 20 * \ln(255/RMSE) \tag{26}$$

When the magnification factor is 2, the PSNR of our method is better than bicubic, and note that SRCNN and we proposed method have similar results. But when the magnification factor becomes 4, our method is better than SRCNN. Besides, our method has only one convolutional layer and one pooling layer and the training steps are much easier and faster.

Table 1 Experimental result of PSNR (DB) and test time between the proposed methods and other methods

Images	Scale	Bicubic		ANR		SRCNN		Proposed	
		PSNR	Time	PSNR	Time	PSNR	Time	PSNR	Time
baby	2	37.07	–	38.44	–	38.30	7.52	38.37	6.05
pepper	2	34.97	–	–	–	36.75	7.71	36.78	6.17
baby	4	31.78	–	32.99	–	32.98	7.46	33.11	6.12
pepper	4	30.60	–	–	–	32.34	7.45	32.47	6.03

Table 1 also shows the running time compared with SRCNN. The programs are executed on the machine with Inter Core 2.8 GHz CPU and 64 GB of RAM. We take approximately 6 s per image, which is contrast with the running time about 7.5 s of SRCNN, no matter with the upscale factor 2 or 4.

5 Conclusions

This paper proposes an efficient method to realize image super resolution based on particle swarm optimization based selective ensemble (PSOSEN) of local receptive fields based extreme learning machine (ELM-LRF). The proposed approach employs pre-processed sub-images to train and test ELM-LRF, and with PSOSEN for better results. For visualization, the output sub-images will be integrated into final whole images. It is demonstrated in experimental results that this method is effective both in training and testing speed and quantitative evaluations. In the future, we will use more high-resolution images and more complicated features to train ELM-LRF for better results, and more details also will be resulted. But on the other hand, through the experiment we found that when we integrate those sub-images into a complete picture, the edge of sub images will affect PSNR. So we will also focus on resolving this problem in the future to get better performance.

Acknowledgments This work is partially supported by the Natural Science Foundation of China (41176076, 51075377, 51379198), the High Technology Research and Development Program of China (2006AA09Z231, 2014AA093410).

References

1. Park, S.C., Park, M.K., Kang, M.G.: Super-resolution image reconstruction: a technical overview. Sig. Process. Mag. IEEE **20**(3), 21–36 (2003)
2. Irani, M., Peleg, S.: Improving resolution by image registration. CVGIP: Graphical Models Image Process **53**(3), 231–239 (1991)
3. Deepu, R., Chaudhuri, S.: Generalized interpolation and its application in super-resolution imaging. Image Vis. Comput. **19**, 957–969 (2001)

4. Tao, H., Tang, X., Liu, J., Tian, J.: Superresolution remote sensing image processing algorithm based on wavelet transform and interpolation. Image Process. Pattern Recog. Remote Sens. **4898**, 259–263 (2003)

5. Surapong, L., Bose, N.K.: High resolution image formation from low resolution frames using Delaunay triangulation. Image process. IEEE Transac. **11**, 1427–1441 (2002)

6. Sina, F., Dirk, R.M., Michael, E., Peyman, M.: Fast and robust multiframe super resolution. Image process. IEEE Transac. on **13**, 1327–1344 (2004)

7. Hardie, R.C., Barnard, K.J., Armstrong, E.E.: Joint MAP registration and high-resolution image estimation using a sequence of undersampled images. Image Process. IEEE Transac. on **6**, 1621–1633 (1997)

8. Atsunori, K., Maeda, S., Ishii, S.: Superresolution with compound Markov random fields via the variational EM algorithm. Neural Netw. **22**, 1025–1034 (2009)

9. Chang H., Yeung, D., Xiong, Y., Super-resolution through neighbor embedding. In: Computer Vision and Pattern Recognition, 2004. CVPR 2004. Proceedings of the 2004 IEEE Computer Society Conference on, Vol. 1 (2004)

10. Freeman, W.T., Pasztor, E.C., Carmichael, O.T.: Learning low-level vision. Int. J. Comput. Vision **40**, 25–47 (2000)

11. Datsenko, D., Elad, M.: Example-based single document image super-resolution: a global MAP approach with outlier rejection. Multidimension Syst. Signal Process. **18**, 103–121 (2007)

12. Yang, J., Wright, J., Huang, T.S., Ma, Y.: Image super-resolution via sparse representation. Image Process. IEEE Transac. on **19**(11), 2861–2873 (2010)

13. An, L., Bhanu, B.: Image super-resolution by extreme learning machine. In: Image processing (ICIP), 2012 19th IEEE International Conference on, pp. 2209–2212 (2012)

14. Dong, C., Loy, C.C., He, K., Tang, X, Learning a deep convolutional network for image super-resolution. Computer Vision–ECCV 2014, pp. 184–199 (2014)

15. Yang, J., Wright, J., Huang, T., Ma, Y.: Image super-resolution as sparse representation of raw image patches. In: Computer Vision and Pattern Recognition, 2008. CVPR 2008. IEEE Conference on, pp. 1–8 (2008)

16. Zeyde, R., Elad, M., Protter, M.: On single image scale-up using sparse-representations. Curves Surf. 711–730 (2012)

17. LeCun, Y., Boser, B., Denker, J.S., Henderson, D., Howard, R.E., Hubbard, W., Jackel, L.D.: Backpropagation applied to handwritten zip code recognition. Neural Comput. **1**(4), 541–551 (1989)

18. Huang, G.B., Zhu, Q.Y., Siew, C.K.: Extreme learning machine: theory and applications. Neurocomputing **70**, 489–501 (2006)

19. Liu, Y., He, B., Dong, D., Shen, Y., Yan, T., Nian, R., Lendasse, A.: Particle swarm optimization based selective ensemble of online sequential extreme learning machine. Math. Probl. Eng. **2015**, 1–10 (2014)

20. Huang, G.B., Bai, Z., Lekamalage, L., Kasun, C.: Local receptive fields based extreme learning machine. Comput. Intell. Mag., IEEE **10**(2), 18–29 (2015)

21. Kennedy, J., Spears, W.M.: Matching algorithms to problems: an experimental test of the particle swarm and some genetic algorithms on the multimodal problem generator. Proceedings of the IEEE International Conference on Evolutionary Computation, pp. 78–83 (1998)

22. Huang, G.B., Zhu, Q.Y., Siew, C.K.: "Extreme learning machine: a new learning scheme of feedforward neural networks", Neural networks, 2004. Proceedings 2004 IEEE International Joint Conference on, vol. 2, (2004)

23. Huang, G.B., Zhou, H., Ding, X.: Extreme learning machine for regression and multiclass classification. Syst. Man Cybern. Part B: Cybern., IEEE Transac. on **42**(2), 513–529 (2012)

24. Zhou, Z.H., Tang, W.: Selective ensemble of decision trees. Rough Sets, Fuzzy Sets, Data Min. Granular Comput. **2639**, 476–483 (2003)

25. Saxe, A., Koh, P.W., Chen, Z., Bhand, M., Suresh, B., Ng, A.Y.: On random weights and unsupervised feature learning. In: Proceedings of the 28th International Conference on Machine Learning (ICML-11), pp. 1089–1096 (2011)
26. Timofte, R., De, V., Gool, L.V.: Anchored neighborhood regression for fast example-based super-resolution. In: Computer Vision (ICCV), 2013 IEEE International Conference on, pp. 1920–1927 (2013)
27. Huynh-Thu, Q., Ghanbari, M.: Scope of validity of PSNR in image/video quality assessment. Electron. Lett. **44**(13), 800–801 (2008)

Sparse Extreme Learning Machine for Regression

Zuo Bai, Guang-Bin Huang and Danwei Wang

Abstract Extreme learning machine (ELM) solves regression and classification problems efficiently. However, the solution provided is dense and requires plenty of storage space and testing time. A sparse ELM has been proposed for classification in [1]. However, it is not applicable for regression problems. In this paper, we propose a sparse ELM for regression, which significantly reduces the storage space and testing time. In addition, we develop an efficient training algorithm based on iterative computation, which scales quadratically with regard to the number of training samples. Therefore, the proposed sparse ELM is advantageous over other ELM methods when facing large data sets for achieving faster training and testing speed, while requiring less storage space. In addition, sparse ELM outperforms support vector regression (SVR) in the aspects of generalization performance, training speed and testing speed.

Keywords Regression · Extreme learning machine (ELM) · Sparse ELM · Quadratic programming (QP) · Support vector regression (SVR)

1 Introduction

Extreme learning machine (ELM) was first proposed as an improvement on classic single-hidden-layer feedforward neural networks (SLFNs) [2] and later was extended to multi-layer networks [3, 4]. Hidden nodes do not need to be tuned and superb performance is provided. Later, a unified ELM is proposed to unify different learning methods [5]. However, the solution of the unified ELM is dense, requiring plenty

Z. Bai (✉) · G.-B. Huang · D. Wang
School of Electrical and Electronic Engineering, Nanyang Technological University,
Nanyang Avenue, Singapore 639798, Singapore
e-mail: zbai1@e.ntu.edu.sg

G.-B. Huang
e-mail: egbhuang@ntu.edu.sg

D. Wang
e-mail: edwwang@ntu.edu.sg

© Springer International Publishing Switzerland 2016
J. Cao et al. (eds.), *Proceedings of ELM-2015 Volume 2*,
Proceedings in Adaptation, Learning and Optimization 7,
DOI 10.1007/978-3-319-28373-9_39

of storage space and testing time. Besides, the unified ELM obtains the solution by matrix inversion with complexity between $O(N^2)$ and (N^3) (N is the number of training samples). To address these issues, Bai et al. propose a sparse ELM for classification, which provides a sparse network and scales quadratically with regard to N.

However, the existing sparse ELM is not capable of dealing with regression problems. Thus, we propose a sparse ELM for regression in this paper and develop a specific, highly efficient training algorithm accordingly. The storage space and testing time are largely reduced and the computational complexity is quadratic to the number of training samples N. Additionally, comparing to support vector regression (SVR) [6], it achieves faster training and testing speed, while obtaining better generalization performance. In general, the proposed method is preferred when facing large-scale regression problems, such as time series prediction, neuroscience, etc.

2   Review of ELM

Since the emergence of ELM, numerous variants of ELM have been suggested. In [5], a unified ELM was proposed, providing a common framework for different learning methods. Later, a sparse ELM was proposed for classification problems in order to reduce the requirement for storage space and testing time [1].

2.1   Unified ELM

In the unified ELM, the problem is solved by minimizing the structural risks and empirical errors. The solution β is calculated:

$$\beta = \mathbf{H}^T \left(\frac{\mathbf{I}}{C} + \mathbf{H}\mathbf{H}^T \right)^{-1} \mathbf{T} = \left(\frac{\mathbf{I}}{C} + \mathbf{H}^T\mathbf{H} \right)^{-1} \mathbf{H}^T\mathbf{T} \tag{1}$$

And kernel form is also applicable when $\mathbf{h}(\mathbf{x})$ is inconvenient to use:

$$f(\mathbf{x}) = \mathbf{h}(\mathbf{x})\beta = \begin{bmatrix} K(\mathbf{x}, \mathbf{x}_1) \\ \vdots \\ K(\mathbf{x}, \mathbf{x}_N) \end{bmatrix}^T \left(\frac{\mathbf{I}}{C} + \mathbf{\Omega}_{\text{ELM}} \right)^{-1} \mathbf{T}$$

$$\mathbf{\Omega}_{\text{ELM}} = \mathbf{H}\mathbf{H}^T : \mathbf{\Omega}_{\text{ELM}i,j} = \mathbf{h}(\mathbf{x}_i) \cdot \mathbf{h}(\mathbf{x}_j) = K(\mathbf{x}_i, \mathbf{x}_j) \tag{2}$$

2.2   Sparse ELM for Classification

In [1], a sparse ELM was proposed for classification, providing a sparse network and reducing the storage space and computational requirement. The solution is:

$$f(\mathbf{x}) = \mathbf{h}(\mathbf{x})\beta = \mathbf{h}(\mathbf{x}) \left(\sum_{s=1}^{N_s} \alpha_s t_s \mathbf{h}(\mathbf{x}_s)^T \right) = \sum_{s=1}^{N_s} \alpha_s t_s K(\mathbf{x}, \mathbf{x}_s) \tag{3}$$

where $\mathbf{x}_s$ is support vector (SV), and $N_s < N$ is the number of SVs. And α_s's are Lagrangian multipliers calculated from the following optimization problem:

$$\text{Minimize: } L_d = \frac{1}{2} \sum_{i=1}^{N} \sum_{j=1}^{N} \alpha_i \alpha_j t_i t_j \Omega_{\text{ELM}i,j} - \sum_{i=1}^{N} \alpha_i \tag{4}$$

$$\text{Subject to: } 0 \leq \alpha_i \leq C, \quad i = 1, ..., N$$

3   Sparse ELM for Regression

3.1   Problem Formulation

In regression problems, the goal is to find a function $f(\mathbf{x})$ that minimizes the expected risk on the probability distribution of the samples [7]. According to structural risk minimization (SRM) [8], the combination of empirical errors $R_{\text{emp}}[f]$ and structural risks $\|\beta\|^2$ is used to approximate the expected risks.

$$\text{Minimize: } R_{\text{reg}} = \frac{\lambda}{2}\|\beta\|^2 + R_{\text{emp}}[f] = \frac{\lambda}{2}\|\beta\|^2 + \frac{1}{N}\sum_{i=1}^{N} c(t_i, f(\mathbf{x}_i)) \tag{5}$$

where the loss function $c(t_i, f(\mathbf{x}_i))$ needs to be convex [7]. In this paper, we choose ε-insensitive loss function because it will lead to a sparse solution. Intuitively speaking, errors will not be considered if they are smaller than ε.

$$c(t_i, f(\mathbf{x}_i)) = \begin{cases} 0 & \text{for } |t_i - f(\mathbf{x}_i)| < \varepsilon \\ |t_i - f(\mathbf{x}_i)| - \varepsilon & \text{for } \text{otherwise} \end{cases} \tag{6}$$

3.2   Optimization

According to (5) and (6), the primal problem of sparse ELM for regression is:

$$\text{Minimize: } L_p = \frac{1}{2}\|\beta\|^2 + C \sum_{i=1}^{N} (\xi_i + \xi_i^*) \tag{7}$$

$$\text{Subject to: } t_i - \mathbf{h}(\mathbf{x}_i)\beta \leq \varepsilon + \xi_i, \ \mathbf{h}(\mathbf{x}_i)\beta - t_i \leq \varepsilon + \xi_i^*, \ \xi_i^{(*)} \geq 0, \quad i = 1, ..., N$$

Thus, the Lagrangian P is:

$$P = \frac{1}{2}\|\beta\|^2 + C\sum_{i=1}^{N}(\xi_i + \xi_i^*) - \sum_{i=1}^{N}\alpha_i\left(\varepsilon + \xi_i - t_i + \mathbf{h}(\mathbf{x}_i)\beta\right)$$

$$- \sum_{i=1}^{N}\alpha_i^*\left(\varepsilon + \xi_i^* + t_i - \mathbf{h}(\mathbf{x}_i)\beta\right) - \sum_{i=1}^{N}\mu_i\xi_i - \sum_{i=1}^{N}\mu_i^*\xi_i^* \tag{8}$$

$$\alpha^{(*)}, \mu_i^{(*)} \geq 0$$

in which $\alpha_i^{(*)}$, $\mu_i^{(*)}$ respectively denotes α_i, α_i^* and μ_i, μ_i^*. With standard optimization method, the dual form is constructed:

$$\text{Minimize: } L_d = \frac{1}{2}\sum_{i=1}^{N}\sum_{j=1}^{N}(\alpha_i - \alpha_i^*)(\alpha_j - \alpha_j^*)\Omega_{\text{ELM}i,j} = -\sum_{i=1}^{N}(\alpha_i - \alpha_i^*)t_i + \varepsilon\sum_{i=1}^{N}(\alpha_i + \alpha_i^*) \tag{9}$$

Subject to: $\alpha_i \cdot \alpha_i^* = 0$ and $\alpha_i, \alpha_i^* \in [0, C]$

in which Ω_{ELM} is the ELM kernel matrix, which has two forms: (1) random hidden nodes form; (2) kernel form. Readers may refer to [1] for details.

For convenience, we substitute $\lambda_i = \alpha_i - \alpha_i^*$ into (9) and λ_i is called Lagrange multipliers. $|\lambda_i| = \alpha_i + \alpha_i^*$ because at least one of α_i and α_i^* is zero:

$$\text{Minimize: } L_d = \frac{1}{2}\sum_{i=1}^{N}\sum_{j=1}^{N}\lambda_i\lambda_j\Omega_{\text{ELM}i,j} - \sum_{i=1}^{N}\lambda_i t_i + \varepsilon\sum_{i=1}^{N}|\lambda_i| \tag{10}$$

Subject to: $-C \leq \lambda_i \leq C$

And the output function of sparse ELM for regression is:

$$f(\mathbf{x}) = \mathbf{h}(\mathbf{x})\beta = \mathbf{h}(\mathbf{x})\left(\sum_{i=1}^{N}\lambda_i\mathbf{h}(\mathbf{x}_i)^T\right) = \mathbf{h}(\mathbf{x})\left(\sum_{s=1}^{N_s}\lambda_s\mathbf{h}(\mathbf{x}_s)^T\right) = \sum_{s=1}^{N_s}\lambda_s K(\mathbf{x}, \mathbf{x}_s) \tag{11}$$

where $\mathbf{x}_s$ is support vector (SV) and N_s is the number of SVs.

Theorem 1 *The dual problem of sparse ELM for regression (10) is convex.*

Proof

$$\frac{\partial}{\partial\lambda_s}L_d = \sum_{i=1}^{N}\lambda_i\Omega_{\text{ELM}s,i} - t_s + \varepsilon\left(\text{sign}(\lambda_s)\right) \tag{12}$$

$$\frac{\partial^2}{\partial\lambda_s^2}L_d = \Omega_{\text{ELM}s,s}$$

Thus, the Hessian matrix $\nabla^2 L_d = \Omega_{\text{ELM}}$.

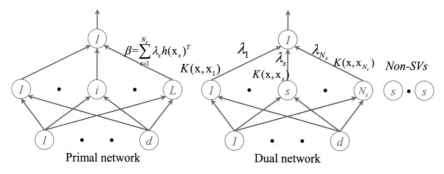

Fig. 1 The primal and dual networks of sparse ELM for regression

(i) Random hidden nodes form:

$$\mathbf{z}^T \left(\nabla^2 L_d \right) \mathbf{z} = \mathbf{z}^T \mathbf{H}\mathbf{H}^T \mathbf{z} = \left(\mathbf{H}^T \mathbf{z} \right)^T \mathbf{I}_{L \times L} \left(\mathbf{H}^T \mathbf{z} \right) \geq 0, \quad \forall \mathbf{z} \in \mathbf{R}^N \qquad (13)$$

Thus, $\nabla^2 L_d$ is positive semi-definite.

(ii) Kernel form: the kernel function K must satisfy Mercer's conditions. Thus $\nabla^2 L_d$ is guaranteed to be positive semi-definite.

Since the Hessian matrix $\nabla^2 L_d$ is positive semi-definite, the first-order derivative $\partial L_d / \partial \lambda_s$ is monotonically increasing. Therefore, the dual problem is convex.

3.3 Sparsity Analysis

If $|t_i - f(\mathbf{x}_i)| < \varepsilon$, α_i and α_i^* should both be zero, making $\lambda_i = \alpha_i - \alpha_i^* = 0$, and corresponding SVs vanish in the expansion of $\boldsymbol{\beta}$ in (11).

The primal and dual networks of sparse ELM for regression are shown in Fig. 1. For the dual network, non-SVs are excluded so that a sparse network is provided. For the primal network, many components are removed from the expansion of $\boldsymbol{\beta}$. And the storage space and testing computations are both proportional to N_s. Therefore, the requirement for storage space and testing time are greatly reduced.

4   Training Algorithm of Sparse ELM for Regression

The dual problem of sparse ELM for regression (10) is in effect a QP problem. In this section, we develop an efficient training algorithm, which divides the large QP problem into a series of smallest possible sub-problems and solves them sequentially.

Unlike SVR, it does not have the sum constraint $\sum_{i=1}^{N} \lambda_i = 0$. Therefore, these smallest possible sub-problems only involves one Lagrange variable and can be calculated analytically.[1]

4.1 Optimality Conditions

In the standard optimization, we derive the relationships that must be satisfied at the optimum. Once they are satisfied, the optimal solution is reached; and *vice versa*. $e_i = t_i - f(\mathbf{x}_i)$ denotes the error between the output and the expected value of sparse ELM.

(i) $\lambda_i = C$:
$$\alpha_i = C, \alpha_i^* = 0 \Rightarrow \mu_i = 0, \xi_i > 0$$
$$t_i - f(\mathbf{x}_i) - \varepsilon - \xi_i = 0 \Rightarrow e_i > \varepsilon \tag{14}$$

(ii) $0 < \lambda_i < C$:
$$\alpha_i \in (0, C), \alpha_i^* = 0 \Rightarrow \mu_i \in (0, C), \xi_i = 0$$
$$t_i - f(\mathbf{x}_i) - \varepsilon = 0 \Rightarrow e_i = \varepsilon \tag{15}$$

(iii) $\lambda_i = 0$:
$$\alpha_i^{(*)} = 0 \Rightarrow \mu_i^{(*)} = C, \xi_i^{(*)} = 0$$
$$|t_i - f(\mathbf{x}_i)| < \varepsilon \Rightarrow |e_i| < \varepsilon \tag{16}$$

(iv) $-C < \lambda_i < 0$:
$$\alpha_i^* \in (0, C), \alpha_i = 0 \Rightarrow \mu_i^* \in (0, C), \xi_i^* = 0$$
$$f(\mathbf{x}_i) - t_i - \varepsilon = 0 \Rightarrow e_i = -\varepsilon \tag{17}$$

(v) $\lambda_i = -C$:
$$\alpha_i^* = C, \alpha_i = 0 \Rightarrow \mu_i^* = 0, \xi_i^* > 0$$
$$f(\mathbf{x}_i) - t_i - \varepsilon - \xi_i^* = 0 \Rightarrow e_i < -\varepsilon \tag{18}$$

4.2 Update Rule

The update rule determines how to decrease the objective function L_d in each step. Assume that λ_c is the chosen Lagrange multiplier, then we have:

---

[1]It is noted to the authors after the previous work [1] that SVM can also be trained without the sum constraint, if satisfying certain conditions [9].

$$L_d = \varepsilon|\lambda_c| - \lambda_c t_c + \frac{1}{2}\lambda_c^2 K_{cc} + \lambda_c z_c^{\text{old}} + W_{\text{const}}$$
$$z_c^{\text{old}} = f_c^{\text{old}} - \lambda_c^{\text{old}} K_{cc}$$

(19)

where W_{const} is a constant term and the superscript "old" indicates previous step. And K_{cc}, f_c^{old} respectively denotes $K(\mathbf{x}_c, \mathbf{x}_c), f(\mathbf{x}_c)^{\text{old}}$ for conciseness.

$$\frac{\partial L_d}{\partial \lambda_c} = \varepsilon\big(\text{sign}(\lambda_c)\big) - t_c + \lambda_c K_{cc} + f_c^{\text{old}} - \lambda_c^{\text{old}} K_{cc}$$

(20)

At the optimal solution $\lambda^\dagger$, the first-order partial derivative $\partial L_d / \partial \lambda_c = 0$.

$$\Rightarrow \lambda_c^\dagger = \lambda_c^{\text{old}} + \frac{1}{K_{cc}}\left(t_c - f_c^{\text{old}} - \varepsilon\big(\text{sign}(\lambda_c^\dagger)\big)\right)$$

(21)

Bound constraint $[-C, C]$ exists for the minimum $\lambda_c^\dagger$. In addition, L_d is not differentiable at $\lambda_c = 0$, causing a discontinuity in $\partial L_d / \partial \lambda_s$. Hence, we need to avoid crossing the value 0 in the update process by imposing more stringent constraint ($[0, C]$ if $\lambda_c^{\text{old}} > 0$; and $[-C, 0]$ if $\lambda_c^{\text{old}} < 0$).

Thus, we can derive the update rule for the following three cases:

(i) $\lambda_c^{\text{old}} = 0$:

$$\lambda_c^\dagger = \frac{1}{K_{cc}}\left(t_c - f_c^{\text{old}} - \varepsilon\big(\text{sign}(t_c - f_c^{\text{old}})\big)\right)$$
$$\lambda_c = \big[\lambda_c^\dagger\big]_{-C}^{C}$$

(22)

(ii) $\lambda_c^{\text{old}} > 0$:

$$\lambda_c^\dagger = \lambda_c^{\text{old}} + \frac{1}{K_{cc}}\left(t_c - f_c^{\text{old}} - \varepsilon\right)$$
$$\lambda_c - \big[\lambda_c^\dagger\big]_{0}^{C}$$

(23)

(iii) $\lambda_c^{\text{old}} < 0$:

$$\lambda_c^\dagger = \lambda_c^{\text{old}} + \frac{1}{K_{cc}}\left(t_c - f_c^{\text{old}} + \varepsilon\right)$$
$$\lambda_c = \big[\lambda_c^\dagger\big]_{-C}^{0}$$

(24)

4.3   Selection Criteria

Which Lagrange multiplier to choose remains an issue. A straightforward approach is to choose the one that decreases the objective function L_d the most.

$$c \in \arg\min_{i=1,\dots,N} \left(L_d(\lambda_i) - L_d(\lambda_i^{\text{old}})\right)$$

(25)

However, it is time consuming to calculate the exact decrease that each Lagrange multiplier brings. Instead, we use the violated degree of optimality conditions as an approximation and choose the one with the highest violated degree:

$$c \in \arg \max_{i=1,\ldots,N} d_i \tag{26}$$

Definition 1 d denotes the degree of violation of the KKT conditions. If $d_i > 0$, it indicates that KKT conditions are not satisfied for λ_i.

(i) $\lambda_i = C$:

$$d_i = \varepsilon - e_i \tag{27}$$

(ii) $0 < \lambda_i < C$:

$$d_i = |e_i - \varepsilon| \tag{28}$$

(iii) $\lambda_i = 0$:

$$d_i = |e_i| - \varepsilon \tag{29}$$

(iv) $-C < \lambda_i < 0$:

$$d_i = |e_i + \varepsilon| \tag{30}$$

(v) $\lambda_i = -C$:

$$d_i = \varepsilon + e_i \tag{31}$$

4.4 Termination Condition

Optimality conditions are not likely to be satisfied exactly since the algorithm is based on iterative update. In fact, they only need to be fulfilled within a tolerance γ. Thus, the training algorithm will be terminated if $\max_{i=1,\ldots,N} d_i < \gamma$.

It was found out that a tolerance equal to the square root of the machine epsilon would present stable results [10]. In our method, we choose $\gamma = 0.001$.

4.5 Convergence Proof

Theorem 2 *The training algorithm for sparse ELM for regression will converge to the global optimal solution in a finite number of iterations.*

Proof The proposed training algorithm of sparse ELM for regression is required to satisfy the following conditions in order to be convergent:

(i) The dual problem (10) is a convex QP one.
(ii) Lagrange multiplier chosen to be updated λ_c violated KKT conditions before the step.
(iii) The update rule guarantees that the objective function L_d will be decreased after the step.
(iv) Lagrange multipliers are all bounded within $[-C, C]^N$.

Therefore, the algorithm is convergent to the global optimum [11].

5   Experiments

We evaluate the performance of the proposed sparse ELM for regression problems in this section. Experimental platform is Matlab R2010b, on Intel i5-2400, 3.1 GHz CPU (except CASP: Matlab R2013a, on Intel Xeon E5-2650, 2 GHz CPU, because of memory issues for the unified ELM and SVR). SVR is realized by SVM and Kernel Methods Matlab Toolbox downloaded from [12] (Table 1).

5.1   Data Sets Description

We adopt different data sets from UCI repository [13]. Each data set is equally split into training and testing sets. For the training set, targets are linearly scaled into [0, 1] and attributes into [−1, 1]; for the testing set, they are respesctively scaled based on the factors used for the training set.

Table 1   Data sets description

Data set	# Train	# Test	# Features
Body fat	126	126	14
Mpg	196	196	7
Housing	253	253	13
Concrete	515	515	8
Mg	693	692	6
Spacega	1554	1553	6
Abalone	2089	2088	8
Wine quality	2449	2449	11
Cpusmall	4096	4096	12
Cadata	10320	10320	8
CASP	22865	22865	9

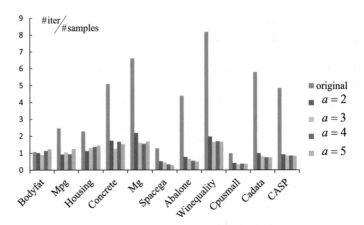

Fig. 2 Number of iterations for original and improved algorithms

Table 2 Detailed training process

Data set	# of λ_i (samples)	# of iterations	# λ_i changed in the process	Average $\frac{\Delta_i(1)}{\Delta_i}$
Bodyfat	126	134	33	2.01
Concrete	515	2622	217	7.88
Abalone	2089	9177	607	12.60
Cpusmall	4096	3998	225	12.34

5.2 Improvements of Convergence Speed

At here, we evaluate the convergence speed of the proposed training algorithm. Gaussian kernel $K(\mathbf{u}, \mathbf{v}) = \exp\left(-\frac{\|\mathbf{u}-\mathbf{v}\|^2}{2\sigma^2}\right)$ is used and parameters are naively set as $C = 1, \sigma = 1$. 20 repetitions are conducted for each experiment. As shown in Fig. 2, the ratios between the number of iterations and the number of training samples (the bars named "original") are much larger than 1.

Let us investigate the detailed training process. Four sets are used for illustration, while the others show similar characteristics. Observing from Table 2, only a part of λ_i's are changed in the process while the total number of iterations is much larger than the number of changed λ_i's. It means that some λ_i's are updated multiple times. $\Delta_i(1)$ denotes change of λ_i in the first time it is updated. And Δ_i is the change that λ_i finally achieves. As the ratios $\frac{\Delta_i(1)}{\Delta_i}$ are larger than 1, we can add a learning rate $\eta = 1 + a \cdot \exp\left(-(\text{TI}_i - 1)\right)$ in the update rule (22)–(24). TI_i denotes which time (1st time, 2nd time or more) the ith Lagrange multiplier λ_i is updated. Therefore, bigger changes can be achieved in the first several times λ_i being updated, so that fewer iterations are required. a is tried with 4 different values: $[2, 3, 4, 5]$. As shown

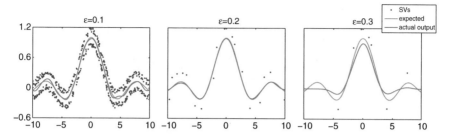

Fig. 3 Expected and actual outputs and SVs with different ε

in Fig. 2, all of them make significant improvements while the difference between each one is trivial. As our focus is not the optimal value of a, we use $a = 3$ for the remaining experiments.

5.3 Influence of ε

At here, we investigate the influence of ε. Gaussian kernel is adopted with naive parameter setting $C = 1, \sigma = 1$ similar to the previous section.

1. *The number of SVs*: synthetic data set "sinc" function is used:

$$y(x) = \begin{cases} \sin(x)/x, & x \neq 0 \\ 1, & x = 0 \end{cases} \tag{32}$$

   where x is uniformly distributed on the interval of $(-10, 10)$. The training and testing sets both include 1000 samples. Uniform noise within $(-0.2, 0.2)$ is added into y of the training data, while testing data is noise-free. ε is tried with 3 values: [0.1, 0.2, 0.3] as illustrated in Fig. 3.
2. *10-fold cross-validation error*: the optimal ε depends on the noise model and level [14], which is usually unknown in real problems. Compared with the unified ELM [5], ε is an additional parameter to be tuned. In order to conduct a fair comparison with the unified ELM later, the value of ε is investigated at here and will be fixed in the overall performance comparison.
   ε is tried with 8 values: [0, 0.02, 0.04, 0.06, 0.08, 0.1, 0.12, 0.14]. And 10-fold cross-validation RMSE is of all training sets are listed in Table 3. Consistent with the expectation, the optimal ε is problem-dependent. $\varepsilon = 0$, 0.02 and 0.06 are acceptable and bigger ε means more sparse network. Thus, $\varepsilon = 0.06$ is fixed for the remaining experiments.

Table 3 10-fold cross-validation RMSE with different ε

	$\varepsilon = 0$	$\varepsilon = 0.02$	$\varepsilon = 0.04$	$\varepsilon = 0.06$	$\varepsilon = 0.08$	$\varepsilon = 0.10$	$\varepsilon = 0.12$	$\varepsilon = 0.14$
Body fat	**0.0608**	0.0676	0.0747	0.0817	0.0882	0.0962	0.0993	0.1041
Mpg	**0.0476**	0.0504	0.0497	0.0525	0.0570	0.0643	0.0684	0.0734
Housing	0.0612	**0.0569**	0.0607	0.0668	0.0756	0.0829	0.0905	0.0990
Concrete	**0.0898**	0.0904	0.0915	0.0938	0.0967	0.0974	0.1000	0.1038
Mg	0.1442	0.1442	0.1441	**0.1408**	0.1410	0.1409	0.1423	0.1445
Spacega	**0.0320**	0.0331	0.0343	0.0374	0.0422	0.0498	0.0589	0.0627
Abalone	0.0781	0.0784	0.0778	**0.0777**	0.0783	0.0800	0.0824	0.0887
Winequality	0.1240	0.1234	**0.1225**	**0.1225**	0.1238	0.1234	0.1230	0.1240
Cpusmall	0.0313	**0.0307**	0.0325	0.0363	0.0428	0.0530	0.0575	0.0626
Cadata	0.1214	0.1210	0.1205	**0.1202**	0.1203	0.1205	0.1211	0.1224
CASP	0.2348	0.2343	0.2334	0.2320	0.2310	0.2300	0.2293	**0.2287**

5.4 Parameter Specification

Linear kernel $K(\mathbf{u}, \mathbf{v}) = \mathbf{u} \cdot \mathbf{v}$ and Gaussian kernel $K(\mathbf{u}, \mathbf{v}) = \exp\left(-\frac{\|\mathbf{u}-\mathbf{v}\|^2}{2\sigma^2}\right)$ are evaluated with 20 values: $[2^{-9}, 2^{-8}, \ldots, 2^{10}]$. In addition, Sigmoid activation function is utilized for the random hidden nodes form, where $L = 2^9$ is selected for all the problems. Optimal parameters are selected based on 10-fold cross-validation accuracy and specified in Table 4.

5.5 Performance Comparison

In this section, the proposed method is compared with the unified ELM and SVR. It is reasonably assumed that the mean RMSE over 20 repetitions provides a credible estimate of the algorithm's performance on each data set.

Tables 5, 6 and 7 list the detailed performance with linear kernel, Gaussian kernel and Sigmoid hidden nodes. For each problem, the smallest testing RMSE and shortest training time are highlighted.

1. *Generalization performance*: sparse ELM is compared with the unified ELM and SVR separately in generalization performance. Wilcoxon signed ranks test is a statistical test suitable for the comparison of two methods [15]. The test value between sparse ELM and SVR with Gaussian kernel is $z = -2.0449 < -1.96$. Thus, sparse ELM achieves better performance at the significance level $\alpha = 0.05$ in this case. Similarly, the test is conducted between sparse ELM and SVR with linear kernel and $z = -1.9560 > -1.96$. Thus, it is highly probable that sparse ELM presents better performance, even though it cannot be stated as statistically significant.

    Furthermore, Wilcoxon test is also conducted between sparse ELM and the unified ELM. And the z values are all bigger than -1.96: -0.9780 for linear kernel, -0.5335 for Gaussian kernel and -1.6893 for Sigmoid hidden nodes. Consequently, the performance of sparse ELM and unified ELM are comparable.

2. *Training and testing speed*: compared with the unified ELM, sparse ELM achieves faster training speed when dealing with large data sets, because it requires complexity of lower order. When the data set is small, sparse ELM costs more training time. However, the training speed is not a big issue for small data sets. In addition, the testing speed of sparse ELM is faster than the unified ELM.

    Comparing to SVR, sparse ELM achieves much faster training and testing speed as easily observed from Tables 5 and 6.

3. *Number of SVs*: as shown in Table 8, the proposed method does provide a sparse network. Therefore, less storage space is needed than the unified ELM.

    However, when comparing to SVR, it is not definite whether the proposed sparse ELM or SVR provides a more compact network.

Table 4 Parameter specifications

Data sets	Sparse ELM				Unified ELM				SVR		
	Linear kernel	Gaussian kernel		Sigmoid nodes	Linear kernel	Gaussian kernel		Sigmoid nodes	Linear kernel	Gaussian kernel	
	C	C	σ	C	C	σ	C	C	C	C	C
Bodyfat	2^0	2^1	2^2	2^2	2^6	2^{10}	2^4	2^6	2^2	2^0	2^2
Mpg	2^2	2^5	2^1	2^3	2^7	2^{10}	2^2	2^{10}	2^0	2^5	2^2
Housing	2^0	2^2	2^1	2^0	2^4	2^{10}	2^5	2^4	2^{-2}	2^3	2^3
Concrete	2^3	2^1	2^0	2^2	2^1	2^7	2^0	2^4	2^{-2}	2^1	2^{-1}
Mg	2^0	2^0	2^0	2^3	2^1	2^3	2^0	2^{10}	2^6	2^1	2^0
Spacega	2^3	2^1	2^1	2^1	2^9	2^9	2^3	2^6	2^3	2^2	2^2
Abalone	2^0	2^0	2^0	2^0	2^3	2^8	2^1	2^{10}	2^2	2^2	2^0
Winequality	2^{-5}	2^{-2}	2^1	2^{-2}	2^6	2^1	2^0	2^5	2^1	2^1	2^2
Cpusmall	2^{-4}	2^2	2^0	2^1	2^3	2^8	2^0	2^{10}	2^{-2}	2^{-1}	2^0
Cadata	2^0	2^0	2^0	2^1	2^5	2^1	2^0	2^6	2^1	2^1	2^0
CASP	2^0	2^0	2^0	2^0	2^2	2^0	2^{-2}	2^{10}	2^0	2^0	2^0

Table 5 Performance of sparse ELM, unified ELM and SVR with linear kernel

Data set	Sparse ELM				Unified ELM				SVR			
	Training RMSE	Testing RMSE	Training time (s)	Testing time (s)	Training RMSE	Testing RMSE	Training time (s)	Testing time (s)	Training RMSE	Testing RMSE	Training time (s)	Testing time (s)
Bodyfat	0.0534	0.0436	0.0066	0.0001	0.0336	**0.0161**	**0.0051**	0.0010	0.0402	0.0368	0.0268	0.0002
Mpg	0.0660	**0.1341**	0.0424	0.0004	0.0602	0.1500	**0.0039**	0.0014	0.0604	0.1452	0.0562	0.0007
Housing	0.0743	**0.1455**	0.0193	0.0004	0.0716	0.1507	**0.0073**	0.0012	0.0723	0.1543	0.0832	0.0008
Concrete	0.1485	**0.1272**	0.3420	0.0012	0.1457	0.1402	**0.0217**	0.0025	0.1462	0.1294	0.7283	0.0034
Mg	0.1674	**0.1570**	0.0840	0.0029	0.1666	0.1582	**0.0259**	0.0044	0.1668	0.1572	1.8135	0.0070
Spacega	0.0477	**0.0406**	0.3424	0.0026	0.0395	0.0435	**0.1390**	0.0291	0.0398	0.0436	4.5092	0.0205
Abalone	0.0802	**0.0771**	**0.3049**	0.0131	0.0806	0.0774	0.3171	0.0502	0.0802	0.0776	19.6114	0.0447
Winequality	0.1345	0.1273	**0.3792**	0.0368	0.1294	**0.1243**	0.4751	0.0562	0.1297	0.1266	46.9333	0.0952
Cpusmall	0.1147	0.1001	**0.5004**	0.0356	0.1036	**0.0961**	2.1691	0.1694	0.1212	0.1009	75.7302	0.0657
Cadata	0.1437	**0.1516**	**10.5743**	0.4547	0.1423	0.1536	15.1593	0.8153	0.1436	0.1537	3548.7487	1.3921
CASP	0.2390	**0.2413**	**102.9173**	3.5126	0.2457	0.2483	116.4548	4.7878	0.2500	0.2529	14400.5332	22.9362

Table 6 Performance of sparse ELM, unified ELM and SVR with Gaussian kernel

Data set	Sparse ELM				Unified ELM				SVR			
	Training RMSE	Testing RMSE	Training time (s)	Testing time (s)	Training RMSE	Testing RMSE	Training time (s)	Testing time (s)	Training RMSE	Testing RMSE	Training time (s)	Testing time (s)
Bodyfat	0.0516	0.0425	0.0066	0.0001	0.0333	**0.0162**	**0.0034**	0.0009	0.0522	0.0439	0.0224	0.0003
Mpg	0.0504	**0.1211**	0.0193	0.0004	0.0405	0.1216	**0.0042**	0.0012	0.0464	0.1228	0.0471	0.0007
Housing	0.0510	**0.1284**	0.0480	0.0008	0.0715	0.1496	**0.0061**	0.0033	0.0700	0.1509	0.0647	0.0012
Concrete	0.0671	**0.1241**	**0.0742**	0.0027	0.0503	0.1394	0.0827	0.0099	0.0583	0.1509	0.7481	0.0058
Mg	0.1307	0.1341	0.0987	0.0093	0.1272	**0.1319**	**0.0359**	0.0166	0.1279	0.1347	1.7136	0.0143
Spacega	0.0412	**0.0385**	**0.1499**	0.0083	0.0358	0.0401	0.1905	0.0880	0.0414	0.0422	4.3292	0.0307
Abalone	0.0753	0.0738	0.3580	0.0460	0.0738	**0.0735**	**0.3428**	0.1477	0.0736	0.0742	18.6725	0.1281
Winequality	0.1262	**0.1200**	0.5233	0.1328	0.1177	0.1204	**0.4816**	0.2035	0.1274	0.1208	45.1656	0.3300
Cpusmall	0.0317	0.0407	**1.2119**	0.0382	0.0248	**0.0352**	1.6629	0.5505	0.0348	0.0383	61.2297	0.1354
Cadata	0.1197	0.1423	**8.9124**	1.5977	0.1203	0.1433	15.3376	3.2928	0.1171	**0.1417**	2813.5930	3.0710
CASP	0.1874	**0.2026**	**51.6688**	15.2476	0.1935	0.2046	121.2519	24.9627	0.2307	0.2354	12269.6436	41.0614

Table 7 Performance of sparse ELM and unified ELM with Sigmoid hidden nodes

Data set	Sparse ELM				Unified ELM			
	Training RMSE	Testing RMSE	Training time (s)	Testing time (s)	Training RMSE	Testing RMSE	Training time (s)	Testing time (s)
Bodyfat	0.0524	0.0448	0.0105	0.0032	0.0331	**0.0198**	**0.0067**	0.0047
Mpg	0.0603	0.1389	0.0583	0.0030	0.0439	**0.1231**	**0.0086**	0.0085
Housing	0.0835	**0.1367**	0.0166	0.0081	0.0717	0.1436	**0.0090**	0.0091
Concrete	0.1224	**0.1374**	0.1630	0.0188	0.1243	0.1407	**0.0534**	0.0465
Mg	0.1555	0.1450	0.5045	0.0286	0.1359	**0.1330**	**0.0400**	0.0370
Spacega	0.0474	0.0450	0.3166	0.0369	0.0374	**0.0401**	**0.1823**	0.1246
Abalone	0.0819	0.0785	**0.3397**	0.0920	0.0756	**0.0741**	0.4459	0.2208
Winequality	0.1318	0.1221	**0.5067**	0.2089	0.1267	**0.1210**	0.6475	0.3216
Cpusmall	0.0345	**0.0373**	1.0375	0.2389	0.0333	0.0376	2.1991	0.8132
Cadata	0.1409	**0.1497**	**13.4122**	1.9611	0.1322	0.1498	21.3416	4.0056
CASP	0.2467	0.2495	**59.3273**	6.7365	0.2314	0.2347	70.7220	9.4233

Table 8 Number of support vectors

Data set	# Total vectors	Sparse ELM			Unified ELM			SVR	
		Linear kernel	Gaussian kernel	Sigmoid nodes	Linear kernel	Gaussian kernel	Sigmoid nodes	Linear kernel	Gaussian kernel
Bodyfat	126	19	15	13.65	126	126	126	16	20
Mpg	196	47	38	35.30	196	196	196	58	47
Housing	253	89	69	83.45	253	253	253	95	90
Concrete	515	313	212	278.25	515	515	515	335	233
Mg	693	497	430	429.55	693	693	693	500	418
Spacega	1554	93	112	152.60	1554	1554	1554	181	174
Abalone	2089	642	659	631.40	2089	2089	2089	703	671
Winequality	2449	1601	1572	1585.55	2449	2449	2449	1574	1602
Cpusmall	4096	858	215	745.80	4096	4096	4096	745	280
Cadata	10320	5684	4798	5305.40	10320	10320	10320	6041	4879
CASP	22865	18232	16442	17892.10	22865	22865	22865	18609	16644

6   Conclusions and Future Work

In this paper, a sparse ELM is proposed to solve regression problems. It provides a sparse network, requiring less storage space and testing time than the unified ELM. Furthermore, an efficient training algorithm, which is based on iterative computation, is developed for proposed method. It has several distinct merits: (1) no sum constraint $\sum_{i=1}^{N} \lambda_i = 0$ and bias b that exist in the SVR, eliminating the inefficiency associated with the sum constraint and bias; (2) no memory issue as it only needs to calculate and store the values encountered in each step, which are quite few; (3) less computational complexity than the unified ELM, so that it achieves faster training speed when dealing with large data sets; (4) sparse network is provided, and thus requiring less storage space and testing time.

In conclusion, the proposed method is preferred when facing large-scale regression problems, such as neuroscience, image processing, time series prediction, etc. In the future, we will try to realize parallel implementation of sparse ELM and to use kernel cache, which was introduced in [16], to further improve the training speed of sparse ELM.

Acknowledgments This work was supported by the Singapore Academic Research Fund (AcRF) Tier 1 under Project RG 80/12 (M4011092).

References

1. Bai, Z., Huang, G.-B., Wang, D., Wang, H., Westover, M.B.: Sparse extreme learning machine for classification. IEEE Trans. Cybern. **44**(10), 1858–1870 (2014)
2. Huang, G.-B., Zhu, Q.-Y., Siew, C.-K.: Extreme learning machine: theory and applications. Neurocomputing **70**, 489–501 (2006)
3. Huang, G.-B., Bai, Z., Kasun, L.L.C., Vong, C.M.: Local receptive fields based extreme learning machine. IEEE Comput. Intell. Mag. **10**(2), 18–29 (2015)
4. Bai, Z., Kasun, L.L.C., Huang, G.-B.: Generic object recognition with local receptive fields based extreme learning machine. Procedia Comput. Sci. **53**, 391–399 (2015)
5. Huang, G.-B., Zhou, H., Ding, X., Zhang, R.: Extreme learning machine for regression and multiclass classification. IEEE Trans. Syst. Man Cybern. Part B **42**(2), 513–529 (2012)
6. Drucker, H., Burges, C.J., Kaufman, L., Smola, A., Vapnik, V.: Support vector regression machines. In: Mozer, M., Jordan, J., Petscbe, T. (eds.) Neural Information Processing Systems 9, pp. 155–161. MIT Press, Cambridge (1997)
7. Smola, A., Schölkopf, B.: A tutorial on support vector regression. Stat. comput. **14**(3), 199–222 (2004)
8. Vapnik, V.: The Nature of Statistical Learning Theory. Springer, New York (1999)
9. Steinwart, I., Hush, D., Scovel, C.: Training SVMs without offset. J. Mach. Learn. Res. **12**, 141–202 (2011)
10. Flake, G.W., Lawrence, S.: Efficient SVM regression training with SMO. Mach. Learn. **46**(1–3), 271–290 (2002)
11. Osuna, E., Freund, R., Girosi, F.: An improved training algorithm for support vector machines. In: Neural Networks for Signal Processing [1997] VII. Proceedings of the 1997 IEEE Workshop, pp. 276–285. IEEE, New York (1997)

12. Canu, S., Grandvalet, Y., Guigue, V., Rakotomamonjy, A.: SVM and kernel methods matlab toolbox. Perception Systmes et Information. INSA de Rouen, Rouen, France (2005)
13. Frank, A., Asuncion, A.: UCI Machine Learning Repository (2010)
14. Smola, A., Murata, N., Schölkopf, B., Müller, K.-R.: Asymptotically optimal choice of ε-loss for support vector machines. In: ICANN 98, pp. 105–110. Springer, Berlin (1998)
15. Demšar, J.: Statistical comparisons of classifiers over multiple data sets. J. Mach. Learn. Res. **7**, 1–30 (2006)
16. Chang, C.-C., Lin, C.-J.: LIBSVM: a library for support vector machines. ACM Trans. Intell. Syst. Technol. **2**(3), 27 (2011)

WELM: Extreme Learning Machine with Wavelet Dynamic Co-Movement Analysis in High-Dimensional Time Series

Heng-Guo Zhang, Rui Nian, Yan Song, Yang Liu, Xuefei Liu and Amaury Lendasse

Abstract In this paper, we propose a fast and efficient learning approach called WELM based on Extreme Learning Machine and 3-D Wavelet Dynamic Co-Movement Analysis to enhance the speed and precision of big data prediction. 3-D Wavelet Dynamic Co-Movement Analysis is firstly employed to transform optimization problems from an original higher-dimensional space to a new lower-dimensional space while preserving the optimum of the original function, and then ELM is utilized to train and forecast the whole process. WELM model is used in the volatility of time series prediction. The forecasts obtained by WELM has been compared with ELM, PCA-ELM, ICA-ELM, KPCA-ELM, SVM and GARCH type models in terms of closeness to the realized volatility. The computational results demonstrate that the WELM provides better time series forecasts and it shows the excellent performance in the accuracy and efficiency.

Keywords Extreme learning machine · Wavelet dynamic Co-Movement analysis · High-dimensional space · Volatility · GARCH models

R. Nian (✉) · Y. Song · Y. Liu
College of Information Science and Engineering, Ocean University of China,
266003 Qingdao, China
e-mail: nianrui_80@163.com

H.-G. Zhang · X. Liu
School of Economics, Ocean University of China, 266003 Qingdao, China

A. Lendasse
Arcada University of Applied Sciences, 00550 Helsinki, Finland

A. Lendasse
Department of Mechanical and Industrial Engineering and the Iowa
Informatics Initiative, The University of Iowa, Iowa City, IA 52242-1527, USA

© Springer International Publishing Switzerland 2016
J. Cao et al. (eds.), *Proceedings of ELM-2015 Volume 2*,
Proceedings in Adaptation, Learning and Optimization 7,
DOI 10.1007/978-3-319-28373-9_40

491

1 Introduction

In the Era of Big Data, how to forecast and model the fluctuation of time series turns out to be one of most challenging topic. Motivated with 3-D Wavelet Dynamic Co-Movement Analysis [1–5], we come up with a fast and efficient Wavelet Analysis approach for the big data via ELM [6, 7]. We first transform optimization problems from an original high dimensional space to a new low dimensional space while preserving the optimum of the original function by 3-D Wavelet Dynamic Co-Movement Analysis, and then employ ELM to forecast the volatility of time series, and the relevant mathematical criterion of the model selection will also be developed for the performance improvements.

The rest of the paper is organized as follows. In Sect. 2, we propose a model called WELM based on Extreme Learning Machine and 3-D Wavelet Dynamic Co-Movement Analysis. Section 3 comes to the conclusions.

2 Fast 3-D Wavelet Dynamic Co-Movement Analysis via ELM

2.1 General Model

The basic idea of the proposed method is that 3-D Wavelet Dynamic Co-Movement Analysis is firstly employed to transform optimization problems from an original higher-dimensional space to a new lower-dimensional space while preserving the optimum of the original function, and then ELM is utilized to train and forecast the whole process. In this paper, employing 3-D Wavelet Dynamic Co-Movement Analysis, we pick out the decisive features which can reflect the whole time or most time domains and cover a wide range of frequencies in frequency domains and have a high strength both in time frequency domains as the input of ELM learning. Let the time series set $X_{1:T,i} = [x_{1i}, \ldots, x_{ti}, \ldots, x_{Ti}]$ as an input in the train set for 3-D Wavelet Dynamic Co-Movement Analysis contain T samples, $t = 1, \ldots, T$. $X_{1:T,i}$ belongs to the original features set $X_t = \{X_{t1}, \ldots, X_{ti}, \ldots, X_{tN}\}$ as an input in the train set or the test set composed of N features for ELM learning, $i = 1, \ldots, N$. After selecting features by 3-D Wavelet Dynamic Co-Movement Analysis, the decisive features can be written as $X_t' = \{X_{t1}', \ldots, X_{tj}', \ldots, X_{tM}'\}$ in a new lower-dimensional space, $t = 1, 2 \ldots, T$ and $1 \leq j \leq M, M \leq N$. Once a set of SLFNs is reasonably established via ELM leaning with the help of 3-D Wavelet Dynamic Co-Movement Analysis, the decisive features set can Let the train set $X_t' = \{X_{t1}', \ldots, X_{tj}', \ldots, X_{tM}'\}$ contain M features, with each feature X_{tj}' composed of $T - c - 1$ samples when $t = 1, 2, \ldots, T - c - 1.c$ is constant. Let the test set $X_t' = \{X_{t1}', \ldots, X_{tj}', \ldots, X_{tM}'\}$ contain M features, with each feature X_{tj}' composed of c samples when $t = T - c, \ldots, T - 1, T$. c is constant. The normalized test input can be written as

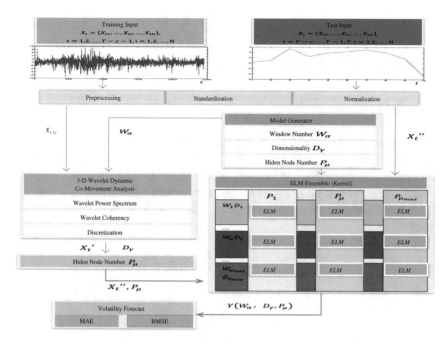

Fig. 1 The flowchart of our proposed approach

$X_t'' = \{X_{t1}'', \ldots, X_{tj}'', \ldots, X_{tM}''\}$, $t = T - c, \ldots, T - 1, T$ and c is constant. The flow chart of our method is shown in Fig. 1. The model selection strategy is further developed in SLFN so as to well fit the simulation process of 3-D Wavelet Dynamic Co-Movement Analysis via ELM ensemble. The decision is mainly made for the following parameters, i.e., the number of the window sizes W_α, $1 \leq W_\alpha \leq T$, the number of the hidden nodes P_μ, $1 \leq P_\mu \leq P_{\mu max}$ and the dimensionality D_γ, $1 \leq D_\gamma \leq N$. In SLFN, the test output of the low dimensional representation in each SLFN can be represented in the D_γ dimensional space as $Y_{test}(W_a, D_\gamma, P_\mu) = [y_{T-c}, \ldots, y_{T-1}, y_T]^T$, $t = T - c, \ldots, T - 1, T$ and c is constant.

2.2 ELM Learning Based 3-D Wavelet Dynamic Co-Movement Analysis

In this paper, we take the ELM learning strategies to perform a fast low-dimensional representation on the basis of 3-D Wavelet Dynamic Co-Movement Analysis and develop the relevant mathematical criterion. Let the decisive features set $X_t' = \{X_{t1}', \ldots, X_{tj}', \ldots, X_{tM}'\}$ range from tr_{min} to tr_{max}, each feature X_{tj}' in X_t' can be mapped into a normalized feature X_{tj}'' by calculating

$X_{tj}^{''} = (X_{tj}^{'} - tr_{\min})/(tr_{\max} - tr_{\min}), t = 1, \ldots, T, j = 1, \ldots, M,$ so that the normalized decisive features $X_t^{''} = \{X_{t1}^{''}, \ldots, X_{tj}^{''}, \ldots, X_{tM}^{''}\}$ can be considered as the train set or the test set for ELM learning.

2.2.1 ELM Learning

In this paper, we try to simply realize the 3-D Wavelet Dynamic Co-Movement Analysis by means of ELM techniques. We propose to facilitate the entire ELM learning in an ensemble at several levels. Let $X_t^{''} = \{X_{t1}^{''}, \ldots, X_{tj}^{''}, \ldots, X_{tM}^{''}\}$ $t = 1, 2, \ldots, T - c - 1, j = 1, \ldots, M$ be the train set for ELM learning. c is constant. In general, ELM is to minimize the training error as well as the norm of the output weights:

$$Minimize: \left\|\tilde{H}\beta^{''} - Y\right\|^2 and \left\|\beta^{''}\right\| \tag{1}$$

where $\tilde{H}$ is the hidden-layer output matrix

$$
\tilde{H} = \begin{bmatrix} h_1(X_{11}^{''}) & \cdots & h_P(X_{1M}^{''}) \\ \vdots & \vdots & \vdots \\ h_1(X_{T1}^{''}) & \cdots & h_P(X_{TM}^{''}) \end{bmatrix}
$$
$$
= \begin{bmatrix} g(a_1 \times X_{11}^{''} + b_1) & \cdots & g(a_P \times X_{1M}^{''} + b_P) \\ \vdots & \vdots & \vdots \\ g(a_1 \times X_{T1}^{''} + b_1) & \cdots & g(a_P \times X_{TM}^{''} + b_P) \end{bmatrix}_{M \times P} \tag{2}
$$

Seen from (2), to minimize the norm of the output weights $\left\|\beta^{''}\right\|$ is actually to maximize the distance of the separating margins of the two different classes in the ELM feature space: $2/\left\|\beta^{''}\right\|$.

The minimal norm least square method instead of the standard optimization method was used in the original implementation of ELM:

$$\beta^{''} = \tilde{H}^\dagger Y \tag{3}$$

where $\tilde{H}^\dagger$ is the Moore–Penrose generalized inverse of matrix $\tilde{H}$. The orthogonal projection method can be used in two cases: when $\tilde{H}^T \tilde{H}$ is nonsingular and $\tilde{H}^\dagger = (\tilde{H}^T \tilde{H})^{-1} \tilde{H}^T$, or when $\tilde{H}^T \tilde{H}$ is nonsingular and $\tilde{H}^\dagger = \tilde{H}^T (\tilde{H}\tilde{H})^{-1}$. If the Number of Training Samples is Not Huge, then we can get

$$\beta^{''} = \tilde{H}^T (\frac{I}{C} + \tilde{H}\tilde{H}^T)^{-1} Y \tag{4}$$

If the Number of Training Samples is Huge and we have:

$$\beta'' = (\frac{I}{C} + \tilde{H}^T \tilde{H})^{-1} \tilde{H}^T Y \tag{5}$$

After 3-D Wavelet Dynamic Co-Movement Analysis, the whole process is further conducted by the ELM ensemble when different types of SLFNs are established with all kinds of variations in the number of the window sizes W_α, the number of the hidden nodes P_μ and the dimensionality D_r. Then learning model of each SLFN in the ELM ensemble can be written as follows,

$$y_{t+1}(W_a, D_\gamma, P_\mu) = f_{P_\mu}(X_t'') = h(X_t'')\beta'' \tag{6}$$

where $t = T - c, \ldots, T - 1, T, 1 \le P_\mu \le P_{\mu max}$. β'' is the vector of the output weights between the hidden layer of P_μ nodes and the output node in the low dimensional space. $h(X_t'')$ is the output (row) vector of the hidden layer with respect to the input X_t'' in the low dimensional space from the training set.

The purpose is to generate multiple versions of ELM learning from different perspective, which will decide the D_r, $P\mu$, D_r in the model selection. A collection of the component SLFNs is organized into the ELM ensemble to seek a better expression of the low-dimensional space. Theoretically, the model selection of the ELM ensemble architecture is as follows.

2.2.2 The Decision of the Dimensionality D_r

In the ELM ensemble, when the W_α and the P_μ are fixed, a set of actual training output $Y = \{y(W_\alpha, 1, P_\mu), \ldots, y(W_\alpha, D_\gamma, P_\mu), \ldots, y(W_\alpha, N_\gamma, P_\mu)\}$ in different kinds of SLFNs, mathematically correlated with the parameters D_r, will be obtained from the time series set $X_{1:T,i} = [x_{1i}, \ldots, x_{ti}, \ldots, x_{Ti}]$, $t = 1, \ldots, T$. In this paper, employing the wavelet power spectrum $|G_{X_{1:T,i}X_{1:T,0}}(\tau, \vartheta)|^2$ and the wavelet coherency $\Gamma^2_{X_{1:T,i}X_{1:T,0}}(\tau, \vartheta)$, we pick out the decisive features. Therefore, the decisive features can determine the dimension D_r.

Wavelet power spectrum. In this paper, we define a wavelet power spectrum $|G_{X_{1:T,i}}(\tau, \vartheta)|^2$ for the time series set $X_{1:T,i} = [x_{1i}, \ldots, x_{ti}, \ldots, x_{Ti}]$, $t = 1, \ldots, T$ is

$$\left|G_{X_{1:T,i}}(\tau, \vartheta)\right|^2 = G_{X_{1:T,i}}(\tau, \vartheta)G^*_{X_{1:T,i}}(\tau, \vartheta) \tag{7}$$

From the CWT of the time series, one can construct the cross wavelet transform (XWT). We can similarly define a cross-wavelet transform of the time series set $X_{1:T,i} = [x_{1i}, \ldots, x_{ti}, \ldots, x_{Ti}]$ and $X_{1:T,0} = [x_{10}, \ldots, x_{t0}, \ldots, x_{T0}]$, $t = 1, \ldots, T$ as $G_{X_{1:T,i}X_{1:T,0}}(\tau, \vartheta) = G_{X_{1:T,i}}(\tau, \vartheta)G^*_{X_{1:T,0}}(\tau, \vartheta)$. The cross-wavelet power spectrum can be a measure of the localized covariance between the time series $X_{1:T,i}$ and $X_{1:T,0}$ for the specified frequency, which is accordingly written as:

$$\left|G_{X_{1:T,i}X_{1:T,0}}(\tau,\vartheta)\right|^2 = \left|G_{X_{1:T,i}}(\tau,\vartheta)\right|^2\left|G^*_{X_{1:T,0}}(\tau,\vartheta)\right|^2 \tag{8}$$

If wavelet power spectrum $\left|G_{X_{1:T,i}X_{1:T,0}}(\tau,\vartheta)\right|^2 \leq 0.5$, the feature $X_{1:T,i}$ will be deleted.

Wavelet coherency. The spectral density functions are $R_{X_{1:T,i}}(\Lambda), i=1,2,\ldots N$, $t=1,\ldots T$ and $R_{X_{1:T,0}}(\Lambda), t=1,2,\ldots,T$, $-\pi\leq\Lambda<\pi$, and the co-spectrum is $CS_{X_{1:T,i}X_{1:T,0}}(\Lambda)$ of the time series $X_{1:T,i}$ and $X_{1:T,0}$. In this paper, we define dynamic correlation for the time series set $X_{1:T,i}=[x_{1i},\ldots,x_{ti},\ldots,x_{Ti}]$, $t=1,\ldots,T$ is

$$\rho_{X_{1:T,i}X_{1:T,0}}(\Lambda) = \frac{CS_{X_{1:T,i}X_{1:T,0}}\Lambda)}{\sqrt{R_{X_{1:T,i}X_{1:T,0}}(\Lambda)R_{X_{1:T,i}X_{1:T,0}}(\Lambda)}} \tag{9}$$

We define the spectral decomposition of the time series $X_{1:T,i}$ and $X_{1:T,0}$:

$$X_{1:T,i} = \int_{-\pi}^{\pi} e^{i\nu t}dI_{X_{1:T,i}}(\Lambda) \quad X_{1:T,0} = \int_{-\pi}^{\pi} e^{i\nu t}dI_{X_{1:T,0}}(\Lambda) \tag{10}$$

where $dI_{X_{1:T,i}}(\Lambda)$ and $dI_{X_{1:T,0}}(\Lambda)$ are orthogonal increment processes.

As is well known, the spectral and cross-spectral density functions of the time series $X_{1:T,i}$ and $X_{1:T,0}$ are related to the above representation:

$$R_{X_{1:T,i}}(\Lambda) = \mathrm{var}(e^{i\nu t}dI_{X_{1:T,i}}(\Lambda)) = \mathrm{var}(dI_{X_{1:T,i}}(\Lambda))$$
$$R_{X_{1:T,0}}(\Lambda) = \mathrm{var}(e^{i\nu t}dI_{X_{1:T,0}}(\Lambda)) = \mathrm{var}(dI_{X_{1:T,0}}(\Lambda)) \tag{11}$$
$$CS_{X_{1:T,i}X_{1:T,0}}(\Lambda) = \mathrm{cov}(dI_{X_{1:T,i}}(\Lambda)dI_{X_{1:T,0}}(\Lambda))$$

If the time series $X_{1:T,i}$ is real, then $dI_{X_{1:T,i}}(\Lambda) = d\bar{I}_{X_{1:T,i}}(-\Lambda)$, so:

$$e^{i\nu t}dI_{X_{1:T,i}}(\Lambda) + e^{-i\nu t}dI_{X_{1:T,i}}(-\Lambda) = 2\cos(\Lambda t)dU_{X_{1:T,i}}(\Lambda) - 2\sin(\Lambda t)dV_{X_{1:T,i}}(\Lambda) \tag{12}$$

where $dU_{X_{1:T,i}}$ and $dV_{X_{1:T,i}}$ are the real and the imaginary part of the orthogonal increment processes $dI_{X_{1:T,i}}$. So we can calculate $\rho_{X_{1:T,i}X_{1:T,0}}(\Lambda)$ according to this.

Similarly, in this paper, we define the wavelet coherency for the time series set $X_{1:T,i}=[x_{1i},\ldots,x_{ti},\ldots,x_{Ti}]$, $t=1,\ldots,T$ is

$$\Gamma_{X_{1:T,i}X_{1:T,0}}(\tau,\vartheta) = \frac{G_{X_{1:T,i}X_{1:T,0}}(\tau,\vartheta)}{\sqrt{G_{X_{1:T,i}}(\tau,\vartheta)G_{X_{1:T,0}}(\tau,\vartheta)}} \tag{13}$$

and

$$\Gamma^2_{X_{1:T,i}X_{1:T,0}}(\tau, \vartheta) = \frac{[\Re G_{X_{1:T,i}X_{1:T,0}}(\tau, \vartheta)]^2 + [\Im G_{X_{1:T,i}X_{1:T,0}}(\tau, \vartheta)]^2}{G_{X_{1:T,i}}(\tau, \vartheta)G_{X_{1:T,0}}(\tau, \vartheta)}$$

$$= \frac{\left|O\big(\vartheta^{-1}G_{X_{1:T,i}X_{1:T,0}}(\tau, \vartheta)\big)\right|^2}{O\Big(\vartheta^{-1}\big|G_{X_{1:T,i}}(\tau, \vartheta)\big|^2\Big) \cdot O\Big(\vartheta^{-1}\big|G^*_{X_{1:T,0}}(\tau, \vartheta)\big|^2\Big)} \quad (14)$$

The functions $\Re G_{X_{1:T,i}X_{1:T,0}}(\tau, \vartheta)$ and $\Im G_{X_{1:T,i}X_{1:T,0}}(\tau, \vartheta)$ in (14) are respectively the real and imaginary parts of $G_{X_{1:T,i}X_{1:T,0}}(\tau, \vartheta)$. Hence the co- and quadrature-wavelet spectra of the time series $X_{1:T,i}$ and $X_{1:T,0}$. O is a smoothing operator. Smoothing is achieved by convolution in time and frequency domain. We define the smoothing operator O as:

$$O(G) = O_s(O_t(G(\tau, \vartheta))) \quad (15)$$

where O_s is smoothing along the wavelet scale axis and O_t smoothing in time. For the Morlet wavelet, a suitable smoothing operator is given

$$O_s(G)|_\vartheta = (G(\tau, \vartheta)*e_1^{\frac{-t^2}{2\vartheta^2}})|_\vartheta$$
$$O_t(G)|_\vartheta = (G(\tau, \vartheta)*e_2\prod(0.6\vartheta))|_t \quad (16)$$

where e_1 and e_2 are normalization constants and $\prod$ is the rectangle function. The statistical significance level of the wavelet coherence is estimated using Monte Carlo methods.

The squared wavelet coherency $\Gamma^2_{X_{1:T,i}X_{1:T,0}}(\tau, \vartheta)$ varies between 0 and 1, with a high value showing strong co-movement between time series, and vice versa. Therefore a graph of wavelet squared coherency will show regions in the time—frequency space. Zero coherency indicates no co-movement while the highest coherency implies the strongest co-movement between time series. If wavelet coherency $\Gamma^2_{X_{1:T,i}X_{1:T,0}}(\tau, \vartheta) \leq 0.5$, the feature $X_{1:T,i}$ will be deleted.

The phase difference $\psi_{X_{1:T,i}X_{1:T,0}}$ which characterizes phase relationship between the time series $X_{1:T,i}$ and $X_{1:T,0}$, in this paper, is defined as:

$$\psi_{X_{1:T,i}X_{1:T,0}} = \tan^{-1}\left(\frac{\mathbb{C}\{O(\vartheta^{-1}G_{X_{1:T,i}X_{1:T,0}}(\tau, \vartheta))\}}{\mathbb{Q}\{O(\vartheta^{-1}G_{X_{1:T,i}X_{1:T,0}}(\tau, \vartheta))\}}\right) \text{ with } \psi_{X_{1:T,i}X_{1:T,0}} \in [-\pi, \pi]$$

$$(17)$$

where $\mathbb{C}$ and $\mathbb{Q}$ are the imaginary and real parts of the smoothed cross-wavelet transform, respectively. If phase difference $\psi_{X_{1:T,i}X_{1:T,0}} \in (0, \pi/2)$ and $\psi_{X_{1:T,i}X_{1:T,0}} \in (-\pi, -\pi/2)$, then the series move in phase (positively co-movement) with $X_{1:T,i}$

leading $X_{1:T,0}$. If $\psi_{X_{1:T,i}X_{1:T,0}} \in (\pi/2, \pi)$ and $\psi_{X_{1:T,i}X_{1:T,0}} \in (-\pi/2, 0)$, then the series move out of phase (negatively co-movement) with $X_{1:T,0}$ leading $X_{1:T,i}$.

Discretization. In this paper, when the time series set $X_{1:T,i} = [x_{1i}, \ldots, x_{ti}, \ldots, x_{Ti}]$, $t = 1, \ldots, T$ with uniform time steps Δt is discrete, the CWT is defined as the convolution of the discrete time series $X_{1:T,i}$ at time t and a scaled α with the Morlet wavelet $\phi_0(t)$:

$$G_{X_{1:T,i}}(\tau, \vartheta) = \sqrt{\frac{\Delta t}{\vartheta}} \sum_{t'=1}^{Q} X_{1:T,i,t'} \phi_0[(t'-t)\frac{\Delta t}{\vartheta}] \tag{18}$$

where Q is the number of points in the time series. By varying the wavelet scale α and translating along the localized time index t', one can construct a picture showing both the amplitude of any features versus the scale and how this amplitude varies with time.

To approximate the continuous wavelet transform, the convolution (18) should be done Q times for each scale, where Q is the number of points. By choosing Q points, the convolution theorem allows us do all Q convolutions simultaneously in Fourier space employing a Discrete Fourier Transform (DFT). The DFT of the discrete time series $X_{1:T,i}$ is:

$$\hat{X}_{1:T,i}(k) = \frac{1}{Q} \sum_{t=1}^{Q} X_{1:T,i} e^{-2\pi ikt/Q} \tag{19}$$

where $k = 1 \ldots Q$ is the frequency index.

In the continuous limit, the Fourier transform of a function $\phi(t/\vartheta)$ is given by $\hat{\phi}(\vartheta\varpi)$. According to the convolution theorem, the wavelet transform is the inverse Fourier transform of the product:

$$G_{X_{1:T,i}}(\tau, \vartheta) = \sum_{k=1}^{Q} \hat{X}_{1:T,i}(k) \hat{\phi}^*(\vartheta\varpi_k) e^{i\varpi_k t\Delta t} \tag{20}$$

where the (*) indicates the complex conjugate, and $\hat{\phi}(\vartheta\varpi_k) = \sqrt{\frac{2\pi\vartheta}{\Delta t}} \hat{\phi}_0(\vartheta\varpi_k)$ and $\sum_{k=1}^{Q} |\hat{\phi}(\vartheta\varpi_k)|^2 = Q$, Q is the number of points. The angular frequency ϖ_k is defined as:

$$\varpi_k = \begin{cases} \frac{2\pi k}{Q\Delta t} & k \leq \frac{Q}{2} \\ -\frac{2\pi k}{Q\Delta t} & k > \frac{Q}{2} \end{cases} \tag{21}$$

Employing (20) and a standard Fourier transform routine, one can calculate the continuous wavelet transform (for a given ϑ) at all t simultaneously and efficiently.

2.2.3 The Choose of the Number of Hidden Nodes P_μ

When the number of the window sizes W_α and the dimensionality D_r are fixed, training ELM is Divided into two kinds of cases.

If the Number of Training Samples is Not Huge. The number of hidden nodes P_μ can be much smaller than the number of training samples, the computational cost reduces dramatically. We have

$$\beta'' = \tilde{H}^T (\frac{I}{C} + \tilde{H}\tilde{H}^T)^{-1}Y \qquad (22)$$

The output function of ELM is

$$y_{t+1}(W_\alpha, D_r, P_\mu) = f(X_t'') = h(X_t'')\beta'' = h(X_t'')\tilde{H}^T(\frac{I}{C} + \tilde{H}\tilde{H}^T)^{-1}Y \qquad (23)$$

where $1 \leq t \leq T - c - 1$, $\quad 1 \leq P_\mu \leq P_{\mu max}$.

The kernel matrix of ELM can be defined as follow. Let $\Theta_{ELM} = \tilde{H}\tilde{H}^T$: $\Theta_{ELMi,j} = h(X_{ti}'')h(X_{tj}'') = K(X_{ti}'', X_{tj}'')$. The output function of ELM can be written compactly as:

$$y_{t+1}(W_\alpha, D_r, P_\mu) = f(X_t'') = h(X_t'')\beta'' = h(X_t'')\tilde{H}^T(\frac{I}{C} + \tilde{H}\tilde{H}^T)^{-1}Y$$

$$= \begin{bmatrix} K(X_t'', X_{t1}'') \\ \cdots \\ K(X_t'', X_{TM}'') \end{bmatrix}^T (\frac{I}{C} + \Theta_{ELM})^{-1}Y \qquad (24)$$

If the Number of Training Samples is Huge. If the number of training data is very large, for example, it is much larger than the dimensionality of the feature space, we have an alternative solution. We have

$$\beta'' = C\tilde{H}^T\xi$$
$$\xi = \frac{I}{C}(\tilde{H}^T)^\dagger\beta''$$
$$\tilde{H}^T(\tilde{H} + \frac{I}{C}(\tilde{H}^T)^\dagger)\beta'' = \tilde{H}^T Y \qquad (25)$$
$$\beta'' = (\frac{I}{C} + \tilde{H}^T\tilde{H})^{-1}\tilde{H}^T\tilde{H}$$

The output function of ELM can be written compactly as:

$$y_{t+1,D_r}(W_\alpha, D_r, P_\mu) = f(X_t'') = h(X_t'')\beta'' = h(X_t'')(\frac{I}{C} + \tilde{H}^T\tilde{H})^{-1}\tilde{H}^T\tilde{H} \qquad (26)$$

where $1 \leq t \leq T - c - 1$, $\quad 1 \leq P_\mu \leq P_{\mu max}$.

2.2.4 The Decision of the Window Sizes W_α

When the number of hidden nodes P_μ and the dimensionality D_r are fixed, the decision of the window sizes W_α depend on the nature of the data and the objective of the research.

2.3 Algorithm

In the following, the WELM algorithm in this paper can be summarized as follows.

Algorithm 1

Input:
   Let the time series set $X_{1:T,i} = [x_{1i}, \ldots, x_{ti}, \ldots, x_{Ti}]$, $t = 1, \ldots, T$ as an input for 3-D Wavelet Dynamic Co-Movement Analysis contain T samples.
Steps:
For $t = 1 : T, i = 1 : N$

(1) Calculate the Discrete Fourier Transform (DFT) of the time series $X_{1:T,i}$
$$\hat{X}_{1:T,i}(k) = \frac{1}{Q} \sum_{t=1}^{Q} X_{1:T,i} e^{-2\pi i k t / Q}$$

(2) Calculate the wavelet transform
$$G_{X_{1:T,i}}(\tau, \vartheta) = \sum_{k=1}^{Q} \hat{X}_{1:T,i}(k) \hat{\phi}^*(\vartheta \varpi_k) e^{i \varpi_k t \Delta t}$$

(3) Calculate the wavelet power spectrum
$$\left| G_{X_{1:T,i} X_{1:T,0}}(\tau, \vartheta) \right|^2 = \left| G_{X_{1:T,i}}(\tau, \vartheta) \right|^2 \left| G^*_{X_{1:T,0}}(\tau, \vartheta) \right|^2$$

   If wavelet power spectrum $\left| G_{X_{1:T,i} X_{1:T,0}}(\tau, \vartheta) \right|^2 \leq 0.5$, the feature $X_{1:T,i}$ will be picked out.
(4) Calculate wavelet coherency
$$\Gamma^2_{X_{1:T,i} X_{1:T,0}}(\tau, \vartheta) = \frac{\left| O(\vartheta^{-1} G_{X_{1:T,i} X_{1:T,0}}(\tau, \vartheta)) \right|^2}{O\left(\vartheta^{-1} \left| G_{X_{1:T,i}}(\tau, \vartheta) \right|^2 \right) \cdot O\left(\vartheta^{-1} \left| G^*_{X_{1:T,0}}(\tau, \vartheta) \right|^2 \right)}$$

   If wavelet coherency $\Gamma^2_{X_{1:T,i} X_{1:T,0}}(\tau, \vartheta) \leq 0.5$, the feature $X_{1:T,i}$ will be picked out.
(5) Delete the feature $X_{1:T,i}$ which has no or little co-movement.
   End
Output:
   Output the decisive features $X'_t = \{X'_{t1}, \ldots, X'_{tj}, \ldots, X'_{tM}\}$ in a new lower-dimensional space, $t = 1, 2 \ldots, T$ and $j = 1, 2, \ldots, M, M \leq N$.

Algorithm 2

Input:
   Let the normalized decisive set $X_t'' = \{X_{t1}'', \ldots, X_{tj}'', \ldots, X_{tM}''\}$ $t = 1, \ldots, T - c - 1$ be the training set and the decisive features $X_t'' = \{X_{t1}'', \ldots, X_{tj}'', \ldots, X_{tM}''\}$ $t = T - c, \ldots, T$ be the testing set.
   Steps:

(1) The features reserved X_t'' and their label t form the set (t, X_t''), which is the input of ELM learning. Hidden node output function $f(a_i, b_i, t)$ and the number of hidden nodes P_μ.

(2) Assign parameters of hidden nodes (a_i, b_i) randomly, $i = 1, \ldots, P_{\mu max}$.

(3) Calculate the hidden layer output matrix $\tilde{H}$:
$$\tilde{H} = \begin{bmatrix} h_1(X_{11}'') & \cdots & h_P(X_{1M}'') \\ \vdots & \vdots & \vdots \\ h_1(X_{T1}'') & \cdots & h_P(X_{TM}'') \end{bmatrix} = \begin{bmatrix} g(a_1 \times X_{11}'' + b_1) & \cdots & g(a_P \times X_{1M}'' + b_P) \\ \vdots & \vdots & \vdots \\ g(a_1 \times X_{T1}'' + b_1) & \cdots & g(a_P \times X_{TM}'' + b_P) \end{bmatrix}_{M \times P}$$

(4) If the Number of Training Samples is Not Huge, the output of ELM learning is:
$$\beta'' = \tilde{H}^T(\frac{I}{C} + \tilde{H}\tilde{H}^T)^{-1}Y$$

(5) If the Number of Training Samples is Huge, the output of ELM learning is:
$$\beta'' = (\frac{I}{C} + \tilde{H}^T\tilde{H})^{-1}\tilde{H}^T\tilde{H}$$

Output:
Input the testing set $X_t'' = \{X_{t1}'', \ldots, X_{tj}'', \ldots, X_{tM}''\}$ $t = T - c, \ldots, T$ into ELM and calculate the output using β''.

3   Conclusion

In this paper, we propose a model called WELM based on a 3-D wavelet dynamic co-movement analysis and extreme learning machine. We randomly select 305 stocks as the input sets of the WELM. As verified by the results, compared to ELM, PCA-ELM, ICA-ELM, KPCA-ELM, SVM and GARCH type models, WELM achieves better generalization performance for regression. WELM (WELM-kernel) not only can turn an original higher-dimensional space into a new lower-dimensional space but also further improve the forecasting accuracy and speed. Using financial time series and SZSE market data samples from CSMAR, the case study results illustrate that the WELM model can be used as a tool by financial market partici-pants. The selection criteria of feature employing a 3-D wavelet dynamic co-movement analysis need further quantitative.

Acknowledgements This work is partially supported by the Natural Science Foundation of P. R. China (31202036), the National Science and Technology Pillar Program (2012BAD28B05), the National High-Tech R&D 863 Program (2014AA093410), the Key Science and Technology Project of Shandong Province (2013GHY11507), the Fundamental Research Funds for the Central Universities (201362030), and the Natural Science Foundation of P. R. China (41176076, 51379198, 51075377).

References

1. Torrence, C., Compo, G.P.: A practical guide to wavelet analysis. Bull. Am. Meteorol. Soc. **79** (1), 61–78 (1998)
2. Rua, A.: Measuring comovement in the time–frequency space. J. Macroecon. **32**(2), 685–691 (2010)
3. Kaiser G.: A Friendly Guide to Wavelets. Springer Science & Business Media (2010)
4. Farge, M.: Wavelet transforms and their applications to turbulence. Annu. Rev. Fluid Mech. **24** (1), 395–458 (1992)
5. Grinsted, A., Moore, J.C., Jevrejeva, S.: Application of the cross wavelet transform and wavelet coherence to geophysical time series. Nonlinear Processes Geophys. **11**(5/6), 561–566 (2004)
6. Huang, G.B., Zhu, Q., Siew, C.K.: Extreme learning machine: theory and applications. Neurocomputing **70**, 489–501 (2006)
7. Huang, G.B., Zhou, H., Ding, X., Zhang, R.: Extreme learning machine for regression and multi-class classification. IEEE Trans. Syst. Man Cybern. **42**(2), 513–529 (2012)

Imbalanced Extreme Learning Machine for Classification with Imbalanced Data Distributions

Wendong Xiao, Jie Zhang, Yanjiao Li and Weidong Yang

Abstract Due to its much faster speed and better generalization performance, extreme learning machine (ELM) has attracted many attentions as an effective learning approach. However, ELM rarely involves strategies for imbalanced data distributions which may exist in many fields. In this paper, we will propose a novel imbalanced extreme learning machine (Im-ELM) algorithm for binary classification problems, which is applicable to the cases with both balanced and imbalanced data distributions, by addressing the classification errors for each class in the performance index, and determining the design parameters through a two-stage heuristic search method. Detailed performance comparison for Im-ELM is done based on a number of benchmark datasets for binary classification. The results show that Im-ELM can achieve better performance for classification problems with imbalanced data distributions.

Keywords Extreme learning machine · Imbalanced data distribution · Imbalanced extreme learning machine · Classification

1 Introduction

Single hidden layer feedforward neural network (SLFN) has been widely used in many fields for its capacity of approximating complex nonlinear processing, which cannot be precisely modelled mathematically using traditional methods, directly from the training samples. Hornik [1] proved that if the activation functions of the SLFNs are continuous, the continuous mappings can be approximated by SLFNs based on the training samples. However, traditional gradient-based training methods of SLFNs are easy to trap in the local minimum and always slow. For this

W. Xiao (✉) · J. Zhang · Y. Li · W. Yang
School of Automation & Electrical Engineering, University of Science
and Technology Beijing, Beijing 100083, People's Republic of China
e-mail: wdxiao@ustb.edu.cn

© Springer International Publishing Switzerland 2016
J. Cao et al. (eds.), *Proceedings of ELM-2015 Volume 2*,
Proceedings in Adaptation, Learning and Optimization 7,
DOI 10.1007/978-3-319-28373-9_41

reason, Huang et al. [2] proposed the extreme learning machine (ELM) as an extension of SLFNs, in which the input weights of SLFNs do not need to be tuned and can be generated randomly, and the output weights are calculated by the least-square method. Due to its faster speed and better generalization performance, ELM has received much attention and many new progresses have been made. For example, Huang et al. proposed the incremental ELM (I-ELM) and its improved versions [3–6], by adopting an incremental construction method to adjust the number of the hidden nodes. Liang et al. [7] proposed the online sequential ELM (OS-ELM) for online learning problems when samples come sequentially. Huang et al. [8] proposed the semi-supervised ELM (SS-ELM) and the unsupervised ELM (US-ELM) to tackle the learning problems when collecting large amount of labeled data is hard and time-consuming. Nowadays, ELM has been used in many fields, such as industrial production [9, 10], human physical activity recognition [11], regression [12], classification [13, 14], etc.

However, ELM rarely involves imbalanced learning, which is used to deal with imbalanced data distribution [15]. Imbalanced data distribution happens in two kinds of situations. In the first situation, the number of majority class is greater than the minority class. While in the second situation, the distribution of the majority class is concentrated and the minority class is relatively sparse. Usually, the first situation has great influence on the performance of the classifier. That is to say, the performance of a classifier can be deteriorated seriously due to the imbalanced data distribution. Dataset with imbalanced data distribution can be found in many areas, such as disease diagnosis, intrusion detection and fraudulent telephone calls, etc. In order to rebalance the data distribution, three main methods are explored: over-sampling to duplicate the minority samples, under-sampling to remove a fraction of the majority samples, and their combination. However, the above methods may bring in redundant samples or remove useful samples.

In order to tackle these drawbacks in the existing classification approaches, we propose a novel imbalanced extreme learning machine (Im-ELM) algorithm for both imbalanced and balanced data distributions. In addition, a two-stage heuristic search method for determining the design parameters of Im-ELM is also presented. The same place between Im-ELM and the Weighted Extreme Learning Machine (W-ELM) in [15] is both of them belong to the cost sensitive learning, the difference of them are: 1) the Im-ELM is implemented by addressing the sum error of each class in the performance index, and the W-ELM adding weight value in the error of each sample; 2) the cost value in the Im-ELM are determined by the two-stage heuristic method, but the weight value in the W-ELM set in advance.

The paper is organized as follows. A brief introduction to ELM theory is given in Sect. 2. The details of Im-ELM, including the effect of imbalanced data distribution on the classifier, the evaluation metrics and the mathematical formulation of Im-ELM for binary classification problems, are described in Sect. 3. Experimental results and analysis are presented in Sect. 4. Finally, conclusions and the future work are given in Sect. 5.

2 A Brief Introduction to ELM

ELM is extended from SLFN. Similar to SLFN, the output of ELM with L hidden nodes can be represented by

$$f_L(x) = \sum_{i=1}^{L} \beta_i G(a_i, b_i, x) \tag{1}$$

where $x \in R^n$ is the input, $a_i \in R^n$ and $b_i \in R^n$ are the learning parameters of the hidden nodes, $\beta_i \in R^m$ is the output weight, and $G(a_i, b_i, x)$ denotes the active function.

For a given dataset for training $\{(x_i, t_i)\}_{i=1}^{N} \subset R^n \times R^m$, where x_i is a n-dimension input vector and t_i is the corresponding m-dimension observation vector, the ELM with L hidden nodes approximating these N training data should satisfy

$$\sum_{i=1}^{L} \beta_i G(a_i, b_i, x) = t_j, \quad j = 1, 2, \ldots, N \tag{2}$$

which can be rewritten compactly as

$$H\beta = T \tag{3}$$

where

$$H(a_1, \ldots, a_L, b_1, \ldots b_L, x_1, \ldots, x_N) = \begin{bmatrix} G(a_1, b_1, x_1) & \cdots & G(a_L, b_L, x_1) \\ \vdots & \ddots & \vdots \\ G(a_1, b_1, x_N) & \cdots & G(a_L, b_L, x_N) \end{bmatrix}_{N \times L} \tag{4}$$

$$\beta = \begin{bmatrix} \beta_1^T \\ \vdots \\ \beta_L^T \end{bmatrix}_{L \times m} \tag{5}$$

$$T = \begin{bmatrix} t_1^T \\ \vdots \\ t_N^T \end{bmatrix}_{N \times m} \tag{6}$$

Here β^T denotes the transpose of the vector β. H is called the hidden-layer output matrix. Parameters a_i and b_i are assigned in advance with random values. (2) becomes a linear system and the output weight β can be estimated by

$$\hat{\beta} = H^\dagger T \tag{7}$$

where $H^\dagger$ is the Moore-Penrose generalized inverse of the matrix H [16]. When $H^\dagger H$ is nonsingular, we have $H^\dagger = (H^T H)^{-1} H^T$.

According to statistical learning theory, the real prediction risk of a learning algorithm consists of empirical risk and structural risk. Usually the empirical risk can be reflected by the sample errors, and the structural risk can be reflected from the distance between the margin separation classes [17, 19, 20].

The above ELM is based on the empirical risk minimization principle [17] and tends to result in over-fitting problem, which can be adjusted by trading off between the empirical risk and the structural risk [2, 18]. A normal operation to do this is to use the weighted sum of them in the performance index by introducing a weight factor C for the empirical risk to regulate the proportion of the empirical risk and the structural risk. Thus, ELM can be described as

$$
\min(\frac{1}{2}\|\beta\|^2 + \frac{1}{2}C\|\varepsilon\|^2)
$$
$$
\text{s.t.}, \sum_{i=1}^{L} \beta_i g(a_i x_i + b_i) - t_j = \varepsilon_j, j = 1, 2, \ldots, N \tag{8}
$$

where $\|\beta\|^2$ stands for the structural risk, and $\varepsilon = [\varepsilon_1, \varepsilon_2, \ldots, \varepsilon_N]$ is the sample errors. β can be calculated from the following expression

$$
\beta = \begin{cases} H^T(\frac{I}{C} + H^T H)^\dagger T, & N < L \\ (\frac{I}{C} + H^T H)^\dagger H^T T, & N > L \end{cases} \tag{9}
$$

where I is the unit matrix.

3   Proposed Approach

In this section, we will firstly analyze the impacts of the imbalanced data distribution on the classification performance, and give a brief introduction to the evaluation metrics we choose. Then, we will present the details of the proposed Im-RELM for binary classification.

3.1   Impacts of Imbalanced Data Distribution on Classifier

Without loss of generality, we take a binary classification problem as an example. Assuming there are lots of negative samples and few positive samples in a binary classification problem. As shown in Fig. 1a, the negative samples are denoted as minus signs and the positive samples are denoted as plus signs. Due to the advantage in quantity, the negative samples tend to push the separation boundary towards the opposite direction to get a better classification result for themselves. From mathematical view, C in (8) is the decisive factor to determine the location of

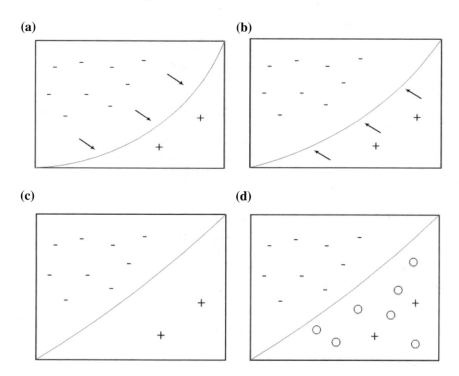

Fig. 1 Impacts of the imbalanced data distribution on classifier: **a** Moving trend of the separation boundary with imbalanced data distribution; **b** Expecting moving trend of the separation boundary; **c** Ideal location of the separation boundary; **d** Rebalancing the proportion of two classes of samples

the separation boundary. When C is too small, the separation boundary is supposed to be close to the positive samples, while when C is relatively larger, the separation boundary may be closer to the negative samples (see Fig. 1b), and the training error is smaller. Surely, the separation boundary is the ideal one when it locates in the middle of the two classes of samples (see Fig. 1c). Therefore, our purpose is to rebalance the proportion of the two classes of samples, and let the separation boundary locate in the middle as near as possible (circle sign in Fig. 1d).

3.2 Evaluation Metrics for Imbalanced Classification

There are many evaluation metrics that can be used to measure the performance of the classifiers in classification problems. Most of them are based on the confusion matrix as shown in Table 1, where TP stands for true positive, TN stands for true negative, FP stands for false positive, and FN stands for false negative. Based on the confusion matrix, the overall accuracy can be defined in (10). However, the

Table 1 Confusion Matrix

	Predicted positives	Predicted negatives
Real positives	True positive (TP)	False negative (FN)
Real negatives	False positive (FP)	True negative (TN)

overall accuracy may be not good enough to evaluate the performance of the classifier in imbalanced classification problem. Also, it is sensitive to the class distribution and misleading in some way [21]. As an example, let us consider a binary classification problem with imbalanced data distribution, which has 998 negative samples and 2 positive samples. According to the above evaluation metrics, the classifier will get 99.8 % accuracy and 0.2 % error by classifying all the samples to the negative class. Superficially, the classifier may get a wonderful result, but in fact it is meaningless to the positive class.

There are three main evaluation metrics to overcome the drawbacks of the above overall accuracy metric in binary classification [22, 23]: G-mean, F-measure and ROC curve. In this paper, we will choose G-mean as the evaluation metric for both binary and multiclass classification problems.

$$accuracy = \frac{TP + TN}{TP + FP + TN + FN} \tag{10}$$

G-mean can be described as follows

$$sensitivity = \frac{TP}{TP + FN} \tag{11}$$

$$specificity = \frac{TN}{TN + FP} \tag{12}$$

$$G - mean = \sqrt{sensitivity * specificity} = \sqrt{\frac{TP}{TP + FN} \times \frac{TN}{TN + FP}} \tag{13}$$

The sensitivity in (11) and the specificity in (12) are usually adopted to evaluate the performance of a classifier in binary classification. The sensitivity is also called the true positive rate or the positive class accuracy, and the specificity is also called the true negative rate or the negative class accuracy. G-mean was proposed based on the above two metrics, which is the geometric mean of sensitivity and specificity [24].

3.3 Im-ELM for Binary Classification

In this subsection, we will propose Im-ELM algorithm for binary classification problem, which is suitable for both balanced and imbalanced data distributions.

3.3.1 Im-ELM

According to the analysis in Sect. 3.1, our purpose is to keep the separation boundary locating in the middle to rebalance the proportion of the two classes. Different from the original ELM, the proposed Im-ELM will set two parameters, C^+ for minority positive samples and C^- for majority negative samples. Im-ELM can be described as

$$\min(\frac{1}{2}\|\beta\|^2 + \frac{1}{2}C^+\|\varepsilon_+\|^2 + \frac{1}{2}C^-\|\varepsilon_-\|^2)$$

$$\text{s.t., } \sum_{i=1}^{L}\beta_i g(a_i x_i + b_i) - t_j = \varepsilon_j, j = 1, 2, \ldots, N \tag{14}$$

where ε_+ stands for the sum error of the positive class and ε_- stands for the sum error of the negative class.

The Lagrangian for (14) can be written as

$$L(\beta, \varepsilon, \varepsilon_+, \varepsilon_-, \alpha, C^+, C^-) = \frac{1}{2}\|\beta\|^2 + \frac{1}{2}C^+\|\varepsilon_+\|^2 + \frac{1}{2}C^-\|\varepsilon_-\|^2 - \sum_{j=1}^{\tilde{N}}\alpha_j(\sum_{i=1}^{\tilde{N}}\beta_i g(a_i x_i + b_i) - t_j - \varepsilon_j)$$

$$= \frac{1}{2}\|\beta\|^2 + \frac{1}{2}C^+\|\varepsilon_+\|^2 + \frac{1}{2}C^-\|\varepsilon_-\|^2 - \alpha(H\beta - T - \varepsilon) \tag{15}$$

According to the KKT condition, by setting the gradient of the Lagrangian with respect to $(\beta, \varepsilon_+, \varepsilon_-, \alpha)$ equal to zero, we have

$$\begin{cases} \dfrac{\partial L}{\partial \beta} = 0 \rightarrow \beta^T = \alpha H \\ \dfrac{\partial L}{\partial \varepsilon_+} = 0 \rightarrow C^+\varepsilon_+^T + \alpha = 0 \\ \dfrac{\partial L}{\partial \varepsilon_-} = 0 \rightarrow C^-\varepsilon_-^T + \alpha = 0 \\ \dfrac{\partial L}{\partial \alpha} = 0 \rightarrow H\beta - T - \varepsilon = 0 \end{cases} \tag{16}$$

As

$$\varepsilon = \varepsilon_+ + \varepsilon_- \tag{17}$$

From the last expression of (16), we have

$$H\beta - T - (\varepsilon_+ + \varepsilon_-) = 0 \tag{18}$$

Substituting the second and the third expressions of (16) in (18), we can compute α as

$$\alpha = -(\frac{1}{\frac{1}{C^+} + \frac{1}{C^-}})(H\beta - T)^T \tag{19}$$

By substituting (19) in the first expression of (16), we can compute β as

$$\beta = \begin{cases} H^T(\frac{I}{C^+} + \frac{I}{C^-} + H^T H)^\dagger T, & N < L \\ (\frac{I}{C^+} + \frac{I}{C^-} + H^T H)^\dagger H^T T, & N > L \end{cases} \tag{20}$$

Similarly, as the matrix for the inversion operation in (20) is with the dimension $L \times L$, where L is the number of hidden nodes, the computational burden for calculating β in Im-ELM is the same as the original ELM.

Finally, we can obtain the following Im-ELM classifier for binary classification from $f(x) = sign\ h(x)\beta$:

$$f(x) = \begin{cases} sign\ h(x)H^T(\frac{I}{C^+} + \frac{I}{C^-} + H^T H)^\dagger T, & N < L \\ sign\ h(x)(\frac{I}{C^+} + \frac{I}{C^-} + H^T H)^\dagger H^T T, & N > L \end{cases} \tag{21}$$

where $h(x) = [h_1(x), \ldots, h_L(x)]$ is the row vector representing the outputs of L hidden nodes with respect to the input x.

3.3.2 Kernel Based Im-ELM

Sometimes, the Im-ELM classifier proposed for the binary classification (in Sect. 3.3.1) may not be possible to separate the samples with data overlapping. Therefore, we will extend (21) by using kernel based ELM [13]. A kernel matrix for ELM is defined as

$$\Omega_{ELM} = HH^T : \Omega_{ELMi,j} = h(x_i) \cdot h(x_j) = K(x_i, x_j) \tag{22}$$

So, the kernel based Im-ELM classifier can be obtained

$$\begin{aligned} f(x)_k &= sign\ h(x)H^T(\frac{I}{C^+} + \frac{I}{C^-} + H^T H)^\dagger T \\ &= sign \begin{bmatrix} K(x, x_1) \\ \vdots \\ K(x, x_N) \end{bmatrix}^T (\frac{I}{C^+} + \frac{I}{C^-} + \Omega_{ELM})^\dagger T \end{aligned} \tag{23}$$

3.3.3 Parameter Determination Strategy in Binary Classification

In Im-ELM, there are three main parameters to be determined, C^+, C^- and the number of hidden nodes L, which can affect the performance of Im-ELM significantly. In this paper, we will propose a two-stage heuristic search method to determine these parameters.

The proposed two-stage heuristic search method consists of the following two stages. In the first stage, the parameter search method in [13, 16] is used to determine the best combinations of C and L for the original ELM in Sect. 2. In the second stage, the number of hidden nodes L is fixed as the result of the stage 1 and the combination of C^+ and C^- is searched by discretizing each of them to discrete values.

3.4 Relevant Definitions in Imbalanced Classification

In binary classification, we define negative samples as the majority class and positive samples as the minority class.

Imbalanced ratio (IR) is defined to represent the imbalanced degree of the dataset

$$IR = \frac{N_+}{N_-}$$

4 Approach Implementation and Experimental Results

In this section, we will implement the proposed Im-RELM for binary classification problems and compare the results with ELM. All the experiments for the algorithms are carried out in Matlab 2012a environment running in an Inter i5 3.2 GHz CPU and 4G RAM.

4.1 Dataset Description and Parameter Settings

In order to verify the validity of our proposed methodology, we have performed experiments on 18 binary datasets from UCI Machine Learning Repository and some of them preprocessed by Fernandez et al. [26]. Table 2 indicates the detailed description of benchmark datasets used in our research work, including the number of attributes, number of categories, as well as IRs. All the experimental results are averaged for 10 runs. According to Table 2, we can find that IRs of the datasets are between 0.0077 and 0.5507. We think that the dataset is highly imbalanced when its IR is below 0.1500 and low imbalanced when its IR is above 0.1500.

Table 2 Description of datasets

Datasets	Attributes	Category	Max-class	Min-class	IR
Abalone19	8	2	4142	32	0.0077
Ecoli1	7	2	259	77	0.2973
Ecoli3	7	2	299	37	0.1237
Glass0	9	2	144	70	0.4861
Glass1	9	2	138	76	0.5507
Glass2	9	2	195	19	0.0974
Iris0	4	2	100	50	0.5000
New-thyroid1	5	2	180	35	0.1944
Pima	8	2	500	268	0.5350
Shuttle-C0_vs_C4	9	2	1706	123	0.0721
Wisconsin	9	2	444	239	0.5383
Yeast3	8	2	1321	163	0.1234

4.2 Experimental Results

We divide the whole datasets into 2 different types, including binary classification with low imbalance and binary classification with high imbalance. For each dataset, we chose 75 % of the samples as the training data and the rest 25 % as the testing data.

Here we will evaluate the performance of Im-ELM for datasets with imbalanced data distributions using G-mean as the evaluation metric. The results are shown in Tables 3 and 4, respectively. With the increasing of the imbalanced degree, the results become worse when using G-mean as the evaluation metric. This is because the conventional overall accuracy evaluation metric cannot reflect the real performance of a classifier.

Table 3 Experimental results of binary classification with low imbalance (G-mean)

Datasets	Sigmoid		Gaussian kernel
	ELM	Im-ELM	Im-ELM
	Testing result (%)	Testing result (%)	Testing result (%)
Ecoli1	87.19	89.65	91.23
Glass0	79.99	82.35	85.59
Glass1	77.36	76.34	80.11
Iris0	100	100	100
New-thyroid1	97.24	98.41	99.18
Pima	68.88	73.37	74.97
Wisconsin	95.31	96.01	96.05

Table 4 Experimental results of binary classification with high imbalance (G-mean)

Datasets	Sigmoid		Gaussian kernel
	ELM	Im-ELM	Im-ELM
	Testing result (%)	Testing result (%)	Testing result (%)
Abalone19	46.55	73.11	72.47
Ecoli3	78.68	90.88	88.16
Glass2	77.42	81.58	83.22
Shuttle-C0_vs_C4	100	100	100
Yeast3	80.56	91.26	93.56

Furthermore, most of datasets in Table 4 (such as Abalone19, Ecoli3 and Glass2, etc.) can get better results when Im-ELM is applied, which indicate that IR has greater influence on the performance of the classifier than the data complexity.

5 Conclusions

In this paper, we propose Im-ELM algorithm for binary classification for both imbalanced and balanced data distributions. We design a two-stage heuristic search method for determining the design parameters. The proposed Im-ELM can achieve better classification performance especially in imbalanced data distribution situation, compared with original ELM.

References

1. Hornik, K.: Approximation capabilities of multilayer feedforward networks. Neural Netw. **4**(2), 251–257 (1991)
2. Huang, G.B., Zhu, Q.Y., Siew, C.K.: Extreme learning machine: theory and applications. Neurocomptuing **70**(1), 489–501 (2006)
3. Huang, G.B., Chen, L., Siew, C.K.: Universal approximation using incremental constructive feedforward networks with random hidden nodes. IEEE Trans. Neural Netw. **17**(4), 879–892 (2006)
4. Huang, G.B., Chen, L.: Convex incremental extreme learning machine. Neurocomptuing **70**(10), 3056–3062 (2007)
5. Huang, G.B., Li, M.B., Chen, L., et al.: Incremental extreme learning machine with fully complex hidden nodes. Neurocomptuing **71**(4), 576–583 (2008)
6. Huang, G.B., Chen, L.: Enhanced random search based incremental extreme learning machine. Neurocomptuing **71**(16), 3056–3062 (2008)
7. Liang, N.Y., Huang, G.B., Saratchandran, P., et al.: A fast and accurate online squential learning algorithm for feedforward networks. IEEE Trans. Neural Netw. **17**(6), 1411–1423 (2006)
8. Huang, G., Song, S., Gupta, J.N., et al.: Semi-supervised and unsupervised extreme learning machines. IEEE Trans. Cybern. **78**(3), 2405–2417 (2014)

9. He, Y.L., Geng, Z.Q., Xu, Y., et al.: A hierarchical structure of extreme learning machine (HELM) forhigh-dimensional datasets with noise. Neurocomputing **76**(3), 407–414 (2014)
10. Zhang, S., Chen, X., Yin, Y.X.: An ELM based online soft sensing approach for alumina concentration detection. Math. Probl. Eng. **2015**, Article ID 268132, 8 pp (2015)
11. Xiao, W.D., Lu, Y.J.: Daily human physical activity recognition based on kernel discriminant analysis and extreme learning machine. Math. Probl. Eng. **2015**, Article ID 790412, 8 pp (2015)
12. Soria-Olivas, E., Gomez-Sanchis, J., Martin, J.D., et al.: BELM: Bayesian extreme learing machine. IEEE Trans. Neural Netw. **22**(3), 505–509 (2011)
13. Rong, H.J., Ong, Y.S., Tan, A.H., et al.: A fast pruned-extreme learning machine for classification problem. Neurocomputing **72**(3), 359–366 (2008)
14. Huang, G.B., Zhou, H., Ding, X., et al.: Extreme learning machine for regression and multiclass classification. IEEE Trans. Syst. Man Cybern. B Cybern. **42**(2), 513–529 (2012)
15. Zong, W.W., Huang, G.B., Chen, Y.Q.: Weighted extreme learning machine for imbalance learing. Neurocomputing **101**(1), 229–242 (2013)
16. Rao, C.R., Mitra, S.K.: Generalized Inverse of Matrices and its Applications. Wiley, New York (1971)
17. Vapnik, V.N.: The Nature of Statistical Learning Theory. Springer, New York (1995)
18. Deng, W.Y., Chen, L.: Regularized extreme learning machine. In: IEEE Symposium on Computational Intelligence and Data Mining, pp. 389–395. IEEE Press, USA (2009)
19. Cristianini, N., Shawe-Taylor, J.: An Introduction to Support Vector Machines. Cambridge University Press, Cambridge (2000)
20. Fung, G., Mangasarian, O.L.:Proximal Support Vector Machine Classifiers. In: Proceedings KDD-2001: Knowledge Discovery and Data Mining, pp. 77–86 (2001)
21. He, H.B., Garcia, E.A.: Learning from imbalanced data. IEEE Trans. Knowl. Data Eng. **21**(9), 1263–1284 (2009)
22. Thai-Nghe, N., Gantner, Z., Schmidt-Thieme, L.: A new evaluation measure for learning from imbalanced data. In: Proceedings of International Joint Conference on Neural Networks, pp. 537–542 (2011)
23. Fawcett, T.: An Introduction to ROC analysis. Pattern Recogn. Lett. **27**(8), 861–874 (2006)
24. Tang, Y.C., Zhang, Y.Q., Chawla, N.C., et al.: SVMs modeling for highly imbalanced classification. IEEE Trans. Syst. Man Cybern. Part B: Cybern. **39**(1), 281–288 (2009)
25. Huang, G.B.: An insight into extreme learning machines: random neurons, random features and kernels. Cogn. Comput. **6**, 376–390 (2014)
26. Fernandez, A., Jesus, M.J., Herrera, F.: Hierarchical fuzzy rule based classification systems with genetic rule selection for imbalanced data-sets. Int. J. Approx. Reason. **3**(50), 561–577 (2009)

Author Index

A
Akusok, Anton, 357, 371, 413

B
Bai, Zuo, 471
Barsaiyan, Anubhav, 285
Björk, Kaj-Mikael, 357, 371, 413

C
Cai, Zhiping, 67
Cao, Yuchi, 55
Chai, Songjian, 263
Chen, Guang, 325
Cheng, Xiangyi, 235
Chen, Hongbiao, 423
Chen, Mo, 249
Chen, Su-Shing, 107
Chen, Yuangen, 345
Choubey, Sujay, 285
Cui, Song, 273

D
Ding, Shi-fei, 93
Dong, Zhaoyang, 189, 263
Duan, Lijuan, 273
Duan, Xiaodong, 295
Du, Hua, 203

E
Eirola, Emil, 385
Er, Meng Joo, 437

F
Feng, Jun, 107

G
Gritsenko, Andrey, 385
Gu, Jason, 1

H
He, Bo, 399, 455
He, Feijuan, 107
Holtmanns, Silke, 371
Huang, Guang-Bin, 471

J
Jiang, Changmeng, 107
Jiang, Changwei, 179
Jiang, Chunfeng, 117
Jia, Xibin, 203
Jia, Youwei, 263

L
Lauren, Paula, 357, 413
Lendasse, Amaury, 357, 371, 385, 413, 491
Li, Chenguang, 325
Li, Fangfang, 423
Li, Feng, 249
Li, Lianbo, 55
Lim, Meng-Hiot, 285
Lin, Jiarun, 67, 317
Lin, Zhiping, 335
Li, Shan, 117
Li, Yanjiao, 503
Liu, He, 117, 131
Liu, Hengyu, 423
Liu, Huaping, 223, 235
Liu, Huilin, 117, 131
Liu, Jun, 43
Liu, Tianhang, 67, 317
Liu, Xinwang, 67
Liu, Xuefei, 491
Liu, Yang, 491
Li, Yao, 131
Lu, Bo, 295
Luo, Minnan, 43
Luo, Xiong, 179

© Springer International Publishing Switzerland 2016
J. Cao et al. (eds.), *Proceedings of ELM-2015 Volume 2*,
Proceedings in Adaptation, Learning and Optimization 7,
DOI 10.1007/978-3-319-28373-9

515

M
Mao, Kezhi, 19
Meng, Ke, 189
Miao, Jun, 273
Miche, Yoan, 357, 371, 413

N
Natarajan, S., 209
Nian, Rui, 357, 399, 413, 455, 491

O
Oh, Beom-Seok, 335
Oh, Kangrok, 335
Oliver, Ian, 371

P
Panigrahi, Bijaya Ketan, 285

Q
Qin, Libo, 325

R
Ramaseshan, Varshini, 209
Ratner, Edward, 385
Ravichander, Abhilasha, 209

S
Schupp, Daniel, 385
Shen, Yue, 399, 455
Shrivastava, Nitin Anand, 285
Song, Yan, 455, 491
Su, Lijuan, 307
Sun, Fuchun, 223, 235
Sun, Kai, 1
Sun, Tingting, 399
Sun, Xia, 107
Sun, Yongjiao, 345
Sun, Zhenzhen, 143

T
Teoh, Andrew Beng Jin, 335
Toh, Kar-Ann, 335

V
Venkatesan, Rajasekar, 437
Vijay, Supriya, 209

W
Wang, Bin, 77, 155
Wang, Danwei, 471
Wang, Dongzhe, 19
Wang, Guoren, 77, 155, 345
Wang, Huan, 179

Wang, Jiarong, 107
Wang, Miao-miao, 93
Wang, Ning, 437
Wang, Xianbo, 169
Wei, Jie, 223
Wong, Pak Kin, 169
Wu, Chengkun, 317
Wu, Lingying, 1
Wu, Q.M. Jonathan, 31
Wu, Zhaohui, 307

X
Xiao, Chixin, 189
Xiao, Wendong, 503
Xie, Haibin, 423
Xu, Jingting, 107
Xu, Xinying, 235
Xu, Yan, 189
Xu, Zhao, 263

Y
Yan, Gaowei, 223
Yang, Dan, 249
Yang, Weidong, 503
Yang, Xiaona, 179
Yang, Yimin, 31
Yang, Zhixin, 169
Yao, Min, 307
Yan, Tianhong, 399, 455
Yin, Jianchuan, 55
Yin, Jianping, 67, 317
Yin, Ying, 325
Yuan, Bin, 273
Yuan, Ye, 295, 345
Yu, Ge, 249
Yu, Yuanlong, 1, 143

Z
Zhang, Dezheng, 179
Zhang, Heng-Guo, 491
Zhang, Tiancheng, 423
Zhang, Xin, 189
Zhao, Jian, 55
Zhang, Jie, 503
Zhao, Rui, 19
Zhao, Yuhai, 325
Zheng, Nenggan, 307
Zheng, Qinghua, 43
Zhong, Jianhua, 169
Zhou, Xun, 189
Zhu, Rui, 77, 155
Zhu, Wentao, 273

Printed in the United States
By Bookmasters